FIFTH EDITION

Greenhouse Operation & Management

Paul V. Nelson

Department of Horticulture
North Carolina State University

Prentice Hall
Upper Saddle River, New Jersey 07458

Library of Congress Cataloging-in-Publication Data

Nelson, Paul V.

Greenhouse operation & management / Paul V. Nelson. —5th ed.

p. cm.

Includes bibliographical references and index.

ISBN 0-13-374687-9

1. Greenhouse management. 2. Floriculture. I. Title.

SB415.N44 1998

635.9'823—dc21 97-24640

CIP

Acquisitions Editor: Charles Stewart
Associate Editor: Kim Gundling
Editorial Production Services: WordCrafters Editorial Services, Inc.
Managing Editor: Mary Carnis
Director of Production and Manufacturing: Bruce Johnson
Prepress Manufacturing Buyer: Marc Bove
Marketing Manager: Melissa Brunner
Editorial Assistant: Kate Linsner
Cover Designer: Marianne Frasco
Printer/Binder: R. R. Donnelley, Harrisonburg, VA

Simon & Schuster/A Viacom Company
Upper Saddle River, NJ 07458

Printed in the United States of America

10 9 8 7 6 5 4 3 2 1

ISBN 0-13-374687-9

Prentice-Hall International (UK) Limited, *London*
Prentice-Hall of Australia Pty. Limited, *Sydney*
Prentice-Hall Canada, Inc., *Toronto*
Prentice-Hall Hispanoamericana, *Mexico*
Prentice-Hall of India Private Limited, *New Delhi*
Prentice-Hall of Japan, *Tokyo*
Simon & Schuster Asia Pte. Ltd., *Singapore*
Editora Prentice-Hall do Brasil, Ltda., *Rio de Janeiro*

Contents

CHAPTER 8 *Watering* *245*

CHAPTER 9 *Fertilization* *285*

CHAPTER 10 *Alternative Cropping Systems* *355*

CHAPTER 11 *Carbon Dioxide Fertilization* *375*

CHAPTER 12 *Light and Temperature* *387*

Preface

Development within the greenhouse industry was relatively slow and steady until shortly after World War II. Then the vast technology of the war effort was redirected toward peacetime applications. Nearly all industries benefited, including the greenhouse industry. This change assumed two faces during the 1950s. The first was technological in nature. That decade saw the advent of aluminum greenhouse frames, polyethylene greenhouse coverings, soilless root substrates, pasteurization of root substrate in lieu of annual replacement, automatic watering, liquid fertilizer application, and many other forms of automation. The second face was one of shifts in domestic production regions fostered by improvements in transport. At first, fresh (cut) flower production shifted from greenhouses around scattered population centers to fields in warm regions. This was followed by development of fresh flower production in greenhouses in the Denver area. These geographical shifts in production brought great hardship to growers in more northern regions.

Technological changes continued their momentum during the 1960s. A large center of greenhouse production of fresh flowers developed in California, further weakening production throughout the remainder of the country. And in the area of marketing, the mass-market outlet for floral products gained sufficient momentum to become a viable force in the marketplace. This market brought floral products into the lives of vast numbers of people who did not previously purchase such items. However, this change created considerable difficulty for the traditional retail florists. The 1970s ushered in even greater change with the importation of fresh flowers from tropical nations into the well-established floral markets of the developed nations. This brought serious pressure on fresh flower production throughout the country. Simultaneously, green (foliage) plant sales grew almost exponentially. This offered an alternative opportunity for domestic fresh flower growers adversely affected by imports, although it was a very painful transition.

During the 1980s, change continued to escalate in many areas. Technology was introduced with increased momentum in such areas as computerized control of the greenhouse environment and plug seedling production. Numerous new fresh flowers and pot crops emerged. In the market arena, large-scale expansion of discount store and supermarket outlets for floral products occurred. By far, the largest changes since World War II occurred during the 1990s. It is a common statement today that more change has occurred in the greenhouse industry in the last 5 years than in the 20 years prior to these. The long-standing auction system of The Netherlands is being threatened by alternative marketing systems. Floral wholesaling, formerly an exclusive function of the traditional floral market nations of Japan, Western Europe, and North America is now shared with tropical floral export nations. Ever increasing numbers of floral crops are being imported by the traditional market nations from a rapidly increasing pool of tropical production nations. As a consequence, production in these former nations is leveling off. Basically, we are experiencing the merger of distinct regional floral industries into one colossal global industry.

The key word that emerges from the history of the greenhouse industry is *change*. This is an industry beset by ever escalating change. Because of this change, great opportunities exist for new as well as established firms. Those who recognize and understand change stand to prosper from it because they will help to shape it and become a part of it. Those who do not assimilate change will probably perish. Today, 57 percent of households in the United States purchase floral products. The 43 percent who do not offer a great opportunity. But how can they be reached? New concepts in product type and marketing methods are needed. These must include innovations to appeal to a society of maturing baby boomers, a society with widening disparity between high- and low-income earners, a society with a perception of less free time, less brand allegiance, and more desire for product variety and ease of access.

The answers to this question of how to meet change in the greenhouse industry will stem from education. Certainly an understanding of the technology for performing greenhouse operations and growing crops is an integral part of this education, but it is not nearly enough. Business management is an equally important component. Whether you own your own business or work for someone else, your well-being will depend on your ability to manage materials, money, and time—both your own and those of others. Without this ability, the application of your technical knowledge of growing crops ultimately will not be profitable and rewarding. Cost accounting and analysis, personnel management, and marketing are all important topics. Because an appreciation of these areas is not common in the young student entering the greenhouse field, this book is written in such a way that these principles are addressed in context with cultural instructions. Additional books and courses would serve this facet of education very well. The third component of a well-rounded education should be an industry perspective. It is important to know the current status of the greenhouse industry and the trends impacting it. From this stage, one can begin to predict future changes. This component of education is gained by attendance at industry conferences; communication with fellow growers, allied industry personnel, and educators; and reading trade literature. A good beginning to this third facet of education is presented in the first chapter of this book.

The outline for this book anticipates the decisions in the order in which they occur for a person entering the greenhouse business. Initially, the decision to enter the field is dealt with in a chapter on the worldwide perspective of floriculture. Decisions involving the physical arrangement of a greenhouse business are taken up in successive chapters considering site selection, greenhouse types, heating and cooling systems. and environmental control systems. Considerations then turn toward the type of root substrate in which the crops are to be grown; pasteurization of the root substrate; maintenance of disease-free conditions in the greenhouse; watering principles and automated systems; fertilizer formulations and methods of application; alternative systems of production such as NFT and rock wool; injection of carbon dioxide gas into the greenhouse atmosphere; light; temperature; chemical growth regulation; pest control; postproduction handling of crops; and cost accounting. These are the main categories of decisions with which you will be faced as you design, build, and operate a greenhouse business.

CHAPTER 1

Floriculture–A Dynamic Industry

The market for floral products in a society is driven by its degree of isolation from flowers and its financial resources. Societies in temperate and frigid climates are cut off from flowers during the winter. Urban societies can likewise be cut off from ornamental settings year round. People, being an organic part of nature, have deep emotional roots in the plant world. Once the basic needs of life are met, they turn to flowers as a natural food for the soul. Flowers and plants reestablish ties with the fundamental nature of our lives, piquing our senses of beauty, peacefulness, and stability. In so doing, they play a therapeutic role. The gift of flowers evokes a powerful message in the recipient by connecting these emotional qualities with the sender's intent, be it gratitude, thoughtfulness, or romance.

The leading markets for floral products in the world coincide with the more economically developed societies. In such societies, people have sufficient extra money to support the floral industry. The largest markets today are Western Europe, the United States and Canada, and Japan. Of the estimated total worldwide retail value of floral products of $37 billion in 1995, approximately $16 billion were consumed in 12 countries of Western Europe, $14 billion in the United States and Canada, and $5.5 billion in Japan.

In previous centuries and the first seven decades of this century, most production of floral products took place in the geographical areas of the retail markets. Suddenly, during the past 25 years, vast floral production areas have appeared in many developing countries, mostly within the warmer regions of the globe. The majority of this new production is exported to European, U.S., Canadian, and Japanese markets. This has brought about shifts of great magnitude in traditional production areas, in the floral product handling industry, and in retail marketing channels. The nature of these changes will be discussed in this chapter. It is a common assessment that more change has occurred in the past 5 years than in the previous 20 years, and that future change will be even faster.

As always, change can be an asset or a detriment. Often, when change can be predicted, it is possible to position oneself in a profitable stance. Change is seldom profitable to those who do not anticipate it.

Origin of the Greenhouse Industry

The greenhouse industry as we know it today probably originated under circumstances similar to those that existed in Holland during its "Golden Age," the 1600s. The stage was set during the first half of the 17th century when The Netherlands became the world's foremost sea power. Its merchant fleet tripled during this period to the point where The Netherlands provided half the world's shipping, and Amsterdam became the world's leading commercial city. The Dutch standard of living was the highest in the world. Milestones during this period occurred in 1602, when the Dutch East India Company was founded, and in 1621, when the Dutch West India Company was founded. Both expanded trade throughout a vast colonial empire. However, The Netherlands was under the proclaimed rule of Spain during this time, a situation that diverted considerable naval strength. Although The Netherlands declared its independence from Spain in 1581, it was not until the end of the Thirty Years' War (1618–1648) that The Netherlands won its cause to become an independent nation and realized its full potential commercial naval strength.

The royal courts of Europe at this time had a taste for elegance and the means to afford it. Spring flowers in the winter and fruit out of season were very enticing. The productive capacity of the large middle class, unique to The Netherlands, and the trade channels of the merchant segment soon gave birth in The Netherlands to what is today the largest greenhouse industry in the world. Grapes were grown along rock walls in western Holland under glass enclosures constructed in a lean-to fashion. These greenhouses conserved the energy of the sun during the winter and permitted early crops of grapes. Today, a vast greenhouse vegetable and cut-flower industry exists, with its center in the Westland area, as a direct descendant of this initial business.

In the region near Amsterdam, field-grown lilac bushes were dug in late fall, prior to the freezing of the ground and were stored outside. Periodically during the winter, bushes were moved into greenhouses where they broke dormancy and flowered (Figure 1-1). The cut blooms graced the palaces of 17th-century royalty in Great Britain, France, Germany, and other countries. Even today this industry persists, although much of this region, near Aalsmeer, is involved in pot-plant culture in general. Today, The Netherlands is the largest producer of floral products in the world, producing almost 25 percent of the total value.

Development of the greenhouse industry in North America followed much later because of its dependence on the economic growth of this new land. Greenhouse technology brought in by immigrants from Europe was used to establish an industry that began to flourish during the 19th century. The first reported greenhouse in the United States was that of James Beckman in 1764, located in New York City (Kaplan 1976). Floriculture first started around the population

(a) (b)

Figure 1-1 **Lilacs were one of the first floral crops grown in The Netherlands, and they are still grown there today. (a) Dormant bushes are dug in the late fall and stored. (b) Periodically during the winter, bushes are brought into the greenhouse for forcing.**

centers of Boston, New York, Philadelphia, and, later, Chicago. In those days, prevailing modes of transportation necessitated production in close proximity to the markets. As the population spread west across the continent, pot- and bedding-plant production followed and became established in each urban area.

United States Production

The United States is the third largest floral producer in the world. Of the total world wholesale production value of about $20 billion, the United States produced $3.27 billion in 1995. The two top producers were The Netherlands with $4.695 billion and Japan with $3.685 billion. Together these three countries produced over 55 percent of the world's floral products. U.S. production fits into five categories. These categories and the percentage of the total production of each in 1995 were: 13.5 percent fresh flowers, 16.4 percent green plants, 22.5 percent flowering pot plants, 43.9 percent bedding plants, and 3.7 percent cut cultivated greens. Each category is undergoing a different change at the present. Some are declining while others are increasing in response to the entry of developing nations into world floral production and the shifting channels within the domestic retail market. Before entering into floral production, it is extremely important to understand the history of change and the forces now shaping future potential in each category. One could be doomed to quick failure by entering the wrong commodity area.

Fresh Flowers

Through the end of the 19th century, floral products were transported by horse-drawn wagons. Because these lacked refrigeration and were rough on the product, transportation was limited to local areas. The development of truck transportation in the early 20th century changed that. Paved roads and faster speeds made it possible to transport greater distances without undue damage to the product. *Fresh flowers,* otherwise known as cut flowers, could be packed in ice. For the first time it was possible to produce a floral commodity in a remote area that lent itself better to profitability than those of numerous local areas. Before looking at the production area shifts that occurred as a result of changing transportation, it is important to understand the three factors that govern the suitability of a given production area: production cost, quality, and transportation cost (Figure 1-2).

Eastern Centers Fresh flowers were the first floral commodity to undergo centralization of production. As trucks became commonplace in the early part of the 20th century, transportation posed less of a problem. The populated areas of eastern Massachusetts, Connecticut, and New York City, particularly Long Island, became major centers for carnation production. Rose production became especially important in these same areas, as well as around Philadelphia and Chicago. From these centers, fresh flowers were transported considerable distances to smaller towns. This early centralization was probably driven by an information infrastructure. These areas had a critical mass of growers and the allied supply industry to share technical information, to foster new innovations of efficiency, and to create the wholesale distribution channels needed. All of these factors led to lower production costs.

Southern Outdoor Production The movement toward centralization suddenly escalated during the 1950s. Air and refrigerated-truck transportation had developed to the extent that shipping fresh flowers to any point was economically feasible. The growing of fresh flowers outdoors in warm climates for shipment to distant markets became a possibility. Production of cut chrysanthemums expanded at a startling rate in Florida, southern California, and, to a lesser extent, Texas. Crops were grown year round under shade fabric supported on inexpen-

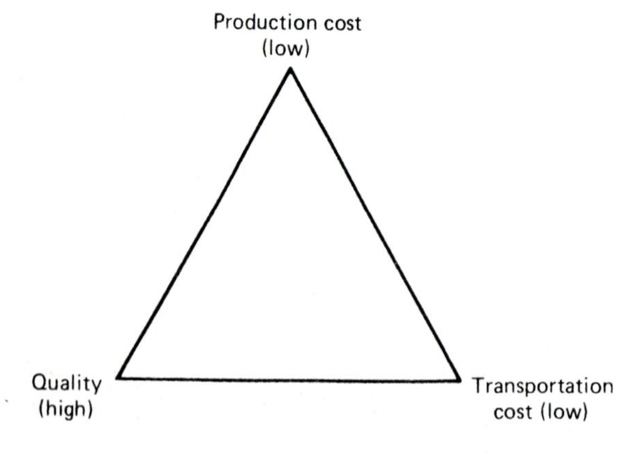

Figure 1-2 Crop production areas often can be evaluated on the basis of three factors: production cost, quality, and transportation cost. If all three factors are ideal, then the area is safe from competition. If conditions are less than ideal, as is usually the case, weakness of any one factor must be offset by strengthening one or both of the others in order for the production area to meet outside competition.

Figure 1-3 **Large areas of crops, particularly fresh flowers such as this chrysanthemum crop in Florida, were grown outdoors under shade fabric in Florida and California.**

sive frames (Figure 1-3). No heating or cooling was required. The increased cost of transportation was more than offset by lower production costs—cheaper growing facilities, no heating expenses, and less expensive labor.

Consequently, the production of stock (a cut-flower crop of secondary importance) essentially came to a halt in northern greenhouses since nearly all the demand was met by southern California growers (Figure 1-4). Northern chrysanthemum growers feared that they too would soon become a relic of the past. Interestingly, chrysanthemum production in the field reached a plateau during the 1960s and came into balance with the rest of the country. The attainment of this position caused many greenhouse growers to turn away from the production of this crop.

Those northern growers who foresaw the trends improved the quality of their product to give themselves a competitive edge over poorer-quality flowers grown during periods of harsh weather conditions in the fields. Particularly in the northern areas, near the ends of the distribution lines from the southern fields, chrysanthemum growers established year-round production schedules to guarantee a steady 52-week supply of flowers. Where it was not possible to meet the competition of southern spray-type chrysanthemums during the winter months, astute growers switched to greenhouse-grown standard chrysanthemums, which at that time did not grow well in the fields.

The equilibrium between field- and greenhouse-grown chrysanthemums was further supported by the lack of control over natural factors in the weather-dependent field environment. Frosts, tropical storms, winds, periods of excessive moisture, and sudden infestations of insects were (and are) all very difficult and

Figure 1-4 **Relatively inexpensive field culture, such as this crop of field-grown stock in California, replaced greenhouse crops in the northern states.** *(Photo courtesy of R. A. Larson, Department of Horticultural Science, North Carolina State University, Raleigh, NC 27695-7609.)*

sometimes impossible to control in the field. When these forces came into play, the market demands for quantity and quality were not met, and the door was opened for controlled-environment (greenhouse) crops. Roses are a good example of this point. They are particularly prone to powdery mildew disease and spider mites as well as to any adversity in handling. Because of the quality factor, field production of roses as a fresh flower has not developed.

Colorado Greenhouse Production Carnation production did not shift to the southern outdoor fields because those climates were too hot. During the late 1950s a shift did occur. Through the efforts of forward-thinking individuals like Professor W. D. Holley of Colorado State University, more satisfactory environments were identified that came closer to fitting the requirements of this crop, which calls for 52°F (11°C) night and 75°F (24°C) day temperatures, high light intensity, and a 12-hour day length. The Denver, Colorado, region offered more temperate summer temperatures and a high light intensity because of its high elevation. From an essentially nonexistent floral industry in 1950, an impressive greenhouse carnation industry grew in Colorado, which accounted in 1965 for 22 percent and in 1975 for 27 percent of the number of blooms produced in the leading 27 states. This production-area shift was supported by the higher product quality that could be produced in the Denver area. Along with carnations, a large greenhouse rose industry also developed in the Denver area.

California Greenhouse Production However, in spite of higher quality production costs, particularly those of fuel, were also high. As a consequence, a shift to greenhouse production rapidly occurred in the San Francisco Bay area. Although light intensity was lower in the Bay area, the shift in quality was more than offset by lower production costs stemming from more temperate winter and summer temperatures and a greater availability of inexpensive labor. Of the total number of U.S. carnation blooms produced, 45 percent were grown in California in 1965, 66 percent in 1975, and 79 percent in 1985. Development of the California industry (Figure 1-5) had its repercussions in that carnation production in Colorado dropped back to 21 percent of U.S. production by 1980. The phenomenal expansion in Colorado and later in California had a devastating effect upon the eastern carnation production areas.

Massive shifts in the production of other fresh flowers into several areas in California also occurred along with that of carnations. The percentages of the total value of fresh-flower production in the United States that were produced in California in the years 1965, 1975, 1985, and 1995 were 27, 36, 56, and 62 percent, respectively. Fifteen percent of the value of all remaining U.S. floral production in 1995 occurred in California. Throughout this period of an initial southward and later westward shift of fresh-flower production, growers in the remainder of the continental United States and Canada were either closing their businesses or shifting to other crops, mainly bedding plants and flowering pot plants.

The shift of fresh-flower production to California was partly in response to the large market developing in the western coastal states. But that alone could not justify the volume of fresh-flower production. Lower production costs due to a milder climate and less expensive labor compared to Colorado and the eastern United States further explained the shift. The final reason was lower trans-

Figure 1-5 **The past three decades have seen a dynamic expansion in greenhouse-grown fresh flowers in California.** *(Photo courtesy of Hall-Manatee Greenhouses, Encinitas, CA.)*

portation costs. The predominant movement of air cargo in 1950 was westward, resulting in partially empty planes returning eastward. To fill these planes, lower rates were offered for eastward transport. This effectively opened the eastern markets to western growers and at the same time protected the western markets against competition from eastern growers. The air freight rate for 1,000 cut carnations (1.67 boxes) from San Francisco to Chicago was $21.47 in 1950 (Table 1-1). This rate fell to a low point of $12.65 in 1965, a strong motivation for air shipment of flowers. Rates then gradually rose. But even in 1980, the rate of $20.75 was still lower than it had been 30 years earlier in 1950.

In 1975, more than 90 percent of fresh flowers were shipped via air from California. Then, as air fares rose and developments in precooling, packaging technology, and truck cooling systems progressed, truck transportation became competitive. In 1980, about 70 to 80 percent of fresh flowers were shipped from California by truck. Since 1980, truck and air freight rates have risen. Which method of transportation is cheaper depends on the destination and the volume being shipped. For example, the air freight rate in 1996 from San Francisco to Chicago was $30 per 1,000 carnation blooms, while the comparable rate to Washington, D.C., a much greater distance, was $25. In 1990, about 60 percent of fresh flowers were shipped from California by air. In 1997 it is estimated that 30 percent are shipped by air.

Foreign Imports In 1969, an intercontinental shift became apparent. Actually the story began in 1966, when two carnation ranges in Bogotá, Colombia, began producing quality carnations at an incredibly low price. They were joined in 1969 by a U.S. firm, and others quickly followed. Today, the majority of fresh flowers consumed in the United States come from Colombia. Greenhouses are primarily covered with polyethylene. Bogotá, at an altitude of 8,660 ft (2,640 m), enjoys a day length close to twelve hours year round because of its location near the equator. Evening temperatures are in the 40–50°F (4–10°C) range, and day-

Table 1-1 **Air Freight Rates from San Francisco to Chicago**[1]

Year	*Rate*
1950	$21.47
1957	$13.03
1965	$12.65
1969	$13.40
1972	$15.15
1980	$20.75
1990	$25.00
1996	$30.00

[1]Per 1,000 cut carnation flowers, 600 flowers per box.

time temperatures are in the 60–70°F (16–21°C) range in all seasons. The area offers high light intensity because of its high altitude. These factors are ideal for high-quality carnation production. Additionally, the cost of labor is low, and there is no expense for heating because flowers are produced in unheated plastic houses in Colombia (Figure 1-6).

Figure 1-6 Carnation production in the Bogotá area of Colombia, South America.

Colombian carnations constituted a modest 0.5 percent of all carnations sold in the U.S. market in 1970, but by the end of 1974 the figure was a stunning 25 percent. In 1978, 1981, 1988, and 1995, imported carnations constituted about 40, 60, 76, and 87 percent, respectively, of total sales in the United States (Figure 1-7a).

A chrysanthemum production industry developed simultaneously in Colombia. At first it was located on the savanna around Bogotá, but later it expanded to the area around Medellín. Medellín is located at a lower altitude of 4,880 ft (1,490 m) in the Andes Mountains northwest of Bogotá. At this lower altitude the average temperatures are about 10°F (6°C) higher, which favors chrysanthemum production. Imported pompon chrysanthemums made up 88 percent of those sold in the United States in 1995, with 79 percent of those imports coming from Colombia (Figure 1-7b). Likewise, 76 percent of standard chrysanthemums were imported, with 62 percent of those imports coming from Colombia (Figure 1-7c).

Rose imports became significant during the late 1970s. The level of imports has grown continuously since then, and in 1995 constituted 66 percent of roses sold in the United States (Figure 1-7d). Rose imports lagged behind carnation and chrysanthemum imports because roses have a short shelf life and are more susceptible to

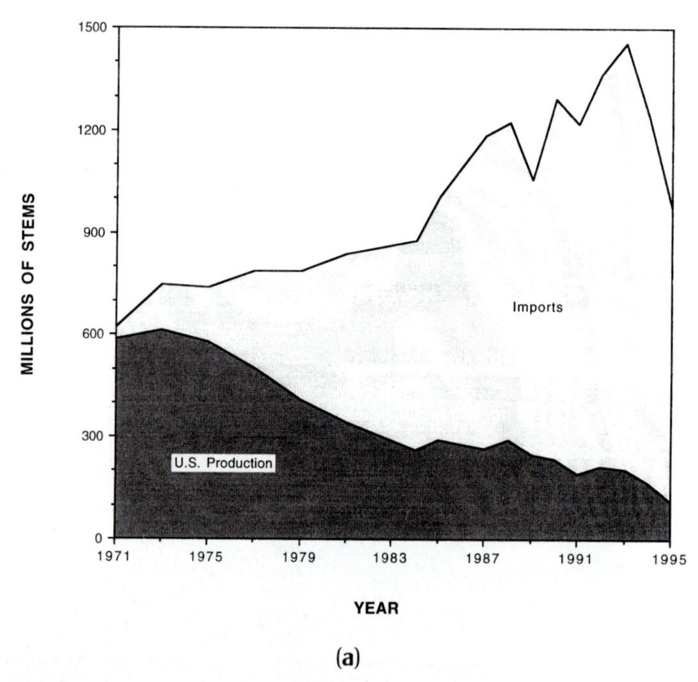

(a)

Figure 1-7 **Number of blooms, or bunches in the case of pompons, of various fresh flowers produced in and imported into the United States per year from 1971 through 1995.** ***(U.S. production data from Agricultural Statistics Board, USDA, 1972 through 1996; import data from Agricultural Marketing Service [1996].)***

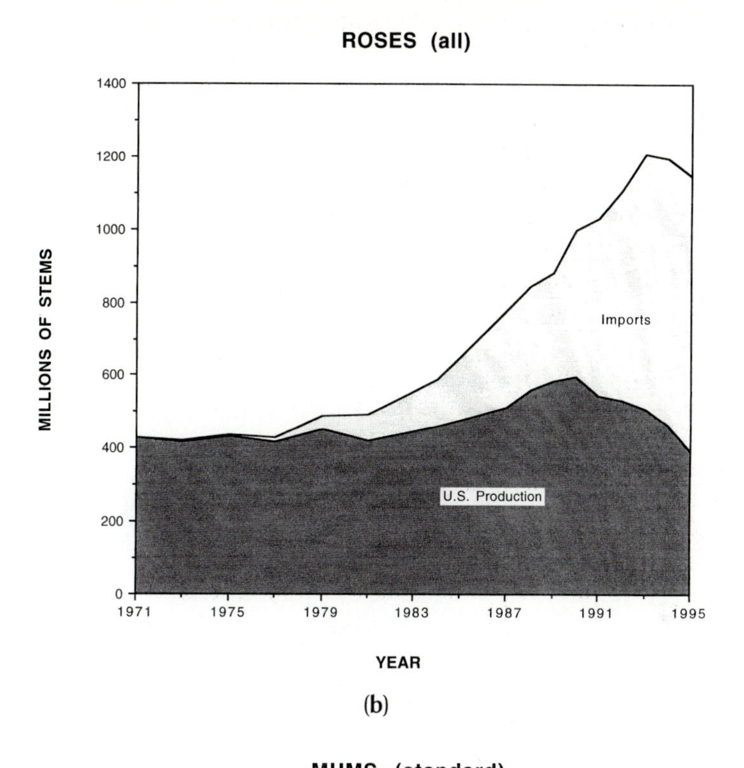

(b)

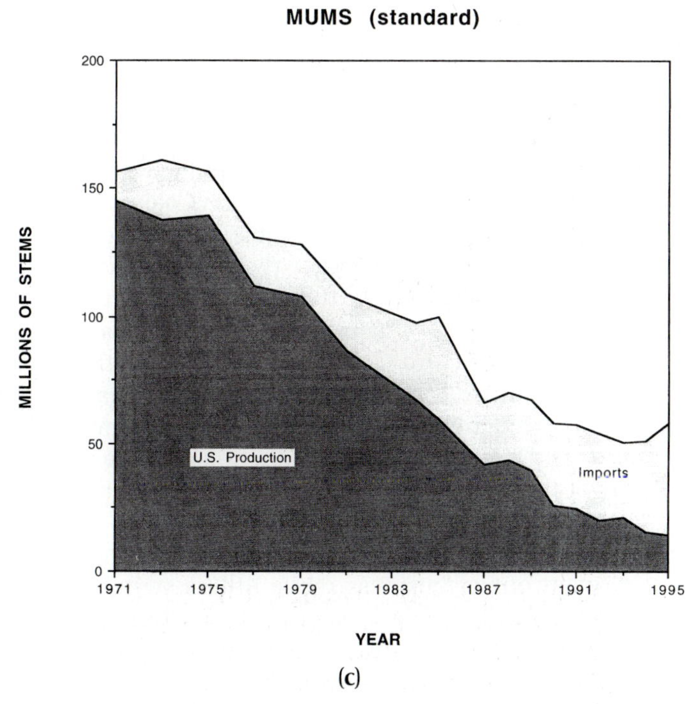

(c)

Figure 1-7 *(continued)*

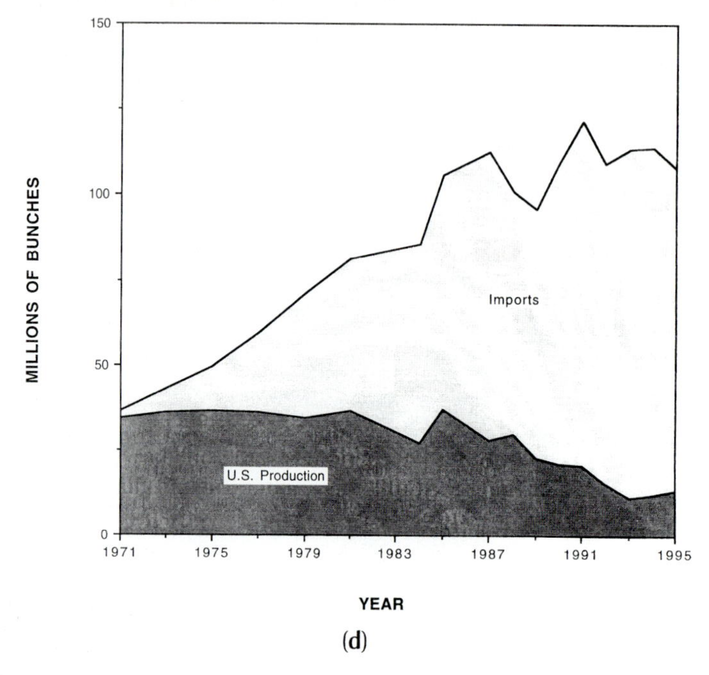

(d)

Figure 1-7 (continued)

mishandling. Roses require more sophisticated production and marketing. The Latin American industry, in the early stage of development, found it difficult to address these needs. However, this was a temporary obstacle, as the growing conditions in countries such as Colombia are conducive to quality production. Sizable quantities of roses began to be imported from Colombia, followed more recently by Ecuador, Guatemala, and Mexico (Table 1-2). Roses from Ecuador are rapidly gaining worldwide acceptance due to their exceptional size and quality.

During the first decade of fresh-flower importing, in the 1970s, the U.S. fresh-flower production industry compensated in part for the decline in domestic production of standard carnations and chrysanthemums by increasing production of miniature carnations and miniature gladiolis. During this period, demand increased in the retail market for smaller sizes of these flowers in arrangements. Due to increased production of these miniature crops and to rising prices of flowers, the total value of U.S. production of fresh flowers continued to increase throughout the decade (Figure 1-8). But averages do not tell everyone's story. Some growers, particularly those operating at a low level of efficiency, went out of business while others switched to containerized crops, mainly green plants (Table 1-3). Greenhouse fresh-flower production in states other than Colorado and California and outdoor production of chrysanthemums in Florida and Texas almost ceased.

During the next decade, the 1980s, a very fortuitous event occurred. The re-

Table 1-2 Countries from Which Fresh Flowers Were Imported into the United States in 1995[1] and the Percentage of Imported Flowers That Came from Each[2]

Country	*Carnation (standard)*	*Carnation (miniature)*	*Chrysanthemum (standard)*	*Chrysanthemum (pompon)*	*Rose (all)*
Colombia	94.23	94.23	97.46	90.42	66.33
Ecuador	3.43	3.43	1.26	1.36	20.99
Costa Rica				7.92	0.53
Guatemala	0.24	0.24	0.47		5.84
Mexico	0.42	0.42	0.39		5.16
The Netherlands	0.29	0.29		0.30	0.37
Bolivia			0.26		0.18
Israel	0.15	0.15	0.12		0.22
Spain	0.90	0.90			
Morocco	0.26	0.26			
Canada					0.27

[1]Only countries that represented more than 0.1 percent of total imports are listed.
[2]From Bureau of the Census (1996).

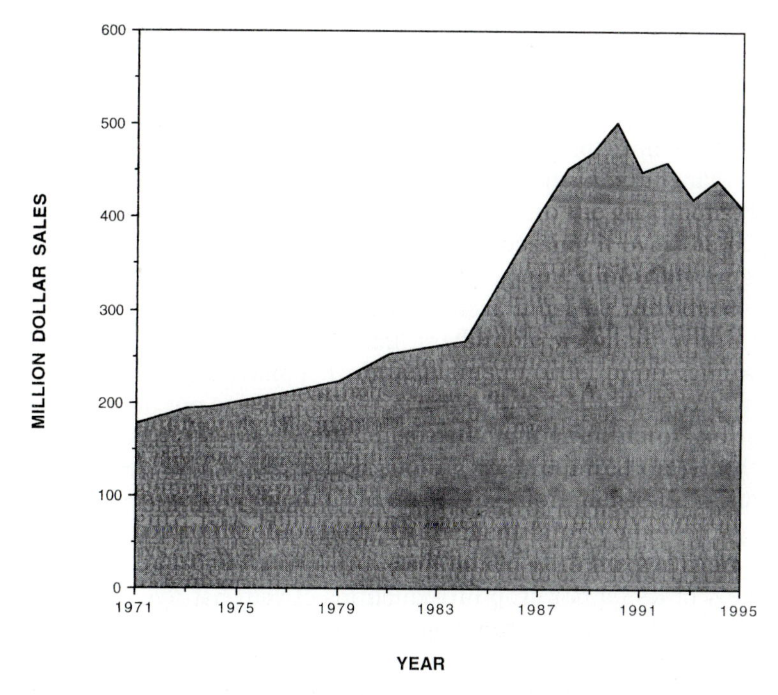

Figure 1-8 Wholesale value of fresh flowers produced in the United States between 1971 and 1995. *(From Agricultural Statistics Board, USDA, 1977 through 1996.)*

Table 1-3 **Total Number of Growers Producing Various Floral Crops in the Leading States Accounting for 90 Percent or More of U.S. Production[1]**

	Number of Growers			
Crop	*1971*	*1979*	*1989*	*1995*
Chrysanthemum, pompon	2,168	999	477	92
Chrysanthemum, standard	2,134	829	397	113
Carnation, standard	1,525	418	254	131
Rose, hybrid tea	323	238	285	175
Pot chrysanthemum	1,394	1,424	1,090	710
Green plants	835	1,687	2,094	1,186
Bedding plants	—	2,819	4,458	2,384
Poinsettia	—	1,977	3,069	2,044

[1]From Agricultural Statistics Board, USDA (1972, 1980, 1990, 1996).

tail floral industry fostered a shift in floral arrangement style. The trend called for freer forms containing new floral materials. A good part of the trend can be credited to the Dutch floral industry, which undertook an ambitious program to export fresh flowers to the U.S. market during the 1980s and met with considerable success. Interestingly, many of the flowers that they exported were new or minor crops in the U.S. market at that time. Such crops included alstroemeria, freesia, cut hybrid lilies, ranunculus, gerbera, and liatris. Introduction of those crops to the consuming public not only met with success but cultivated a demand for more of the same and for more new flowers. What started as a perceived disaster for the domestic floral production industry turned out to have a silver lining.

This demand for new flowers came at a very opportune time. During the 1980s, production of gladiolis and pompon chrysanthemums took a downturn. If not for the shift on the part of domestic growers to the new minor fresh-flower crops, total fresh-flower production might have declined during that decade. As it turned out, total production rose (Figure 1-8). New crops included alstroemeria, anemone, freesia, gerbera, liatris, lilies, and several others. Again, not all growers were successful in this shift. Some went out of business and others shifted to containerized crops, mainly flowering pot plants and green plants.

The value of fresh-flower production in the United States reached its peak in 1990 and has been in a general decline since. This is due in part to a decline in rose production after 1991 (Figure 1-7d). The selling price of roses reflects the high supply level, thus reducing profitability and ultimately the production level. The decline in domestic fresh-flower production is also a result of adoption of many of the minor crop introductions of the past two decades in the newly developing floral production areas of the world. As demand has been established in the developed markets of the world, these new crops have been adopted in the new production areas (see Table 1-4 for a list of crops exported from Colombia to the United States). Thus, crops that served to prevent a downtrend in production during the 1980s now face very strong competition during the 1990s.

The current shift in domestic fresh-flower production consists of the adoption of specialty crops to meet the continually changing acceptance of new flowers in arrangements. Many of these crops are being grown outdoors, especially in warmer climates. These crops include annuals as well as perennials. Some were previously wildflowers and others garden flowers. They include such plants as *Achillea,* ageratum, aster, *Aquilegia,* calendula, carnations from the garden cultivars, *Carthamus, Celosia,* cosmos, *Craspedia,* delphinium, *Echinacea, Eryngium,* godetia, *Gomphrena,* kalanchoe, *Lisianthus, Lupinus,* rudbeckia, sedum, verbena, and veronica. This list is not nearly complete. The optimism in such a list lies in the large number of potential plants not yet commercialized and in the wide range of climatic conditions required for growth. Growers in all regions should be able to find crops well adapted to their conditions in the future.

The import competition felt in the fresh-flower arena in the United States has had its counterparts in Europe and Japan. Within Europe a large increase occurred in the production of fresh flowers in the warmer southern countries of Spain, France, and Italy. Also, flowers are now being imported from African and Southeast Asian countries. These changes are similar to the shifts in the United States, first to southern fields, then to greenhouses in Colorado and California, and finally to foreign imports.

Table 1-4 **Fresh-Flower Crops Exported from Colombia to the United States in 1995**[1]

Crop	*Value (in U.S. dollars)*	*Percentage of Total Export Value*	*Number of Stems or Bunches*
Rose	116,260,940	31.6	
Carnation, standard	86,986,133	23.7	
Chrysanthemum, pompon	48,851,678	13.3	
Carnation, miniature	34,493,229	9.4	
Chrysanthemum, standard	7,098,635	1.9	
Other	73,927,214	20.1	
Alstroemeria			142,793,000
Daisies			344,000
Freesia			8,462,000
Gerbera			36,952,000
Gladiolus			65,000
Gypsophila (bunches)			6,094,000
Iris			402,000
Lily			994,000
Orchid			4,000
Statice (bunches)			3,189,000
Tulip			221,000
Other			228,429,000

[1]Data on value of crops are from Floriculture International (1996); data on number of stems and bunches are from Agricultural Marketing Service (1996).

Flower imports originate primarily from parts of the world along established trade routes. The Middle East and Africa are the likely origins of floral imports into Western Europe because of the established trade between these two parts of the world. Similarly, North America and South America are logical trade partners, as are Japan and the other Asian countries. Well-established means of transportation exist along each of these routes. The cost of shipping in these channels is inexpensive relative to shipping between parts of the world in different channels. While there are well-established trade channels between Japan and the United States, as well as Europe and the United States, the United States is not the recipient of significant quantities of floral imports from Japan or Europe (Table 1-5). The reason is that production and marketing costs do not vary

Table 1-5 **Global Distribution of Fresh Flowers Exported from Colombia in 1995 (Expressed as a Percentage of Total Exports)[1]**

Country	*Percentage of Total Exports*
United States	77.31
United Kingdom	6.95
Germany	2.45
The Netherlands	1.94
Canada	1.95
Sweden	1.33
Spain	1.33
Puerto Rico	1.13
Austria	0.99
South Ireland	0.74
Italy	0.65
Argentina	0.63
France	0.52
Finland	0.44
Japan	0.36
Antilles	0.36
Hong King	0.28
Switzerland	0.17
Martinique	0.11
Norway	0.11
Guadalupe	0.10
Kuwait	0.07
Greece	0.04
Denmark	0.03
Trinidad	0.01
Total export value	$475,782,614

[1]From Floriculture International (1996).

enough between these areas to offset the shipping costs. Labor is expensive in both areas and energy inputs are high.

This does not mean that no imports will travel along these channels. There are always niche markets. For instance, the United States imports proteas from South Africa, cut bulb crops from The Netherlands, and orchids from Thailand because each of these countries has production advantages not yet found in the Latin American countries that are our trading partners.

Green Plants

Green plants, plants sold in a pot and valued more for their foliage than for their flowers and also commonly referred to as foliage plants, made up 16.4 percent of U.S. floral production in 1995. They include philodendrons, dracena, ficus, croton, ferns, a wide range of hanging-basket plants, and many others. The size of this crop was very stable through 1970, with a wholesale value of about $25 million. After that, it exploded to a value of $282 million in 1978 (Figure 1-9). The area in production expanded 308 percent from 1968 through 1978, while the wholesale value increased 968 percent! In 1978, the market began to saturate, prices leveled off, and many marginally efficient producers perished. This stress fostered change through the surviving innovative growers. Attention to quality,

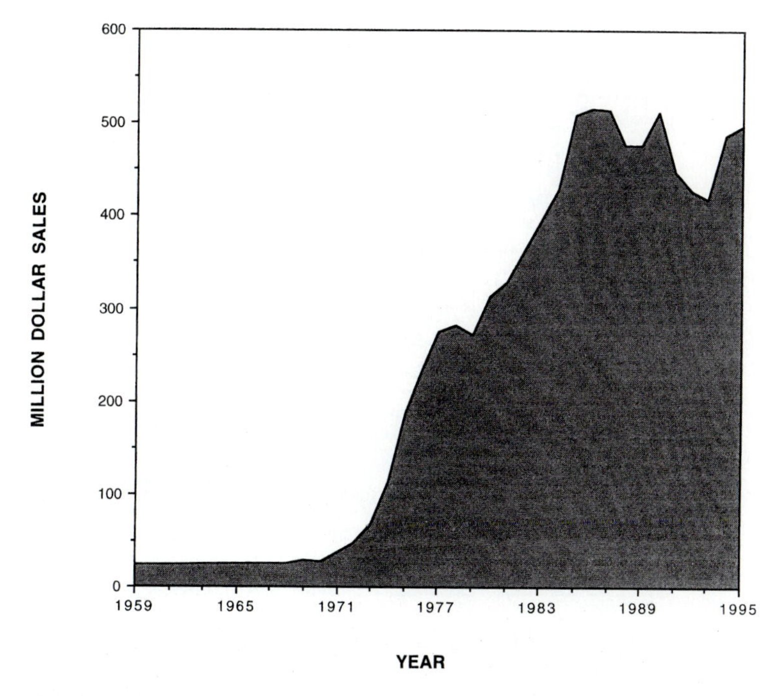

Figure 1-9 **Wholesale value of green (foliage) plants produced in the United States from 1959 through 1995.** ***(From Agricultural Statistics Board, USDA, 1960 through 1996.)***

acclimatization of plants to better guarantee their survival in the consumer environment, and sensitivity to the changing desires of the consuming public for smaller as well as larger pot sizes and new crops all led to new increases in demand for the years 1980 through 1985. The wholesale value climbed to $508 million in 1985. In 1986, the green-plant industry entered another period of adjustment and has had essentially zero growth since. One contributing factor was the large number of growers who were attracted to this commodity by the former boom period (Table 1-3).

Of the U.S. production value of green plants, 62.4 percent was grown in Florida, 17.2 percent in California, 4.5 percent in Texas, and 2.0 percent in Hawaii in 1995. These are the four leading states in production. Only a modest number of green plants, 13.9 percent of the total value, are grown in the other 46 states. Many of the green plants are of tropical origin and can be produced more economically in subtropical areas.

Northern areas, where heated greenhouses are required, have found a future in green plants, although it is more modest than that of the subtropical regions. Premium hanging-basket plants, being large in volume and cumbersome to handle, are expensive to ship. They are best grown close to their terminal markets. The production costs of hanging-basket plants are low, since many of the fixed costs such as greenhouse depreciation and heat are shared with another crop on the benches or the ground below. Hanging baskets offer a means for utilizing nearly 100 percent of the equivalent floor space of a greenhouse (Figure 1-10).

Figure 1-10 The expense involved in producing hanging-basket plants is shared with other crops since the baskets occupy space over the walks that was not formerly used. It is possible to achieve 100 percent utilization of the equivalent floor area of a greenhouse with such a production program.

Many greenhouse enterprises in temperate regions purchase green plants in the final stage of development from the subtropical regions and hold them in their greenhouses until they can sell them to the network of retail outlets they service. Other green-plant crops are purchased in various intermediate stages of development and are grown to the finished market stage in the temperate-region greenhouses.

The rapid rise in green-plant sales began one year after fresh-flower imports began their rapid ascent. Further, one of the two major fresh-flower production areas initially stressed by imports was the field production areas of Florida, California, and Texas. The demise of one production segment within the same areas in which the rise of a second segment took place presented an opportunity for some growers. These were growers who realized what was happening and who had the fortitude to change.

Flowering Pot Plants

Notable among the *flowering pot plants* are African violet, azalea, begonia, gloxinia, hydrangea, kalanchoe, lilies (including Easter and hybrid types), poinsettia, pot mums (potted chrysanthemum for indoor use), and spring flowering bulbs. The production value of this group of crops increased steadily in the United States through 1992 and then declined for the first time in over two decades (Figure 1-11a). Flowering pot plants made up 22.5 percent of the value of U.S. floral production in 1995. The long steady increase in production was due to an increase in consumer demand for the traditional plants—a renewed interest in crops that had waned during the middle of this century, including calceolaria, cineraria, cyclamen, and kalanchoe, as well as for new crops such as *Clerodendron,* exacum, gerbera, and Rieger begonia. However, many of these crops are not new to Europe, where a much wider variety of plants are grown commercially. The recent drop in production, is due in great part to a decline in pot mum production since this is one of the most valuable crops within the group (Figure 1-11b). Production of the most valuable crop, poinsettia, and of lilies has continued to rise (Figures 1-11c, 1-11d).

The decline in overall flowering pot-plant production may also be a reflection of imported pot plants. Nearly all imported flowering pot plants come from Canada. Approximately 4.9 million pot mums were imported from Canada in 1995. These represented 13.3 percent of pot mums sold in the United States. At the same time, 1.46 million poinsettias were imported from Canada. These constituted 2.5 percent of the poinsettias sold in the United States.

The future still looks optimistic for flowering pot-plant production. However, one must look carefully at the trends of individual crops. One concern for the future is that of quarantines. Quarantine 37, aimed at preventing the influx of disease and insect problems, has inadvertently served to keep the import of pot plants to a minimum. Exemptions from such regulations are being sought by European and Latin American producers. Partly because of the relatively inert, soilless root substrate used today, such exemptions may increase. It was possible

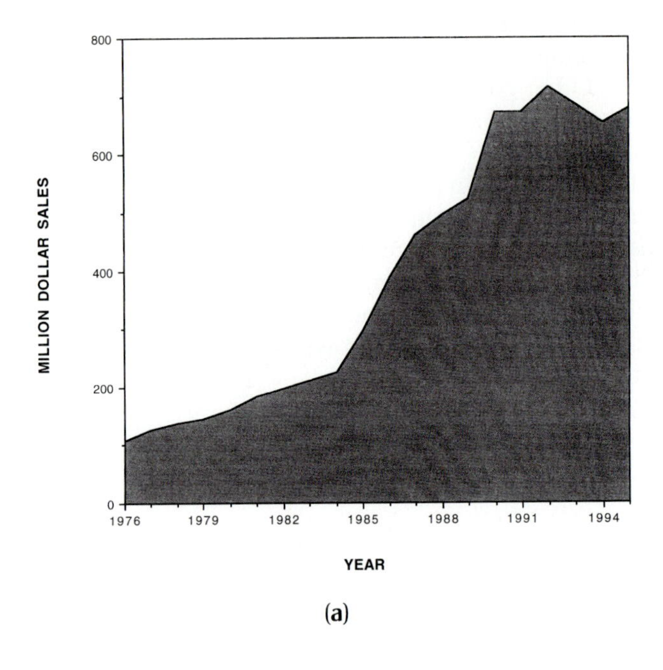

(a)

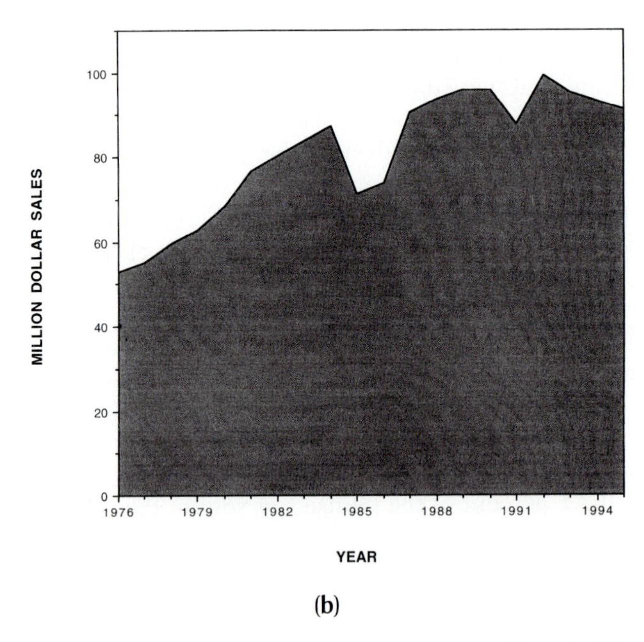

(b)

Figure 1-11 Wholesale value of (a) all flowering pot crops, (b) potted florist chrysanthemum, (c) poinsettia, and (d) all types of pot lilies produced in the United States from 1976 through 1995. *(From Agricultural Statistics Board, USDA, 1977 through 1996.)*

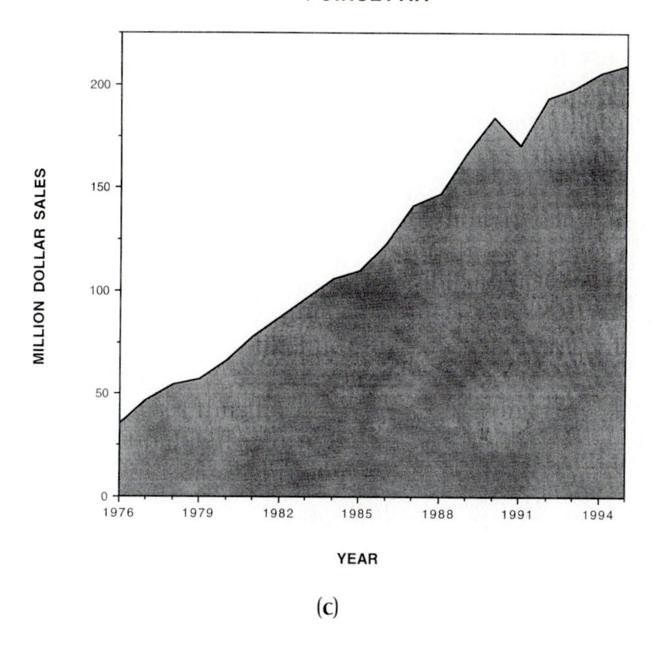

(c)

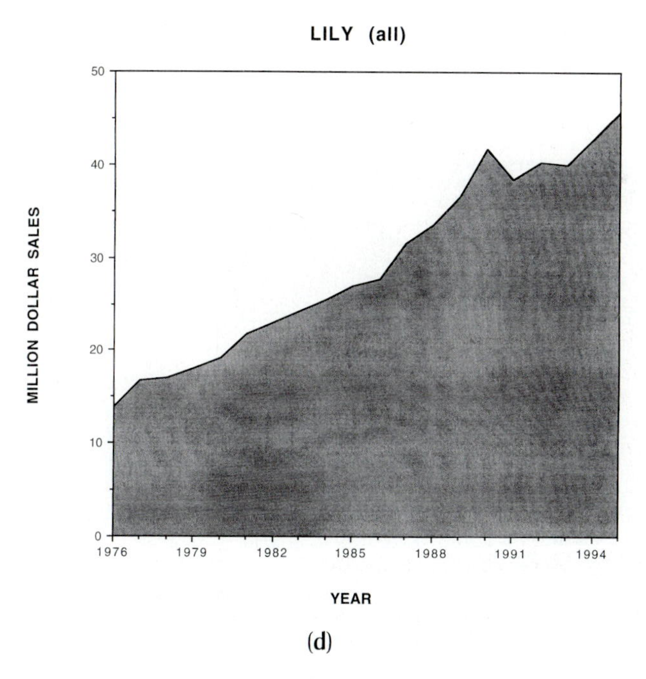

(d)

Figure 1-11 *(continued)*

for fresh-flower growers to shift to green and flowering pot plants and bedding plants when import competition increased. Such fertile fields do not exist for pot-plant producers. It is becoming more important for domestic growers to increase their production and marketing efficiency. In short, we are moving steadily toward a single international floriculture industry in which production of each commodity category will go to that group of growers who can best manage costs.

Bedding Plants

Bedding plants are a unique group of plants (Figure 1-12). Over 100 plant species are grown. The bedding plant group includes annual and perennial floral plants as well as vegetable plants in flats and small pots for planting in the garden, hanging baskets and bowls of flowering plants, geraniums from seeds or cuttings in flats and pots, and hardy garden chrysanthemums. The wholesale value of bedding-plant production in the United States in 1995 was $1.325 billion, or 44 percent of total floral production. Total bedding-plant production value (Figure 1-13) and bedding-plant production value as a percentage of total floral production has increased steadily for the past two decades. Bedding plants have provided an opportunity for growers forced out of fresh-flower and green-plant production in the United States, Europe, and Japan. The future looks excellent for bedding plants.

A significant change in greenhouse technology during the past fifteen years has been the production of seedlings as "plugs." While this method of production is used extensively in the bedding-plant industry, it is not restricted to it. Within

Figure 1-12 A bedding-plant crop at Rockwell Farms, Rockwell, NC.

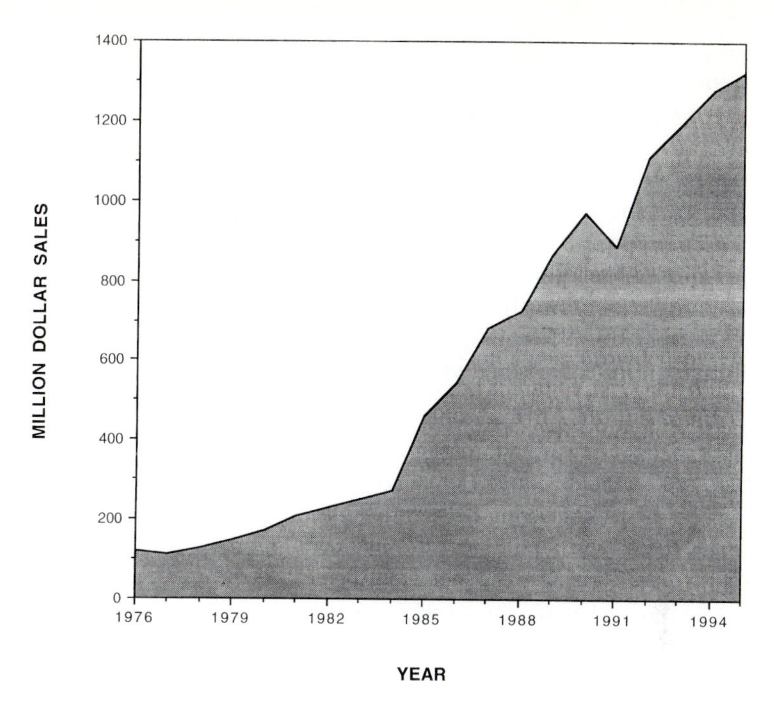

Figure 1-13 **Wholesale value of bedding plants produced in the United States from 1976 through 1995.** *(From Agricultural Statistics Board, USDA, 1977 through 1996.)*

the floral industry, a number of fresh-flower, green-plant, and flowering pot-plant crops are also established this way. In this system, seeds are mechanically sown in trays containing from less than 100 to more than 800 small cells. One plant is produced in each cell, as opposed to broadcasting seed in an open flat and later digging the seedlings out for transplanting. Customarily, plug seedlings are germinated in ideally controlled growth rooms. From there, the plug trays are moved to the greenhouse, where they remain at a high density for several more weeks, depending upon the plant species. Automatic spray booms deliver water and fertilizer, and again conditions are idealized for this second stage of growth. Such plug seedlings may be used by their propagator or sold to smaller firms that cannot afford the equipment and facilities for producing plugs. The highest labor input into bedding-plant production is the transplanting of seedlings. Mechanical transplanters are available that will remove plug seedlings from trays and plant them into flats that have been automatically filled with root substrate.

Plug technology is revolutionizing the bedding-plant industry and greatly aiding growers of flowering and green pot-plant crops that are propagated from seed. The advantages include lower overhead costs (because the seedlings can be held at high densities for a considerable time); a better chance to establish ideal cultural conditions during the early stages of growth for the purpose of reducing crop time; less transplanting shock, which leads to faster finish-plant production;

and reduction in sowing and transplanting labor. An equally great advantage that plug seedling technology has brought to the floral industry is the door that has been opened to other plant industries. Most tobacco seedlings, many forest-tree seedlings, and numerous vegetable seedlings for commercial outdoor production are now produced by plug technology.

As with most technological advancement, the firms that first adopt it are able to translate the advantage into large profits. As the technology is more widely adopted, the savings are passed on to the consumer in terms of lower prices. This stems from those producers who are trying to gain a greater local market share by lowering their prices below those of their competitors. After this point is reached, the remaining producers must adopt the technology simply in order to maintain their current profitability. Individuals entering the bedding-plant field should study the currently changing plug technology in order to assimilate it into their business.

Floral Markets

Western Europe, Japan, and the United States collectively make up slightly over 90 percent of the world's $37 billion floral retail market. The U.S. portion is approximately $12.5 billion, or 34 percent. Each region arrived at its current profile of market channels at different times and in different ways. However, the current profile of each and the impending changes facing each are very similar.

The Traditional Florist

Traditionally, Americans purchased their green plants, flowering pot plants, and fresh flowers at full-service retail flower shops. These shops also provide the services of floral arrangement and delivery. Traditional florists were characterized in the past by nonspontaneous purchases. Eighty-five percent of fresh-flower sales went into funeral and wedding orders in the early 1950s. Traditional holiday sales accounted for another heavy proportion of fresh-flower and pot-plant sales, including such demands as Easter lilies at Easter, poinsettias at Christmas, and roses on Valentine's Day. Apart from the strongly motivated wedding, funeral, and floral holiday customers, routine customers represented only a small percentage of the American public—about 25 percent.

This historic profile has changed considerably. During the mid- to late 1960s, floral mass-market channels began to develop, such as in supermarkets and discount stores. While the mass markets reached many new customers, they also reached some of the traditional florists' customers. This created a period of serious economic challenge for traditional florists. Changes brought on by this pressure have included offering simpler floral bouquets in addition to arrangements, selling individual flowers by the stem, including smaller plants in the line of pot plants, relocating to higher pedestrian traffic areas, and embracing a higher degree of self-service. Over 40 percent of the total value of floral sales pass through traditional florist shops today.

A distinguishing feature of traditional florists is the services they sell in addition to the floral material. Fresh flowers are professionally arranged beyond the competence of the average consumer. Pot plants are wrapped in foil and tied with a bow. These products carry the prestige of the flower shop's name. The convenience of home delivery is an important part of the service. Traditional florists also find support through the services of organizations referred to as *wire services,* such as 1-800-FLOWERS, AFS Florists, America's Florist, Florists' Transworld Delivery, National Flora, and Teleflora. Customers can access a wire service directly by telephone or through a local florist affiliated with a given wire service and place an order for flowers or plants that will be arranged and delivered nearly anywhere in the world by a florist in that locality. Wire services account for about 6 percent of total floral sales.

Services cost the retail florist dearly in trained personnel and in physical overhead such as buildings and trucks. These costs must be passed on to the consumer. It is not uncommon for floral products to be marked up to a retail price three or four times the wholesale price paid for them. While this markup may seem exorbitant, it is not. It is fair in light of the overhead expenses of the flower shops and is acceptable to the customers who frequent these outlets. The emotional effect of a gift of flowers received by a mother some 3,000 miles away from her children on her birthday is very significant. The impact of a florist's van pulling up to the home of a hostess to deliver flowers for an elegant evening gathering is as great as that of the flowers alone.

The Mass Market

Not everyone places emphasis on prestige and service, nor can everyone afford the prices that a full-service florist must charge. As is the case with so many other commodities, new methods have increased the availability and decreased the retail price of floral products, making repeat purchases possible for a larger proportion of the public—some of the 75 percent of the population who were not reached in the past.

Mass marketing is the sale of floral products through outlets in high-traffic locations such as supermarkets, large discount stores, garden centers, and plant boutiques in shopping malls, on busy street corners, and in airport terminals (Figure 1-14). The objective is a high-volume business. Since purchases are usually spontaneous, prices must be sufficiently low to attract a customer's discretionary dollars (money remaining after the necessities of life are purchased). Thus, costly services such as floral arranging and delivery are often avoided. Plants such as Easter lilies would be sold in plastic pots, whereas in traditional flower shops they might be in more expensive clay pots. Other pot plants might be in the same size pot that is sold by the traditional florist, but the plant was allotted less space in the greenhouse during production. The result is a moderately

Figure 1-14 This streetside outlet for fresh flowers in San Francisco is one of the forms mass marketing has taken in recent years. Others include no-service cash-and-carry shops in supermarkets, airports, and shopping malls. *(Photo courtesy of J. C. Raulston, Department of Horticultural Science, North Carolina State University, Raleigh, NC 27695-7609.)*

smaller plant produced at a lower cost. Since the price received is different, growers often alter their production techniques depending on the market they are servicing. For example, a poinsettia in a 6.5-inch (17-cm) pot might sell for $4.50 to a traditional florist but only $2.75 to the mass market.

Floral mass marketing was developed long ago in Europe, where it is now highly perfected. It developed later in the United States and Canada. The potential of mass marketing, however, was recognized well before that time. The results of the Northeast Regional Marketing Project NEM-8, conducted in the 1950s (Zawadzki, Larmie, and Owens, 1960), indicated optimistic prospects for mass marketing of fresh flowers through nonfloral outlets regularly patronized by consumers. Initial development was slow, partly because of the initiative and resources needed to bring about such a change in production and marketing and partly because of resistance within the industry. Many conventional florists in the 1950s and 1960s boycotted growers who sold in the mass-market channels. Taken at face value, mass marketing appeared to be a great threat to the continued existence of the conventional florist market system. Lower prices and high volume struck a chord of fear, in spite of optimistic predictions coming out of market studies. Such studies indicated that flowers purchased from mass-market outlets would be for a different purpose than those purchased through the already established florists' outlets. These studies further indicated that mass-market sales would help to create a greater appreciation and desire for flowers in the United States and thereby enhance sales in the conventional florists' channel. These predictions have come true.

Eighty-two percent of supermarkets in the United States have a floral department. This is up from about half in 1989. Over 20 percent of the value of floral sales passed through supermarkets in 1995, compared to 16 percent in 1989. Fresh flowers, green plants, and flowering pot plants are sold in these outlets. The outlook is very optimistic for supermarkets. A current trend within supermarkets is to offer more services. Twenty-five percent of supermarkets now include a full-service florist offering custom arrangements. One chain has its own wire service.

Another major category of mass markets is the discount stores such as K-Mart, Target, and Wal-Mart, plus the home-improvement stores such as Home Depot and Lowes. Large volumes of green plants, flowering pot plants, and bedding plants are sold in this channel. It is a rapidly growing market. The discount/home-improvement stores have presented a special challenge to growers. Because of their large size relative to growers, they have been able to dictate price to a large extent. This has made it nearly impossible for small growers and difficult for midsize growers to deal in this channel. Even the larger growers with 25 acres (10 ha) or more of greenhouses find it difficult to develop the required levels of production and marketing efficiency necessary to maintain a reasonable profit. This situation has prompted the development of larger greenhouse businesses to better realize an economy of scale through automation. In 1995, the Fremont Group made an attempt to buy into a partnership with many of the larger bedding-plant firms in the United States, in order to gain a controlling share of this commodity in the country. Had it succeeded, this new firm would have been in a position to equitably negotiate prices and sales terms with colossal discount/home-improvement firms such as Wal-Mart. Currently, Color Spot Nurseries, Inc. of California is acquiring bedding-plant production firms in California with the goal of achieving 50 percent of the market share in bedding plants on the West Coast. In mid-1996 the company had about 500 acres (200 ha) in production, with 100 acres (40 ha) of greenhouses and 60 acres (24 ha) of shade houses.

The mass-market channels offer daily exposure of floral products to greater numbers of people than would have been possible through traditional florists. This in turn leads to new customers who are not tradition bound. These new customers purchase new crops; accept new colors, forms, and sizes; and are receptive to new outlets.

Future Market Potential

Floral production in the world has caught up with market demand. This problem for the producer is further compounded by world floral demand that leveled off during 1994 and 1995. More needs to be done to stimulate market growth. The potential is there for growth. Only 57 percent of households in the United States purchased floral products in 1995.

Promotional programs have met with great success in the floral industry. During a recent Christmas season the wire service company 1-800-FLOWERS promoted their new item, the Holiday Flower Tree. The tree is actually an arrangement made from a treelike base of boxwood and evergreens decorated with flowers. They planned to sell 20,000 units, but when the season ended, 340,000 trees had been sold. The Society of American Florists (SAF), serving as a parent organization to all segments of the industry, including growing, transportation, wholesaling, and retailing, has made great efforts through the American Florists' Marketing Council (AFMC) to collect voluntary contributions from all segments for the purpose of promoting floral products nationally. This program has had a great impact on expanding sales.

Unfortunately, not enough money has been invested in floral promotion. When one takes into consideration the whole floral industry in the United States, probably less than 1 percent of retail value is spent on promotion. No other viable industry would attempt to promote itself on such a weak basis. More must be done nationally. Fortunately, this need is recognized and is being pursued. In 1994, the PromoFlor program was signed into law in the United States with the charge to develop a national generic promotion and information program to maintain, expand, and develop markets for fresh flowers and cut greens. Originally, PromoFlor was proposed to cover the broad area of floral production. That proposal did not receive a sufficient vote from the floral industry. A scaled-down proposal to cover fresh flowers and cut greens passed. All wholesalers of these commodities with gross sales totaling more than $750,000 per year pay an assessment. The program collects about $12 million per year. The spokescharacter for PromoFlor is Buzz the Bumblebee, who appears on television, radio, and posters, as well as in person at various events.

Equally important to funding promotional programs is the vision of where potential markets exist. Current research addressing this question identifies the following profiles of consumers. With the aging of the baby boom generation, nearly 40 percent of Americans are age 50 or older. This segment has considerable resources to purchase floral products. But these purchases will reflect the age of the group. A situation that baby boomers and younger people share is lack of time. Outlets must be quickly accessible, near the workplace or in established shopping areas. The population is much more individualistic in its choice of products and places less value on product or company loyalty than in the past. This opens up the door for continual change in types of plants or arrangements, sizes, colors, and end uses of the purchases.

Another marketing technique has arisen out of the fact that the proportion of people working out of their homes is rising rapidly. Many of these individuals use computers for their work. Such individuals are not commuting to work and as such are not in contact with floral outlets as much as others. There is an excellent opportunity here to reach them through the computer. When 1-800-FLOWERS launched a World Wide Web site to market flowers for a three-month trial period, they received 15,000 to 30,000 visits per day to their Internet site. About 1,000 orders were completed, with an average value of $60. The total sales through this channel were anticipated to be $15 million for the first year.

Another trend that bears scrutiny is the widening gap between high- and low-wage earners. Middle-income America is expected to continue diminishing. This will increase the proportion of people purchasing higher-end products. On the other hand, it will impose changes in the mass market, which currently reaches the middle-income group. Opportunities abound for those retailers and growers who meet these challenges.

Production Opportunities Near Developed Markets

Small Growers

Small growers will always have a place in the floral industry. Niches that can best be handled by these growers include superior quality, new crop introductions, low-volume specialty crops, and integration of production and retailing. It is imperative to provide the level of quality sought in a given market. But the level of quality sought varies in different markets. Most people purchasing standard chrysanthemums for an arrangement in their home would be satisfied with 5-inch (13-cm) -diameter blooms. While they would accept larger blooms for the same price, they probably would not pay extra for them. The 5-inch bloom meets their perceived need, given the dimensions of the room in which they will display these flowers. On the other hand, a five-star hotel ordering flowers for its lobby perceives a need for larger flowers and expects to pay more. Beyond a given point, increases in quality are met with diminishing demand. It is difficult for large production firms to offer ultrahigh quality levels. Chrysanthemums, for example, require a longer growing time and greater space allocation. A small grower is better positioned to meet the smaller demands for high quality.

A period of time is required to develop the full market potential after a new crop is introduced. The demand during this developmental stage is too small to lend itself to the high level of automated production required by larger firms. This window of time offers an advantage for small growers because it is often possible to command higher prices than will be possible later when the market is saturated. Within the fresh-flower category, numerous new introductions are possible if one simply looks at the availability of annual and perennial flowers, vegetables, and wild plants. Recent introductions in the marketplace have included ornamental kale on long 30-inch (76-cm) stems and long stems of pepper plants bearing fruit but not leaves. Given time, plants with large market potential will be adopted by the larger domestic growers and then may even shift to the developing countries.

A number of specialty crops have a small demand even after the market potential has been met. Since it is difficult for a large grower to produce large numbers of low-volume crops, these crops are best left to the small growers. Such crops could include bonsai plants, terrarium plants, aquatic plants for fish tanks, plants with unusual fragrances, rare plants for plant collectors, and collections of a given category of plant such as begonia, geranium, or carnivorous plants.

Integration of production and retailing is another vehicle that small growers can use to carve out a niche for themselves. Value can be added to greenhouse products in several ways that would support higher sales prices. Many people have never been in a production greenhouse and would enjoy the experience enough to consider it an outing. Taking this one step further, blocks of plants ready for market could be designated for self-selection. Customers would select their own plants and carry them to the cashier. This could be done for fresh flowers as well as pot plants. Customers would pick their own flowers and pay by the stem. To foster such a fresh-flower program, a point might be given for every dollar's worth of purchases. A given number of points would entitle the customer to a class in floral design. A class in houseplant care might be offered for pot-plant purchases. Other services sometimes found in production/retail greenhouses include free root substrate, fertilizer solution, or repotting. Each of these services can translate into higher sales prices above the retail market average.

Large Growers

Large production firms will continue to gain a greater proportion of the market for main crops, at the expense of smaller growers. These crops include flowering pot plants such as Easter lilies, geraniums, gloxinia, poinsettia, pot mums, the principal bedding-plant annual and perennial species, and major green-plant crops. These firms will continue to become larger to justify more automation to lower production costs and to become more efficient in marketing.

Midsize Growers

Midsize growers may be in the most difficult situation. Their production level of the main crops may not lend itself to the levels of automation possible for the large growers, and many small niche items enjoyed by the small growers may not provide the required revenue for their business scale.

However, midsize growers should be in a good position to service traditional florists. The sizes of orders delivered to each florist are small compared to deliveries to discount stores. Large trailer trucks often cannot be accommodated at florist shops, yet these trucks are mandatory for the level of marketing economy required of large producers. The cost of marketing is higher to traditional florists than to the mass market, and the selling price is generally higher. The economics of this situation does not suit the large grower very well.

Cooperative arrangements among several midsize growers are another option. In such arrangements, each grower could specialize in a given crop or category of crops. The larger quantity of each crop grown would foster automation and lower production cost. Marketing could be carried out more efficiently by a single department representing all growers in the cooperative.

Midsize growers marketing to the mass market will have to be extremely efficient. Analysis of production and marketing costs will be essential. Those crops that return a profit should be increased and those that do not could be purchased in the finished stage for satisfying customer demand. Another option to consider is the purchase of prefinished plants from other larger growers. These are plants that are grown to an intermediate stage by one grower and then sold to a second *finish grower,* who carries them through their final stage of production. An example would be pot mums that have received supplemental long-day treatment, have been pinched, have been treated with a chemical height retardant, and are now ready for the final stage of production at a cooler temperature. Another example would be bedding plants already transplanted into flats. In this case, the finish grower does not need to purchase seeds, root substrate, and flats, or provide the labor of transplanting, which is the greatest labor input into the production of this crop. These items are provided by a larger grower who can purchase at a much lower discounted price and can afford expensive transplanting equipment.

Careers in Greenhouse Management

The greenhouse industry offers a wide variety of career opportunities. The range of employers traverses greenhouse firms, wholesalers and retailers, allied supply and facilities companies, private and governmental associations, private and governmental extension services, high schools and universities, industrial and governmental research labs, publishers, and others. A wide mix of career interests can supplement a horticultural background, such as business, statistics, journalism, education, computer science, basic biology, and plant breeding. The world greenhouse industry has a bright future that will expand as a function of economic development of countries and population growth.

Individuals interested in plant production can set their goals on conservatory management or commercial greenhouse production. Those interested in conservatory management should supplement their training in the areas of plant identification and ecology. Anyone planning a career with a horticultural production company or an agency offering advice relative to crop production should seek additional training in business. This is very important whether one plans to own his or her own business or assume a management position within a firm. In either event, it is imperative that the person be capable of assisting the firm in generating profits that more than offset his or her own salary.

There are many careers in the wholesale marketing of greenhouse products. These may be found in auctions, wholesale houses, or brokerage firms around the world. Likewise, many other opportunities exist in the retail channel. These may be in florist shops, garden centers, plant departments of discount stores, or wire service companies. Numerous companies supply the greenhouse industry with production supplies, agricultural chemicals, seeds, plant material, equipment, and greenhouse facilities. Many positions within these companies require knowledge of greenhouse management.

Private plant consultant firms and governmental extension services also employ people with greenhouse training. A B.S. degree is generally required for county-level positions, an M.S. degree for directorship of county offices, and a Ph.D. degree for statewide positions. Along these same lines of education, the teaching profession is open to individuals trained in greenhouse management. Those with a B.S. degree can teach horticulture courses in high schools. An M.S. degree is usually sufficient for teaching in technical and junior colleges, and a Ph.D. degree is the norm for teaching in universities.

The field of research offers careers as well. Graduates with B.S. degrees are often employed as research technicians for companies and governmental institutions. Research project leaders generally hold M.S. degrees in smaller companies and Ph.D. degrees in larger companies. These companies may be producers of seeds, plant material, pesticides, organisms for biological pest control, or a wide variety of other products.

Other less obvious careers include a wide array of possibilities, including plant quarantine, horticultural statistics, writing for avocational or professional magazines, and management of growers' and marketing associations.

It cannot be emphasized too strongly that technical knowledge in horticulture is only half of what it takes to succeed in this industry. You must become equally well versed in principles of business management and marketing. While the foundation for this education lies in formal courses, ideally it will be expanded by an educational process that you will maintain throughout your career. You can do this by reading books on the subject, subscribing to business periodicals, and most important, establishing communications with people in other phases of the greenhouse industry. Knowledge of the responsibilities of these people is your responsibility, even though you are not directly involved in their work. In addition to attending your local greenhouse growers' association meetings and short courses, you should become accustomed to participating in conferences held by associations representing other areas such as transportation, wholesaling, and retailing.

No knowledge is of value unless there is a mind to assemble it into a plan and a spirit to activate it. If you feel a glow when you handle a plant, or a thrill when you walk into a greenhouse, then you are in the right field of endeavor. Make up your mind now that this is your field and that you will succeed. Do not waste further energy doubting yourself. Instead, direct your efforts toward learning how to succeed and formulating a plan. Take the time now to write down on paper what it is you want to achieve in life. This book will take you through many of the decisions you face. As you study the facts, reflect on how they can fit into your plan. Seek further knowledge to fill in your plan by perusing the references listed at the end of each chapter, attending local growers' short courses, establishing a relationship with commercial people, and, best of all, taking a job in the area of your choice—perhaps part-time during your school year or during the summer. Sooner or later, you must gain practical experience to supplement your book learning if you are going to be able to apply it. It is not your background or your current aptitude that will bring you success, but rather your knowledge of what you want out of life, your belief in yourself, and the persistent effort you are willing to put forth to achieve your goal.

References

Excellent information about international floriculture can be obtained from the monthly periodical *FloraCulture International,* available from P.O. Box 9, Batavia, IL 60510-0009, and the annual *AIPH Yearbook of International Horticultural Statistics,* produced by the International Association of Horticultural Producers, P.O. Box 930099, 2509 AB's-Gravenhage, The Netherlands.

1. Agricultural Marketing Service. 1996. *Fresh fruits, vegetables, and ornamental plants: weekly summary—shipments and arrivals.* Agr. Mkt. Ser., Fruit and Vegetable Div., P.O. Box 96456, Washington, DC 20090-6456.
2. Agricultural Statistics Board. 1996. *Floriculture crops: 1995 summary.* USDA, National Agr. Statistics Ser. Sp. Cir. 6-1 (96). USDA, Washington, DC. (Available annually except in 1983 and 1984.)
3. Ball, V. 1976. Early American horticulture. *Grower Talks* 40(3):1–56.
4. Bureau of the Census. 1996. *U.S. imports of merchandise: cut flowers and flower buds nesoi, fresh.* U.S. Dept. of Commerce, Bureau of the Census, Foreign Trade Div., Washington, DC 20233.
5. FloraCulture International. 1996. 1995 Colombian export overview. *FloraCulture International* 6(11):42–43.
6. Johnson, D. C. 1990. *Floricultural and environmental horticulture products: a production and marketing statistical review, 1960–88.* USDA, Economic Res. Ser., Commodity Economics Div., Statistical Bul. 817.
7. Kaplan, P. 1976. Origins of commercial floriculture in U.S. found to predate Declaration of Independence. *Florist* 10(2):39–46.
8. Kiplinger, D. C., and R. W. Sherman. 1962. *Florist crops for mass market outlets.* Ohio Agr. Exp. Sta. Res. Bul. 928.
9. Staby, G. L., J. L. Robertson, D. C. Kiplinger, and C. A. Connover. 1976. *Proc. National Floricultural Conference on Commodity Handling.* Ohio Florists' Association, 2130 Stella Ct., Suite 200, Columbus, OH 43215-1033.
10. Zawadzki, M. I., W. E. Larmie, and A. L. Owens. 1960. *Selling flowers in supermarkets.* Univ. of Rhode Island Agr. Exp. Sta. Bul. 355.

CHAPTER 2

Greenhouse Construction

In the United States, the term *greenhouse* refers to a structure covered with a transparent material for the purpose of admitting natural light for plant growth. The structure is usually heated artificially and differs from other growing structures, such as cold frames and hotbeds, in that it is sufficiently high to permit a person to work from within. The European definition of a greenhouse differs in that it refers to a structure that receives little or no artificial heat; the term *glasshouse* is used in Europe to refer to an artificially heated structure. Quite frequently, two or more greenhouses in one location are referred to as a *greenhouse range.* A building associated with the greenhouses that is used for storage or for operations in support of growing of plants, but is not itself used for growing plants, is referred to as a *headhouse* or *service building.* Greenhouses are to be found in many designs, including the conventional A-shaped, Quonset, and gutter-connected types. The transparent coverings are as varied as the designs. Originally, glass was used, but now film plastics, fiberglass-reinforced plastic (FRP), acrylic panels, and polycarbonate panels and sheets are used as well. The future holds promise of new covering materials that will reduce the burden of heating and cooling and also of new frame designs that will be more economical.

Location

The first consideration in establishing a greenhouse range is that of location. Several factors to be considered follow.

Room for Expansion

A parcel of land larger than the grower's immediate needs should be acquired. The ultimate size of the range should be predicted. Area should then be added to this predicted figure to accommodate service buildings, storage, access drives, and a parking lot. Additionally, extra space should be allotted to cover unforeseen needs. To meet environmental codes of some municipalities, it is necessary to use holding ponds for water effluent from the range in order to reduce nutrient release into streams. Doubling the area covered by greenhouses would constitute a bare minimum land requirement.

The floor area of service buildings required for small firms equals about 13 percent of the greenhouse floor area. This requirement diminishes with increasing firm size, 7.5 percent of the growing area for large firms with 400,000 ft^2 (37,000 m^2) of greenhouse area. On the average, service buildings occupy 10 percent of the growing area (Brumfield et al. 1981).

Topography

The service building and greenhouses should be on the same level for easy movement of personnel and materials and to permit maximum automation. Thus, the building site should be as level as possible to reduce the cost of grading. The site should be well drained. Because of the extensive use of water in greenhouse operations, providing a drainage system is always advisable. Where drainage is a problem, it is wise to install drainage tile below the surface prior to constructing the greenhouses. It is also advisable to select a site with a natural windbreak, such as a treeline or hill, on the north and northwest sides. In regions where snow is expected, trees should be 100 ft (30.5 m) away in order to keep drifts back from the greenhouses. To prevent shadows on the crop, trees located on the east, south, or west sides should be set back a distance of 2.5 times their height.

Land-Use Prediction

Local zoning and tax laws are subject to changes brought on by development pressures. Such changes have brought about the termination of many greenhouse businesses, as witnessed by the extensive disbandment of the once-vast greenhouse industry on Long Island. The past development of the location in question should be carefully studied in order to assess its future direction. Some local governments classify greenhouses as agricultural businesses to protect them from prohibitive property taxation due to zoning shifts. Others, in order to change the occupants within a zone, have denied expansion permits to floral production businesses.

Climate

As indicated earlier, climatic conditions have dictated worldwide geographical shifts in horticulture. Such forces are also at work within local regions. The primary limiting factor to crop production in greenhouses is low light intensity during the winter. Areas where there is frequent fog, inclement weather, or shadows from the north slope of tall mountains are poor for crops in general. The better light intensity of higher altitudes is advantageous for crops in general and particularly for the high-light-requiring crops such as carnation and rose. The advantage is much lower for crops with a requirement for low light intensity, such as African violet, begonia, gloxinia, and most green plants. The greenhouses of one carnation range that was forced to terminate operations on Long Island were disassembled and trucked to carefully selected high-elevation sites in the Appalachian Mountains and then reconstructed. These sites were located well above the customary morning fog layer of this region and enjoyed high light intensity and cool summer temperatures—conditions ideal for carnation growth. Subsequent production records testified to the successful selection of these sites.

Labor Supply

Present and future labor needs should be assessed and should be in accord with the labor supply of the area. Procurement of a labor supply has been a perennial problem in the horticulture industry. While the solution has appeared to rest on locating close to an urban area, this does bring on a problem of wage level. Traditionally, greenhouse wages have been low, which has given the labor-recruitment advantage to the more technologically advanced industries. The solution appears to lie in meeting the competition directly through higher wages. Higher wages can be offset by automation, which reduces the number of employees but increases the productivity of each.

Accessibility

A site should be selected where shipping routes are easily accessible. Marketing of floral crops costs approximately one-quarter of the gross wholesale return (Brumfield et al. 1981). Minimization of shipping costs by close proximity to the target markets will go a long way toward alleviating this burden. At the same time, local carrier costs for goods received will be reduced.

Site location is often the deciding factor in the type of fuel used. In some regions, natural gas is a cheaper source of energy than other fuels. Some greenhouse ranges are not able to take advantage of this factor because their location is at a prohibitive distance from the gas line, while the competition located near the line can enjoy this advantage. In one situation where a greenhouse range was built at a high altitude to take advantage of light conditions, the remoteness of the location necessitated the transfer of oil from large tank trucks to smaller trucks during delivery, thus raising the cost of the oil.

Water

Water is one of the most frequently overlooked resources in the establishment of a greenhouse business. Before a site is purchased, the available water source should be tested for quality (see Chapter 8) and quantity. There have been several cases in which businesses located in coastal and riverbed regions were compelled to move to new locations to obtain water of suitable quality. The cost of removing ions such as sodium, chloride, and bicarbonate can reduce profitability, but failure to do so results in plant injury. Water quantity is equally important, since as much as 2 quarts of water can be applied to 1 ft^2 (20 L/m^2) of growing area in a single application. Well water is the desired source, since municipal water is often too costly and may contain harmful fluoride (as discussed in the "Water Quality" section of Chapter 8). Pond or river water is subject to disease organisms and may require expensive sterilization.

Orientation

Shadows are cast by the greenhouse structure. The magnitude of the shadows depends on the angle of the sun and thus on the season of the year. The effect can be most detrimental to growth in the winter, when the sun remains closer to the horizon and shadows are longer.

Single greenhouses located above 40°N latitude in the Northern Hemisphere should be built with the ridge running east to west so that low-angle light of the winter sun can enter along a side rather than from an end where it would be blocked by the frame trusses. Below 40°N latitude, the ridge of single greenhouses should be oriented from north to south, since the angle of the sun is much higher. Ridge-and-furrow and gutter-connected greenhouses (greenhouses connected to one another along their length) at all latitudes should be oriented north to south. This north-south arrangement avoids the shadow in a greenhouse that would occur from the greenhouse lying immediately south of it in an east-west arrangement. Although the north-south orientation has a shadow from the frame trusses, it is much smaller than the shadow that would be cast from a whole greenhouse located to the south.

L. G. Morris made the calculations presented in Table 2-1 in England at a latitude of about 50°N. They leave little doubt that the ridge of a single greenhouse at higher latitudes should run from east to west. The difference in light intensity due to orientation is not great during the summer, when the angle of the sun is high. A difference shows up in the winter, when light is an important issue.

Table 2-1 Effect of Greenhouse Orientation on Light Transmission in Midsummer and Midwinter at a Latitude of Approximately 50°N

	% Transmission	
Orientation	*Midsummer*	*Midwinter*
North-south	64	48
East-west	66	71

Floor Plan

It is important to develop a greenhouse floor plan that allows for more future expansion than will likely occur. This consideration best ensures that an efficient operation will always be possible. If there is a slope to the land, construction should begin at the midpoint (average elevation). In this way, the soil excavated will provide the fill needed during each expansion. In the end, there will be one final elevation for the entire greenhouse range. A plan as pictured in Figure 2-1 allows for additions to the service building and greenhouses without removal of previous buildings or the accumulation of multiple service buildings. The service building is centrally located in a nearly square design of the firm, which minimizes distances that plants and materials need to be moved. The whole firm is on one elevation and is internally connected so that an internal transport system can be used (Figure 2-2).

Doors between the service building and the greenhouse should be wide enough to facilitate full use of the corridor width. Doors at least 10 ft (3.1 m) wide by 9 ft (2.7 m) high are common. The drive through the corridor and greenhouse should be at least 8 ft (2.4 m) wide in larger greenhouse ranges to allow two trains to pass each other. It would also be good to have the greenhouse gutters at least 12 ft (3.7 m) above the floor to accommodate automation and thermal blankets and still leave room for future innovations. Many greenhouses are being built with

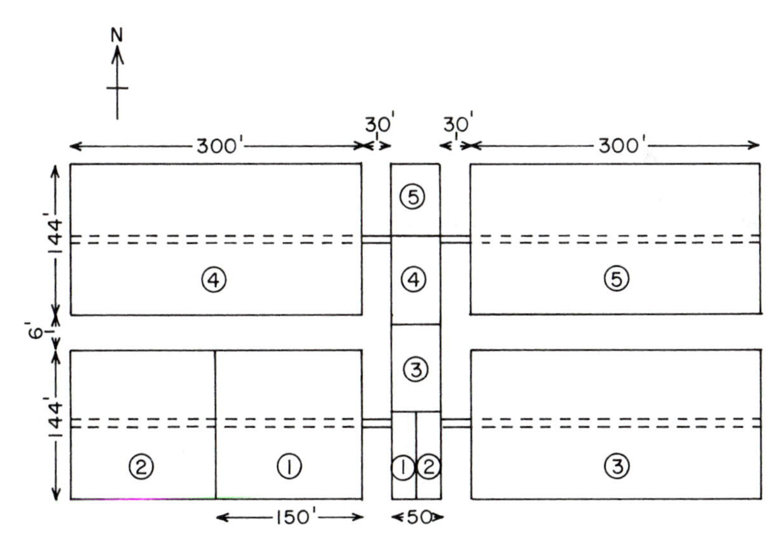

Figure 2-1 **This floor plan for a greenhouse firm allows for construction in five phases. The design employs a single central service building. Building phases are numbered consecutively. The adjacent greenhouse or service-building walls between each building phase are removed. Greenhouses are separated from the service building by transparent corridors 30 ft (9.1 m) long to prevent shadows in the growing area.**

Figure 2-2 **An internal transport system at Rockwell Farms in Rockwell, North Carolina, consisting of a train of shelved, self-tracking trailers and an electric cart for pulling the train. Main aisles in each greenhouse block need to be wide enough to accommodate two trains passing each other.**

14- and 15-foot (4.3- and 4.6-m) -high gutters. The greenhouse gutters need to be oriented north to south. A greenhouse gutter length of 144 ft (44 m) coincides well with the greenhouse brands using 12- and 24-foot (3.7- and 7.3-m) gutter modules, while a 140- or 150-foot (43- or 46-m) gutter length works well for brands with 10-foot (3.1-m) gutter modules. All of these gutter lengths are within the maximum effective summer cooling distance of approximately 150 ft (46 m). The service building should have at least 16-foot (4.9-m) eaves to allow for doors 12 ft (3.7 m) wide by 14 ft (4.3 m) high, which are needed to accommodate trailer trucks for both receipt of goods and shipping of plants. Eighteen-foot (5.5-m) eaves are better where two levels are desired within the service building, which is the situation when a storage area is located above the office area. All plant loading should be carried out from a central point inside the service building.

Service buildings are constructed from a variety of materials, with steel buildings being the most common. Economics generally dictates the type. Some growers use a few bays of the greenhouse itself for the service building because of the 25 to 50 percent lower cost. In this case, the service area may be covered with the same transparent material as the growing area or with standard building materials such as board and shingle or corrugated metal. Facilities within the service building include offices, restrooms, a lunch/break area, a pesticide storage and handling area, a fertilizer room, a workshop, storage, a truck loading dock, a staging area for market-bound plants and flowers, possibly a central boiler area, and extra space for staging various jobs.

Glass Greenhouses

Only glass greenhouses existed prior to 1950. Glass greenhouses have an advantage of greater interior light intensity over plastic panel and film-plastic-covered greenhouses. The greater light intensity is due to the single glass covering during the day compared to the double plastic covering that remains on all the time. The heat requirement can be the same in a glass greenhouse as in a double-layer film plastic greenhouse if a thermal screen is installed in the glass greenhouse for use during the night and the glass greenhouse is located in a mild climate where little daytime heating is needed. Glass greenhouses tend to have a higher air infiltration rate, which leads to lower interior humidity. This is advantageous for disease prevention. On the other hand, glass greenhouses have a higher initial cost than double-layer film plastic greenhouses. Recently, the prices of glass greenhouses, particularly the low-profile type, have come much closer to film plastic greenhouse prices. When comparing the price of a glass greenhouse to a film plastic greenhouse, one needs to take into account the initial purchase price of each as well as the cost of re-covering the film plastic greenhouse every three to four years. The typically higher long-term cost of the glass greenhouse is justified by some growers by the higher crop yield. In 1995 the proportion of greenhouse covering types in use in the United States was 14.3 percent glass, 22.7 percent FRP and rigid plastic panels, and 63.0 percent film plastic (Agricultural Statistics Board, 1996).

Several styles of glass greenhouses are designed to meet specific needs. A *lean-to* design is used when a greenhouse is placed against the side of an existing building (Figure 2-3a). This design makes the best use of sunlight and minimizes the requirements for roof supports. It is found mostly in the retail industry. An *even-span* greenhouse is one in which the two roof slopes are of equal pitch and

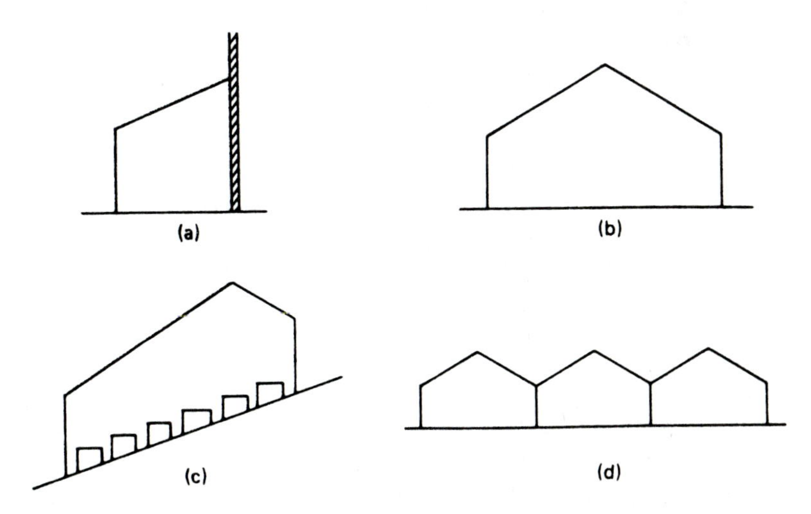

Figure 2-3 **Four basic greenhouse styles: (a) lean-to, (b) even-span, (c) uneven-span, and (d) ridge-and-furrow.**

width (Figure 2-3b). By comparison, an *uneven-span* greenhouse has roofs of unequal width, which make the structure adaptable to the side of a hill (Figure 2-3c). This style is seldom used today because such greenhouses are not adaptable to automation.

Finally, a *ridge-and-furrow* design uses two or more A-frame greenhouses connected to one another along the length of the eave (Figure 2-3d). The eave serves as a furrow or gutter to carry rain and melted snow away. The side wall is eliminated between greenhouses, which results in a structure with a single large interior. Consolidation of interior space reduces labor, lowers the cost of automation, improves personnel management, and reduces fuel consumption because there is less exposed wall area through which heat can escape. The snow load must be taken into account in the frame specifications of these greenhouses. Snow cannot slide off the roofs, as in the case of individual free-standing greenhouses, but must melt away. Heating pipes are generally located beneath the gutters for this purpose. In spite of snow loads, ridge-and-furrow greenhouses are effectively used in the northern countries of Europe and in Canada.

Basically, three frame types have been used in greenhouses. *Wood frames* were used for greenhouses under 20 ft (6.1 m) in width. Side posts and columns were constructed of wood without the use of a truss. Wider houses required sturdier frames. *Pipe frames* served well for greenhouses up to a width of about 40 ft (12.2 m) (Figure 2-4a). The side posts, columns, cross-ties, and purlins were constructed from pipe. Again, a truss was not used. The pipe components did not all interconnect but depended on attachment to the sash bars for support. Some greenhouses under 50 ft (15.2 m) in width and most over this width were built on *truss frames* (Figure 2-4b). Flat steel, tubular steel, or angle iron is welded together to form a truss encompassing the rafters, chords, and struts. Struts are support members under compression, while chords are support members under tension. Angle-iron purlins running the length of the greenhouse are bolted to each truss. A frame thus constructed can stand without support of sash bars. Columns are used only in very wide truss-frame houses of about 70 ft (21.3 m) and wider.

Today, glass greenhouses are primarily of the truss-frame type. Truss-frame greenhouses are best suited to prefabrication, which has made the construction of greenhouses more economical over the years. Automation has also fostered wider houses, which require the strength of truss frames.

The glass on the greenhouse is attached to sash bars. In earlier days, sash bars were made exclusively of wood, primarily cypress and redwood. The wood required periodic painting to protect it against rot. Ideally, exteriors were painted every two years and interiors every five to seven years. This practice was costly. Aluminum sash bars and ventilators were introduced in the early 1950s. The resultant all-metal greenhouses were very expensive at the outset but quickly became competitive with houses having wooden sash bars. All-metal greenhouses proved cheaper to maintain since they required no painting. Today, virtually all glass-greenhouse construction is of the metal type.

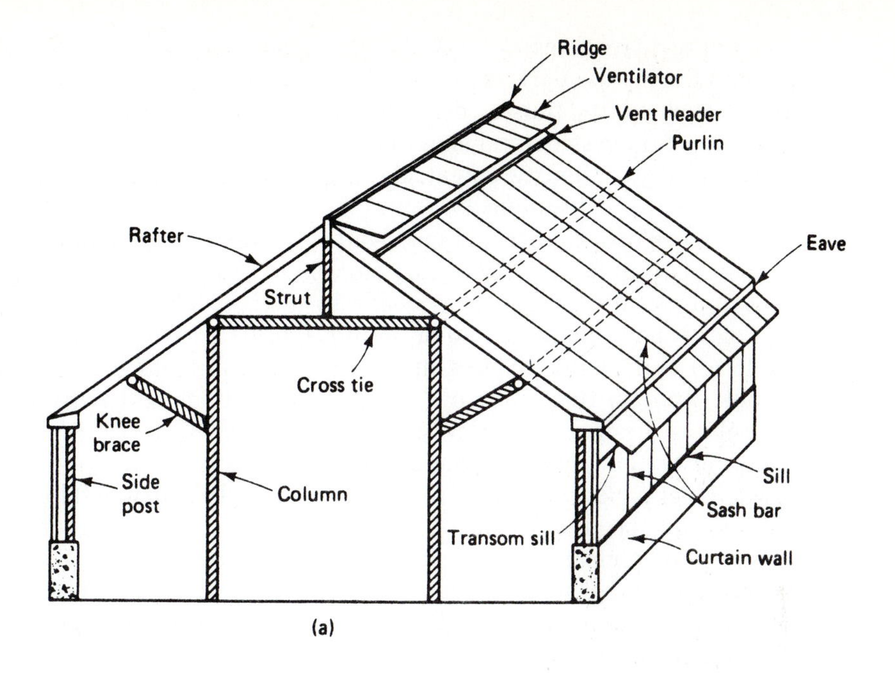

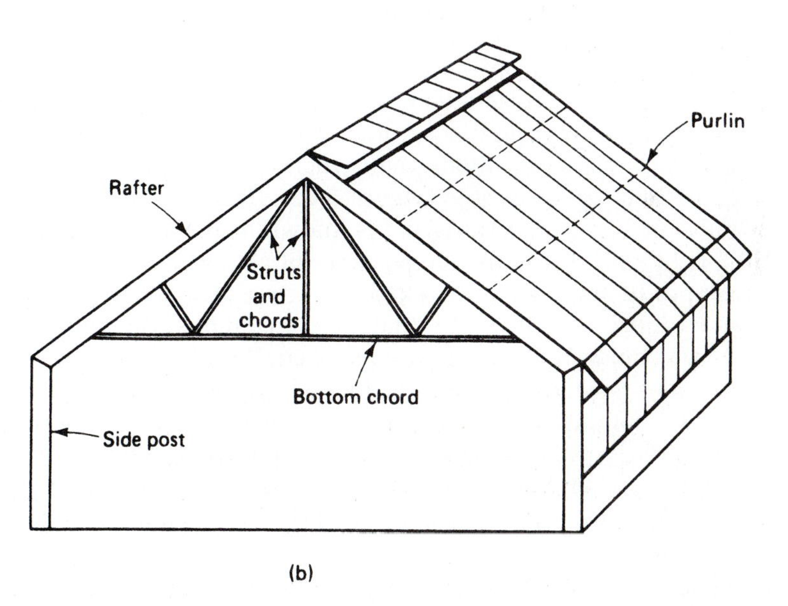

Figure 2-4 Structural components of (a) a pipe-frame greenhouse and (b) a truss-frame greenhouse. In the house in (b), the side posts, rafter, chords, and struts are one unit known as a *truss.*

The structural members of the greenhouse cast shadows that reduce plant growth during the dark months of the year. Aluminum sash bars are stronger than wooden ones; thus, wider panes of glass can be used with aluminum bars. The reduction in structural material plus the reflectance of aluminum have given these metal greenhouses a great advantage over wooden greenhouses in terms of higher interior light intensity.

The original 16-inch (41-cm) width of glass has evolved over the years to widths of 20 in (51 cm), 24 in (61 cm), 29 in (73 cm), 36 in (91 cm), and 39 in (1 m), a width that is now available in Europe. The original length of 18 in (46 cm) is now commonly 30 or 36 in (76 or 91 cm), but can range up to 57 in (1.45 m) in the United States and 65 in (1.65 m) in Dutch greenhouses. Mostly double-strength float glass of 1/8-inch (3-mm) thickness is used in American greenhouse designs. It currently sells for about \$0.85/ft^2 (\$9.15/m^2). The larger panes of Dutch greenhouses are tempered glass of 5/32-inch (4-mm) thickness. Even larger panes of 5/32-inch (4 mm) tempered glass up to 6 ft by 13 ft (1.8 m by 3.9 m) have been used since 1985. These panes cover the roof from gutter to ridge. Being tempered, they can be bent to fit a curved roof (Figure 2-5). Tempered glass in the smaller panes sells for about \$1.10/ft^2 (\$11.84/m^2) and in the very large panes for \$1.85/ft^2 (\$19.91/m^2). Hammered glass (glass with a rough, uneven surface) has been used to a moderate degree in Dutch-design greenhouses. This glass scatters the light so that intensity is more uniform across the inside of the greenhouse, which leads to more uniform crop growth. Hammered glass is seldom used today because of its high rate of breakage during shipping and handling. Low-iron-content glass is often used in rose greenhouses because of its higher light transmission, which is 90 to 92 percent versus 88 percent for float glass. Low-iron glass costs about \$1.00/ft^2 (\$10.76/m^2).

Today's glass-greenhouse construction can be categorized as *high profile* (Figure 2-6a) or *low profile* (Figure 2-6b). The low-profile greenhouse is most popular in The Netherlands and is known as the *Venlo* greenhouse. Eaves are 10.5 ft (3.2 m) apart, and single panes of glass extend from eave to ridge. The lower profile slightly reduces exposed surface area, thereby reducing the heating cost. It is suggested, however, that these greenhouses are more expensive to cool in warm climates where fans are required. Ventilator cooling during intermediate seasons is not as effective due to the lower height from ground to ventilator; thus, fan cooling may be used more extensively. In North America, both low- and high-profile greenhouses are currently popular. Since heating and cooling differences are very small, price appears to be the dominant factor in the relative popularity of these greenhouse types. The price advantage has shifted back and forth in the United States.

High-profile greenhouses are available with special sash bars that hold two or even three layers of glass. This layering produces one or two dead-air spaces to cut heat loss (see Table 3-1 for heat-loss specifications). A problem with this design is the unsealed junction between pieces of glass in the inner layer. Moisture and dust can get between the layers and reduce light transmission. It is expensive to remove and clean the glass.

Figure 2-5 (a) A range of gutter-connected greenhouses being covered with tempered glass panes 6 ft (1.8 m) wide by 10 ft (3.1 m) long, each extending from gutter to ridge and bent to fit the arched roof shape. (b) The resulting covering is exceedingly strong and has very high light transmission. *(Photos courtesy of Westbrook Greenhouse Systems Ltd., P.O. Box 99, Grimsby, Ontario L3M 4G1, Canada.)*

(a)

(b)

Figure 2-6 (a) A high-profile ridge-and-furrow greenhouse. (b) A low-profile ridge-and-furrow greenhouse.

Film Plastic Greenhouses

Role

Flexible plastic films, including polyethylene, polyester, and polyvinyl chloride, have been used for greenhouse coverings. Polyethylene is principally used today. Film plastic is currently the leading greenhouse covering in the United States for two reasons. First, film plastic greenhouses with permanent metal frames cost less than glass greenhouses. Even greater savings can be realized when film plastic is applied to less permanent frames, such as Quonset greenhouses. Second, film plastic greenhouses are popular because the cost of heating them is approximately 40 percent lower compared to single-layer glass or FRP greenhouses. Granted, a thermal screen can be installed inside a glass greenhouse that will lower the heat requirement to approximately that of a double-layer film plastic greenhouse, but this further increases the cost of the glass greenhouse.

Polyethylene film was developed in the late 1930s in England, and its use as a greenhouse covering was pioneered around the middle of this century. The use of polyethylene for greenhouses has increased rapidly and continues to do so. In the United States alone, about 40 acres (16 ha) of plastic greenhouses were in use in the mid-1950s, about 2,300 acres (920 ha) by the mid-1960s, and 4,800 acres (1,920 ha) in 1977. In 1995, 63 percent of all greenhouses in the United States were film plastic, 22.7 percent were FRP or rigid plastic panel, and 14.3 percent were glass, according to the US Agricultural Statistics Board (1996). That report places the total area used by film plastic greenhouses at 7,060 acres (2824 ha). While the percentage is probably quite accurate, the area is greatly underestimated. Polyethylene film greenhouses constitute the largest portion of new greenhouse construction in Canada and the United States today. Their popularity, however, is not very great in northern Europe where glass remains popular.

Some disadvantages occur along with the advantages of film plastic. These covering materials are short-lived compared to glass and plastic panels. The highest-quality, ultraviolet (UV) light–resistant, 6-mil-thick (.006 in, 0.15 mm) polyethylene films last four years. UV light from the sun causes the plastic to darken, thereby lowering light transmission, and to become brittle, which subjects it to breakage in the wind. While the time required to cover a 30-foot-by-100-foot (9.1-m-by-30-m) Quonset-design greenhouse is minimal (about eight labor hours), the task is never-ending and carries implicit costs of management and the use of equipment. However, under proper management, the savings in fuel as well as the lower initial purchase price givc the film plastic greenhouse a lower cost than a glass greenhouse.

Types of Film Plastic

Polyethylene Polyethylene has always been and still is the principal choice of film plastic for greenhouses in Canada and the United States. Nearly all current greenhouses have two layers. The outer layer is customarily 6 mil (0.15 mm) thick, while the inner layer only needs to be 4 mil (0.1 mm) thick. All polyethylene used for covering year-round production greenhouses has a UV inhibitor in it; otherwise, it would last for only one heating season. Three-year and four-year products are on the market. UV-grade polyethylene is available in widths up to 50 ft (15.2 m) in flat sheets and up to 25 ft (7.6 m) in tubes. Standard lengths include 100, 110, 150, 200, and 220 ft (30.5, 33.5, 45.7, 61.0, and 67.0 m). Several companies provide custom lengths up to a maximum of 300 ft (91.5 m).

A polyethylene covering is colder in the winter than the air inside the greenhouse. When warm, moist greenhouse air contacts the cold polyethylene, it cools. As a result, water vapor condenses on the polyethylene surface. The surface is repellent to water; thus, the water forms into beads. With time, the water beads increase in size to a point where they drop off to the plants below. The wet foliage fosters disease development, while the constantly wetted soil becomes waterlogged and oxygen deficient. If the plastic surface were not as repellent to water, the condensing water would spread out better over the surface of the plastic, without forming droplets large enough to drop off. Ultimately this water would flow down along the surface of the plastic to the gutter, where it would be collected. A liquid surfactant, Sun Clear, is available that can be sprayed on the inner surface of film plastic and rigid-panel greenhouses to give this benefit of lower surface tension. The materials cost for this treatment is $0.0033/ft^2 ($0.035/m^2) of surface treated and needs to be applied periodically. Today, polyethylene film as well as rigid FRP, acrylic, and polycarbonate panels are available with an antifog surfactant built into the film or panel. It is advisable to use an antifog product because in addition to the water-dripping problems, this condensation also reduces light intensity within the greenhouse.

Warm objects such as plants, the greenhouse frame, and soil radiate infrared (radiant) energy to colder bodies, such as the sky at night. This condition creates a large source of heat loss in greenhouses. Polyethylene is a poor barrier to radiant heat. Polyethylene with infrared (IR)-blocking chemicals formulated into it during manufacture will stop about half of the radiant heat loss. On cold clear nights, as much as 25 percent of the total heat loss of a greenhouse can be prevented in this way. On cloudy nights, only about 15 percent is saved.

Light that is utilized in photosynthesis is termed *photosynthetically active radiation* (PAR) and comprises the wavelengths from 400 to 700 nm. Transmission of PAR through polyethylene can vary with the brand of polyethylene used and the chemical additives it contains (Table 2-2). UV-stabilized polyethylene, on average, transmits about 87 percent of PAR. IR-absorbing polyethylene, which reduces radiant heat loss, transmits about 82 percent of PAR. The amount of light passing through two layers of a greenhouse covering is approximately the square of the decimal fraction of the amount passing through one layer. Where 87 percent (0.87) passes through one layer of UV-inhibited polyethylene, only 76 percent (0.87×0.87) passes through two layers. PAR transmission through two layers of IR-absorbing polyethylene is 67 percent.

While two layers of polyethylene transmit less light than one layer of glass, it is questionable how much less light there is in a polyethylene greenhouse. Kozai, Gourdriaan, and Kimura (1978) developed a simulation model for a glass-

Table 2-2 **Light-Transmission Values for Various Greenhouse Coverings**

Covering	*Number of Layers*	*% Transmission*[1]
Glass (double-strength float, 3.2-mm)	1	88
	2	77
Glass (low-iron, 3.2-mm)	1	90–92
	2	81–85
FRP (clear, 0.64-mm)	1	88
	2	77
Polyethylene (4- or 6-mil, 0.10- or 0.15-mm, UV-stabilized)	1	87
	2	76
Polyethylene (4- or 6-mil, 0.10- or 0.15-mm, IR-absorbing)	1	82
	2	67
Vinyl, clear[2]	1	91
Vinyl, hazy[2]	1	89
Polyvinyl fluoride film (4-mil, 0.10-mm)	1	92
	2	85
ETFE (Tefzel T^2)[2]	1	95
	2	90
Acrylic panels (8- or 16-mm)	2	83
Polycarbonate panels (6- or 8-mm)	2	79

[1]Light-transmission values for photosynthetically active radiation (PAR, 400–700 nm) and for single sheets are from American Society of Agricultural Engineers (1995), unless otherwise indicated. Transmission values for two layers were computed by squaring the single-layer values.

[2]Manufacturers' specifications.

covered high-profile greenhouse located at 30°41′N latitude. Although they indicated a light transmissivity of 86 percent at a zero angle of incidence for the glass itself, the light transmissivity for the entire greenhouse varied from 50 percent to 60 percent, depending on season and orientation of the greenhouse. The difference between 60 percent and 86 percent was due to the angle of incidence of light, sash bars, and structural members. Polyethylene greenhouses have less structural material than high-profile glass greenhouses. This could compensate in part for the lower light transmissivity through a double layer of polyethylene compared to a single layer of glass. The surface area of structural members is very similar in the low-profile glass and film plastic greenhouses. The glass greenhouse has sash bars 39 in (1 m) apart and no trusses, while the film plastic greenhouse has bows 3 or 4 ft (0.9 or 1.2 m) apart. In this comparison, light intensity is much higher in the glass greenhouse.

The latest technology in polyethylene production makes use of the co-extrusion process. Three liquid resins are extruded simultaneously such that a single layer of film can have three different chemistries across it. In the present tri-extruded films, the inner core contains the antifog surfactant. This chemical is not entirely compatible with polyethylene. The repelling forces cause it to slowly bleed out of the core through the overlying zones of polyethylene that do not contain it. In this way, it can last a few years.

The IR-blocking chemical is also placed in the core. Typically, this chemical weakens polyethylene. If polyethylene containing it is stretched, it can turn cloudy. By confining the IR block to the core, a zone of unusually strong and clear polyethylene can be developed over each side of the core, which alleviates much of the clouding problem. It also gives the film an overall strength greater than that of standard three-year film of the same thickness. Some tri-extruded film is warranted for four years when it is 6 mil (0.15 mm) thick and is on either the outer or the inner covering of the greenhouse. Four-mil (0.1-mm) film can also be warranted for four years if it is used under a 6-mil, UV-inhibited film. The 4-mil tri-extruded film is sufficiently strong to equal the life of the current 6-mil, three-year film.

Average prices for 6-mil (0.15 mm), three-year polyethylene film in quantities sufficient to cover one acre of greenhouse are \$0.06/ft^2 (\$0.65/m^2) for UV-inhibited, \$0.07/ft^2 (\$0.75/m^2) for UV-inhibited plus antifog, and \$0.08/ft^2 (\$0.86/m^2) for UV plus IR block. Several greenhouse construction companies can be hired to re-cover polyethylene greenhouses. Prices vary from \$0.12 to \$0.23/ft^2 (\$1.29 to \$2.47/m^2) of ground covered. The lower price applies to greenhouses covered with a tube of plastic, since both layers of plastic can be applied in one step. Gutter-connected greenhouse designs also frequently contribute to the lower price because less surface area per unit of ground area is exposed, as compared to Quonset greenhouses. The use of channel locks for attaching the plastic, as opposed to staples and batten strips, can further reduce the price.

Vinyl UV light–resistant vinyl (polyvinyl chloride) films of 8- and 12-mil (0.20- and 0.30-mm) thicknesses are guaranteed for four and five years, respectively. This guarantee was a decided advantage years ago when polyethylene lasted for only one or two years. With the recent advent of four-year polyethylene, the advantage is nearly gone. The cost of 12-mil (0.30-mm) vinyl is \$0.215/ft^2 (\$2.31/m^2), which is three times that of 6-mil (0.15-mm) polyethylene. Although vinyl film is produced in rolls up to 50 in (1.27 m) wide, any width can be purchased, since the supplier can seal strips of vinyl together. The vinyl films tend to hold a static electrical charge, which attracts and holds dust. This in turn reduces light transmittance until the dust is washed off. Vinyl films are seldom used in the United States. However, in Japan 95 percent of greenhouses are covered with film plastic and within this group 90 percent are covered with vinyl film.

Polyester For a time, Mylar brand polyester film offered the strong advantage of durability. Films of 5-mil (0.13-mm) thickness were used for roofs and lasted four years, while 3-mil (0.08-mm) films were used on vertical walls and had a life expectancy of seven years. Although the cost of Mylar was higher than that of polyethylene, it was offset by the extra life expectancy. Other advantages included a level of light transmittance equal to that of glass and freedom from static electrical charges, which collect dust. Other industrial uses were found for Mylar in the mid-1960s, and soon its price increased out of the practical realm for floriculture. Polyester is still used frequently, however, in heat retention screens because of its high capacity to block radiant energy.

Tefzel T^2 (ETFE) The most recent category of greenhouse film plastic covering to appear is Tefzel T^2 film (ethylene tetrafluoroethylene). Actually, this film has had an application as the transparent covering on solar collectors for many years. The anticipated life expectancy is 20 years or longer. The light transmission is 95 percent and is greater than that of any other greenhouse covering. A double layer has a light transmission of 90 percent. Tefzel T^2 film is more transparent to infrared radiation than other film plastics; thus, less heat is trapped inside the greenhouse during hot weather. As a result, less cooling energy is required. On the negative side, the film is available only in 50-in (1.27-m) -wide rolls. This would require clamping rails on the greenhouse about every 4 ft (1.2 m). Efforts are underway to produce wider sheets of the film. Tefzel T^2 film currently sells for about $1.20/ft^2 ($12.92/m^2). If reasonable widths become available, this price would not be excessive because a double layer covering would still cost less than a polycarbonate-panel covering with its aluminum extrusions, would last longer, and would have much higher light intensity inside the greenhouse.

Film Plastic Greenhouse Designs

Wood-Frame Greenhouses When polyethylene first entered the horticultural scene, the cost of establishing a business was significantly reduced. People who didn't have the funds to set up an expensive glass greenhouse range could now enter the greenhouse industry. Accordingly, inexpensive frames were sought. Pine wood was commonly used. Various frames were designed through the 1950s and 1960s. The A-frame (Figure 2-7) was one of the more popular. The scissors-truss frame (Figure 2-8) was particularly strong. These and other designs were used for greenhouses ranging from 20 to 30 ft (6.1 to 9.1 m) wide.

A single layer of film plastic was generally used until the early 1960s, when fuel costs entered into the picture. Double-layer coverings of film plastic were desired to try to achieve a potential 40 percent savings in fuel cost. At first, the second layer was applied from the inside of the greenhouse. This task was difficult in greenhouses with columns, because holes had to be cut in the inner layer of plastic to maneuver it around these columns. The holes were generally not sealed and thus left avenues of entry for warm air into the dead-air space between the plastic layers, which reduced its insulating property. Stronger truss designs were sought to eliminate columns.

Next, the width of the dead-air space became a consideration. Ideally, the

Figure 2-7 An inexpensive but temporary A-frame film plastic greenhouse, a type which was very popular in the early days of film plastic greenhouses.

Figure 2-8 A scissors-truss film plastic greenhouse designed at Virginia Polytechnic Institute. This is a particularly strong design.

dead-air space should be 0.5 to 4 in (1.3 to 10 cm) thick (U.S. Housing and Home Finance Agency 1954). When it exceeds 4 in (10 cm), air currents can become established inside and reduce the insulating property of this space. Warm air immediately above the inner covering rises up into contact with the outer covering, and here it gives up heat. As it cools, the air becomes heavy and drops back to the inner covering to pick up more heat. This loss does not become very significant until a space of 18 in (46 cm) is reached. Below 0.5 in (1.3 cm), the insulating property again diminishes, and when the two layers touch, the insulation value is totally lost. A-frame and scissors-truss designs necessitated dead-air spaces several feet thick. To avoid this problem, the Gothic-arch greenhouse was developed at Virginia Polytechnic Institute (Figure 2-9). Greenhouses up to 30 ft (9.1 m) wide with a frame thickness of 4.5 in (11 cm) could then be constructed without columns.

The greenhouses thus far discussed were constructed with short-life wood, which required frequent painting to prevent rotting. White paint was usually used to increase interior light intensity. Today, as in the past, when paint is applied in or on a greenhouse, a mercury-based paint should be avoided. Mercury will volatilize from the paint for a considerable length of time and thus cause damage to the crop. Paints sold specifically as greenhouse paints are safe.

Posts and other wood in contact with the ground should be treated with a wood preservative. Treated wood may be purchased for this purpose, or the wood may be treated at the time of use. Several treatments are available, but not all are safe. Pentachlorophenol and creosote should not be used. Creosote in contact with roots and foliage can burn them. Pentachlorophenol produces fumes

Figure 2-9 A Gothic-arch greenhouse of the type designed at Virginia Polytechnic Institute. The trusses used are fabricated during the construction of the greenhouse. This greenhouse offers a pleasing appearance and is devoid of internal columns.

that can last for more than a year and are toxic to plants. Entire crops can be killed by moving them into a new greenhouse with treated posts. A single treated board can cause abnormal growth throughout the greenhouse. A very suitable wood preservative is copper naphthenate, which is sold under several trade names. Generally used as a 2% solution of copper naphthenate, it can be sprayed, dipped, or applied with a brush. It is an excellent preservative for frame members as well as for wooden benches and plant flats. Lumber that has been pressure-treated with Wolman salts is safe for greenhouse use (Beese 1978).

Quonset Greenhouses With time, the price of wood became objectionably high relative to metal. The cost of continual painting was an added burden. By 1970, film plastic greenhouse designs were mainly of two styles. The first, and least expensive, which persists to this day, is the Quonset-style greenhouse (Figure 2-10). Quonset greenhouses can be purchased prefabricated or can be fabricated on the site. Many excellent designs are available from nearly every manufacturer of greenhouses, in widths up to 36 ft (11 m). Quonset greenhouses fit into two principal niches. First, they are typically less expensive than the gutter-connected greenhouses; therefore, they are a popular choice among individuals entering the industry on a limited budget. This is especially true for growers fabricating their own greenhouses. Second, they are handy when a small, isolated cultural area is needed, such as for cool holding of azaleas.

Fabrication of a Quonset greenhouse is not very difficult. Often, the trusses are constructed from water pipe that is bent to fit a 180° arc, modified for somewhat more vertical sides. In greenhouses 20 ft (6.1 m) wide, 0.75-inch (1.9-cm)

Figure 2-10 **A metal-frame, Quonset-style greenhouse, which is very popular today with users of film plastic. This greenhouse is inexpensive, does not require painting, and is well suited to a double covering of film plastic.**

pipe is used; 1-inch (2.5-cm) pipe is used for a 30-foot (9.1-m) greenhouse width. Aluminum electrical conduit should not be used, since it does not have sufficient strength to support snow loads. Slightly larger pipe, into which the pipe arches are inserted for support, is driven into the ground. A 2-in-by-8-in (5-cm-by-20-cm) wooden plank is attached to the base of the pipe arches so that it runs along the ground, partially buried. This provides a basal point of attachment for the film plastic. The pipe arches, or trusses, are supported by pipe purlins running the length of the house. Trusses are spaced 36 to 48 in (91 to 122 cm) apart. The width of film plastic required to cover a Quonset greenhouse of given width can vary according to the height and shape of the trusses. A 20-foot (6.1-m) -wide greenhouse generally requires a 32-foot (9.8-m) -wide sheet of plastic. The covering width for a 30-foot (9.1-m) -wide Quonset greenhouse varies greatly; the more common widths are 40 and 42 ft (12.2 and 12.8 m). Polyethylene greenhouses are constructed 2 to 3 feet (61 to 91 cm) shorter than the length of the roll of plastic used to cover them. The plastic is metered out onto the roll by weight and may vary in length by 1 or 2 percent. Also, some extra plastic is required to reach over the ends of the greenhouse for clamping onto the end walls.

Quonset houses are either constructed in a freestanding style or arranged in an interlocking ridge-and-furrow manner, as depicted in Figure 2-11. In this latter case, the trusses overlap sufficiently to place a bed of plants between the overlapping portions of adjacent houses. A single large interior thus exists for a set of houses, an arrangement that is better adapted to the movement of labor and to automation.

Figure 2-11 **An interconnecting arrangement of Quonset greenhouses offering a single large interior for several greenhouses. This greenhouse arrangement is in harmony with the current needs for automation and efficiency of movement.**

Gutter-Connected Greenhouses The gutter-connected house is the most efficient film plastic greenhouse design (Figure 2-12). It is cheaper, and thus more feasible, to automate the single consolidated space inside a gutter-connected greenhouse than the multiple equivalent spaces in several Quonset greenhouses. For instance, a heat-retention screen in a 1-acre (0.4 ha) gutter-connected green-

(a)

(b)

Figure 2-12 (a) Exterior view of a gutter-connected polyethylene greenhouse range. (b) Interior view of a gutter-connected polyethylene range at Metroliner Greenhouse in Huntersville, North Carolina.

house costs about 1.25/ft^2 (\$13.45/m^2) and about \$3.00/ft^2 (\$32.28/m^2) in an equivalent area of 20-foot (6.1-m) -wide Quonset greenhouses. Management of personnel is more efficient when they are all in one room with the supervisor, as opposed to being scattered about in multiple locations without supervision. Movement of materials and product into and out of the greenhouse requires less labor in a single large space than in numerous small spaces. The heating cost is less in a gutter-connected greenhouse, because there is less exposed surface area. A 20-foot-by-98-foot (6.1-m-by-30-m) Quonset greenhouse with a 32-foot (9.8-m) -wide covering has 1.65 square feet of exposed surface per square foot of floor space, while a 1-acre (0.4-ha) gutter-connected greenhouse with a 10-foot (3.1-m) gutter height has 1.50 square feet of exposed surface per square foot of floor area. There is 10 percent less exposed surface per unit of floor area in the gutter-connected greenhouse. The modest increase in the price of a gutter-connected greenhouse compared to a Quonset greenhouse is quickly returned with interest. Even if the first unit of a gutter-connected greenhouse is not suitably large for full automation, subsequent additions can make it so. When additions are made, the film plastic is removed from an existing side wall and the new houses are connected at that point without any resulting discontinuity. In this way, a modest initial investment, unadaptable for automation, can be developed through expansions into a structure well suited to automation.

It is more difficult to reapply film plastic coverings on the end walls than on the roofs. For this reason, many owners of gutter-connected film plastic greenhouses use double-layer polycarbonate panels on the end walls. Panel thicknesses of 6 or 8 mm (0.24 or 0.32 in) are common. Considering the small additional cost and the greater strength of the 8-mm (0.32-inch) panel, it appears to be the wiser choice. Since there is less light load on these vertical walls than on the roof, this covering can be expected to last for 20 years. Single-layer corrugated polycarbonate is also becoming popular for partition walls within the greenhouse.

The height of gutters above the ground has been increasing over the years to accommodate the continuing evolution of climate-control equipment and automation devices. The original gutter-connected greenhouses typically had an 8-foot (2.4-m) gutter height. Today, 12 ft (3.7 m) is the minimum recommended height, and 14 ft (4.3 m) is becoming common. Gutters may be constructed from galvanized steel or from aluminum. Freedom from rust justifies the additional cost of aluminum gutters. Some gutters have film plastic attachment channels and a condensate drip collector molded into them. These are desirable features. The gutters in ridge-and-furrow glass greenhouses and gutter-connected film plastic greenhouses should slope to carry water away. A slope of 6 in per 100 ft (0.5 cm/m) is common. Often the floor is sloped at the same angle. If the floor is not sloped, it is important to have adequate drainage built into it.

The distance between gutter rows depends on the greenhouse brand purchased. This distance ranges from 10.5 to 40 ft (3.2 to 12.2 m). A number of greenhouse designs, up to and including 18 ft (5.5 m) between gutters, offer a truss frame that permits placement of gutter-supporting columns under every second or sometimes every third gutter. Although these extra-strong truss greenhouses can cost more, they offer the advantage of larger column-free spaces inside the greenhouse. This facilitates the use of automation and can reduce its cost. Greenhouses with wider spaces between column rows are inherently weaker unless stronger, more costly trusses are used. When selecting a greenhouse, it is important to know the wind load of the structure and the live and snow loads of the roof before comparing prices. Greenhouses with spacings between gutters of 12, 17, 21, 22, and 30 ft (3.7, 5.2, 6.4, 6.7, and 9.1 m) can be covered by film plastic sheets 14, 20, 24, 25, and 36 ft (4.3, 6.1, 7.3, 7.6, and 11.0 m) wide.

Another option in gutter-connected greenhouses is the contour of the roof. Traditionally they had a relatively flat Quonset-arch shape. Today, they can be obtained with a Gothic-arch shape or a peaked roof. The Gothic arch has a higher ridge than the Quonset, which results in steeper slopes to the roof. The peaked roof rises straight from the gutters to the ridge, in the typical shape of a glass roof. The steeper slopes of both the Gothic-arch and peaked designs facilitate downward flow of water condensate along the inner surface of the film plastic to the drip collector on the gutter. This reduces condensate drip on the crop and root substrate. Advantages include higher light transmission through the plastic, less disease on the crop, and prevention of waterlogging of the root substrate. The Gothic-arch and peaked designs also facilitate the use of roof ventilators on film plastic greenhouses.

The final choice to make when selecting a gutter-connected greenhouse design is that of active versus passive cooling. Greenhouses are now offered that have roll-up side curtains that can be installed on two or all four walls (Figure 2-13) and roof ventilators (Figure 2-14). In lieu of roof ventilators, in some de-

Figure 2-13 **A gutter-connected greenhouse range with roll-up sides on four sides. Note that the polyethylene sides have been lowered about one-third of their height.**

Figure 2-14 **A range of gutter-connected greenhouses with roof ventilators for passive cooling. Note that the end walls are covered with polycarbonate panels and the roofs are covered with polyethylene film. Louvers in the gables are for cold air intake for the winter cooling system.** ***(Photo courtesy of Westbrook Greenhouse Systems Ltd., P.O. Box 99, Grimsby, Ontario L3M 4G1, Canada.)***

signs the whole roof is retractable (Figure 2-15). The purpose of the side-curtain-plus-roof-ventilator system or the retractable roof with or without roll-up side curtains is to replace high-energy-consuming fan-and-pad cooling systems. These passive cooling systems work well in hot and cold climates. They were initially used for cold-temperature-tolerant crops such as bedding plants, garden chrysanthemum, many hanging-basket plants, and foliage plants. More recently, they are being used for less tolerant crops such as poinsettia. The main limitation appears to be the tolerance of the crop to full light intensity when the roof is open. This would adversely affect plants such as African violet and gloxinia.

One factor must be weighed against the savings in cooling energy. Most of the roll-up side curtains consist of a single layer of film plastic, and the retractable roof is also a single layer. The higher heat transmission of these single coverings and the possibility of cold air infiltration at the points of closure, if these are not carefully constructed, raise the heating costs. This is probably a temporary disadvantage. Undoubtedly, double-layer venting systems will be developed in the near future. In the meantime, a thermal screen can be installed inside the greenhouse to be used at night during the heating season to rectify this disadvantage.

The gutter-connected greenhouse brings us full circle to the category of permanent metal-frame greenhouses, of which the glass greenhouse is a member.

Figure 2-15 A gutter-connected range of retractable-roof greenhouses covered with a single layer of reinforced polyethylene film at Smith Gardens, Marysville, WA. The greenhouse is open with the roof film gathered at each truss. *(Photo courtesy of Cravo Equipment, Ltd., RR 1, Brantford, Ontario N3T 5L4, Canada.)*

The major portion of new greenhouse construction in the United States today, whether for temporary or permanent use, is film plastic.

Double-Layer Covering

Today, virtually all film plastic greenhouses make use of the air-inflated system. Two layers of film plastic are applied directly on top of each other from the outside of the greenhouse and are held apart by a cushion of air maintained at low positive pressure. This air-inflated system offers the easiest method for covering a greenhouse with two layers of film plastic. It insures a longer life expectancy of the film because the outer layer of plastic rests on a cushion of air. Plastic applied by techniques other than the air-inflated system is constantly chafed against the trusses by the lifting and dropping action of the wind. This greatly reduces its life expectancy.

If the width of the film sheet needed to cover the greenhouse is 25 ft (7.6 m) or less, a roll of tube plastic can be used. This roll has effectively two sheets of plastic in it. Thus, only one roll needs to be applied to the roof to obtain two layers of plastic. If the covering width needs to be wider than 25 ft (7.6 m), two rolls of single-sheet plastic will have to be applied. This increases labor time. During covering, the roll of plastic is often suspended in the air on a spindle, often attached to a tractor. The roll is situated just off one end of the greenhouse and is level with and perpendicular to the ridge of the greenhouse. Each of the leading corners of the plastic is drawn by an individual. The two individuals walk along the opposite sides of a Quonset greenhouse, pulling the plastic out over the whole length of the greenhouse. For a gutter-connected greenhouse, the two individuals walk in the gutters on either side of one bay of the greenhouse to the opposite end. The ends of the sheet overlap the greenhouse ends by a few inches and are attached at that point. No attachment is made to the trusses. Two sheets of plastic are attached to each end of the greenhouse as well.

On a Quonset greenhouse, the plastic roof covering sheets are attached to a frame member running along the ground on either side. On a gutter-connected greenhouse, the plastic sheets are attached to a clamping channel located in the gutter. In either type of greenhouse, the ideal method of attachment is a clamping channel (Figure 2-16). Such channels may be purchased independently of the greenhouse or may be obtained as an integral part of the greenhouse. The layers of film plastic are laid over the channel, and then a metal rod or extrusion is placed over the film plastic and pushed into the channel, locking the plastic in place. Sometimes on self-fabricated Quonset greenhouses, the plastic is attached by placing a batten strip over it and stapling through it into the wooden member running along the ground on either side of the greenhouse. The batten strip is usually a thick plastic strip about 1 inch (2.5 cm) wide, which can be obtained

Figure 2-16 **A clamping rail for attaching plastic film to greenhouses. The two layers of plastic are laid over an aluminum extrusion that is fixed permanently to the greenhouse frame. A second aluminum extrusion is pressed over the first extrusion and is locked in place with thumb screws, thereby locking the plastic in place.**

from greenhouse-supply companies. Most prefabricated greenhouses are equipped with metal channel locks.

The tension under which the plastic is installed is important, since film plastics contract and expand to a considerable degree with temperature shifts. When it is applied on a cold day, the film should be pulled taut. On a hot day, with temperatures above 80°F (27°C), about 2 to 3 in (5 to 8 cm) of slack should be left in the covering all the way along one side of a Quonset greenhouse 20 ft (6.1 m) wide to permit contraction over the truss when cold weather comes. If this slack is not allowed, the film will tear loose from the points of attachment when it contracts during cold weather. Conversely, if it is not pulled taut when it is applied on a cold day, excess slack will occur during warm weather, resulting in an excessive air space between the two layers.

A small squirrel-cage fan is installed inside the greenhouse to inflate the space between the two film plastic layers (Figure 2-17a). Air is maintained at between 0.2 and 0.3 inch (5.1 and 7.6 mm) of water-column pressure. Higher pressures are used during heavy winds—up to 0.5 inch (13 mm)—to prevent excessive movement and breakage of the film by the wind. The high pressure should not be maintained because the plastic will stretch. The fan should have an adjustable door on the air inlet for adjusting the pressure between the two films. For a greenhouse measuring 26 ft by 96 ft (8 m by 29.3 m), a fan deliver-

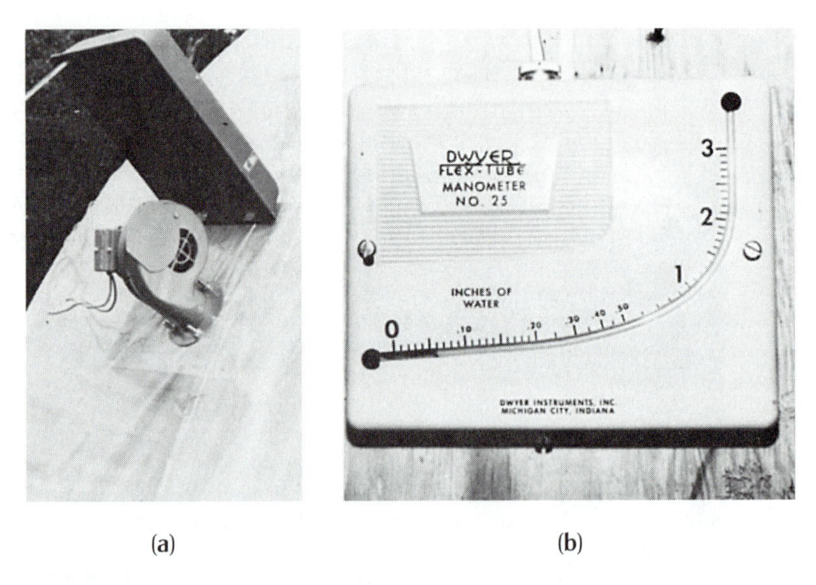

Figure 2-17 (a) A squirrel-cage fan used to inflate the space between two layers of plastic. The plate on the side of the fan can be moved to adjust the air supply to the fan and, consequently, the pressure between the two coverings on the greenhouse. (b) A manometer used to measure the air pressure between the two plastic covers.

ing air at 200 to 400 cubic feet per minute (cfm) (5.7 to 11.3 cubic meters per minute [cmm]) at a static water pressure of 0.5 inch (13 mm) (about 1 A, 115 W) is sufficient. During a snowstorm when snow sticks to the roof, it may be necessary to turn off the fan. This will allow the two polyethylene layers to come together, thus eliminating the insulating effect of the dead-air space. More heat will escape through the covering to melt snow and clear it from the roof.

Air pressure between the plastic layers is monitored with a manometer, which can be purchased from greenhouse-supply companies (Figure 2-17b). Conversely, a manometer can be easily fabricated by the grower as follows:

1. Bend a 2-foot (61-cm)-long piece of clear plastic tubing into the shape of a U and attach it to a board.
2. Make an X-shaped cut in the inner layer of plastic and insert one end of the plastic tube.
3. Seal the film plastic to the tube with plastic tape.
4. Put about 8 in (20 cm) of water in the tube such that it settles at the bottom of the U. Leave both ends of the tube open.
5. Attach a ruler vertically to the board behind or alongside the plastic tube.

Pressure between the layers of plastic will push the water down on the film plastic side of the U and up on the opposite side of the U. A rise in water level of 0.2 to 0.3 inch (5.1 to 7.6 mm) indicates the desired pressure. Coloring the water will help make it more visible.

The fan is generally mounted on the end wall of the greenhouse. A hole is cut in the end wall adjacent to the fan so that air feeding the fan is drawn in from outside. Outside air is colder than the air between the plastic layers. As the cold air warms in the roof cavity, it dries. This helps to control condensation between the layers of plastic. Such condensation leads to light reduction as well as corrosion problems. If warm moist air from inside the greenhouse were used, it would cool and water would condense in the roof cavity.

A flexible tube, such as that used for a clothes dryer, is installed between the fan and the inner layer of plastic to be inflated. A X-shaped cut is made in the inner layer of plastic, and the tube is inserted through it. The four points of plastic resulting from the cut are pulled out over the tube and taped to it to make an airtight seal. Air is conducted from the fan to the inner space through this tube. This system is sufficient to inflate the entire roof of a Quonset greenhouse. Generally, the two layers of plastic pull tight at the ridge of an A-frame greenhouse, thus separating the roof into two inflatable portions. In this case, air from the fan can be divided in a 4-inch (10-cm) stovepipe tee and introduced to each side of the roof through flexible tubing immediately below the ridge. Side or end walls can be inflated as well without adding additional fans. Flexible connectors are sold for this purpose (Figure 2-18). Alternatively, pieces of garden hose can be inserted between the layers of plastic to connect the roof cavity to the end- or side-wall cavities.

Figure 2-18 A flexible jumper tube for connecting the air cavity between the two film plastic layers on the roof with the cavity in the side wall. This permits one fan to pressurize the two cavities.

Rigid-Panel Greenhouses

Polyvinyl Chloride

Polyvinyl chloride (PVC) rigid panels have, for the most part, been dropped from use. Initially, they showed promise as an inexpensive covering (about 40 percent of the cost of long-lasting FRP). They had a life expectancy of five years or better at a time when polyethylene lasted one year. Commercial use of these panels soon indicated that this life expectancy was much shorter, sometimes as little as two years. This was unacceptable because the cost of PVC panels was four to five times that of polyethylene film and because they required much more time to install. Rigid PVC, like its film plastic counterparts, was subject to the deteriorating effect of UV light, which caused it to turn dark and become brittle. At first, light transmission was reduced; later, the panels would break apart. Rigid PVC was purchased in corrugated panels 26 or 28 in (66 or 71 cm) wide and 8, 10, or 12 ft (2.4, 3.1, or 3.7 m) long. The panels were available in various colors; however, clear panels were used for general greenhouse culture.

Fiberglass-Reinforced Plastic

Role Fiberglass-reinforced plastic (FRP) was more popular as a greenhouse covering in the recent past than it is today. As with PVC, corrugated panels were used because of their greater strength. Flat panels are occasionally used on the end and side walls where the load is not as great. Panels are available in 51.5-inch (1.3-m) widths, lengths up to 24 ft (7.3 m), and a variety of colors. The panels are flexible enough to conform to the shape of Quonset greenhouses, which makes FRP a very versatile covering material.

FRP can be applied to the inexpensive frames of film plastic greenhouses (Figure 2-19) or to the more elaborate frames of glass-type greenhouses (Figure 2-20). In the former case, the price of the FRP greenhouse lies between that of a film plastic greenhouse and that of a glass greenhouse, but the cost is offset by elimination of the need for replacement of film plastic. In the latter case, the FRP greenhouse costs about the same as the glass greenhouse.

FRP and glass greenhouses each have advantages and disadvantages, and growers are divided as to their preferences. FRP is more resistant to breakage by factors such as hail or vandals. Sunlight passing through FRP is scattered by the fibers in the panels, with the result that light intensity is rather uniform throughout the greenhouse by comparison with a glass covering. Plants on the north sides of beds, and particularly in the north beds as a whole, grow much better. Hammered glass, however, offers a similar light-scattering benefit.

There are disadvantages as well. The acrylic surface of FRP panels is subject to etching and pitting by dust abrasion and chemical pollution. Thus, glass fibers become exposed and subject to fraying, and they begin to collect dust as well as to harbor algae. The resultant effect is a darkening of the panels and a subsequent reduction in light transmission. The situation can be corrected by scrubbing the FRP surface clean with a stiff brush or steel wool and then painting on a new surface of acrylic resin. The material is inexpensive, but the labor is extensive. The

Figure 2-19 **A Quonset greenhouse being covered with sheets of corrugated FRP.**

Figure 2-20 A permanent iron-frame greenhouse with FRP covering.

need for refinishing varies with the grade of FRP purchased. Some grades do not carry a guarantee and may last only five years or so. Other grades carry guarantees of various lengths of time up to 20 years. Those with a UV light–resistant protectant hold the longest life expectancy. The guarantee generally protects the level of light transmission and compensates for the unused portion of the term of the guarantee. Many growers in high-light areas such as Denver count on 10 years of use and then replace the FRP to regain maximum light transmission. By contrast, glass can last as long as a grower's life or longer. The decision between glass and FRP is not clear-cut. Some northern growers have been known to cover only the north slope of a glass greenhouse with FRP to increase the light intensity within. A portion of the sun's rays impinging upon the north roof are then transmitted inward, rather than being deflected off.

Light Transmission The total quantity of light transmitted through clear FRP is roughly equivalent to that transmitted through glass (as shown in Table 2-2) but diminishes in relation to its color. For greenhouse crops in general, only clear FRP permits a satisfactory level of light transmission (88 to 90 percent). Colored FRP has found a limited use in greenhouses used for growing some houseplants that require low light intensity and in display greenhouses used for holding plants during the sales period.

Heat Transmission FRP has the distinct advantage over glass of being easier to cool. In an experiment conducted at Colorado State University with two greenhouses of identical size and style, one was covered with clear FRP and the other with glass. The length of time that the cooling fans operated in each greenhouse was recorded for 13 consecutive months. Fewer hours of cooling were required in the FRP greenhouse month by month. At the end of the 13 months, a total of 2,066 hours of cooling had been required in the glass greenhouse versus only 1,668 hours in the FRP greenhouse. This represented a reduction of 19 percent. The winter heat requirement of corrugated FRP greenhouses is about equivalent to that of structurally tight glass greenhouses.

Construction FRP greenhouses require fewer structural members than glass greenhouses since sash bars are not needed. The construction labor input is accordingly lower for an FRP greenhouse. The price of FRP is similar to that of glass. The thickness of FRP is measured in terms of weight per square foot. Where a snow load is expected, 5-oz weights (37 mil, 0.94 mm thick) are used on peaked roofs. The 4-oz weight (30 mil, 0.76 mm thick) is common on arched roofs and vertical walls. Trusses are spaced 8 to 10 ft (2.4 to 3.1 m) apart and purlins 4 ft (1.2 m) apart. High-quality, 5-oz FRP with a warranty of 20 years costs about \$1.00/ft^2 (\$10.76/m^2). There is an FRP with an extra-thick coating of plastic resin over the glass fibers to guard against fraying of the fibers, also with a 20-year warranty, that costs about \$1.25/ft^2 (\$13.46/m^2). FRP panels are 50.5 to 52.6 in (1.28 to 1.34 m) wide but, with overlap, have an effective covering width of 48 in (1.22 m).

The greenhouse must be constructed to be as airtight as possible. Corrugated plastic closures (Figure 2-21) are available for insertion between the FRP panel and frame components such as the eave and the sill, to seal off outer air. Flashing is used at the ridge to cover the exposed ends of the FRP panels for the purpose of preventing water entry. The flashing can be constructed from aluminum or corrugated FRP. The FRP panels are attached to the purlins by aluminum screw nails or by aluminum wood screws. These nails and screws have a rubber washer immediately beneath the head to seal the hole made by the shaft.

When condensation flows along the inner surface of FRP, it does so along the corrugation valleys. If the FRP panels are attached directly to the purlins, the corrugation valleys are in contact with the purlins. Condensation, upon reaching this point, flows onto the purlin and drips from its lower edge, thus causing harm to plants beneath. The FRP panel must be elevated away from the purlin. Metal U-shaped supports are placed between the purlin and the corrugation ridges of the FRP panel. The nail or screw attaching the panel to the purlin passes through the support (Figure 2-22).

Fire Hazard Many greenhouse structures are insured. One cause of destruction is fire, which is not a significant danger in glass greenhouses but is a very definite concern in FRP greenhouses. The glass fibers themselves do not burn, but the polyester and acrylic resins binding them together do. A few years ago, a fire believed to have originated from a faulty electrical wire beneath a sheet of black shade cloth spread to the FRP covering of a ridge-and-furrow range on a windy night. More than an acre of greenhouses was consumed within 20 minutes. Insurance rates are assessed according to the risk involved, which is greater for FRP greenhouses. Fire-retardant FRP panels are available and carry the best

Figure 2-21 **A corrugated plastic closure strip in place, sealing off the outer air at the point of attachment of a corrugated FRP panel to the frame member.**

rating for building materials (Class I). These panels offer no support for sustainment to flame even when they are directly attacked. Benefits associated with standard greenhouse FRP are not associated with fire-retardant FRP by manufacturers; thus, these panels are not commonly used for greenhouses.

Acrylic and Polycarbonate

Acrylic and polycarbonate double-layer rigid panels have been available for about 15 years for greenhouse use. A number of research institutions have installed these panels. There has been a steady increase in acceptance in the commercial industry. The heaviest commercial use thus far has been for glazing

Figure 2-22 **U-shaped metal supports are placed between the purlin and the FRP panel to provide a space so that condensation water can flow along the inner surface of the panel to the ground or to a gutter. The pot label was inserted to demonstrate this space.**

side and end walls on film plastic greenhouses (Figures 2-14 and 2-15) and for retrofitting old glass and FRP greenhouses. A moderate number of new buildings have been completely glazed with these panels. While the acrylic panels are highly flammable, the polycarbonate panels are not flammable. The acrylic panels are particularly popular with research institutes due to their higher light transmission and longer life. The polycarbonate panels are by far the most popular choice for commercial greenhouses due to lower price, flame resistance, and greater resistance to hail damage.

Acrylic panels are available in thicknesses of 16 mm (0.64 inch) and 8 mm (0.32 inch). The thicker panels cannot be bent, but the thinner panels can be bent to fit curved-roof greenhouses. Panels are available with a coating to prevent condensation drip. The two acrylic layers of these panels are held apart by ribs spaced approximately 0.6 to 0.9 inch (16 to 24 mm) apart. The panels have a width of 47.25 in (120 cm) and come in lengths up to 39 ft (11.9 m). The effective covering width of the panel is 48 in (122 cm), since the metal support member takes up space between panels. Heat-loss (U) values for 8-mm and 16-mm panels are 0.65 and 0.58 Btu per hour per square foot per degree Fahrenheit (Btu/hr/ft^2/°F) temperature differential from inside to outside the greenhouse (3.7 and 3.3 W/m^2/K), respectively. The heat-loss value for glass is 1.13 Btu (6.40 W), which is nearly double this. PAR light transmission is 83 percent. The thinner panels carry a limited warranty against loss of more than 3 percent light transmission in 10 years. The 8-mm panels sell for \$2.10/ft^2 (\$22.60/m^2), while the 16-mm panels sell for approximately \$2.90/ft^2 (\$31.22/m^2).

Polycarbonate panels come in thicknesses of 4, 6, 8, 10, and 16 mm (0.16, 0.24, 0.32, 0.40, and 0.64 inch). The thinner panels can be bent to fit curved-roof greenhouses, while the thicker ones cannot. Their skin thicknesses and rib dimensions are similar to those of acrylic panels. Depending on the manufacturer, panels are available with a coating to prevent condensation drip and also with an acrylic coating for extra protection from UV light. Panels are available in widths from 4 to 8 ft (1.2 to 2.4 m). Lengths are available up to 32 ft (9.8 m). Heat-loss values are 0.72, 0.65, and 0.58 Btu/hr/ft^2/°F (4.1, 3.7, and 3.3 W/m^2/K) for the 6-, 8-, and 16-mm panels, respectively. PAR transmission is 79 percent. The loss in light transmission is anticipated at 1 percent per year in the thin panels. The 6-mm and 8-mm panels sell for \$1.60/ft^2 and \$1.70/ft^2 (\$17.22/m^2 and \$18.30/m^2), respectively.

Benches and Beds

Fresh Flowers

The first choice in growing fresh flowers is whether to grow them in raised benches or in ground beds. If the crop is of moderate height, such as chrysanthemum and snapdragon, raised benches can be used; however, these benches should be located close to the ground to keep the plants at a practical level for disbudding, spraying, and harvesting. Rose plants are grown for about five years and become exceedingly tall during this time. Most are grown in ground beds to minimize height. Carnations are grown from one to two years and also become very tall. Years ago, they were commonly grown in ground beds without bottoms, but the occurrence of a bacterial wilt disease nearly destroyed this business in the northeastern United States, and since then they have been grown in raised benches. (It was not possible to pasteurize the root substrate deep enough in the bottomless ground beds, and the disease continually recurred.)

If ground beds are selected, they should be constructed in a manner that isolates the root substrate contained within from external soil. In this way, the root substrate can be thoroughly pasteurized on a routine schedule, thus reducing the possibility of disease. Concrete is very suitable for ground beds (Figure 2-23). The bottom should be V-shaped, with the longitudinal center at least 1.5 in (4 cm) lower than the sides. A half tile should be placed in the center over the V, and the bed sloped 1 inch per 100 ft (1 cm per 1,200 cm) to ensure drainage of water. The bottom of the bed should be filled level with gravel to ensure lateral movement of water into the tile. A hole to permit drainage of water should be located at the point where the tile contacts the lower end of the bed. The drainage tile serves another valuable purpose, since steam introduced through the tile will percolate up through the root substrate and pasteurize it.

Other less expensive ground beds can be constructed as well (Figure 2-24). Side walls can consist of treated wood or cement blocks. The wall should be at least 8 in (20 cm) deep and extend down to a well-drained foundation substance, such as a sandy subsoil. If the base substance is not well drained, drainage tile

Figure 2-23 **A concrete ground bed used for cut-flower production. The bed is sloped and has a V-shaped bottom. A half tile runs the length of the bed at the lowest point to conduct water to a drain hole at the end of the bed. A concrete trough running across the greenhouse collects water from all beds and carries it out of the greenhouse. Steam can be injected into the drain hole for pasteurization of the bed between crops.**

should be installed in this substance below each bed. Walks should be filled with gravel; paved walks should be sloped for drainage. It is important that walks be separated from beds to ensure that (1) the soil in them, easily contaminated by soil carried in on shoe bottoms, does not spread into the beds, and (2) water remains where it is applied, rather than running off into the walks.

Ground beds are well suited for tall crops, such as rose and carnation. If raised beds are preferred for cut flowers, they should be situated close to the ground. An 8-inch (20-cm) concrete block serves as a good post to separate the bench from the ground. The bottom should have abundant drainage holes along its length. Raised bottoms should be as level as possible to prevent wet and dry areas. Benches are most commonly constructed from concrete or treated wood. Concrete benches can be poured in place or assembled from precast concrete boards. One board is used for each side; several boards, running lengthwise, are used for the bottom. The bottom boards have a 1/2-inch (1.3-cm) space between them for drainage. Galvanized iron brackets are used to bolt the sides to a pipe frame or concrete cross-support beneath the bench floor.

The preferred woods for bench construction are cypress, redwood, locust, and cedar because of their resistance to decay. Wooden benches should be painted with a copper naphthenate preservative. The natural preservative in redwood is corrosive to iron and steel; therefore, nails, screws, or bolts should be made of other types of metals such as aluminum, brass, or zinc.

The preferred widths of cut-flower benches and beds are 3.5 and 4.0 ft (1.1 and 1.2 m). Roses are conveniently grown in 4-foot (1.2-m)-wide beds because

(a)

(b)

Figure 2-24 (a) Drainage tiles embedded in gravel beneath a ground bed. (b) Ground beds with treated wood sides. The root substrate is placed on the gravel base containing the drainage tile.

bushes are planted 1 foot (30 cm) apart in each direction. This permits four plants across the bed. The other cut-flower crops may be found in either width of bed. Except in very wide greenhouses, benches run the length of the greenhouse. The beds and benches should be 8 in (20 cm) deep to accommodate 7 in (18 cm) of root substrate. Rose beds, which should be 1 foot (30 cm) deep, are an exception. Eighteen-inch (46-cm) walks should be used between all benches except in the center of the greenhouse, where a 2-foot (61-cm) walk should be established. This longitudinal arrangement of benches allows for the use of about 67 percent of the floor area for growing.

Pot-Plant Crops

Raised benches are generally used for pot-plant crops. They should be 32 to 36 inches (81 to 91 cm) high for convenience of working. Benches should not exceed a 3-foot (91-cm) width if they are against a wall, or a 6-foot (1.8-m) width if they are accessible from both sides. It is difficult to handle plants in the center of wider benches, and labor becomes inefficient. It is important to have air circulation around each plant to reduce the incidence of condensation on foliage and thus the possibility of disease. Pot plant benches should not have sides. The floor of the bench should be as open as possible. Spruce or redwood lath in woven wire, similar to snow fencing but manufactured more precisely for benches, makes excellent bench floors and is sold for this purpose. The spruce lath can be supported with a 2-inch-by-4-inch (5-cm-by-10-cm) wooden frame (Figure 2-25)

Figure 2-25 **A raised pot-plant bench using spruce lath for the floor and a 2-inch-by-4-inch (5-cm-by-10-cm) wood frame. Cement blocks are used for legs.**

or by a pipe frame. The frame itself is often supported by concrete blocks. One-inch (2.5-cm)-square, 14-gauge welded-wire fabric and expanded metal also make excellent bench floors. These benches permit proper circulation of air.

A special category of pot-plant benches is used for ebb-and-flow culture. These benches are watertight to accommodate periodic flooding with fertilizer solution and are plumbed to a tank below them for holding the solution when it is not in use. (See Chapter 9 for details.)

Cut-flower benches generally run lengthwise in a greenhouse to minimize the number of end posts needed for supporting plants and the time necessary to attach and tighten support wires. Since support is not a consideration in pot plant benches, these benches usually run across the greenhouse to minimize handling of heavy pots. A 3-to-4-foot (0.9-to-1.2-m)-wide center aisle is provided along the length of smaller greenhouses to permit motorized carts to be used for transporting plants and materials. In larger greenhouse ranges, an 8-foot (2.4-m) center drive should be provided for larger internal transport equipment. Side walks should be 18 in (46 cm). Benches may be located at the ends of the walks. Benches in this arrangement are known as *peninsular benches* and can result in as much as 80 percent growing area, as opposed to 67 percent in the longitudinal arrangement.

Movable-bench systems can increase production space up to about 90 percent of the floor space. By turning a crank at the end of the bench or by simply pushing the bench, the bench platform can be moved to either side. As a bench is moved from right to left, an aisle on the left side closes and a new aisle opens up on the right side (Figure 2-26). When several movable benches are used, only one aisle is needed, which can be shifted to any position.

The number of benches permitted per aisle is a difficult question. In a stationary-bench arrangement, a production operation could be carried out in each aisle simultaneously. This would be a benefit for crops requiring constant attention, such as frequent respacing, disbudding, pinching, or selection of plants for market. A crop such as Easter lily or poinsettia, which does not require as

Figure 2-26 **An aisle-eliminator bench system. (a) The bench on the right is in its extreme right position. (b) The bench on the right has been moved to its left position, thus shifting the aisle to the right of this bench.** ***(Photos courtesy of Simtrac, Inc., Skokie, IL 60076.)***

many production operations and is marketed over a short period, is well adapted to movable benches. As many as five benches may be used per 2-foot (61-cm) aisle for such a crop.

Benches are an expense that is often forgotten in the pricing of a greenhouse firm. Prefabricated benches are available in many designs and cost \$2.50 to \$3.50/ft^2 (\$26.91 to \$37.67/m^2) of bench. An additional cost of \$0.50 to \$1.00/ft^2 (\$5.38 to \$10.76/m^2) is required to have them installed.

A different concept for ridge-and-furrow ranges with a large single interior calls for paving the floor with porous asphalt or concrete and growing pot plants directly on the floor (Figure 2-27). Water percolates through the pavement to a gravel bed beneath, while weeds are unable to grow through this layer. Standard asphalt paving with a reduced quantity of binder can be used, or porous concrete made from a mixture of 2,800 lb (1 yd^3, 0.76 m^3) of $^3/_8$-inch (10-mm) dust-free gravel, 5.5 bags (94 lb, 43 kg each) of cement, and 23.4 gal (88 L) water can be used (Aldrich and Krall 1978; Aldrich and Bartok 1994). Porous concrete is generally poured in a layer 4 in (10 cm) thick. It will withstand a working compressive strength test of 600 pounds per square inch (psi) (4,137 kPa). Light vehicles may be driven over the floor for setting up and removing crops. This system makes it possible to use 90 percent of the floor area for growing. A disadvantage occurs with crops requiring extensive hand labor operations, because working at ground level is fatiguing and takes its toll on labor efficiency. Bedding plants, azalea liners, some green plants, and poinsettias are well suited to this system. A concrete floor costs approximately \$2.00/ft^2 (\$21.53/m^2) for materials and labor.

Figure 2-27 **A gutter-connected range in which pot plants are grown on a pavement of water-porous asphalt. Growing space is maximized in this greenhouse, and tractors or trucks can be used for moving plants and materials.**

A recent evolution of the paved floor concept is found in the *flood floor.* This is a nonporous concrete floor that slopes from the sides to the center of each greenhouse bay. The perimeter of the floor in each bay has a side wall to contain water. An inlet/drain pipe is situated under the low point at the middle of the bay. When water or fertilizer solution is required, the floor is flooded for about 10 minutes. Water moves into the root substrate of the pot by capillary action. The floor is then drained (see Chapter 9 for more details). A complete flood floor costs between \$3.50 and \$4.00/ft^2 (\$37.67 and 43.06/m^2) for the concrete floor, plumbing, tank, pumps, and labor.

Cost of Greenhouse Construction

Presented in Table 2-3 is the range of 1996 commercial construction prices for 1 acre (0.4 ha) of various types of greenhouses. Included in the basic structure category are the total frame, ventilators for the cooling pads, greenhouse ends, doors, and covering. Labor includes placement of the unit heating and cooling systems. The heating systems are of the forced-air unit heater type. The cooling systems include a pressurized convection tube for winter cooling and a cross-fluted cellulose pad-and-fan system for summer. Greenhouse prices can vary more than is shown in Table 2-3. The strength of structural components can be reduced where no snow is expected, or increased in areas of abnormally high snow and wind.

The basic price of a polyethylene greenhouse can be as reasonable as \$1.50/ft^2 (\$16.14/m^2). This figure, however, does not include the covering, end walls, erection labor, heating, cooling, wiring, and plumbing. When these items are added, a polyethylene greenhouse can cost \$6.75 to \$7.75/ft^2 (\$72.66 to \$83.42/m^2). If benches and thermal screens are desired, the price goes up to \$10.50 to \$12.50/ft^2 (\$113.02 to \$134.55/m^2). Missing yet are the prices of land, grading, service buildings, access drives, and parking areas, which could easily add another \$3.00/ft^2 (\$32.28/m^2) of greenhouse, for a total of \$13.50 to \$15.50/ft^2 (\$145.31 to \$166.84/m^2).

Low-profile glass greenhouses cost from \$2.50 to \$4.00/ft^2 (\$26.91 to \$43.06/m^2) more than polyethylene greenhouses. High-profile greenhouses are not priced in the table because at the present time they are considerably higher in price than low-profile greenhouses. Much of their current application is for research and educational institutions. Other greenhouse coverings are also not included in the table. Acrylic and polycarbonate panels on permanent frames could cost more than glass greenhouses. FRP on permanent frames would be priced similarly to glass greenhouses. Quonset polyethylene greenhouses might be as low as half the basic price of a gutter-connected polyethylene greenhouse.

Selection of a greenhouse should not be based solely on the total purchase price. Maintenance, such as re-covering polyethylene every 3 years or FRP every 10 to 20 years, must be assigned a cost. The 40 percent fuel savings in a double-layer film plastic greenhouse, or the nearly 50 percent fuel savings of acrylic and polycarbonate panels compared to single-layer glass, must enter into the decision.

Table 2-3 Range of Prices ($\$/ft^2$ of floor area) for 1 Acre (0.4 ha) of Gutter-Connected, Double-Layer Polyethylene and Low-Profile Glass Greenhouses[1]

Item	Polyethylene	Glass
Structure with cover	$2.00–3.00	$4.50–6.50
Erection labor	$1.00	$1.50
Heating system[2]	$0.50–1.00	$0.50–1.00
Cooling system[2]	$1.25–0.75	$1.25–0.75
Plumbing	$0.75	$0.75
Wiring	$1.25	$1.25
Subtotal	$6.75–7.75	$9.75–11.75
Benches	$2.50–3.50	$2.50–3.50
Thermal screen	$1.25	$1.25
Total	$10.50–12.50	$13.50–16.50
Polycarbonate ends[3]	$0.30	—
Central heating system[3]	$2.00	$2.00

[1]Prices are derived from a broad range of greenhouse suppliers in the United States in 1996. $\$1.00/ft^2 = \$10.76/m^2$.

[2]The lowest heating and highest cooling prices represent firms in warm regions, while the opposite combination represents cold regions. Heating systems are of the forced-air unit heater type. Cooling systems include the summer fan-and-pad system plus the winter convection-tube system.

[3]Add these prices to the total to upgrade the greenhouse to these features.

The predominant choice of the industry today is film plastic. There are, however, valid arguments in favor of glass. The higher light intensity inside modern glass greenhouses compared to double-layer polyethylene greenhouses more than offsets the higher price of the former for some growers. With the numerous options available and the differences in prices for greenhouse frames, coverings, heat-conservation systems, and heating systems, it is extremely important that a greenhouse operator study the available information and perform the appropriate cost analysis. This is an easy time to go bankrupt from the purchase of too much cost-saving technology.

Summary

1. Greenhouse location is as important as the greenhouse design itself. Factors to be sought in a location are as follows:
   - **a.** Room for expansion.
   - **b.** A level, well-drained site.

c. Reasonable tax structure at present and in the future.

d. A climate favorable for the crop intended.

e. Available labor.

f. Reasonable proximity to utilities and shipping routes.

g. A plentiful supply of high-quality water.

2. The greenhouse-business floor plan is crucial to efficiency. Greenhouses should be consolidated into a square block to minimize distances for material movement, increase ease of personnel management, and reduce heating and automation costs. There should be a single service building centrally situated. The whole range should be on a single elevation.

3. Glass greenhouses are permanent and can last as long as the owner's life or longer. The material expense and labor of periodically replacing the covering is eliminated with glass, but the overall cost of a glass structure is higher. There are two general styles: high-profile greenhouses, which can be free-standing or connected in a ridge-and-furrow fashion, and low-profile Dutch-type greenhouses, which are constructed in a ridge-and-furrow style only because of their narrow bay width of 10.5 ft (3.2 m).

4. Film plastic greenhouses are the least expensive to build. They lend themselves well to temporary business ventures, businesses operated for only one season of each year, and locations where there is a tax advantage for nonpermanent structures. Film plastic greenhouses offer an inexpensive means of entering the flower-growing business. However, film plastic ranges can be built on permanent, metal, gutter-connected frames, permitting the full degree of automation and efficiency of any glass or FRP range. Polyethylene is the most common film plastic in use and is usually applied as an air-inflated double layer. The insulating property of the double layer reduces fuel consumption by about 40 percent over a greenhouse with a single covering of polyethylene, glass, or FRP, which makes the double-layer polyethylene greenhouse less expensive to purchase and operate in spite of the periodic labor and the cost of replacing the plastic. Film plastic greenhouses constitute the greatest portion of new greenhouse construction.

5. A third type of greenhouse is the FRP (fiberglass-reinforced plastic)-panel greenhouse. FRP panels can be bent to fit most film plastic greenhouse frames. This reduces the labor of replacing film plastic since FRP, depending on grade, will last 5 to 20 years. FRP is also used on permanent metal-frame greenhouses. In this latter case, the overall structure generally costs about the same as a glass greenhouse. The FRP covering does not last as long as glass, but it is more resistant to breakage, is cheaper to cool in the summer, and has a more uniform light intensity throughout the greenhouse. The highest concentration of use is in California.

6. Acrylic and polycarbonate double-layer panels are gaining popularity. Thick 16-mm (0.64-inch) panels can reduce heat loss by 50 percent compared to single-layer glass. Heat savings are moderately less with thinner panels. The

thinner panels of 6 and 8 mm (0.24 and 0.32 inch) can be bent to fit Quonset designs. A large proportion of polycarbonate panels are being used for side and end walls on film plastic greenhouses and for retrofitting old glass and FRP greenhouses.

7. Fresh-flower crops are grown in either ground beds or raised benches. Such beds are either 3.5 or 4 ft (1.1 or 1.2 m) wide and are generally 8 in (20 cm) deep, but 1 foot (30 cm) is best for rose beds. Fresh-flower beds are oriented along the length of the greenhouse with 18-inch (46-cm) aisles between them. This arrangement of beds allows for 67 percent utilization of floor space for growing.

8. Pot plants can be grown on raised benches or directly on the floor. Raised benches have open bottoms constructed from wire hardware cloth, expanded metal, spruce lath, or treated boards with at least a ½-inch (1.3-cm) space between them. Sides are either not used or are low. Benches are usually 5 to 6 ft (1.5 to 1.8 m) wide and are arranged in a peninsular style. A central aisle, 3 ft (91 cm) wide or wider, runs the length of the greenhouse. Benches and smaller aisles radiate out from the central aisle to either side. Such an arrangement makes more efficient use of floor space—up to 80 percent growing area—and minimizes hand carrying of plants. Some pot crops are grown directly on floors paved with porous asphalt or concrete. Water penetrates the floor, while weed growth is inhibited. Others grow on nonporous concrete ebb-and-flood floors. These latter two systems permit use of up to 90 percent of the floor space.

References

Various greenhouse manufacturers offer literature concerning products and technical information.

1. Agricultural Statistics Board. 1996. *Floriculture crops: 1995 summary.* USDA, National Agr. Statistics Ser., Sp. Cir. 6-1 (96). Washington, DC.
2. Aldrich, R. A., W. A. Bailey, J. W. Bartok, Jr., W. J. Roberts, and D. S. Ross. 1976. *Hobby greenhouses and other gardening structures.* Pub. NRAES-2. Northeast Reg. Agr. Eng. Ser., Cornell Univ., 152 Riley-Robb Hall, Ithaca, NY 14853.
3. Aldrich, R. A., and J. W. Bartok, Jr. 1994. *Greenhouse engineering,* 3rd rev. Pub. NRAES-33. Northeast Reg. Agr. Eng. Ser., Cornell Univ., 152 Riley-Robb Hall, Ithaca, NY 14853.
4. Aldrich, R. A., and T. J. Krall. 1978. Compression strength of porous concrete. *Pennsylvania Flower Growers' Bul.* 307:1–6.
5. American Society of Agricultural Engineers. 1995. Commercial greenhouse design and layout: ASAE EP460 July 93. In: *American Society of Agricultural Engineers Standards 1995,* 42nd ed., pp. 596–602. St. Joseph, MI: Amer. Soc. of Agr. Engineers.
6. Bartok, J. W., Jr. 1984. Greenhouse startup and expansion. *Greenhouse Manager* 3(1):57–78.
7. Beese, E. J. 1978. Wood preservatives and treated lumber for use in landscape construction. *Illinois State Florists' Association Bul.* 377(May–June):20–21.

8. Brumfield, R. G., P. V. Nelson, A. J. Coutu, D. H. Willits, and R. S. Sowell. 1981. *Overhead costs of greenhouse firms differentiated by size of firm and market channel.* North Carolina Agr. Res. Ser. Tech. Bul. 269.
9. Courter, J. W. 1965. *Plastic greenhouses.* Univ. of Illinois Coop. Ext. Ser. Cir. 905.
10. Godbey, L. C., T. E. Bond, and H. F. Zornig. 1979. Transmission of solar and long-wavelength energy by materials used as covers for solar collectors and greenhouses. *Trans. Amer. Soc. Agr. Engineers* 22(5):1137–1144.
11. Gray, H. E. 1956. *Greenhouse heating and construction.* Florists' Publishing Co., 343 S. Dearborn St., Chicago, IL.
12. Kozai, T., J. Gourdriaan, and M. Kimura. 1978. *Light transmission and photosynthesis in greenhouses.* Wageningen, The Netherlands: Center for Agr. Publishing and Documentation.
13. National Greenhouse Manufacturers Association. 1994. *Standards: design loads in greenhouse structures, ventilating and cooling greenhouses, greenhouse heat loss, and greenhouse retrofit.* National Greenhouse Mfg. Association, 7800 S. Elati, Suite 113, Littleton, CO 80120.
14. Robbins, F. V., and C. K. Spillman. 1980. Solar energy transmission through two transparent covers. *Trans. Amer. Soc. Agr. Engineers* 23(5):1224–1231.
15. U.S. Housing and Home Financing Agency. 1954. *Thermal insulation value of air space.* Housing Res. Paper 32. U.S. Housing and Home Financing Agency, Div. of Housing Res., Washington, DC.
16. Wiebe, J., and R. E. Barrett. 1970. *Plastic greenhouses.* Pub. 40. Ontario Dept. of Agr. and Food.

CHAPTER 3

Greenhouse Heating

Heat is measured by the *British thermal unit* (Btu), defined as the amount of heat required to raise 1 lb of water 1°F. When the number of Btu's becomes large, as in heating greenhouses, it is more convenient to use the larger unit *horsepower* (hp). One boiler horsepower is equivalent to 33,475 Btu. To convert from Btu to boiler horsepower, divide Btu's by 33,475. In the metric system, a *calorie* (cal) is defined as the amount of heat required to raise 1 gram (g) of water 1°C. One *kcal* equals 1,000 cal or 3.968 Btu. In international units, the *joule* (J) is used, which is equivalent to 0.239 cal or 0.00095 Btu. Reciprocally, 1 Btu equals 252 cal or 1,055 J. One *watt* (W) is equal to 1 J per second.

Types of Heat Loss

The requirements for heating a greenhouse reside in the task of adding heat at the rate at which it is lost. Most heat is lost by *conduction* through the covering materials of the greenhouse. Different materials, such as aluminum sash bars, glass, polyethylene, and cement curtain walls, vary in conduction according to the rate at which each conducts heat from the warm interior to the colder exterior. For instance, aluminum sash bars conduct heat faster than wood, which results in more rapid loss of heat. (Since the upkeep of wood, however, is much greater, its use is not justified.) Listed in Table 3-1 are heat-loss (U) values for several greenhouse coverings. A greenhouse covered with one layer of polyethylene, for example, loses 1.20 Btu of heat through each square foot of covering every hour when the outside temperature is 1°F lower than the inside. When a second layer of polyethylene is added, only 0.7 Btu is lost. This is a reduction of almost 40 percent of the heat loss.

There are limited ways of insulating the covering material without blocking light transmission. As previously mentioned, a dead-air space between two coverings appears to be the best system. Forty percent of the heat requirement can be saved when a second covering is applied. The savings diminish when the air space between the two coverings increases to the point where air currents can be established in the space—generally with a spacing of 18 in (46 cm) or greater—and the insulation value is completely lost when the two layers touch each other.

Although thermopane glass panels (two layers of glass factory-sealed with a dead-air space) significantly reduce heat loss, they have been too expensive to justify. Sash bars that hold two and even three layers of glass are available from some manufacturers. This establishes one or two dead-air spaces between the layers of glass and yields overall U values of 0.70 and 0.47 Btu (3.97 and 2.66 W), respectively. Double-layer rigid panels of either acrylic or polycarbonate plastic also utilize the concept of dead-air space for heat conservation. Although more expensive than conventional coverings, these materials in 16-mm-thick panels have a lower U value of 0.58 Btu (3.29 W).

A second mode of heat loss is that of air *infiltration*. Spaces between panes

Table 3-1 Heat Loss through Various Greenhouse Coverings

	Heat Loss (U)[1]		*Radiation Loss*[2] *(% of total)*
Covering Material	*Btu*	*W*	
Glass, single-layer	1.13	6.40	4.4
Glass, double-layer	0.70	3.97	—
Film (single-layer) over glass	0.85	4.82	—
Film (double-layer) over glass	0.60	3.40	—
PVC (rigid)	0.92[3]	5.21	—
FRP (corrugated)	1.20	6.80	1.0
Acrylic or polycarbonate (16-mm panels)	0.58	3.29	—
Acrylic or polycarbonate (8-mm panels)	0.65	3.69	—
Polycarbonate (6-mm panels)	0.72	4.08	—
Polyethylene film, single-layer	1.20	6.80	70.8
Polyethylene film, double-layer	0.70	3.97	—
Polyvinyl fluoride, single-film (Tedlar)	—	—	30.0[4]
Polyester film, single-layer	1.05[3]	5.95	16.2
Corrugated cement asbestos board	1.15	6.51	
Concrete, 4-inch (10-cm)	0.78	4.42	
Concrete, 8-inch (20-cm)	0.58	3.29	
Concrete block, 4-inch (10-cm)	0.64	3.63	
Concrete block, 8-inch (20-cm)	0.51	2.89	

[1]Btu/hr/ft^2/°F, W/m^2/K; U is the combined loss of heat due to conduction and radiation. From National Greenhouse Manufacturers' Association (1994), unless otherwise indicated.

[2]Radiation loss is the amount of radiant heat passing through the covering expressed as a percentage of the total radiant heat beaming upon it. From Duncan and Walker (1973).

[3]Manufacturers' specifications.

[4]From Whillier (1963).

Table 3-2 **Air Infiltration in Greenhouses**[1]

Greenhouse Type	*Air Exchanges per hour*
Double-layer film plastic	0.5–1.0
Acrylic or polycarbonate panels[2]	1.0
Glass, new construction or new FRP	0.75–1.5
Glass, old construction, good condition	1–2
Glass, old construction, poor condition	2–4

[1]From American Society of Agricultural Engineers (1995).
[2]From National Greenhouse Manufacturers Association (1994).

of glass or FRP and around ventilators and doors permit the passage of warm air outward and cold air inward. A general assumption holds that the volume of air held in a greenhouse can be lost as often as once every 60 minutes in a double-layer film plastic or polycarbonate panel greenhouse; every 40 minutes in an FRP or a new glass greenhouse; every 30 minutes in an old, well-maintained glass greenhouse; and every 15 minutes in an old, poorly maintained glass greenhouse (see Table 3-2). About 10 percent of the total heat loss from a structurally tight glass greenhouse occurs through infiltration loss.

A third mode of heat loss from a greenhouse is that of *radiation.* Warm objects emit radiant energy, which passes through air to colder objects without warming the air significantly. The colder objects become warmer. Glass, vinyl plastic, FRP, and water are relatively opaque to radiant energy (do not readily permit the passage of radiant heat), whereas polyethylene is not (Table 3-1). Polyethylene greenhouses can lose considerable amounts of heat through radiation to colder objects outside, unless a film of moisture forms on the polyethylene to provide a barrier.

Heating Systems

The heating system must provide heat to the greenhouse at the same rate at which it is lost by conduction, infiltration, and radiation. There are three popular types of heating systems for greenhouses. The most common and least expensive is the *unit heater* system. In this system, warm air is blown from unit heaters that have self-contained fireboxes. Heaters are located throughout the greenhouse, each heating a floor area of 2,000 to 6,000 ft^2 (186 to 558 m^2). A second type of system is *central heat,* which consists of a central boiler that produces steam or hot water, plus a radiating mechanism in the greenhouse to dissipate the heat. The third type of system is *radiant heat.* In this system, gas is burned within pipes suspended overhead in the greenhouse. The warm pipes radiate heat to the plants. There is a fourth possible type of system, although it has gained almost no place in the greenhouse industry: the *solar heating* system. Solar heating is still too expensive to be a viable option.

In all greenhouse heating systems, it is important that the exhaust not contact the crop. When the fuel source is of high purity and is thoroughly combusted, only carbon dioxide and water vapor are produced—but it is rare that fuels are completely combusted. Products of incomplete combustion, including ethylene gas, are injurious to plants (Figure 3-1). Ethylene gas can cause a distorted, corkscrew type of stem growth, curling of leaves, narrow leaves, and abortion of buds. Fuels also contain impurities. Sulfur is commonly found in coal, oils, and gases. Upon combustion, it is released as sulfur dioxide gas (SO_2). Sulfur dioxide gas dissolves into moisture films on the plant surfaces and is converted to sulfurous acid and, after oxidation, sulfuric acid, which burns the cells it contacts (Figure 3-2). Small tan spots appear, or in severe cases, the entire leaf may die.

Unit Heater Systems

Unit heaters are often referred to as *forced-air heaters.* The price of a unit heater system varies with the climate in which it is located. The typical cost including installation labor is \$0.50 to \$1.00/ft^2 (\$5.40 to \$10.76/m^2) of greenhouse floor. The low initial investment for the unit heater system is suitable for greenhouse firms that start small and expand steadily, purchasing heaters as needed.

Figure 3-1 Ethylene gas injury to chrysanthemums caused by fumes escaping from an improperly vented unit heater inside the greenhouse. Leaves are distorted and abnormally narrow, and the terminal bud has aborted. *(Photo courtesy of J. W. Love, Department of Horticultural Science, North Carolina State University, Raleigh, NC 27695-7609.)*

Figure 3-2 Sulfur dioxide injury on Rieger begonia foliage appearing as circular, tan spots. Improperly vented heaters can emit the gas. Carbon dioxide generators burning fuel with an undesirably high sulfur content also produce toxic levels of this gas inside the greenhouse.

Unit Heaters These heaters consist of three functional parts, as illustrated in Figure 3-3. Fuel is combusted in a firebox to provide heat. The heat is initially contained in the exhaust, which rises through the inside of a set of thin-walled metal tubes on its way to the exhaust stack. The warm exhaust transfers heat to the cooler metal walls of the tubes. Much of the heat is removed from the exhaust by the time it reaches the stack through which it leaves the greenhouse. A fan in the back of the unit heater draws in greenhouse air, passing it over the exterior side of the tubes and then out the front of the heater to the greenhouse environment again. The cool air passing over hot metal tubes is warmed. In short, the metal tubes serve as heat exchangers, absorbing heat from the hot exhaust passing through the inside of them and transferring it to the cool greenhouse air passing over the outside of them.

Generally, the fuel supply and fan are controlled by a temperature sensor located in an appropriate area of the greenhouse. Heat is supplied only as needed. Unit heaters burn a variety of fuels, including No. 2 oil, kerosene, LP gas, and natural gas. Fuel types, however, cannot be changed without alteration to the unit heater.

Unit heaters come in vertical as well as horizontal designs (see Figure 3-4) based upon the direction in which the heated air is exhausted from the heater. Vertical heaters take in air from the ridge area of the greenhouse and expel it downward toward the floor. These heaters are purchased in a size capable of

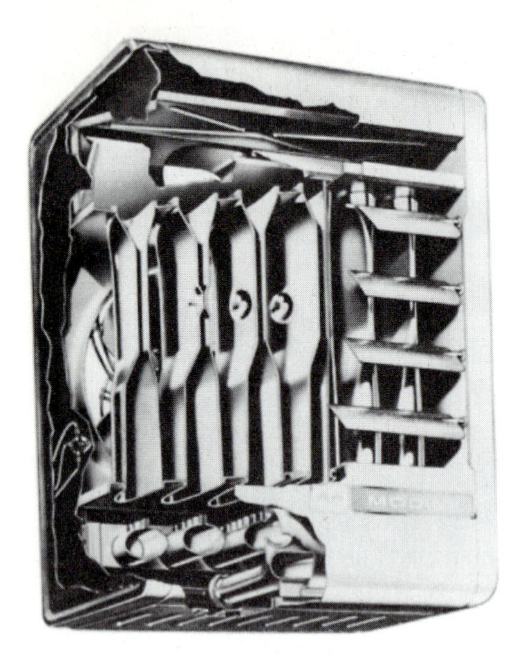

Figure 3-3 Interior view of a horizontal unit (forced-air) heater. Fuel is combusted in the chamber at the bottom. Hot fumes rise inside the heat-exchanger tubes, giving up heat to the walls of the tubes. Smoke exits at the top rear into a stack. A fan behind the unit forces cool greenhouse air over the outside of the tubes, where it picks up heat. *(Photo courtesy of Modine Manufacturing Company, Racine, WI 53401.)*

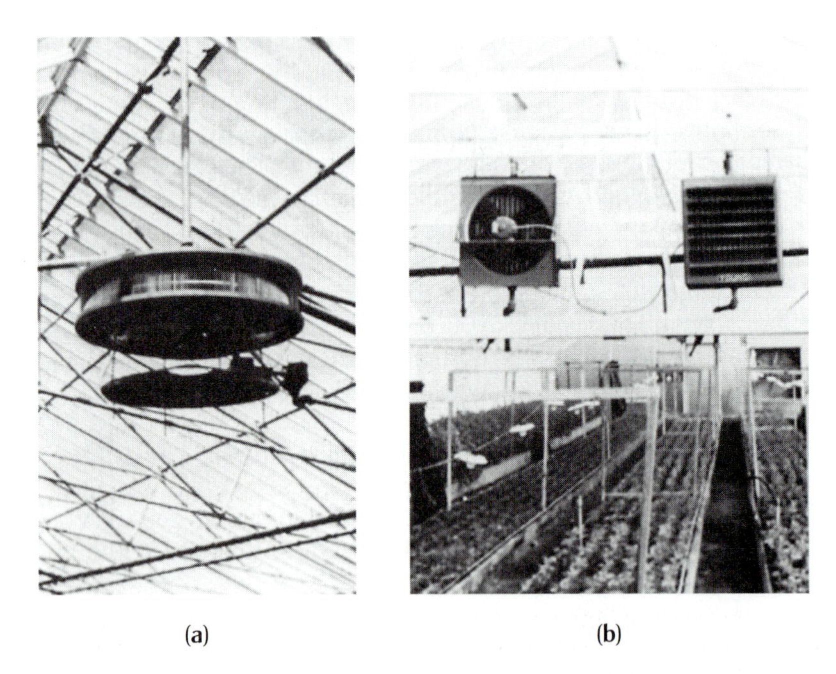

(a) (b)

Figure 3-4 (a) A vertical unit heater typical of the early types used for greenhouse heating. (b) A horizontal unit heater commonly used today.

heating a square area having sides equal to the width of the greenhouse. They are suspended from the ridge of the greenhouse, well above head height, and are spaced along the length of the greenhouse at intervals equal to its width. When unit heaters first became popular in the 1940s, the vertical type was believed best for greenhouse application. Uneven temperatures and drying of the soil sometimes occurred, which resulted in non-uniform growth. Horizontal unit heaters are the standard heaters used today. The uneven temperature and drying problems are reduced with horizontal air distribution. It is possible to use fewer but larger heaters, thus reducing the initial cost of the heaters as well as the labor of installation. Horizontal heaters are also adaptable to the newer integrated systems of heating, cooling, and horizontal airflow.

Whenever fuel is combusted, oxygen is consumed. Old glass greenhouses may or may not have sufficient air leaks to provide the needs of the firebox. Plastic greenhouses are tighter, and there have been many cases where burners have gone out during the night after consuming the available oxygen, causing the crop to freeze. A shortage of oxygen often leads to the formation of odorless carbon monoxide gas before the flame goes out. *An employee entering such a greenhouse could lose his or her life.* As a general rule, 1 in^2 of opening from the outside should be provided near the heater for every 2,500 Btu capacity of the heater (1 cm^2/114 W). A stovepipe, tile, or flexible clothes dryer tube may be placed near the burner intake, extending outside. It is frequently buried for convenience. An 8-inch (20-cm) -diameter pipe would provide the 50 in^2 required for a 125,000 Btu (5,700 W) heater. The end of the tube should be covered with a screen to prevent the entry of animals.

The exhaust stack on unit heaters must be sufficiently tall to develop an updraft to draw fumes out of the heater and must be high enough above the greenhouse roof to permit dissipation of the smoke without reentry into the greenhouse. The stack should extend 8 to 12 feet (2.4 to 3.7 m) above the firebox to ensure a proper air draft.

Heat Distribution: Convection Tubes In very small greenhouses, the fan in the unit heater may be all that is required to distribute the heat uniformly throughout the greenhouse. Most commercial greenhouses are too large for such simplicity. For these, two warm-air distribution systems exist: *convection tubes* and *horizontal airflow* (HAF). In the convection-tube system, a transparent polyethylene tube is connected to the air outlet of the unit heater (Figure 3-5). The polyethylene tube is installed along the length of the greenhouse above plant height and is sealed at the distant end. Round holes 2 to 3 inches (5 to 8 cm) in diameter are located in pairs at opposite sides of the tube every few feet (0.5 to 1.0 m) along the tube length. Warm air from the heater moves through the tube and out the side holes. The warm air comes out at a high velocity in a jet stream and quickly mixes with the surrounding air. This system ensures that heat is distributed from one end of the greenhouse to the other. When neither heating nor cooling is required, many growers keep the fan in the unit heater running without heat so that air from the greenhouse is continuously circulated through the tube. Air circulation gives more uniformity of temperature in the greenhouse,

Figure 3-5 **A horizontal unit heater connected to a transparent polyethylene convection tube with holes along either side for uniform distribution of heat.**

conserves heat, and reduces the occurrence of disease by reducing condensation on plant foliage. The polyethylene tube is also used to bring in cold air and distribute it when cooling is needed during the winter.

Care must be taken to locate unit heaters and air distribution tubes below any thermal screens and photoperiodic shade blankets that may be used in the greenhouse. Some firms have installed the air distribution tubes beneath benches. This is feasible only where long benches are situated in such a manner that the tubes do not need to cross aisles.

Heat Distribution: Horizontal Airflow (HAF) A more recent system for establishing uniform temperature in greenhouses is the horizontal airflow (HAF) system developed at the University of Connecticut. This system uses small horizontal fans for moving the air mass (Figure 3-6).

The greenhouse may be visualized as a large box containing air. It is difficult to start the air moving, but once it is moving in a circular pattern, like water in a bathtub, it is easy to keep it moving. The horizontal airflow pattern of the HAF system also results in the movement of warmer air from the gable to plant height, thereby reducing heating costs. Temperatures at plant height are more uniform with the HAF system than with other systems.

Minimum and maximum airflow velocities for this system are 50 and 100 feet per minute (fpm) (0.25 and 0.5 m/sec). Below this level, airflow is erratic, and uniform mixing of air cannot be assured. A velocity of 50 fpm (0.25 m/sec)

Figure 3-6 **Horizontal air fans in a gutter-connected greenhouse range. These fans are used to distribute warm air from the heaters, incoming cold air during winter cooling, and interior greenhouse air when neither heating nor cooling is on.**

causes slight leaf movement on plants with long leaves, such as tomato. This system should move air at 2 to 3 cfm/ft^2 (0.6 to 0.9 cmm/m^2) of floor space. Fans of $1/30$ to $1/15$ hp (31 to 62 W) and a blade diameter of 16 in (41 cm) are sufficient. Commercial, continuous-duty motors should be used. With approximately one fan per 50 ft (15.2 m) of greenhouse length, fans should be aimed directly down the length of the greenhouse and parallel to the ground. The first fan should be installed no closer than 10 to 15 ft (3.1 to 4.6 m) from the end of the house; the last one should be placed, 40 to 50 ft (12.2 to 15.2 m) from the end toward which it is blowing.

Specifications for the HAF system are shown in Figure 3-7 and are described as follows.

1. For individual houses, install two rows of fans along the length of the greenhouse, each row one-quarter of the width of the greenhouse in from the side wall. The row of fans on one side of the greenhouse should blow air opposite to the direction of the row of fans on the other side of the greenhouse to form a circulating pattern. Fans should be 2 to 3 feet (0.6 to 0.9 m) above the plants. A unit heater serves as the first fan in one row of small greenhouses of approximately 2,000 ft^2 (186 m^2). In larger greenhouses, particularly in cold climates, two unit heaters are installed in opposite corners of the greenhouse such that each heater substitutes for the first fan in each row of fans.

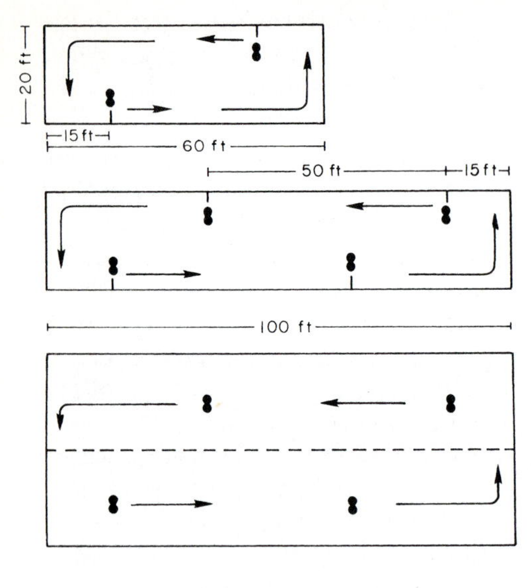

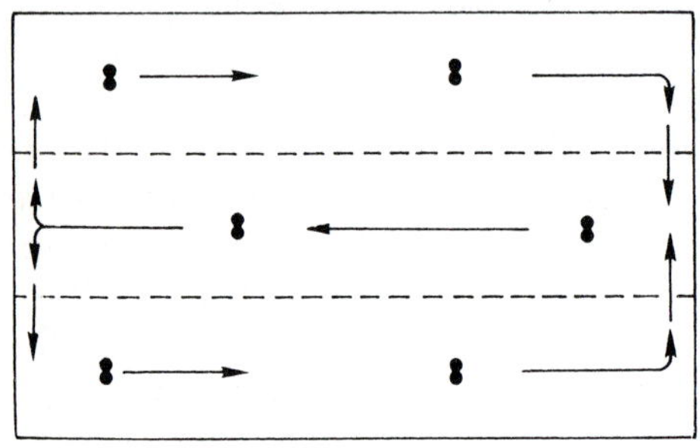

Figure 3-7 Fan arrangements for a horizontal airflow (HAF) system in various greenhouse sizes. Fans are located one-quarter of the width of the greenhouse in from the side walls in the first two single greenhouses illustrated. They are located under the ridge in the ridge-and-furrow greenhouse diagrams. *(Adapted from Aldrich and Bartok [1994].)*

This places the heat source in the path of the airflow. The fan in the unit heater serves to circulate the air.

2. For ridge-and-furrow houses, install a row of fans down the center of each greenhouse. A unit heater should be substituted for the first fan in each greenhouse. If the block contains an even number of greenhouses, move air down one house and back in the adjacent greenhouse. In this way, each pair of greenhouses has a circulating air pattern. Connecting gutters must be sufficiently high to permit air movement beneath them. If the block contains an odd number of greenhouses, move air in the same direction in the first and third houses and back the opposite way in the second house. Again, the first unit in each greenhouse is a unit heater.

Central Heating Systems

A central heating system consists of one or more boilers in a central location that provide steam or hot water to the various greenhouses. The typical cost of a central boiler system, including the heat distribution components in the greenhouse and installation, is \$2.50 to \$3.25/ft^2 (\$27 to \$35/m^2) of greenhouse floor space. The materials cost for the boiler with controls and pumps is between \$0.40 and \$0.75/ft^2 (\$4.30 and \$8.07/m^2).

The extra \$2 to \$3/ft^2 (\$21 to \$32/m^2) spent on a central heating system compared to a unit heater system must be made up somewhere. This is accomplished in five ways over the long term. Boilers can burn cheaper fuels than unit heaters and radiant heaters. Wood chips, logs, coal, and the heavier No. 4, 5, and 6 grades of oil can be burned in boilers. Unit heaters are restricted to gas and No. 2 or lighter oil. Wood costs only 20 to 25 percent of the cost of oil. Larger firms realize further savings in the cheaper maintenance of one or two large boilers compared to numerous unit heaters, and in the longer life expectancy of the boilers. When the boilers are used to supply hot water to floor heating systems, it is possible to keep the greenhouse air temperature 5 to 10°F (3 to 6°C) cooler, thereby reducing heat loss from the greenhouse. Unlike unit heater systems, a portion of the heat from central boiler systems is delivered to the root and crown zone of the crop. This can lead to improved growth of the crop and to a higher level of disease control. Each of these factors translates into higher monetary returns from the crop. Over the long haul, central heating systems pay for themselves; otherwise, there wouldn't be as many in use in the industry today.

The first choice to be made after deciding to use a central heating system is whether it will be a hot-water or a steam system. Today, hot water is the system of choice worldwide. Traditionally, European greenhouses were equipped with hot water, while larger American greenhouses used steam for heating. Those who chose steam probably did so because of the faster response time from the boiler when heat was needed, because less heating pipe was required in the greenhouse, and because hot water circulating pumps were not required in the steam system. Hot water is now the most popular medium for carrying heat from the boiler to the greenhouse for several reasons. The uniformity of temperature across the greenhouse and over time is greater with hot water. There is a larger reserve of heat in a hot-water system in the event of boiler failure. The temperature of hot water can be sufficiently low to use it to heat pipes located in the greenhouse floor or in pipes suspended within the foliage canopy of fresh flowers or vegetable plants, whereas steam would be too hot.

The Boiler Boilers, like unit heaters, consist of three principal components (Figure 3-8). The first is a firebox where fuel is burned. The second is a flue that provides a passage for the smoke from the firebox to a chimney that vents the smoke to the outer air. The third component is a heat exchanger consisting of a network of metallic tubes either filled with or surrounded by water. The first part of the flue conducts heat laden smoke over or through these tubes. Heat in the smoke is transferred to the tubes. The tubes, in turn, pass heat on to the water. If it is a hot-water boiler, the heated water is pumped to the greenhouse, where it passes through a heat exchanger that releases the heat to the greenhouse air. If it is a steam boiler, the water is heated to a temperature sufficiently high to allow it to turn to steam—at or above 212°F (100°C). The steam is under pressure,

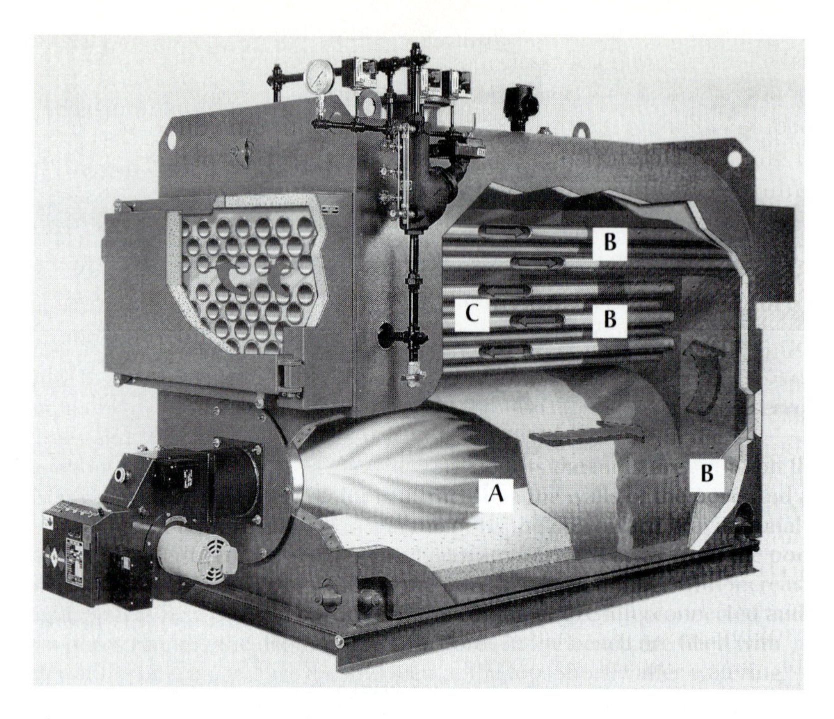

Figure 3-8 **A cut-away view of a firetube design boiler. The burner mounted on the lower front of the boiler provides a flame inside the firebox "A". The firebox and firetubes are jacketed with water "B". The hot exhaust from the flame moves to the rear of the firebox where it rises and enters the firetubes "C". The exhaust moves through the firetubes to the front of the boiler at which point it enters the upper firetubes and passes back to the rear of the boiler where it exits into the chimney flue "D". Water over the firebox and surrounding the firetubes picks up heat from the exhaust and is used to heat the greenhouse.** ***(Photo courtesy of Boilersmith, Ltd., P.O. Box 70, Seaforth, Ontario N0K 1W0, Canada.)***

which propels it through pipes to the greenhouse, where its heat is also released to the air through a heat exchanger.

There are two general types of hot-water boilers used in greenhouses today. The first is the traditional *high-mass* boiler, which is constructed with steel or cast-iron tubes and also known as a *fire-tube* boiler. This is a large boiler; it is the type that must be used if the fuel is wood, coal, or oil. The second is the newer *low-mass* boiler, which contains copper *fin tubes.* It is also known as a *compact boiler.* The compact boiler burns only natural gas or manufactured gas, as soot from oil or coal could plug the narrow spaces between the fins of the heat-exchanging copper tubes. The compact boiler is typically cheaper to purchase, and occupies considerably less space. A 3-million-Btu (877,193 W) compact boiler would contain only about 10 gallons of water, compared to several hundred gallons in a ferrous metal boiler of equal heat output. Because of this, compact boilers are more efficient at either end of the heating season, when heat is required for short periods with long nonheating intervals between them. Larger quantities of heat need to be put into the high-mass ferrous boilers to bring them to operational temperature. This difference in efficiency diminishes as the season turns colder and the greenhouse heat requirement becomes more constant. Growers who have natural gas available tend to find an economic advantage in compact boilers. Those who don't would have to purchase propane, which is considerably more expensive. This could more than negate the lower purchase price of the compact boiler.

The boiler may be located in the greenhouse or in the service building. In either location, heat that escapes from the boiler jacket, the pipes carrying steam or hot water from the boiler, and the return lines carrying condensate or cool water back to the boiler serves a useful purpose at that location. However, when the boiler is located in the greenhouse, the high humidity results in corrosion and premature breakdown of switches, pumps, and motors. Most growers today consider it more desirable to locate the boiler in the service building, since the atmosphere is drier.

Attention should be paid to the placement and height of the smokestack in a central system. The stack should be sufficiently tall so that shifting winds cannot sweep emitted gases into the greenhouses, where they can cause plant injury. It is best to place the stack in a position such that the prevailing winds will carry the smoke away from the range and also such that the stack does not cast a shadow on the crop. The north side and the northeast corner, for instance, would be good locations for the boiler and stack under conditions of prevailing winds from the west.

Heat Distribution Heat is delivered from the boiler to the greenhouse in either steam or hot water. In the greenhouse, heat is then exchanged from the steam or hot water to the crop through pipe coils, unit heaters, or a combination of the two.

Pipe Size and Quantity. Hot water has been customarily supplied at a temperature of 180°F (82°C) in 2-inch (51-mm) iron pipe in American greenhouses and at a temperature of 203°F (95°C) in 2-inch (51-mm) iron pipe in Dutch greenhouses. Steam systems, on the other hand, usually supply steam at a temperature of 215°F (102°C), which is 3°F (2°C) above the temperature at which water turns to steam and is possible because the system is under a low pressure of 5 psi or so. Since there is less resistance to the flow of steam, smaller iron pipes of 1.25- or 1.5-inch (32- or 38-mm) diameter are used in the greenhouse coil. More recently, aluminum pipes have been used as well. These are typically smaller-diameter pipes with fins for increasing heat release to compensate for their smaller diameter.

The amount of pipe needed in a greenhouse coil can be determined by referring to the heat-supply values listed in Table 3-3 for various types of pipe. A greenhouse requiring 160,000 Btu/hr would need 1,000 linear ft (305 m) of 2-inch (51-mm) hot-water pipe to provide this heat. This was determined by dividing the total heat requirement for the greenhouse by the amount of heat that 1 linear foot

Table 3-3 Heat Available from Various-Diameter Pipes Heated by Hot Water or Steam[1]

Heat Source	*Pipe Diameter*	*Heat Supplied*	
		Btu/hr/ft	*W/m*
Steam, 215°F (102°C)	1.25 in (32 mm)	180	173
Steam, 215°F (102°C)	1.5 in (38 mm)	210	202
Hot water, 180°F (82°C)	2 in (51 mm)	160	154
Hot water, 203°F (95°C)	2 in (51 mm)	200	192

[1]The inside air temperature of the greenhouse is 60°F (16°C).

(0.3 m) of pipe can provide. In this case, 160,000 Btu/hr is divided by 160 Btu per linear foot of 2-inch hot-water pipe, which yields an answer of 1,000 ft of pipe. If a 1.5-inch (38-mm) system of steam pipes were used instead, the need would be 160,000 Btu/hr divided by 210 Btu/hr, or 762 ft (232 m) of pipe.

Wall Pipe Coils. Placement of heating pipes is very important. Considerable heat is lost through the side walls of the greenhouse. In addition, warm plants radiate heat energy to colder objects outside the greenhouse. The result is a disproportionately high cooling effect in the outer beds of plants. To counteract this heat loss, pipes are installed on the inside of the four perimeter walls of the greenhouse. Side pipes should have a few inches of clearance on all sides to permit the establishment of air currents and should be located low enough to prevent blockage of light entering through the side walls. They are generally attached to the opaque curtain wall. The side wall coil of pipes should have a heat-supplying capacity equal to the heat loss through the walls of the greenhouse.

When several pipes are stacked above one another, their effectiveness is reduced. Additional pipes must be added to compensate. Table 3-4 shows the effect. For two pipes, the effect is insignificant. Five pipes in a stack, however, are only as effective as four pipes placed apart from one another. In a heating design where the heat of four pipes is needed in the side coil, five pipes would have to be installed. Overhead pipes are spaced sufficiently far apart to avert the problem.

If more than one row of pipe is required in a side-wall stack, *fin pipe* can be used so that only one pipe is required (Figure 3-9). Fin pipe is a conventional pipe with numerous thin metal plates radiating outward from it to increase the surface area of the pipe and thus the rate at which it transfers heat from the hot water or steam contained inside to the surrounding air. Depending on the design, 1 linear foot (0.3 m) of fin pipe can be equivalent to 4 or more linear feet (1.2 m) of conventional pipe. It should be remembered that heat released from fin pipe is much more intense than heat from conventional pipe. It is therefore important to distribute fin pipes evenly throughout a greenhouse. If a single continuous coil of fin pipe is not needed around the entire greenhouse, then the fin pipe should be alternated with conventional pipe at equidistant intervals.

Table 3-4 **Heat-Supply Relationship of Pipes in a Vertical Stack Compared to Pipes Located Separately from One Another[1]**

Number of Pipes in Vertical Stack	*Number of Individual Pipes Giving an Equivalent Amount of Heat*
1	1
2	2
3	2.67
4	3.33
5	4
6	4.33
8	5

[1]From Gray (1956).

Overhead Pipe Coils. The wall pipe coil counteracts heat lost through the four perimeter walls. Heat lost through the roofs and gables is supplied through an overhead or in-bed coil of pipes that is situated across the entire greenhouse (Figure 3-10a). The overhead coil is not the most desirable source of heat because it is located above the plants. Heat rises from the coil to the top of the greenhouse, where it serves no function and is quickly lost to the outside. Energy needs to be expended to drive the heat down to the plant zone. However, overhead coils are popular because they place the pipe out of the way of pedestrian traffic and automation.

Figure 3-9 **Fin pipe installed in the overhead pipe coil of a high-profile glass greenhouse.**

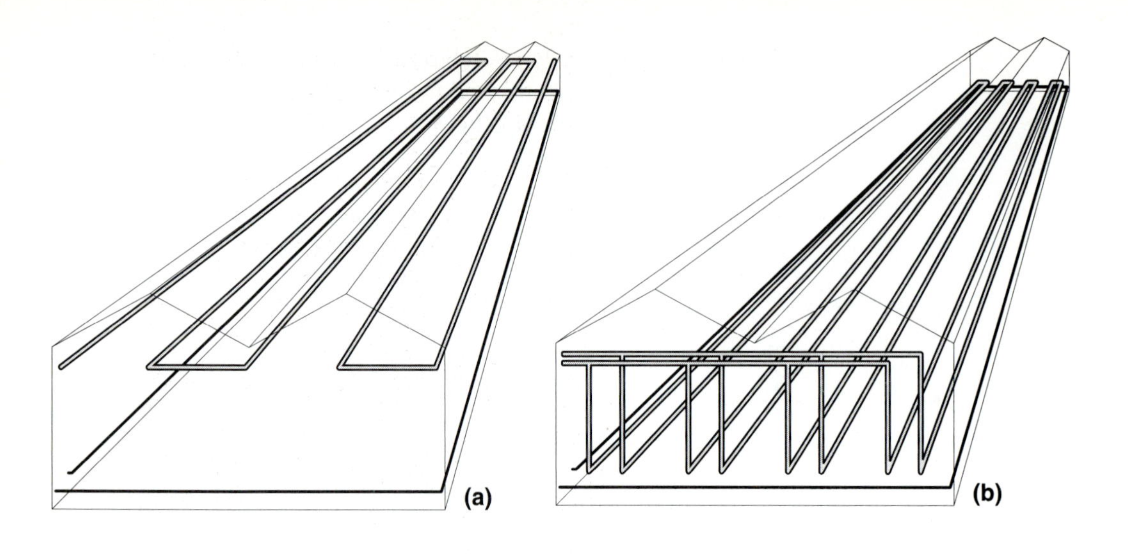

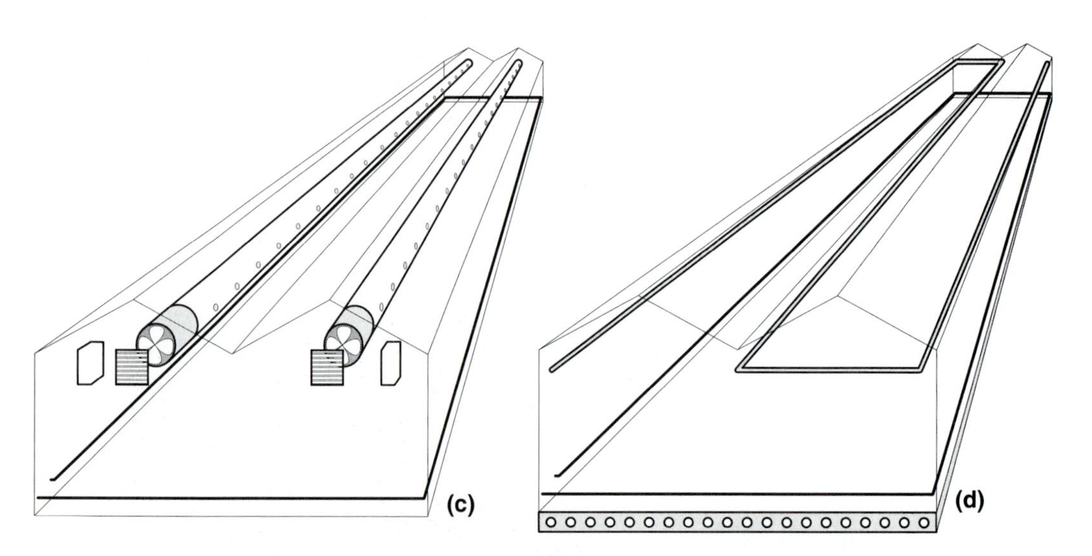

Figure 3-10 **Sketches showing the arrangement of heating pipe coils in a ridge-and-furrow greenhouse heated by (a) a wall coil and overhead coil, (b) a wall coil and an in-bed pipe coil, (c) a wall coil and overhead unit heaters, and (d) a wall coil, a limited overhead coil, and an in-floor coil.**

In-Bed Pipe Coils. A better pipe arrangement, when the greenhouse layout allows it, is the in-bed coil (Figure 3-10b). By placing the heating pipes near the base of the plants, the roots and crown of the plants are heated better than in the overhead-coil system. This leads to improved growth and greater disease control. Also, heat is kept lower in the greenhouse (where it is needed), resulting in bet-

Figure 3-11 **A 2-inch (51-mm) hot-water heat pipe supported by the lower frame of a movable pot-plant bench.**

ter energy efficiency. For pot crops on benches, pipes are installed in the framework of the bench beneath the tabletop (Figure 3-11). This arrangement is also possible for movable benches because the frame remains fixed in place. Fresh-flower and vegetable beds can likewise be heated with hot-water pipes, which are suspended by flexible rubber hoses from overhead mains (Figure 3-12). The heat pipe is confined to the bed and does not cross aisles. For crops such as roses, these pipes may be located in the bed of plants, while for others they are located on either side of each bed. Hot water is used in these systems because temperatures lower than that of steam are required to avoid burning the plants. Hot water also ensures a uniform temperature throughout the greenhouse. To facilitate removal of plants, root substrate pasteurization, and replanting, the heating pipes can be lifted and tied overhead without disconnecting them.

Hot water heating pipes are also being installed under slabs of rock wool in a more recent cultural system to ensure that the roots are warmed. The nutrient solution itself is being heated in some nutrient film technique (NFT) hydroponic systems to deliver heat directly to the roots, rather than only to the air above the plants. (Rock wool and other closely related cultural systems are described in Chapter 10.)

Box vs. Trombone Pipe Coils. Pipe coils can be arranged in two styles, either box or trombone (Figure 3-13). Box coils are used in hot-water systems. Hot water entering the greenhouse through the pipe main is distributed in a header, or branch tee, to several smaller pipes, through which it passes simultaneously to the opposite end of the greenhouse. There, it combines and returns to the boiler to be reheated. There is a resistance to the flow of water in the pipe. The box coil

Figure 3-12 **The hot-water pipes heating this tomato crop growing in rock wool are located between the plant rows and just above the floor for maximum efficiency of heat distribution. The hot-water pipes are suspended from overhead mains by flexible rubber hoses. This permits the pipes to be raised overhead for cleaning purposes when the crop is finished.**

minimizes this resistance by reducing the length of pipe through which any given portion of the water must flow and by increasing the cross-sectional area of the combined pipe through which the water passes.

Trombone coils are used for steam systems. Resistance to flow is not a problem for steam, but the rapid drop in pressure and temperature along the pipe is. If a box coil were used for steam conduction, the entry end would be hot and the exit end much cooler, resulting in an intolerable temperature gradient in the greenhouse. A continuous pipe is used in a trombone coil. Steam enters at the top of the coil and passes to the distant end of the greenhouse. It returns to the entry end in the second pipe down and then back to the distant end in the third pipe down. This arrangement continues until, at the end of the coil, water condensate and steam enter a trap that permits the return of water, but not steam, to the boiler. No temperature gradient exists along the length of the coil. The gradient exists from the top to the bottom of the coil and is of no consequence. The overhead pipe coil is usually a trombone coil, whether hot water or steam is used. In the case of a hot-water system, two overhead trombone coils are used to reduce resistance.

Enhanced Heat Distribution Form Coils. While some greenhouse firms rely on the innate distribution of heat from the wall plus overhead or in-bed pipe coils

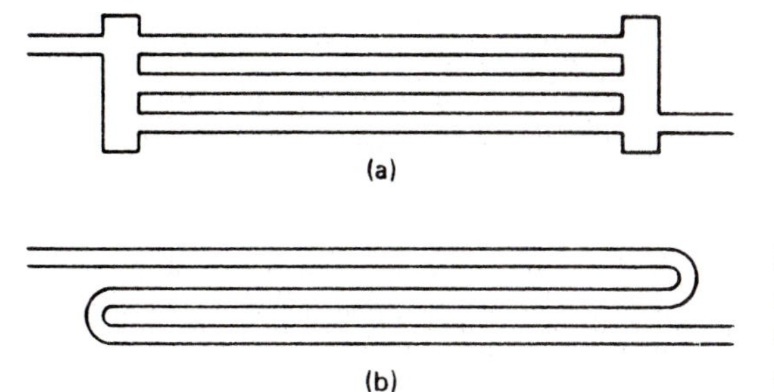

Figure 3-13 (a) A box coil used to distribute hot water through a greenhouse. (b) A trombone coil used in a steam system of heating.

just described, others further enhance the distribution. Advantages of improved temperature uniformity in the greenhouse can include reduced heating bills due to the return of heat from the top of the greenhouse to the bottom, more uniform maturation of the crop, and less disease as a result of the elimination of condensation on plants in cold spots in the greenhouse. Two systems used for distributing heat released from pipe coils are convection tubes and HAF.

Convection tubes were described earlier, in the "Unit Heater Systems" section. In that context, convection tubes were connected to unit heaters to distribute heat from the unit heater. Since heat in this current system is derived from pipe coils, the unit heater is not used. A clear polyethylene convection tube is installed overhead at the center of each greenhouse in such a way that it runs the length of the greenhouse. A pressurizing fan (Figure 3-14) is located in the inlet

Figure 3-14 A newly installed horizontal unit heater on the right and convection-tube housing on the left. Note the pressurizing fan in the cylindrical housing. The convection tube will be attached to the outside of the cylindrical housing.

end of the tube and the opposite end is tied shut. This fan takes in heated air and forces it down the tube. The heated air exits through pairs of holes on opposite sides of the tube along the length of the greenhouse. This system stirs the air in the whole greenhouse and combats stratification of warm air at the top of the greenhouse. The convection tubes are used for distribution of heat, for recirculating air when neither heating nor cooling is on, and for winter cooling.

The alternative, and recently more popular, system for distributing heat is HAF (Figures 3-6 and 3-7). This system was also described in the "Unit Heater Systems" section. It is set up in the same manner as described, except that unit heaters are not used. As in the case of convection tubes, HAF is used for distribution of heat, recirculation of air when neither heating nor cooling is on, and for winter cooling.

Unit Heaters. Some firms substitute unit heaters for overhead or in-bed pipe coils (Figure 3-10c). The wall pipe coil is still used. These unit heaters differ from those previously described in that they do not contain a firebox. The heater consists of a steam or hot-water fin-pipe coil and a fan. Steam or hot water from the boiler passes through the coil while the fan passes cool greenhouse air over the coil to heat the air. Two examples of these unit heaters can be seen in Figure 3-4. Hot air emitted from these unit heaters can be circulated through convection tubes. Alternatively, the unit heaters can be placed in line with HAF fans for heat circulation.

Floor Pipe Coil. Thus far we have seen that there are two distribution systems for heat from central boilers: the wall coil combats heat loss through the walls, and the overhead coil, the in-bed coil, or overhead unit heaters compensate for heat lost through the roofs and gables. In this final option, a portion of the heat generally supplied through the overhead coil or overhead unit heaters is redirected to an in-floor pipe coil (Figures 3-10d and 3-15). Floor heating is more effective than in-bed pipe coil heating and therefore enhances the three advantages previously cited for in-bed coils: increased yield, greater disease control, and less fuel consumption. A fourth advantage associated only with floor heating is the ability to dry the floor quickly. This is essential when flood floors are used for irrigation/fertilization (Chapter 8). In this system, plants are set on the floor, which makes drying the floor difficult. Unless the floor is dried quickly after watering, the humid environment in the plant canopy fosters disease development.

In-floor pipe coils often consist of 0.75-inch (19-mm) -diameter pipes buried in the floor 6 to 12 inches (15 to 30 cm) apart, depending upon the heat requirement. The floor may consist of porous or solid concrete. Standard polyethylene pipe is not used because of the possibility of breakage. PVC pipe can be used, but is not popular because of its inflexibility and cost. Polybutylene has been a popular choice for its flexibility, strength, and high temperature tolerance. More recently, cross-linked polyethylene (PEX) pipe has gained popularity because of its lower cost and freedom from breakage. EPDM flexible tubing (synthetic rubber) is also highly effective and withstands high temperatures.

Hot water, generally at a temperature of 90 to 120°F (32 to 49°C), is circu-

(a) (b) (c)

Figure 3-15 **An installation sequence for an in-floor hot water heating system. (a) Three-quarter-inch (19-mm) hot-water tubing in place on the subsoil that will support the concrete floor. (b) Attachment of the hot-water tubing to the inlet and outlet manifold pipes at one end of the greenhouse. (c) Placement of concrete over the hot-water tubing.** ***(Photos courtesy of Greenlink, 22 S. Pack Square, Suite 404, Asheville, NC 28801.)***

lated through the pipe to maintain the desired temperature in the plant canopy. During periods of maximum heat requirement, the water temperature can be as high as 140°F (60°C). Hot water is pumped the length of the floor and then back to the inlet end to provide a bidirectional flow for the purpose of uniform heat distribution along the length of the greenhouse. In general, heat is applied at the rate of 20 Btu/hr/ft^2 (63 W/m^2) of floor. During periods of high heat demand, these systems may be called upon to supply 30 Btu/hr/ft^2 (95 W/m^2). Since the root zone and plant area are heated first, the air temperature above the plants is commonly 5 to 10°F (3 to 6°C) lower than in conventionally heated greenhouses, with no loss in plant growth. As in the case of radiant heating, this lowers the temperature differential across the greenhouse covering and thereby cuts fuel costs. The supply of hot water to the floor is generally tied into a soil-temperature sensor rather than an air-temperature sensor. In larger computer-controlled greenhouse firms, the temperature of the water circulating in the floor is determined by the computer. As the rate of demand for heat increases to hold the soil at the set point, the temperature of the water circulated in the floor is increased. The water-temperature decision can be based on the rate of decline in soil temperature along with the outside temperature. Circulating water temperatures may be as low as 100°F (38°C) and as high as 140°F (60°C). By using the minimum temperature of water necessary to accomplish the task, the efficiency of the boiler is raised.

Heat is first supplied via the floor in floor heating systems, and only when this is insufficient are other coils or unit heaters used. With experience, an air temperature can be found that allows the floor heating system to maintain the desired soil temperature. The air temperature is usually 5 to 10°F (3 to 6°C) lower than the air temperature customarily recommended for the crop. A floor heating system can provide all of the required heat during the fall and spring. On cold winter days, supplemental heating, such as an above-ground pipe coil or unit heaters, will be required. Over the whole year, floor heating may provide from 20 to 50 percent of the total need and may average out to 25 percent. This percentage is highest in warm climates. If the floor is covered with a crop of pot plants or bedding-plant flats, a high percentage of the total heat need will be met because the plants are near the heat source and tend to hold the heat down. When plants are grown on benches, the efficiency of this system is reduced. Higher air temperatures are required at the elevation of the plants. Also, heat is able to escape more freely from the uncovered floor to the greenhouse gable, where it is not desired. Hanging baskets reduce the efficiency even more. The total heat requirement cannot be supplied through the floor in cold climates because the amount of heat supplied to the floor would result in excessive cement and plant temperatures. The total system, including a porous concrete floor with heating pipes, the perimeter wall pipe coil, the overhead pipe coil, a hot-water boiler, controls, and installation labor, costs about \$4.25 to \$5.00/ft^2 (\$45.75 to \$53.82/m^2) of greenhouse floor.

Another recent method for heating the root zone is available in various commercial packages (Figure 3-16). Flexible EPDM tubing can be placed on the floor or on or beneath a bench surface. Tubes are usually spaced 2 in (5 cm) apart along the length of the floor or bench, but may be spaced closer together or farther apart to meet local heat needs. The inlet and outlet mains for the tubing are located on the same end of the floor or bench to provide bidirectional flow. Tubing is 5/16 inch (8 mm) in outside diameter. The tubing is sufficiently strong to withstand placement of pots placed directly on it as well as people walking on it. As in the case of the in-floor hot-water coil, control of the system is generally dependent on soil-temperature sensors. Similar water temperatures are used.

Figure 3-16 **EPDM hot-water heat tubing used on the surface of a greenhouse bench for heating a crop of bedding plants. Note that the flats (or pots) are placed directly on the heating tubes.** *(Photo courtesy of Biotherm Co., 421 Second St., Petaluma, CA 94952.)*

Many firms without central hot water boilers have installed EPDM tube heating systems for specialized purposes such as plant propagation in small zones within their greenhouse range. Smaller hot-water heaters independent of the primary heat source of the firm are used in these cases for the tube heating system.

Radiant Heater Systems

Grower reports on fuel savings suggest a 30 to 50 percent fuel bill reduction with the use of low-energy infrared radiant heaters, as compared to the unit heater system (Figure 3-17). These heaters emit infrared radiation, which travels in a straight path at the speed of light. Objects in the path absorb this electromagnetic energy, which is immediately converted to heat. The air through which the infrared radiation travels is not heated. After objects such as plants, walks, and benches have been heated, they then will warm the air surrounding them. It is the soil and plant temperatures that are important to growth. Air temperatures in infrared radiant–heated greenhouses can be 5 to 10°F (3 to 6°C) cooler than in conventionally heated greenhouses with equivalent plant growth. In the conventional system, the air is heated first; the air then heats the plants. Thus, air temperatures tend to be higher than plant temperatures at night. This encourages condensation on plant surfaces. Disease is discouraged by the lesser amount of condensation in infrared radiant–heated greenhouses.

Infrared radiant heaters used in greenhouses are available in sizes from 20,000 to 120,000 Btu/hr (5,860 to 35,160 W) in 20,000 Btu/hr increments. The distance between heaters can be 30 to 40 ft (9.1 to 12.2 m). They are placed in tandem overhead along the length of the greenhouse. Above the line of heaters, running the length of the greenhouse, is a deep-dish metal reflector to direct all rays down toward the plants and to give the proper uniformity of heat across the production area. The composition of this reflector is important to ensure that maximum reflectivity is achieved. A high-quality metal for this use is aluminum. Each heater mixes air from the greenhouse with fuel and injects it into a 4-inch (10-cm) steel pipe. Fuel is ignited by a direct-fire ignition system rather than by

(a)

(b)

(c)

Figure 3-17 (a) A greenhouse installation of an infrared radiant heating system. (b) Burners installed in two adjacent infrared radiant heating lines. (c) The exhaust fan and outlet port serving two infrared radiant heating lines. *(Photos courtesy of Growth Zone Systems, 1735 Cedardale Rd. D-6, Mt. Vernon, WA 98274.)*

using a pilot light or a spark plug. The pipe heats to a temperature around 900°F (480°C). This is not sufficiently hot to cause the pipe to emit visible red light, which would interfere with the photoperiodic timing of some crops (see Chapter 12). Actually, the temperature can be varied by the manufacturer to suit specific greenhouse spatial needs. The pipe extends the length of the greenhouse, where it exits to the outside. Fumes are drawn along the length of the pipe through a vacuum developed by a pump in the end of the pipe. A vacuum of 2 in (5 cm) of water column is developed in the pipe. A 0.5-hp (370-W) pump can handle up to 16 smaller heaters. Since the pipe in the vicinity of each heater can be 900°F (480°C), it is important that plants not be placed within 5 ft (1.5 m) of the pipe. Radiant heaters today can heat a width of plant surface up to two times the height of the heaters above the plant surface.

Reasons for fuel savings fall into two categories. First, fuel gases in this system exit at less than 150°F (65°C) as opposed to 400 to 600°F (204 to 315°C) in conventional greenhouse heaters. Thus, more heat is derived from the combusted fuel. The efficiency of combustion is claimed to be about 90 percent. Second, cooler air temperatures in the greenhouse ensure a smaller temperature differential from outside to inside. Therefore, less heat is lost from the greenhouse. It is important that high-velocity air circulation as generated by convection tubes not be used in infrared radiant–heated greenhouses. Air currents set up by such fans would cool the plants and carry the air warmed by the plants and floor to the cold greenhouse covering. A horizontal airflow (HAF) system (discussed earlier in this chapter), having a gentler airflow, works well with this system. Another advantage of radiant heating is the reduction of about 75 percent in electrical consumption over a conventional unit heater system. The only motor required in the infrared radiant heating system is in the exhaust fan.

Installation and materials for an infrared radiant heating system can range from about $1.50/ft^2 ($16.14/m^2) in warm southern states to about $2.25/ft^2 ($24.21/m^2) in northern states and provinces. Although capital costs of this system are higher than those of the conventional unit heater system, the fuel savings could pay for the additional cost over a few years. However, one should be aware that in recent years unit heaters and boilers have become available with low stack temperatures (300°F, 150°C), which allows them to share in part of the benefit that formerly belonged exclusively to radiant heaters. Second, the advantage of the lower inside-to-outside temperature differential across the greenhouse covering offered by radiant heating systems can be achieved in floor heating systems (discussed earlier in this chapter). Finally, radiant heaters burn either natural gas or manufactured gas. While natural gas can be competitive with other fuels, manufactured gases have become expensive relative to other fuels in recent years. The grower who must use manufactured gases could be at a disadvantage.

A choice of radiant heating products exists. Factors to consider in selecting one include thermal efficiency, emissivity, reflectivity, fixture efficiency, and pattern efficiency. *Thermal efficiency* is the ratio of heat potential in the fuel consumed to the energy released in the heater. *Emissivity* is a measure of the capacity of the heater tubes to release infrared energy. *Reflectivity* is a measure of the ability of the reflector to redirect energy. *Fixture efficiency* refers to the amount of infrared energy that is absorbed by the heating fixture and converted to heat, ultimately to be convected away. This amount should be as low as possible. *Pattern efficiency* is a measure of the ability of the heater to distribute radiant energy to the space in a manner consistent with the needs of the space. The overall efficiency of the system is a combination of all of these factors.

Solar Heating Systems

Solar heating is often pondered as a partial or total alternative to fossil-fuel heating systems. Few solar heating systems exist in greenhouses today. As will be seen, the economics of such a system bear scrutiny. In this section, the fundamental principles and components of solar heating will be considered. The components (Figure 3-18) consist of (1) a collector, (2) a heat storage facility, (3) an exchanger to transfer the solar-derived heat to the greenhouse air, (4) a backup heater to take over when solar heating does not suffice, and (5) a set of controls.

Collector Various solar heat collectors are possible, but the type that has received greatest attention is the flat-plate collector. This consists of a flat black plate (rigid plastic, film plastic, or board) for absorbing solar energy. The plate is covered on the sun side by two or more transparent glass or plastic layers and on the back side by insulation. The enclosing layers serve to hold the collected heat within the collector. Water or air is passed through or over the black plate to remove the entrapped heat and carry it to the storage facility.

A greenhouse itself is a solar collector. Some of its collected heat is stored in the soil, plants, greenhouse frame, walks, and so on. The remaining heat can be excessive for plant growth and is therefore vented to the outside. The excess vented heat could just as well be directed to a rock bed for storage and subsequent use during a period of heating. Heat derived in this manner could provide up to half of the total heat requirement for greenhouses in the southern United States and perhaps 10 to 20 percent of the total requirement in northern states.

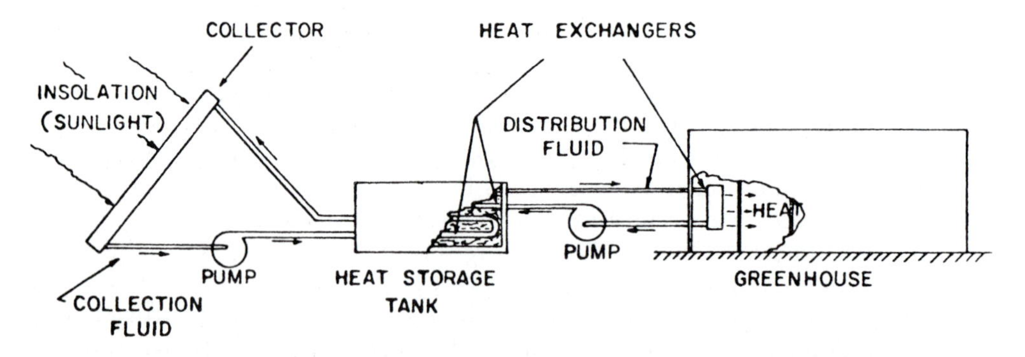

Figure 3-18 A typical solar heating system for greenhouses. *(From D. H. Willits, Department of Biological and Agricultural Engineering, North Carolina State University, Raleigh, NC 27695-7625.)*

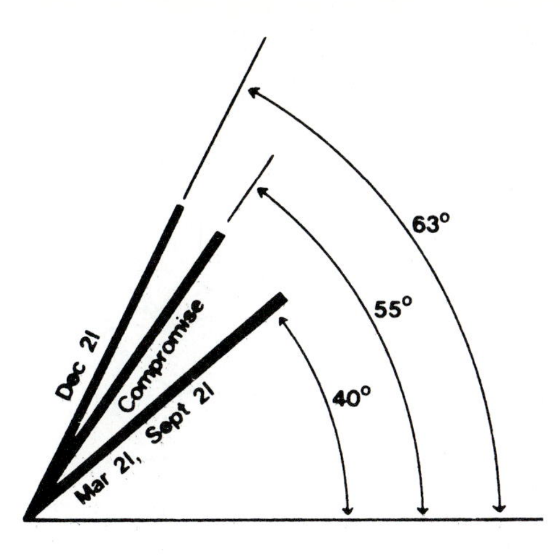

Figure 3-19 **The best angle of tilt with respect to the ground for a solar collector at 40°N latitude (Philadelphia, Denver) is 40° on March 21 and September 21. During the 91 days from September 21 to December 21, it increases by 23° to 63°. After December 21, it decreases continually to a value of 40° by March 21. A stationary collector is generally oriented at a compromise angle equal to the latitude plus 15°, or 55° in this example.**

Collection of heat by flat-plate collectors is most efficient when the collector is positioned perpendicular to the sun at solar noon. The required angle of tilt with respect to the ground is equal to the latitude on March 21 and September 21 (the spring equinox and fall equinox). The angle should be gradually increased to a maximum of the latitude plus 23° on December 21 (the winter solstice) and then decreased thereafter. Since movable collectors add considerable expense, a stationary compromise angle of the latitude plus 15° is often used (Figure 3-19).

The amount of solar radiation reaching the earth's surface varies with such factors as weather conditions and elevation. Average daily quantities of solar radiation striking a square foot of horizontal surface during July and January are presented in Figure 3-20. While an average solar input of 600 Btu/ft^2 (1,625 kcal/m^2, 6,800 kJ/m^2) of surface per day is expected in the Washington, D.C., area (38°N latitude), not all can be trapped by a solar collector. At solar noon, a flat-plate collector using water can have an efficiency of 65 percent, but the efficiency diminishes at either side of that point to 0 percent in the early morning and late afternoon. Considering an overall efficiency of 40 percent, 240 of the 600 Btu impinging on a square foot of collector in a day can be trapped for heating a greenhouse. Based on a heat output of about 100,000 Btu per gallon of oil, 417 ft^2 of collector would be required to equal in one day the heating capacity of 1 gal of oil (10.25 m^2 of collector/L oil). At least ½ square foot of collector surface is required per square foot of greenhouse floor area, and in northern areas 1 square foot may be needed.

Heat absorbed by the black plate inside the collector is often removed by water or air. The black plate may be a sheet of black plastic tubes fused together. In this case, water can be passed through the interior of these tubes. If it is a solid black sheet, such as polyethylene, water may be passed over its surface. Water picks up the heat and is then transferred to a storage tank. Air may likewise be passed through or over the black plate to remove heat from it. Water collectors require a flow rate of 1 to 3 gallons per minute (gpm) per 100 ft^2 (0.4 to 1.2

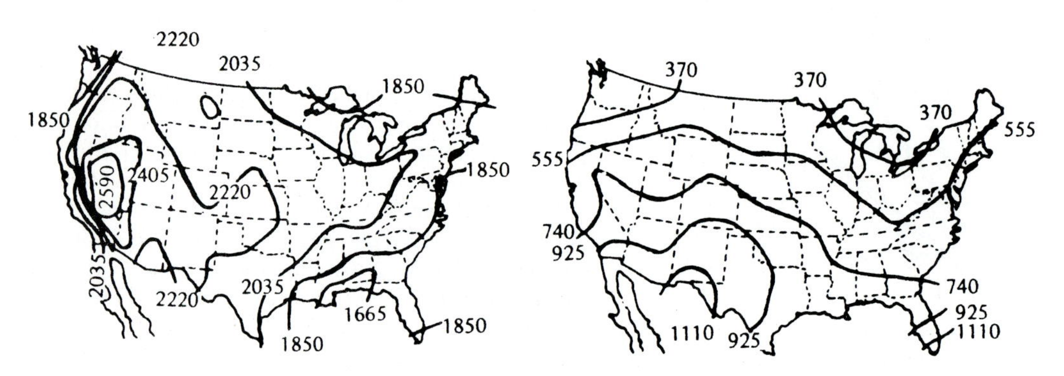

Figure 3-20 Average daily solar radiation received on a square foot of horizontal surface throughout the United States in July and in January. One Btu/ft² is equivalent to 2.7 kcal/m² or 11.4 kJ/m². *(From Ross et al. [1978].)*

L/min/m²) of collector surface. Correspondingly, air-heating collectors require a flow rate of 5 to 15 cfm/ft² (1.5 to 4.6 cmm/m²) of collector.

Heat and Storage Exchange Water and rocks are the two most common storage materials for heat in the greenhouse at the present. One pound of water can hold 1 Btu of heat for each 1°F rise in temperature (4.23 J/g water/°C). Thus, its specific heat is 1. Rocks can store about 0.2 Btu per pound for each 1°F rise in temperature (0.83 J/g rock/°C). The specific heat in this case is 0.2. To store equivalent amounts of heat, a rock bed would have to be three times as large as a water tank. A rock storage bed lends itself well to an air-collector and forced-air heating system. In this case, heated air from the collector, along with air excessively heated inside the greenhouse during the day, is forced through a bed of rocks. The rocks absorb much of the heat. The rock bed may be located beneath the floor of the greenhouse or outside the greenhouse, assuming that it is well insulated against heat loss. During the night, when heat is required in the greenhouse, cool air from inside the greenhouse is forced through the rocks, where it is warmed and then passed back into the greenhouse. A clear polyethylene tube with holes along either side serves well to distribute the warm air uniformly along the length of the greenhouse. Conventional convection tubes (discussed earlier in this chapter) can be used for distributing solar-heated air.

A water storage system is well adapted to a water collector and a greenhouse heating system making use of a pipe coil or a unit heater with a water coil contained within. Heated water from the collector is pumped to the storage tank during the day. As heat is required, warm water is pumped from the storage tank to a hot-water or steam boiler or into the hot-water coil within a unit heater. Although the solar-heated water will be cooler than the thermostat setting on the boiler, heat will be saved, since the temperature of this water will not have to be raised as high to reach the output temperature of water or steam from the boiler.

Low-temperature solar systems have been the most popular for greenhouses thus far because of their lower price. Solar input during the daytime can cause a storage-unit temperature rise in these systems of up to 30°F (17°C) above the evening baseline temperature. Each pound of water can thus supply 30 Btu of heat, and each pound of rock 6 Btu, as it cools 30°F. A 20-foot-by-100-foot (6.1-m-by-30.5-m) double-layer polyethylene greenhouse has been reported to lose about 3,500 Btu/hr/°F (1,848 W/K) of temperature differential between inside and outside. If an inside temperature of 60°F (16°C) and an average outside night temperature of 35°F (2°C) are experienced and the heating period is considered to be 13 hours long, about 1.1 million Btu (1.17 million kJ) of heat will be required. This would require a 4,400-gallon (16,600-L) water storage tank. (Note that a 1.1-million-Btu heat requirement divided by [(Btu/lb × °F) × 30°F × (8.3 lb/gal water)] equals 4,400 gal.) To store the same quantity of heat, about 2,000 ft^3 (57 m^3) of rock would be required.

The water or rock storage unit occupies a large amount of space and a considerable amount of insulation if the unit is placed outdoors. Placing it inside the greenhouse offers an advantage in that escaping heat is beneficial during heating periods. It is detrimental when heating is not required. Rock beds can pose a problem in that they must remain relatively dry. Water evaporating from these beds would remove considerable heat.

Backup Heater Today, a solar heating system is considerably more expensive than a conventional system. Current strategy calls for sizing a solar system to meet the average winter needs. A conventional fossil-fuel backup system is installed to meet the additional heating needs of the coldest nights. This compromise increases the chances of justifying the cost of a solar heating system.

Controls To illustrate typical controls in a solar-heated greenhouse, a water system is considered. The first control activates when the water in the collector becomes 10°F (6°C) warmer than the water in the storage tank and cuts off when the differential is 5°F (3°C). Water is pumped from the collector to the top of the storage tank. Cooler water at the bottom of the storage tank returns to the collector. A second control activates the storage tank to the greenhouse heat-exchanger pump when the greenhouse air temperature drops and turns it off when the desired temperature is reached. A third control turns on the backup heater at a temperature 2°F (1°C) below the desired air temperature in the event that the solar system fails to hold the desired temperature. A fourth control empties water from the collector into an underground tank when the collector temperature approaches freezing and refills it when the collector temperature rises.

Economics High-capacity collectors capable of raising the storage unit temperature more than 30°F (17°C) have the advantage of requiring less collector area and storage capacity. High-capacity systems are very expensive; thus, low-capacity collectors are more typically used in greenhouses. Costs in the mid 1980s for a low-capacity system were about \$4 to \$5/ft^2 (\$43 to \$54/m^2) for the collector and \$8 to \$10/ft^2 (\$86 to \$108/m^2) for the total system.

Even at \$8/ft^2 (\$108/m^2) of collector for the total system, the price per acre for a solar heating system is \$348,500 (\$871,000/ha). This is assuming a ratio of 1 ft^2 of collector per square foot of greenhouse floor area. Such a system might meet total heat requirements in southern regions where 1 gallon of oil is consumed per square foot of floor area per year (41 L/m^2/yr) for single-glazed greenhouses. The annual savings in fuel based on \$0.60/gal for oil (\$0.16/L) would be \$26,100/acre (\$64,500/ha). Taking into account interest on invested capital, repairs, electrical consumption, and implicit costs, the payoff period for this system would be well beyond 20 years and highly questionable.

Interest in solar heating waned over the first half of the 1990s, because tax incentives for installing these systems were taken away and the cost of fuel fell relative to other costs. However, solar heating systems do exist in commercial greenhouse firms. Generally, these firms are small, and the owner may have been satisfied to overlook portions of the true cost of the system. The owner may have constructed the system personally without placing a value on his or her labor. The firm may have financed the system out of prior profits and failed to calculate an interest cost for the money. The profits could otherwise have been invested and yielded interest. The lost interest is a real opportunity cost, which should be added into the total cost of the solar system.

The advent of solar heating does not appear to be on the horizon at this time. Factors that could set the stage would include a return to disproportionately high fuel prices, more efficient collectors, and/or an inexpensive, high-capacity heat-storage medium.

Greenhouse Temperature Sensing

Cooling and heating systems are controlled by temperature sensors. Since temperature gradients exist in greenhouses with even the best of heating systems, placement of the sensor is very important. Its location should reflect the average temperature in the greenhouse. If it is placed in a location near the heater or in a direct flow of warm air, the heater will turn on and off according to conditions in that warm spot and the remainder of the greenhouse will run colder than desired. Consequently, the majority of the crop might be delayed. Sensors need to be placed in an average temperature location, usually near the center of the greenhouse. The height of the sensor placement is also very important with respect to the vertical temperature gradient. The sensor should be located at the height of the growing points of the plants. For pot-plant crops, this height is usually 6 to 12 in (15 to 30 cm) above the pot rim. For cut flowers, the height varies, and the sensor should be attached to a post on which it can be raised or lowered.

Direct or indirect rays of sunlight will raise the sensor temperature well above the air temperature. This will prevent operation of the heater on cold but bright winter days, when heat is needed. The sensor should therefore be shielded from the sun's rays. The sensor housing pictured in Figure 3-21a is a homemade

(a)

(b)

Figure 3-21 Greenhouse sensor housings. (a) A homemade housing consisting of an aspirated box that houses the heater thermostat, a low-temperature alarm thermostat, and a thermometer. The box has a reflective outer surface, louvered ends, and a fan to provide a minimum airflow of 600 fpm (3 m/sec). It is located at the height of the growing points of the plants. (b) A commercial sensor housing, less than 6 in (15 cm) in diameter, which contains an aspirating fan. *(Photo courtesy of Q-Com, Inc., 17782 Cowan Ave., Irvine, CA 92614.)*

unit that has the following features. The outer surface of the box is painted in a reflective color, such as white or aluminum, to reduce heat buildup. The ends of the box have louvers to permit air passage but prevent entry of the sun's rays. A fan is installed to provide a minimum airflow through the box of 600 fpm (3 m/sec). This ensures that a large mass of air is continually monitored by the sensor and that the temperature inside the box does not rise. The sensor station should have an alarm capability. If the temperature drops to a low set point, such as 50°F (10°C), an alarm in the manager's or owner's home would be activated. The alarm system should be powered by a battery or a standby generator to ensure that it operates during an electrical power failure. When the alarm is to be located a long distance from the greenhouse, it is possible, working through the local telephone company, to use their existing lines. More often than not, firms purchase sensor housings from companies that sell sensors and computers. These are typically small, plastic, aspirated units that can contain multiple sensors, as pictured in Figure 3-21b.

Thermostats have historically been used for temperature sensing and control in the greenhouse. Common thermostats operate around a bimetallic strip. The strip curves to conform to the air temperature because two metals fused together in the strip have different thermal expansion coefficients. Such a thermostat generally has a switch built into it. It might be a mechanical switch activated by contact with the end of the bimetallic strip as the strip curves, or it could be a mercury switch attached to the end of a bimetallic coil. A second type of thermostat in use today makes use of a thin metal tube filled with liquid or gas. The tube is shaped into a coil. As the liquid or gas inside changes volume in response to temperature, the end of the tube moves in such a manner that it activates a switch. Depending on the number of settings, such thermostats can cost $200 or more. These thermostats are not highly accurate, nor are they reproducible over time. They need to be calibrated regularly. One problem has been the variation from one thermostat to another within a brand. Even an individual thermostat may slip upward out of calibration one time and downward the next time.

More accurate and reproducible temperature sensing is obtained by growers who use thermocouples and thermistors. Of the two, the thermistor is the more common. Temperature sensing for dedicated microprocessor and computer control of greenhouses is most often accomplished with thermistors. A thermistor is a solid-state chip that changes its voltage output according to the temperature. This sensor requires a circuit to carry the signal to a switch. The switch may be a conventional one for smaller equipment or a relay switch for larger equipment. The circuit can be adjusted to activate switches at specific voltage (temperature) settings. The response is exclusive to temperature, and no other factors are integrated.

The thermocouple consists of two wires of dissimilar metal attached together. Current flow through the junction of the two metals is measured. It is necessary to have a reference temperature in order to convert the current flow rates at each thermocouple into temperature. The reference temperature is usually measured with a thermistor. Thus, when temperature is measured in one zone, a thermistor provides the less expensive alternative. When several temperature sensors are required, it can be cheaper to use thermocouples because they are considerably cheaper than thermistors.

Emergency Heaters and Generators

The risk of electrical power failure is always present. If a power failure occurs during a cold period, such as a heavy snow or ice storm, crop loss due to freezing is likely. Heaters and boilers depend on electricity. Solenoid valves controlling fuel entry, safety-control switches, thermostats, and fans providing air to the firebox all depend on electrical energy.

Power failure can be equally devastating during the summer. Temperature control in greenhouses lacking ventilators is dependent upon electrical exhaust fans. It is likely that the temperature will rise to more than 120°F (49°C) in a closed greenhouse on a clear summer day if a ventilation system is not in effect. High temperatures cause delay in flowering of many crops and, if prolonged for several days, can cause flower bud abortion. Many other types of equipment used in growing crops, including the water pumps, depend upon electrical power. For these reasons, it is important that a standby electrical generator be installed (Figure 3-22).

Figure 3-22 A standby electric generator (left) used in the event of power failure to maintain operation of the boiler (right), cooling system, and possibly a portion of the lights used for photoperiodic timing of the crop. *(From J. W. Love, Department of Horticultural Science, North Carolina State University, Raleigh, NC 27695-7609.)*

The generator can be wired into the greenhouse circuit in such a way that it automatically turns on in the event of a power failure. Some thought should be given to the types of equipment that will be run in this situation. It is rare that the cost can be justified for a generator to handle all power needs. Lights used during the night for control of crop flowering draw considerable power and often cannot be handled by available generators. As will be discussed in Chapter 12, it is possible to use cyclic (flash) lighting, in which the crop is divided into three to five zones. Only one zone is lighted at a time, thus reducing the load demand. During the summer, if the entire cooling system cannot be handled, a proportion of the fans should be maintained to prevent excessive temperatures.

A standby electrical generator is essential to any greenhouse operation. It may never be used, but if required for even one critically cold night, it becomes a highly profitable investment. Generators are available from a number of used-equipment sources, such as government-surplus stores. A firm with a tractor might consider an electrical generator that is powered by the power takeoff on the tractor. A minimum of 1 kilowatt (kW) of generator capacity is required per 2,000 ft^2 (186 m^2) of greenhouse floor area.

It is equally likely that the heating system will fail. Temperatures can drop rapidly in a greenhouse if the insulating properties of the coverings are poor. The rate of temperature decline is increased by lower outside temperatures and increases in wind velocity. Frequently, time is insufficient to seek assistance or to repair the heater before the inside temperature reaches the freezing point. In northern latitudes, this period can be as short as three or four hours. Greenhouse owners using a central boiler system sometimes purchase two boilers to do the job of one. If one fails, the other can still maintain temperatures above freezing. In greenhouses heated by unit heaters with self-contained fireboxes and by infrared radiant heaters, there is little to fear since there are many heaters. It is unlikely that more than one or two could fail at any one time. Situations where there is only one heater in a given greenhouse or only one central boiler require that a backup heating system be available.

Some growers have installed natural-gas or LP-gas burners on flexible fuel lines in the greenhouse (Figure 3-23). When needed, these burners can be moved out into aisles from their storage places under benches or along walls. Because they are already connected to a fuel source and are ready to light manually, no electricity is required.

The Salamander radiant heater is a popular and inexpensive backup. A kerosene supply is maintained in a pot at the bottom of a stovepipe. It is combusted within the bottom part of the vertical stovepipe. The fumes rise up the pipe and out the top into the greenhouse. For this reason, a door should be opened or a ventilator should be opened about ½ inch (1.3 cm) to prevent concentration of the fumes. The stovepipe turns red and radiates considerable quantities of heat. One heater can raise the temperature of 12,000 ft^3 (340 m^3) of air 25 to 30°F (14 to 17°C) and is considered adequate emergency heat for up to 1,500 ft^2 (140 m^2) of greenhouse floor area. The heater burns 0.5 to 1 gallon (1.9 to 3.8 L) of kerosene per hour. One-gallon (3.8-L) cans have been used as well for emergency heat. The top is removed, and two 1-inch (2.5-cm) holes are cut

Figure 3-23 **A 250,000-Btu/hr portable propane heater used for heating the greenhouse during periods of heater failure.**

in opposite sides 2 to 3 in (5 to 8 cm) down from the top to provide air circulation. The can is half-filled with alcohol and ignited. Many other systems are feasible. It is important that one be available.

Fuel

Solid, liquid, and gaseous fuels represented by wood, coal, oil, and gas are used for greenhouse heating. Each has advantages and disadvantages. The choice can be influenced by antipollution regulations. The use of coal and high-sulfur-content oils has been disallowed in some areas.

Heating efficiency is an important factor for any greenhouse firm when selecting a fuel and a heater or boiler. It is a very elusive factor since it has many forms. Combustion efficiency is of little value, since it is merely the percentage of heat supplied in the consumed fuel that is released in the firebox of the heater or boiler. Most of the lost heat remains in the unburned fuel. Thermal efficiency is far more important, because it is the percentage of heat supplied in the consumed fuel that leaves the boiler in the hot-water or steam. The difference between combustion efficiency and thermal efficiency is due mainly to heat lost up the smokestack, and to a lesser degree to heat lost from the jacket of the boiler. A far more valuable measurement would be seasonal efficiency. This is the average of the thermal efficiency over the total heating season. Thermal efficiency is highest when a boiler is operating near full capacity, as in the middle of winter. At the beginning and end of the heating season, efficiency is lowest since heat is supplied infrequently, and thus a higher percentage of heat is consumed in warming and maintaining the boiler temperature. Unfortunately, few seasonal efficiency figures are available for the greenhouse application. Thermal efficiencies in the range of 82 to 88 percent can be expected for oil- and gas-fired heaters and boilers. Thermal efficiencies for coal are much lower and range from about 55 percent for sub-bituminous coal to 70 percent for anthracite coal. Wood boilers can have a thermal efficiency of 65 percent. The preceding values are reasonable at this time but do vary with style and brand of heater or boiler.

Natural gas is the most desirable fuel because the initial installation of a natural gas system is cheaper, storage tanks are not required, and the gas burns clean, which reduces the labor of adjusting and cleaning the boiler. In some parts of the world, natural gas is cheaper than oil. Propane and butane gases have many of the advantages of natural gas, but are much more expensive.

Oil is generally the next choice. An oil-fueled system is easily automated, but storage tanks are necessary and considerably more ash and soot result. The boiler exhaust passages, or tubes, must be cleaned often, and the burner needs to be adjusted at least annually. Fuel oils are available in five grades, designated No. 1, 2, 4, 5, and 6. No. 1 is slightly heavier than kerosene and is generally used to heat private homes. The oil becomes heavier (more viscous) as the number increases. No. 6 oil must be preheated before ignition, or it will not flow through the nozzle in the burner. No. 2 oil is used in small greenhouse heaters, and the heavier grades are used in large boilers. The heavier oils cost slightly less and have a higher heat content. Large central boilers, which can burn heavier grades, offer a fuel cost advantage.

Coal is available in many grades. The terms *anthracite* and *bituminous* refer to hard and soft coals, respectively. Many intermediate kinds exist, with no distinct lines of demarcation. Materials softer than bituminous also exist, ranging all the way to peat. All are the compacted remains of plant material. Coal requires considerable above-ground storage space and more handling labor than oil, and yields large volumes of ash, which must be removed and disposed of.

Boilers are commercially available for burning wood. These systems can be completely automated. Owners of moderate-sized greenhouses requiring a boiler of 100 hp (980 kW) output and larger could consider this option. A few have done so and are realizing a considerable savings in their heating costs. Fuel can consist of green chips made from entire trees, green chips intended for paper pulp, or sawdust. Green chips have a heat content of about 4,500 Btu/lb (10.5 kJ/g) and a moisture content of about 40 to 50 percent, depending on tree species. Dried wood has a heat content of about 8,500 Btu/lb (19.8 kJ/g) and a moisture content of 18 percent. Taking into account a thermal efficiency of 65 percent for wood boilers and 85 percent for oil boilers, a price of $1/ton for green wood chips is equivalent to 2.0¢/gal for No. 2 oil ($1/metric ton of wood = 0.48¢/L oil). The current price of $15/ton for green whole-tree chips is equivalent to an oil price of 30¢/gal (7.9¢/L). This is half of the current 60¢/gal price of oil.

Not all of the fuel price differential is profit, since the price of chips cited here is FOB the chipping plant and a more complex fuel handling system is required for wood. A storage shed is needed to keep rain off the wood. Remember, 1 ton of wood is required for every 50/gal of oil normally consumed (1 metric ton of wood/206 L oil). A silo is required to continuously supply wood to an auger, which feeds it into the boiler. An existing coal boiler can often be converted to burn wood. A tractor is needed for moving wood around in the storage shed and into the silo. Finally, a large bin is needed to collect ash from the boiler. Although the system can be automated, additional labor is required to remove ash from the boiler and dispose of it. In spite of these costs and others, one large greenhouse firm was able to pay back the additional capital cost over an oil system in less than two years and has since realized considerable savings in heating costs. Modern systems burn wood clean enough to meet federal air-pollution standards. The key to successful burning of wood lies in having a steady source of wood. Often, this is not available.

During the 1970s and 1980s, log-burning boilers made their way into the greenhouse industry. They can be purchased for heating requirements as small as 200,000 Btu/hr (56,600 W). The firebox can accommodate 6-foot (1.8-m) logs, which are loaded by tractor. These open-system hot-water boilers generate no pressure. As such, they are free of governmental inspection requirements for pressurized boilers. The burning efficiency is about 65 percent. Cracked logs, undesirable sizes, and species normally left behind after harvesting the forest can be used for fuel in these boilers. Taking into account the burning efficiencies, 1 cord (128 ft^3) of green wood provides the heat output of approximately 125 gallons of No. 2 oil (1 m^3 of wood = 126 L of oil). At the current price of $40/cord, this would be equivalent to buying oil at 32¢/gal (8.5¢/L). For other conversion purposes, it is handy to know that 1 cord of green wood can range in weight from 3,200 lb/cord (400 kg/m^3) for white pine to 5,700 lb/cord (714 kg/m^3) for white oak. Log-burning boilers cost much less than boilers used for burning oil or gas. Logs can be left out in the rain prior to burning; therefore, storage sheds are not required. Labor is required to cut the logs and to load them two to three times daily into the firebox. For this latter reason, many firms have discontinued the use of log-burning boilers during the past 10 years.

The quantity of fuel required for one night, or for any given period of time, can be predicted by knowing the heat value of the fuel to be used, the thermal efficiency of the heater burning the fuel, and the heat required in the greenhouse. The heat requirement can easily be calculated (as will be seen later in this chapter). The heat values of common greenhouse fuels are listed in Table 3-5, along with some common thermal efficiencies.

The data in Table 3-5 can be used to determine that the heater in a greenhouse requiring 100,000 Btu of heat per hour would burn 11.9 lb of anthracite coal, or 0.85 gal of No. 2 oil, or 118 ft^3 of natural gas. All are equivalent in heat value. Each is determined by multiplying the heat value of the fuel by the decimal value of the thermal efficiency of burning that fuel to obtain the heat output of the fuel. The heat output is then divided into the Btus of heat required in the greenhouse. The thermal efficiencies were assumed to be 65 percent for coal and 85 percent for oil and gas. In the case of anthracite coal, the heat value of 12,910 Btu/lb was multiplied by 0.65 (the decimal fraction of the 65 percent thermal ef-

Table 3-5 **Typical Heat Contents for Various Types of Fuel Used for Greenhouse Heating**[1]

Fuel	*Heat Value*	
Moist Coal—Mine Run	**Btu/lb**	**kJ/g**
Anthracite (hard)	12,910	30.0
Semi-anthracite	13,770	32.0
Low-volatile bituminous	14,340	33.3
Medium-volatile bituminous	13,840	32.2
High-volatile bituminous	10,750–13,090	25.0–30.4
Sub-bituminous	8,940–9,150	20.8–21.3
Fuel Oils	**Btu/gal**	**kJ/mL**
No. 1	132,900–137,000	37.1–38.2
No. 2	135,800–141,800	37.9–39.6
No. 4	140,600–153,300	39.2–42.8
No. 5	148,100–155,900	41.3–43.5
No. 6	149,400–157,300	41.7–43.9
Gases	**Btu/ft^3**	**kJ/dm^3**
Natural	1,000	37.3
Manufactured	550	20.5
Propane[2]	2,570	95.7
Butane	3,225	120.1
Wood	**Btu/lb**	**kJ/g**
Green chips	4,500	10.5
Dried pellets	8,500	19.8

[1]The heat value is the amount of heat contained in the fuel. The useful amount of heat in the fuel can be determined by multiplying the heat content by the decimal fraction of the thermal efficiency of the fuel in a given boiler. Current thermal efficiencies can be 55 percent for softer coals to 70 percent for hard coal, 85 percent for oil and gas, and 65 percent for wood. Efficiency will vary from one boiler design to another.

[2]One gallon of propane has a heat value of 91,690 Btu (25.6 kJ/mL), while 1 gallon of butane has a heat value of 102,000 Btu (28.5 kJ/mL).

ficiency that can be achieved when burning this coal) to obtain a heat output of 8,392 Btu/lb. The 100,000 Btu required in the greenhouse was then divided by 8,392 Btu, which is the heat output of 1 pound of coal, resulting in a need for 11.9 pounds of coal.

The cost of fuel is a strong factor in its selection. Equivalent costs of three types of fuel are listed in Table 3-6. The five figures on any line in the table are equivalent. That is, a given amount of heat would cost the amount shown for each of the three fuels within a given line. Taking the first line, for example, 2.0¢ per kilowatt hour (kWh) is equivalent to paying 46.9¢ per gallon of No. 2 oil or

Table 3-6 **Comparative Costs of Electricity, Oil, and Gas**[1]

An Electric Rate of:	Is the Same As Heating with			
	Fuel Oil at:		Gas at:	
¢/kWh	¢/gal[2]	¢/L[2]	¢/therm	¢/m³ (natural gas)[2]
2.0	46.9	12.4	38.6	4.1
2.2	51.6	13.7	42.5	4.6
2.4	56.3	14.9	46.4	5.0
2.6	61.0	16.1	50.2	5.4
2.8	65.7	17.4	54.1	5.8
3.0	70.4	18.6	58.0	6.3
3.2	75.1	19.8	61.9	6.6
3.4	79.8	21.1	65.7	7.1
3.6	84.5	22.3	69.5	7.5
3.8	89.2	23.6	73.4	7.9
4.0	93.9	24.9	77.3	8.3
4.4	103.3	27.3	85.1	9.2
4.8	112.7	29.8	92.7	10.0
5.2	122.0	32.3	100.5	10.9
6.0	140.8	37.2	115.9	12.4
6.8	159.6	42.2	131.4	14.1

[1]Adapted from a table by Clifford M. Tuck and Associates, Athens, GA 30604.
[2]Heat values: kWh = 3,416 Btu; gal = 139,000 Btu; therm = 100,000 Btu.
Thermal efficiency of boiler = 85 percent for No. 2 fuel oil and for gas.

38.6¢ per therm of gas. One should check local prices for fuel. If oil is available for 46.9¢ per gallon and electricity for 3¢ per kWh, it is much cheaper to heat with oil. On the other hand, if gas costs 40¢ per therm, each Btu of heat would cost less from gas than oil.

Calculation of Heat Requirements

A-Frame Greenhouse

In order to determine the heat requirement, the surface of an A-frame greenhouse must be divided into four components, as illustrated in Figure 3-24. They are the roof, gable, wall, and curtain wall. Heat lost under standard conditions through each of these areas can be found in Tables 3-7 and 3-8. All values in the

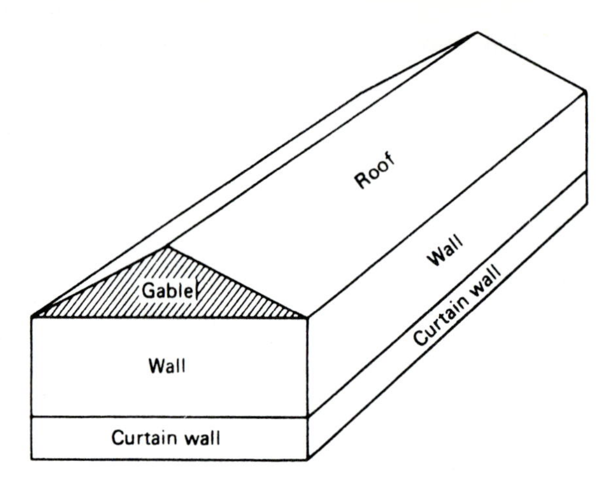

Figure 3-24 **Diagram of an A-frame greenhouse showing component areas needed to determine the heat requirement of this greenhouse.**

tables are listed as thousands of Btus/hr (MBtu). A figure of 5 in the table, for example, means 5,000 Btu/hr. One MBtu/hr is equivalent to 293 W or 252 kcal/hr. The gable and roof losses can be found in Table 3-7. There are two wall components: (1) the wall, covered with a transparent covering and (2) the curtain wall below it, which has a nontransparent covering such as asbestos-cement or concrete block. The heat loss from each is determined separately in Table 3-8. The wall length in each case refers to the total perimeter of the greenhouse, since the wall extends around four sides of the greenhouse.

All heat losses thus far determined are for standard conditions, which include a 70°F (39°C) temperature difference from the outside to the inside and an average wind velocity of 15 miles per hour (mph) (6.7 m/sec). It is likely that you will have different temperature and wind conditions or a different type of greenhouse construction. You can change the heat-loss values in Tables 3-7 and 3-8 by multiplying them by two correction factors. First, determine the difference in temperature between your desired inside night temperature and the coldest outside temperature you expect to encounter during the winter. (Local temperature probabilities can be obtained from the nearest U.S. Weather Bureau office or by purchasing the most recent "Weather Bureau Climatological Data" pamphlet from the Superintendent of Documents, Government Printing Office, Washington, DC 20042.) Next, determine the average wind velocity for your area. For most areas, 15 mph (6.7 m/sec) will suffice. (However, this too can be checked out with the nearest U.S. Weather Bureau office.) Select a climate factor (K) from Table 3-9 for your particular temperature difference and wind velocity, and multiply each heat-loss value from Tables 3-7 and 3-8 by the factor. Select a construction factor (C) from Table 3-10 for the type of greenhouse you have, and multiply this by the heat-loss values for the gable, roof, and wall (trans-

Table 3-7 **Standard Heat-Loss Values for Gables and Roofs of A-Frame Greenhouses**[1]

	Greenhouse Width in ft (m)														
	16 (4.9)	*18 (5.5)*	*20 (6.1)*	*22 (6.7)*	*24 (7.3)*	*26 (7.9)*	*28 (8.5)*	*30 (9.1)*	*32 (9.8)*	*34 (10.4)*	*36 (11.0)*	*38 (11.6)*	*40 (12.2)*	*50 (15.2)*	*60 (18.3)*
	Gable Loss (both) in MBtu/hr[2]														
Greenhouse Length in ft (m)	*5*	*6*	*8*	*10*	*11*	*13*	*15*	*18*	*20*	*23*	*26*	*29*	*32*	*50*	*72*
	Roof Loss (both) in MBtu/hr														
5 (1.5)	7	8	9	10	11	12	12	13	14	15	16	17	18	22	26
10 (3.0)	14	16	18	19	21	23	25	27	28	30	32	34	35	45	54
20 (6.1)	28	32	35	39	42	46	50	53	57	60	64	67	71	88	106
30 (9.1)	42	48	53	58	64	69	74	80	85	90	96	101	106	133	160
40 (12.2)	57	64	71	78	85	92	99	106	113	120	127	135	142	177	212
50 (15.2)	71	80	89	97	106	115	124	133	142	151	159	168	177	222	266
60 (18.3)	85	96	106	117	127	138	149	159	170	181	191	202	212	265	318
70 (21.3)	99	112	124	136	149	161	173	186	198	211	223	235	248	310	372
80 (24.4)	113	127	142	156	170	184	198	212	227	241	255	269	283	354	424
90 (27.4)	127	143	159	175	191	207	223	239	255	271	287	303	319	398	478
100 (30.5)	142	159	177	195	212	230	248	266	283	301	319	336	354	443	532
200 (61.0)	283	319	354	390	425	460	496	531	567	602	637	673	708	885	1,062
300 (91.4)	425	478	531	584	637	690	743	797	850	903	956	1,009	1,062	1,328	1,594
400 (121.9)	566	637	708	779	850	920	991	1,062	1,133	1,204	1,274	1,345	1,416	1,770	2,124
500 (152.4)	708	797	885	974	1,062	1,150	1,239	1,328	1,417	1,505	1,593	1,682	1,770	2,213	2,666

[1]Tables 3-7 through 3-12 adapted from Bohanon, Rahilly, and Stout (1993).
[2]One MBtu/hr = 239 W or 252 kcal/hr.

Table 3-8 Standard Heat-Loss Values for Greenhouse Walls[1]

Wall Length in ft (m)	Wall Height in ft (m)						
	2 (0.61)	4 (1.22)	6 (1.83)	8 (2.44)	10 (3.05)	12 (3.66)	14 (4.27)
	Wall Loss in MBtu/hr[2]						
5 (1.5)	1	2	2	3	4	5	6
10 (3.0)	2	3	5	6	8	10	11
20 (6.1)	3	6	9	13	16	19	22
30 (9.1)	5	9	14	19	24	29	34
40 (12.2)	6	13	19	26	32	38	45
50 (15.2)	8	16	24	32	40	48	56
60 (18.3)	9	19	28	38	47	58	67
70 (21.3)	11	22	33	44	55	67	78
80 (24.4)	13	25	38	51	63	77	90
90 (27.4)	14	28	43	58	71	86	101
100 (30.5)	16	32	47	64	79	96	112
200 (61.0)	32	63	95	128	158	192	224
300 (91.4)	47	95	142	192	237	288	336
400 (121.9)	63	127	190	256	316	384	448
500 (152.4)	79	158	237	320	395	480	560

[1]From Bohanon, Rahilly, and Stout (1993).
[2]One MBtu/hr = 293 W or 252 kcal/hr.

parent covering only). Determine a curtain-wall construction factor (CW) from Table 3-11 and multiply the curtain-wall heat-loss value by this factor. All greenhouse component heat losses have now been corrected. The four corrected values should be added together to determine the total heat input required to heat the greenhouse for 1 hour.

If the heating system is located inside the greenhouse, your calculation is finished. Purchase a boiler with a net rating equal to the heat requirement calculated. If a central heating system is located in a separate building, an additional quantity of heat will be necessary to compensate for heat losses from the delivery and return lines to and from the greenhouse. An engineer should be consulted to determine what this loss is, and it should be added to the heat requirement calculated for the greenhouse.

Table 3-9 Climate Factors (*K*) for Various Average Wind Velocity and Temperature Conditions[1]

Inside-to-Outside Temperature Difference in °F (°C)	*Wind Velocity in mph (m/sec)*				
	15 (6.7)	*20 (8.9)*	*25 (11.2)*	*30 (13.4)*	*35 (15.6)*
30 (16.7)	.41	.43	.46	.48	.50
35 (19.4)	.48	.50	.53	.55	.57
40 (22.2)	.55	.57	.60	.62	.64
45 (25.0)	.62	.65	.67	.70	.72
50 (27.8)	.69	.72	.74	.77	.80
55 (30.6)	.77	.80	.83	.86	.89
60 (33.3)	.84	.88	.91	.94	.98
65 (36.1)	.92	.96	.991	1.03	1.07
70 (38.9)	1.00	1.04	1.08	1.12	1.16
75 (41.7)	1.08	1.12	1.17	1.21	1.25
80 (44.4)	1.16	1.21	1.26	1.30	1.35
85 (47.2)	1.25	1.30	1.35	1.40	1.45
90 (50.0)	1.33	1.38	1.44	1.49	1.54

[1]Standard heat-loss values from Tables 3-7, 3-8, and 3-12 are multiplied by a factor (K) to correct them for local wind and temperature conditions. From Bohanon, Rahilly, and Stout (1993).

Table 3-10 Greenhouse Construction Factors (*C*) for the Common Types of Greenhouses in Use Today[1]

Type of Greenhouse	*C*
All metal (tight glass house—20- or 24-inch [51- or 61-cm] glass width)	1.08
Wood and steel (tight glass house—16- or 20-inch [41- or 51-cm] glass width—metal gutters, vents, headers, etc.)	1.05
Wood house (glass with wooden bars, gutters, vents, etc.—up to and including 20-inch [51-cm] glass spacing)	
Good tight house	1.00
Fairly tight house	1.13
Loose house	1.25
FRP-covered wood house	.95
FRP-covered metal house	1.00
Double glass with 1-inch (2.5-cm) air space	.70
Plastic-covered metal house (single thickness)	1.00
Plastic-covered metal house (double thickness)	.70

[1]Standard heat-loss values for transparent components of greenhouses such as gables and roofs in Table 3-7, transparent side walls in Table 3-9, and ends as well as covering in Table 3-12 are multiplied by a factor (C) to correct them for the type of construction. From Bohanon, Rahilly, and Stout (1993).

Table 3-11 **Curtain-Wall Construction Factor (*CW*) for Various Types of Coverings Used in the Nontransparent Curtain Wall**[1]

Type of Covering	*CW*
Glass	1.13
Asbestos-cement	1.15
Concrete, 4-inch (10-cm)	.78
Concrete, 8-inch (20-cm)	.58
Concrete block, 4-inch (10-cm)	.64
Concrete block, 8-inch (20-cm)	.51

[1]The standard heat-loss value for the curtain wall from Table 3-8 is multiplied by this factor to correct it for the type of covering. From National Greenhouse Manufacturers Association (1994).

EXAMPLE PROBLEM

The following steps are taken to determine the heat requirement for an all-metal, glass-covered greenhouse measuring 30 ft (9.1 m) wide by 100 ft (30.5 m) long. The curtain wall is 2 ft (0.61 m) high and is constructed of 4-inch (10-cm) concrete block. The glass wall above the curtain wall is 6 ft (1.83 m) high. An average wind velocity of 15 mph (6.7 m/sec) is expected. A 60°F (33°C) temperature difference is expected between the outside low temperature of 0°F (–17°C) and the inside temperature of 60°F (16°C).

1. Set up a chart as illustrated here:

Greenhouse Component	*Standard Heat Loss (MBtu/hr) (from Table 3-7 or 3-8)*	*K (from Table 3-9)*	*C or CW (from Table 3-10 or 3-11)*	*Corrected Heat Loss (MBtu/hr)*
Gable			(*C*)	
Roof			(*C*)	
Wall (transparent)			(*C*)	
Curtain wall			(*CW*)	______
		Total heat requirement		

2. In Table 3-7, find the appropriate heat-loss value for both gables combined, immediately below the figure for the greenhouse width. For a 30-foot (9.1-m) width, it is 18 MBtu (18,000 Btu) per hour.
3. In Table 3-7, find the heat-loss value for the combined roofs, at the point where the 30-foot (9.1-m) greenhouse-width column and the 100-foot (30.5-m) greenhouse-length row intersect. In this case, it is 266 MBtu/hr.

4. Figure the length of the side wall. It is equal to the perimeter of the greenhouse, which equals 100 + 30 + 100 + 30 ft, or 260 ft (79.3 m). Find the heat-loss figure for the transparent wall measuring 6 ft (1.83 m) high and 260 ft (79.3 m) long and for the curtain wall measuring 2 ft (0.61 m) high and 260 ft (79.3 m) long in Table 3-8. Since there are no figures in the table for a wall length of 260 ft (79.3 m), look up the values for 200 ft (61.0 m) and for 60 ft (18.3 m) and add them together to arrive at the answer. For the transparent wall, 95 MBtu/hr are lost through a 200-foot (61.0-m) wall and 28 MBtu/hr more are lost through an additional 60 ft (18.3 m) of the wall. The total loss is equal to 95 + 28, or 123 MBtu/hr. The curtain-wall heat loss is equal to 32 + 9, or 41 MBtu/hr.

5. Determine the *K* factor from Table 3-9 for a wind velocity of 15 mph (6.7 m/sec) and a temperature difference of 60°F (33°C). The *K* value is 0.84, which lies at the intersection of the wind-velocity column and the temperature-difference row. Enter this value in the chart in the appropriate spaces after each of the four greenhouse components.

6. Determine the *C* factor from Table 3-10 for the type of greenhouse construction. The example greenhouse is constructed with a metal frame and a glass covering and has a *C* factor of 1.08. Enter this value in the appropriate spaces after the gable, roof, and transparent-wall components. These are the three components constructed with the given materials.

7. Find the *CW* factor for the curtain wall in Table 3-11 and enter it in the chart in the appropriate space in the curtain-wall row. For a 4-inch (10-cm) concrete-block wall, the *CW* factor is 0.64.

8. Correct each of the standard heat-loss values in the chart by multiplying each by the *K* factor and then, in turn, multiplying each answer by the *C* or *CW* factor in the same row. Enter these four values in the chart.

9. Add the four corrected heat-loss values together to arrive at the total heat loss. This value is the amount of heat that must be applied to the greenhouse each hour to maintain the desired temperature if the heater is located in the greenhouse. For the example greenhouse, a heater or boiler with a net rating of 391,273 Btu/hr is needed.

Greenhouse Component	*Standard Heat Loss (MBtu/hr) (from Table 3-7 and 3-8)*		*K (from Table 3-9)*		*C or CW (from Table 3-10)*		*Corrected Heat Loss (MBtu/hr)*
Gable	18	×	0.84	×	1.08	=	16.330
Roof	266	×	0.84	×	1.08	=	241.315
Wall (transparent)	123	×	0.84	×	1.08	=	111.586
Curtain wall	41	×	0.84	×	0.64	=	22.042
				Total heat requirement			391.273

10. If the heater is located in a building apart from the greenhouse, the loss from the boiler, the steam or hot-water mains, and the return lines must be determined and added to the preceding figure.

11. In a mild climate, all heat could be provided by an overhead unit heater system. In a cold climate, a wall coil of pipes should provide an amount of heat equal to the loss through the transparent wall plus the curtain wall. In this example, the requirement would be 111.586 + 22.042, or 133.628 MBtu/hr. The remaining heat, gable plus roof (257.645 MBtu/hr), is provided by the overhead system.

12. If desired, the fuel consumption could be calculated for an hour during the night described. Divide the total heat requirement by the heat output of the fuel used:

$$\text{Anthracite coal: } \frac{391{,}273 \text{ Btu/hr}}{8{,}392 \text{ Btu/lb coal}} = 46.6 \text{ lb/hr}$$

$$\text{No. 2 oil: } \frac{391{,}273 \text{ Btu/hr}}{97{,}000 \text{ Btu/gal oil}} = 4.0 \text{ gal/hr}$$

Quonset Greenhouse

Determination of the heat requirement for a Quonset greenhouse requires a few modifications because of the difference in shape, as diagrammed in Figure 3-25. Quonset greenhouses are covered with film plastic, FRP, or polycarbonate, and a curtain wall is rarely used. The transparent covering usually extends to the ground. Two surface areas are considered in the heat calculation: (1) the two ends collectively, and (2) the covering that extends for the length of the greenhouse, which covers the roof and walls but not the ends. Heat-loss values under standard conditions through these two components are found in Table 3-12.

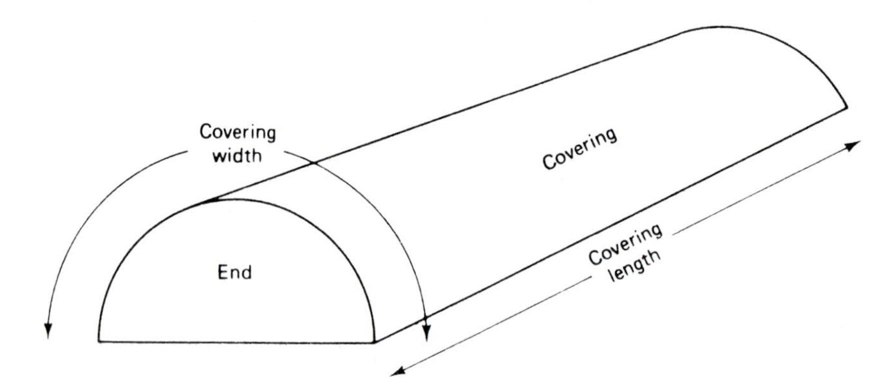

Figure 3-25 **Diagram of a Quonset greenhouse showing component areas needed to determine the heat requirement of this greenhouse.**

Table 3-12 Standard Heat-Loss Values for Quonset-Type Greenhouses for the Combined Ends and for the Entire Covering along the Length of the Greenhouse[1,2]

	Covering Width in ft (m)											
	18 (5.5)	*20 (6.1)*	*22 (6.7)*	*24 (7.3)*	*26 (7.9)*	*28 (8.5)*	*30 (9.1)*	*32 (9.8)*	*34 (10.4)*	*36 (11.0)*	*38 (11.6)*	*40 (12.2)*
	End Loss in MBtu/hr[2]											
House Length in ft (m)	*8*	*10*	*12*	*15*	*17*	*20*	*23*	*26*	*29*	*33*	*36*	*40*
	Covering Loss in MBtu/hr[2]											
5 (1.5)	7	8	9	9	10	11	12	13	13	14	15	16
10 (3.0)	14	16	17	19	21	22	24	25	27	28	30	32
20 (6.1)	28	32	35	38	41	44	47	51	54	57	60	63
30 (9.1)	43	47	52	57	62	66	71	76	81	85	90	95
40 (12.2)	57	63	70	76	82	89	95	101	103	114	120	127
50 (15.2)	71	79	87	95	103	111	119	127	134	142	150	158
60 (18.3)	85	95	104	114	123	133	142	152	161	171	180	190
70 (21.3)	100	111	122	133	144	155	166	177	188	199	211	222
80 (24.4)	114	127	139	152	164	177	190	202	215	228	240	253
90 (27.4)	128	142	157	171	185	199	214	228	242	256	271	285
100 (30.5)	142	158	174	190	206	221	237	253	269	285	301	316
200 (61.0)	285	316	348	380	411	443	475	506	538	570	601	633
300 (91.4)	427	475	522	569	617	664	712	759	807	854	902	949
400 (121.9)	570	633	696	759	822	886	949	1,012	1,075	1,139	1,202	1,265
500 (152.4)	712	791	870	949	1,028	1,107	1,187	1,265	1,345	1,424	1,503	1,582

[1]These values are for standard conditions, including a 70° F(39°C)difference from outside to inside temperature and an average wind velocity of 15 mph (6.7m/sec). From Bohanon, Rahilly, and Stout (1993).

[2]One MBtu/hr = 293 W or 252 kcal/hr.

The values must be corrected for your own conditions in the same way that the heat values for an A-frame greenhouse were corrected. The same *K* and *C* factors are located in Tables 3-9 and 3-10, respectively. The end and covering heat-loss values are multiplied by each of these factors to determine the corrected heat-loss values. The two corrected heat-loss values are added together to arrive at the heat requirement for the greenhouse.

EXAMPLE PROBLEM

Listed next are steps to follow in calculating the heat requirement of a metal-frame Quonset greenhouse measuring 30 ft (9.1 m) wide by 100 ft (30.5 m) long and covered with two layers of polyethylene, each measuring 40 ft (12.2 m) wide. A temperature difference of 60°F (33°C) and an average wind velocity of 15 mph (6.7 m/sec) are expected.

1. Locate the heat-loss value for the two ends combined in Table 3-12. It is the value immediately below the covering width of 40 ft (12.2 m), or 40 MBtu (40,000 Btu) per hour.
2. Locate the heat-loss value for the covering in Table 3-12. It is the value located at the intersection of the column below the covering width of 40 ft (12.2 m) and the row for a greenhouse length of 100 ft (30.5 m). In this example, the covering heat loss is equal to 316 MBtu/hr.
3. Determine the *K* factor from Table 3-9 for a wind velocity of 15 mph (6.7 m/sec) and a temperature difference of 60°F (33°C). The *K* factor is 0.84.
4. Find the *C* factor from Table 3-10 for this metal-frame greenhouse covered with a double layer of polyethylene. The *C* factor is 0.70.
5. Multiply each of the standard heat-loss values by the *K* factor and then by the *C* factor to determine the corrected heat-loss values:

$$40 \times 0.84 \times 0.70 = 23.520 \text{ MBtu/hr}$$

$$316 \times 0.84 \times 0.70 = 185.808 \text{ MBtu/hr}$$

6. Add the two corrected heat-loss values together to determine the heat requirement of the greenhouse. This is the net load of the heater when it is located within the greenhouse.

Greenhouse Component	*Standard Heat Loss (MBtu/hr) (from Table 3-12)*		*K (from Table 3-9)*		*C (from Table 3-10)*		*Corrected Heat Loss (MBtu/hr)*
Combined ends	40	×	0.84	×	0.70	=	23.520
Covering	316	×	0.84	×	0.70	=	185.808
			Total heat requirement				209.328

$$\text{Total required kcal/hr} = 209.328 \text{ MBtu/hr} \times 252 = 52{,}750 \text{ kcal/hr}$$

$$\text{Total required W} = 209.328 \text{ MBtu/hr} \times 293 = 61{,}333 \text{ W}$$

Gutter-Connected Greenhouse

A gutter-connected greenhouse generally has three components in terms of heat-requirement computation. They are the roof, the gables, and the walls (Figure 3-26). Standard heat loss from the walls is determined from Table 3-8. The wall height is the distance from ground to gutter, while the wall length is the perimeter of the greenhouse. The heat loss from each roof is determined from Table 3-12. The heat loss calculated for one roof is multiplied by the number of roofs in the greenhouse. The heat-loss values for ends listed in Table 3-12 yield values too large for the loss from gables. The end of a Quonset greenhouse equates to the gable plus part of the side wall of a gutter-connected greenhouse. Gable heat loss is best determined by calculating the surface area of a gable and then figuring 0.08 MBtu/hr standard heat loss for every 1 ft^2 (2.52 W/m^2, 2.17 kcal/hr/m^2). The gable area can be satisfactorily estimated by multiplying the gable height by the gable width and finally by 0.55:

$$\text{height} \times \text{width} \times 0.55 = \text{area of one gable} \tag{3-1}$$

Actually, a different equation is needed for each manufacturer's design, but this equation will come close enough for all. Be sure to multiply the area of one gable by the number of gables. There are two gables per roof. When the standard heat losses have been determined for the roofs, walls, and gables, each must be multiplied by the appropriate K value from Table 3-9 and C value from Table 3-10.

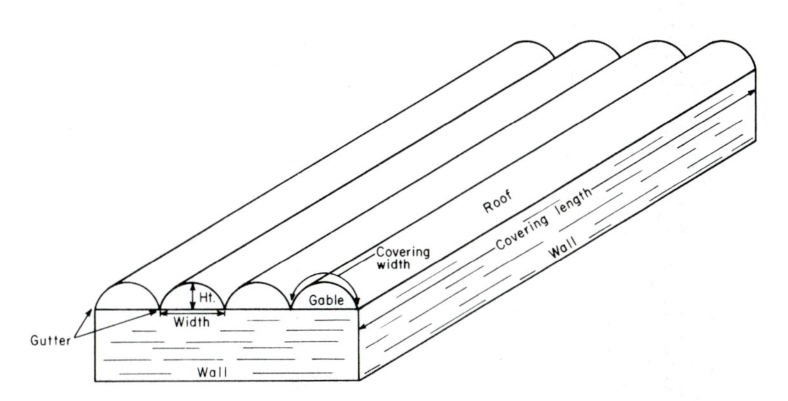

Figure 3-26 **Diagram of a gutter-connected greenhouse showing component areas needed to determine the heat requirement of this greenhouse.**

The sum of the three corrected heat-loss values is the total heat loss for the greenhouse.

EXAMPLE PROBLEM

The following steps are taken to determine the heat requirement for a gutter-connected greenhouse with four bays (as shown in Figure 3-26) each measuring 21 ft (6.4 m) wide by 100 ft (30.5 m) long. The walls are 12 ft (3.66 m) high, the gables are 5 ft (1.5 m) high, the covering width of each bay is 24 ft (7.3 m), the greenhouse frame is metal, and the entire greenhouse is covered with a double layer of polyethylene. A temperature difference of 65°F (36°C) and a wind velocity of 20 mph (8.9 m/sec) is expected.

1. Figure the length of the side wall. It is equal to the perimeter of the greenhouse, which equals 100 + (4 bays × 21 ft) + 100 + (4 bays × 21 ft) = 368 ft (112.2 m). Find the heat-loss value for the wall measuring 12 ft (3.66 m) high and 368 ft (112.2 m) long in Table 3-8. Since there are no figures for 368 ft (112.2 m), look up values for 300 ft (91.4 m) and 70 ft (21.3 m) and add them together to arrive at an answer that is sufficiently accurate. The total loss is 355 MBtu/hr, the sum of 288 MBtu/hr for the 300-foot (91.4-m) length and 67 MBtu/hr for the 70-foot (21.3-m) length.
2. Determine the heat loss through the four combined roofs. In Table 3-12, the heat loss through a roof covering width of 24 ft (7.3 m) and length of 100 ft (30.5 m) is 190 MBtu/hr. This is multiplied by 4 (for the four greenhouse bays) to arrive at a total of 760 MBtu/hr.
3. Calculate the heat loss through the eight gables. First determine the area of these gables. Each gable has an area equal to 0.55 × base × height, or 0.55 × 21 ft × 5 ft which equals 57.75 ft^2. The total for the eight gables is 462 ft^2. Multiply this value by 0.08, since there is a standard heat loss of 0.08 MBtu/hr for each square foot of surface area. The heat loss for the eight gables is 37 MBtu/hr.
4. Determine the *K* factor from Table 3-9 for a wind velocity of 20 mph (8.9 m/sec) and a temperature difference of 65°F (36°C). The *K* factor is 0.96.
5. Find the *C* factor from Table 3-10 for this metal-frame greenhouse covered with a double layer of polyethylene. The *C* factor is 0.70.
6. Construct a heating chart and enter the standard heat-loss values for the walls, roofs, and gables. Multiply each by the appropriate *K* and *C* factors just determined. Add the three resulting corrected heat-loss values to determine the required net load of the heater when it is located in the greenhouse.

Greenhouse Component	Standard Heat Loss (MBtu/hr) (from Tables 3-8 and 3-12)		K (from Table 3-9)		C (from Table 3-10)		Corrected Heat Loss (MBtu/hr)
Walls	355	×	0.96	×	0.70	=	238.56
Roofs	760	×	0.96	×	0.70	=	510.72
Gables	37	×	0.96	×	0.70	=	24.86
				Total heat requirement			774.14

Significance of *K, C,* and *CW* Factors

Standard heat-loss factors are multiplied by *K, C,* and *CW* factors to correct them for local conditions. When local conditions are the same as the standard conditions under which the heat-loss values of Tables 3-7, 3-8, and 3-12 were determined, these factors are equal to 1. Obviously, multiplication by 1 does not change the heat-loss values. Note in Table 3-9 that the *K* factor has a value of 1 for an average wind velocity of 15 mph (6.7 m/sec) and a temperature difference of 70°F (39°C).

If the wind velocity remains at 15 mph (6.7 m/sec) and the inside temperature is reduced by 10°F (6°C) so that the temperature difference is now 60°F (33°C) rather than 70°F (39°C), less heat will be required in the greenhouse. This can be seen in the *K* factor, which would become 0.84. Since 0.84 is 16 percent less than 1, there is a 16 percent savings in heat realized. When a factor less than 1, such as this factor of 0.84, is multiplied by the standard heat-loss value, the heat loss, or in other words the heat requirement, diminishes. The most highly resistant coverings to heat transmission, those that retain heat best in the greenhouse, have the lowest *C* factors in Table 3-10. The same is true for the curtain-wall covering materials in Table 3-11.

Heat Conservation

Greenhouse Design

Fuel economy can be designed into a greenhouse firm. Heat loss is a function of the amount of exposed greenhouse surface area. Quonset greenhouses can have close to 1.65 ft^2 of exposed surface per square foot of floor area, while large blocks of gutter-connected greenhouses can approach a ratio of 1.50 ft^2 of exposed area per square foot of floor area. This is a 10 percent reduction in surface area. Thus, a sizable decrease in heat loss can be realized through greenhouse design.

Double Covering

A double-layer polyethylene greenhouse will consume about 40 percent less fuel than an equivalent single-layer glass, FRP, or polyethylene-covered greenhouse. Polycarbonate panels will reduce the heat requirement by approximately 50 percent, compared to a single-layer greenhouse.

Thermal Screens

Greenhouses with a maximum distance between supporting post rows lend themselves more economically to the installation of *thermal screens.* A thermal screen is a curtain of material such as polyethylene, polyester film, aluminized polyester film strips, or polyester cloth that is drawn from eave to eave or gutter to gutter as well as around the inner perimeter of the greenhouse each night to box in the crop. It is drawn off in the morning by a motorized mechanism. Polyester is superior to polyethylene because it blocks radiant heat better (see Table 3-1). Often, the curtain has an aluminized surface on one side to further reflect radiant heat back to the soil and plants at night so that it cannot leave the greenhouse. Thermal screens serve also to block convection of heat, keeping it around the plants and away from the greenhouse covering. Less heat is lost because the temperature differential across the greenhouse covering is lower. Heat curtains can reduce fuel consumption by 20 to 60 percent, with 40 percent being a realistic mean value.

Thermal-screen systems, including installation, are available at a cost of $1 to $3/ft^2 ($10.76 to $32.29/m^2) of floor area covered. The price differential relates to the number of obstacles within the greenhouse (such as supporting post rows), the number of zones to be independently covered, whether the greenhouse was designed to accommodate a screen, whether the screen is drawn from truss to truss or eave to eave (the latter being cheaper), and the type of screen material. This system is not nearly as practical for Quonset greenhouses because the higher price would apply due to the small installation needed for each greenhouse. The composition of screen materials varies depending on the roles the screen is intended to play. There are three roles: (1) heat retention on winter nights, (2) partial sun screening on bright summer days, and (3) total exclusion of light for lengthening the night in summer for photoperiodic crops (see Chapter 12 for the latter two roles). Screens are available to perform any one of these functions singly. Screens are also available for the combination of heat retention plus sun screening or for the combination of heat retention plus photoperiodic control. If all three functions are required in a greenhouse, two automatic screen-pulling systems will have to be installed, which is entirely possible.

Firms that grow photoperiodically timed crops (discussed in Chapter 12), which require covering with shade cloth to lengthen the night, might be more inclined to install a thermal screen. The mechanism that automatically pulls the thermal screen during the winter is also used to pull the shade cloth from spring to fall. One curtain material can be used for both functions. Thus, a single investment can be recouped in two ways. (Such a system can be seen in Figure 12-12.)

Thermal screens result in a lower greenhouse-covering temperature, which reduces the tendency to melt snow. There is a greater risk of collapse from snow load, which can be remedied by leaving the screen open during a snowstorm. A snow-sensing device should be installed on the roof for this purpose. Insurance rates may be higher for greenhouses equipped with thermal screens. Some growers have a problem with condensation collecting on the thermal screen, but porous screens are available to solve this problem. Finally, the rush of cold air on the plants when the screen is drawn open in the morning troubles some growers. To get around this problem, many screens have recently been installed immediately below the roof covering material in a way that the screens are drawn from truss to truss. There is less of a shading problem with this arrangement.

Lower Air Temperature

Heating systems that permit lower air temperature result in considerable heat conservation. Heated floors and radiant heaters allow for at least a 5°F (3°C) reduction in air temperature. As seen in Table 3-9, this can result in an 8 percent reduction in heat requirement if the air temperature is lowered from 60°F (16°C) to 55°F (13°C). (The K value for 15 mph wind velocity and a temperature differential of 60°F is 0.84 while for a 15 mph wind velocity and a temperature differential of 55°F it is 0.77. The latter K value is 8.3 percent lower than the former.)

Radiant Heat

Low-energy radiant heaters can also be designed into the heating system for a fuel savings of 30 percent or more where natural gas is available. In high-ridge greenhouses, it is possible to lower the radiant heating system and to install a thermal screen above the radiant heaters.

Wall Insulation

Little benefit is derived from scattered light entering through the north wall of a greenhouse. A 5 to 10 percent savings in fuel can be realized by constructing a solid, insulated north wall with a reflectorized inner surface. Another 3 to 6 percent savings can be gained by installing insulation over the curtain walls of the greenhouse. Further savings are possible by installing insulation 12 in (30 cm) into the ground around the perimeter of the greenhouse.

Sealing Air Leaks

A number of techniques for sealing air leaks can be applied to existing energy-inefficient greenhouses. Several commercial glass greenhouses have been covered with two layers of air-inflated polyethylene for a fuel savings of 40 to 60 percent.

One problem that goes along with this system is the reduction in light transmission. In an Ohio State University study, a solar radiation reduction of 35 percent was measured within a conventional glass greenhouse as a result of glass, sash bars, and frame. An additional 18 percent reduction occurred from a double layer of polyethylene over the glass. For high-light-requiring crops, this situation may be intolerable, but for many others it appears to be acceptable. Considerable heat can be lost through cracks between overlapping panes of glass or sheets of FRP. Aging causes these cracks to open up and may also cause the glazing compound to become brittle and fall away from the area between the glass and the sash bar. Cracks may occur in the glass, with corners falling out, or some panes of glass may slide, opening up holes. Eventually, reglazing of a glass greenhouse becomes necessary. Reglazing should be done as soon as the need becomes evident to prevent heating costs from rising. Under the present fuel situation, it could be false economy to put off a reglazing job.

Windbreaks

The climate factors in Table 3-9 give a good indication of the effect of wind on the heat requirement. For every 5-mph (2.2-m/sec) rise in average wind velocity above 15 mph (6.7 m/sec), there is a 4 percent increase in heat loss from the greenhouse. The velocity of wind striking a greenhouse can be reduced by providing windbreaks of trees. Fast-growing evergreen trees, such as hemlock or Leyland cypress, serve well. In some cases, trees are already growing prior to construction of the greenhouse range. Care should be taken to leave these where they can perform a strategic role. While windbreaks are important, they must never cast a shadow over the growing area. This would result in loss of productivity, which would be more costly than the fuel saved by the windbreak. Windbreaks on the east, west, or south side should be located away from the greenhouse a distance equal to 2.5 times the height of the windbreak to prevent winter shadows from interfering with crop growth. In general, a 5 to 10 percent fuel savings can result from windbreaks.

High-Efficiency Heaters

High-efficiency heaters are available today that have more extensive heat exchangers than previous models. As a consequence, more heat is removed from the exhaust gases. Exhaust temperatures of 600°F (315°C) and higher can now be reduced to about 300°F (150°C). Where one is operating an inefficient boiler, the efficiency can be increased by installing a chimney heat reclaimer in the furnace flue pipe. This consists of a heat exchanger through which air in some models and water in other models is passed. The warmed air may be used to heat the service building or part of the greenhouse, while the warm water may be used for irrigation. Care should be taken not to lower the stack temperature below the manufacturer's recommendation. At low temperatures, water, acids, and other corrosive compounds can condense and cause deterioration of the chimney.

Heater Maintenance

Heaters will consume fuel at varying efficiencies, depending upon adjustment of the fuel-to-air ratio. For this reason, heaters should be maintained in good condition. Omission of a periodic service call can cost far more in increased fuel consumption. Soot may build up in the flue passageways of boilers, providing insulation on those iron surfaces that are in actuality the heat exchanger of the boiler. Less heat is transferred to water and more goes up the smokestack, thus increasing fuel consumption. A soot layer 1/8 inch (3 mm) deep can cause a heat loss of up to 15 percent, and a 3/16-inch (5-mm) layer can cause a 21 percent loss in heat captured by the boiler. Boilers should be cleaned on a regular basis. Special materials for coating flue tubes can reduce the tendency for soot to adhere to the surface, allowing more to pass out in the smoke effluent. On the average, these tubes are more efficient heat exchangers, assuming that a cleaning schedule is still maintained.

Thermostat Maintenance

Many other maintenance possibilities exist for reducing heat loss. Thermostats or sensors should be accurately calibrated so that higher-than-desired temperatures are not maintained. This maintenance needs to be done periodically (about every six months) against a calibrated thermometer. Highly precise thermostats should be used. Bimetallic-strip thermostats generally activate a heater at the desired temperature setting and turn it off when a higher temperature is reached. The interval between is known as the *dead load.* A dead load of 2°F (1°C) is quite acceptable for these thermostat types. The dead load can be 6°F (3°C) or more in a malfunctioning thermostat. Considerable heat is wasted each time the thermostat activates. Such a thermostat should be replaced.

Cool-Temperature Crops

Within some crops are cultivars that can be satisfactorily produced at lower temperatures than others. This is particularly true for poinsettias and chrysanthemums. Greenhouse crops as a whole can be produced at lower temperatures than are generally recommended, but the cropping time is increased. Arguments have been set forth for and against this procedure in reference to fuel conservation. In some cases, the fuel savings are lost in forms such as overhead costs and fuel consumption during the period of extended growth. Before adopting this form of heat conservation, one should test it out and keep accurate records.

Combined Economics

It should be obvious that if several heat-saving options are adopted, the total heat savings will not be equal to the sum of the savings of each option. If a second layer of polyethylene is applied to a plastic greenhouse, a savings of 40 percent might be realized on the original heat bill. The fuel consumption now equals 60 percent of the original. Further installment of a thermal screen, predicted to save 40 percent of the fuel consumption, would not lower the fuel consumption to 20 percent of the original value. It would reduce the fuel consumption after installing the second layer of polyethylene by 40 percent. This would be a 24 percent fuel reduction ($0.40 \times 0.60 = 0.24$). The price of energy conservation must always be measured against the value of the energy conserved.

Summary

1. Heat must be supplied to a greenhouse at the same rate with which it is lost in order to maintain a desired temperature. Heat can be lost in three ways—by conduction, by infiltration, and by radiation. Heat is conducted directly through the covering material in conduction loss. In infiltration loss, heat is lost as warm air escapes through cracks in the covering. In radiation loss, heat is radiated from warm objects inside the greenhouse through the covering to colder objects outside.
2. A unit heater system, in which each heater has a firebox, is the cheapest and consequently the most popular system in America. Heat is distributed from the unit heaters by one of two common methods. In the convection-tube method, warm air from unit heaters is distributed through a transparent polyethylene tube running the length of the greenhouse. Heat escapes from the tube through holes on either side of the tube in small jet streams, which rapidly mix with the surrounding air and set up a circulation pattern to minimize temperature gradients. The second method of heat distribution is horizontal airflow (HAF). In this system, fans located above plant height are spaced about 50 ft (15 m) apart in two rows such that the heat originating at one corner of the greenhouse is directed down one side of the greenhouse to the opposite end and then back along the other side of the greenhouse. Both of these distribution systems can be used for circulating air when neither heating nor cooling are used and for introducing cold outside air during winter cooling.
3. A central heating system can be more efficient than unit heaters, especially in large greenhouse ranges. They are particularly popular in European greenhouses. In this system, two or more large boilers are in a single location. Heat is transported in the form of hot water or steam (mainly hot water) through pipe mains to the growing area. There, heat is exchanged from the hot water in a pipe coil on the perimeter walls plus an overhead pipe coil located across the greenhouse or an in-bed pipe coil located in the plant zone. Some greenhouses have a third pipe coil embedded in a concrete floor. A set of unit heaters obtaining heat from hot water or steam from the central boiler can be used in lieu of the overhead pipe coil.

4. Low-intensity infrared radiant heaters can save 30 percent or more in fuel over more conventional heaters. Several of these heaters are installed in tandem in the greenhouse. Lower air temperatures are possible since the plants and root substrate are heated directly.
5. Solar heating systems are found in hobby greenhouses and small commercial firms. Both water and rock storage systems are used. The high cost of solar systems has discouraged any significant acceptance by the horticulture industry to date.
6. Emergency equipment is a necessity and should include a heat source as well as an electrical generator. The generator can be installed to start automatically upon power failure. The need for heat should be signaled by a thermostat-activated alarm system in the manager's or owner's home.
7. Temperature sensor placement is very crucial. The sensor should be at the height of the growing point of the plants and in a location typical of the average temperature of the greenhouse. It should be in a light-reflecting box that is aspirated at a minimum airflow rate of 600 fpm (3 m/sec). The aspirated box should also contain other temperature-sensing controls and a thermometer for testing and correcting the sensors.
8. Relatively easy procedures have been outlined for calculating the heat requirement of greenhouses. Information necessary for determining the heat requirement for an A-frame greenhouse is contained in Tables 3-7 through 3-11 and for a Quonset greenhouse in Tables 3-9 through 3-12. Calculations for a gutter-connected greenhouse make use of a combination of all tables.
9. The heat requirement of a greenhouse can be reduced by installing double greenhouse coverings; by using a greenhouse design with minimal exposed surface area; by using thermal screens; by repairing broken glass and tightening existing glass; by using a windbreak of trees to reduce wind velocity; by using high-efficiency (low-stack-temperature) heaters and boilers; by periodically adjusting and cleaning heaters, boilers, and thermostats; and possibly by using cool-temperature-tolerant varieties of plants.

References

Various manufacturers of heating equipment offer literature concerning products and technical information.

1. Agricultural Development and Advisory Service. 1976. *Greenhouse heating systems.* Ministry of Agriculture, Fisheries and Food, Mechanization Leaflet 27. London: Her Majesty's Stationery Office.
2. Aldrich, R. A., W. A. Bailey, J. W. Bartok, Jr., W. J. Roberts, and D. S. Ross. 1976. *Hobby greenhouses and other gardening structures.* Pub. NRAES-2. Northeast Reg. Agr. Eng. Ser., Cornell Univ., 152 Riley-Robb Hall, Ithaca, NY 14853.

3. Aldrich, R. A., and J. W. Bartok, Jr. 1994. *Greenhouse engineering, 3rd rev.* Pub. NRAES-33. Northeast Reg. Agr. Eng. Ser., Cornell Univ., 152 Riley-Robb Hall, Ithaca, NY 14853.
4. American Society of Agricultural Engineers. 1995. Heating, ventilating, and cooling greenhouses: ASAE EP406.2 Feb. 95. In: *American Society of Agricultural Engineers Standards 1995,* 42nd ed., pp. 559–566. St. Joseph, MI: Amer. Soc. of Agr. Engineers.
5. Badger, P. C., and H. A. Poole. 1979. *Conserving energy in Ohio greenhouses.* Ohio Coop. Ext. Ser. Bul. 651. The Ohio State Univ., Columbus, OH.
6. Blom, T., F. Ingratta, and J. Hughes. 1982. *Energy conservation in Ontario greenhouses.* Pub. 65. Ontario Ministry of Agr. and Food.
7. Bohanon, H. R., C. E. Rahilly, and J. Stout. 1993. *The greenhouse climate control handbook.* Form C7S. Muskogee, OK: Acme Eng. and Manufacturing Corp.
8. Boyette, M. D., and R. W. Watkins. 1988. *Getting into hot water.* North Carolina Agr. Ext. Ser. Bul. AG-398. North Carolina State Univ., Raleigh, NC.
9. Duncan, G. A., and J. N. Walker. 1973. *Poly-tube heating-ventilation systems and equipment.* AEN-9. Dept. of Agr. Eng., Univ. of Kentucky, Lexington, KY.
10. Gray, H. E. 1956. *Greenhouse heating and construction.* Florists' Publishing Co., 343 S. Dearborn St., Chicago, IL.
11. Jacobson, J. S., and A. C. Hill, eds. 1970. *Recognition of air pollution injury to vegetation: a pictorial atlas.* Informative Report No. 1. Pittsburgh, PA: Air Pollution Control Association.
12. Jahn, L. G. 1985. *Wood energy guide for agricultural and small commercial applications.* North Carolina Agr. Ext. Ser. Bul. AG-363. North Carolina State Univ., Raleigh, NC.
13. Koths, J. S. and J. W. Bartok, Jr. 1985. Horizontal air flow. *Connecticut Greenhouse Newsletter* No. 125 (special issue). Dept. of Agr. Pub., Univ. of Connecticut, Storrs, CT 06268.
14. National Greenhouse Manufacturers Association. 1994. Standards: Design loads in greenhouse structures, ventilating and cooling greenhouses, greenhouse heat loss, statement of policy—greenhouse retrofit. Natl. Greenhouse Manufacturers Association, 7800 S. Elati, Suite 113, Littleton, CO 80120.
15. Poole, H. A., and P. C. Badger. 1980. *Management practices to conserve energy in Ohio greenhouses.* Ohio Coop. Ext. Ser. Bul. 668. The Ohio State Univ., Columbus, OH.
16. Roberts, W. J. 1996. *Soil heating for greenhouse production.* Pub. E208. Rutgers Coop. Ext. Ser., New Brunswick, NJ.
17. Roberts, W. J., J. W. Bartok, Jr., E. E. Fabian, and J. Simpkins. 1989. *Energy conservation for commercial greenhouses,* 3rd rev. Pub. NRAES-3. Northeast Reg. Agr. Eng. Ser., Cornell Univ., 152 Riley-Robb Hall, Ithaca, NY 14853.
18. Sherry, W. J. 1983. Which greenhouse cover is for you? *Greenhouse Manager* 2(2):126–132.
19. U.S. Housing and Home Financing Agency. 1954. *Thermal insulation value of air space.* Housing Res. Paper 32. U.S. Housing and Home Financing Agency, Div. of Housing Res., Washington, DC.
20. Walker, J. N., and G. A. Duncan. 1975. *Estimating greenhouse heating requirements and fuel costs.* AEN-8. Dept. of Agr. Eng., Univ. of Kentucky, Lexington, KY.
21. ———. 1974. *Greenhouse heating systems.* AEN-31. Dept. of Agr. Eng., Univ. of Kentucky, Lexington, KY.
22. Whillier, A. 1963. Plastic covers for solar collectors. *Solar Energy* 7(3):148–151.

CHAPTER 4

Greenhouse Cooling

Greenhouses require two distinctly different forms of cooling, one for summer and the other for winter. Most localities, with the general exception of those in higher elevations, experience periods of summer heat that are adverse to greenhouse crops. Temperatures inside the older standard ventilator-cooled greenhouses are frequently 20°F (11°C) higher than those outside, in spite of open ventilators. This system of cooling is known as *passive cooling.* Detrimental effects of high temperatures are typified by loss of stem strength, reduction of flower size, delay of flowering, and even bud abortion.

Evaporative cooling systems, developed to reduce the excess-heat problem, work well for summer cooling. The evaporative cooling systems are based on the process of heat absorption during the evaporation of water. The two evaporative cooling systems in use today are *fan-and-pad* and *fog*. Excess heat can likewise be a problem during the winter. Even when the outside temperature is below the desired inside temperature, the entrapment of solar heat can raise the inside temperature to an injurious level if the greenhouse is not ventilated. The challenge during winter cooling is to temper the excessively cold incoming air before it reaches the plant zone. Otherwise, hot and cold spots in the greenhouse will lead to uneven crop timing and quality. Two active winter cooling systems that have been developed to solve this problem are convection-tube cooling and horizontal airflow (HAF) cooling. The cost of these four active summer and winter cooling systems, which have been developed over the past 40 years, has recently stimulated major improvements in passive ventilator cooling of greenhouses.

The fan-and-pad evaporative cooling system has been available since 1954 and is still the most common summer system in greenhouses (Figure 4-1). Along one wall of the greenhouse, water is passed through a pad that is usu-

(a)

(b)

Figure 4-1 An installation of (a) a 4-inch (10-cm) -thick cross-fluted cellulose evaporative pad and (b) exhaust fans in the opposite wall used for evaporative cooling of a greenhouse during the summer.

ally placed vertically in the wall. Traditionally, the pad was composed of excelsior (wood shreds), but today it is commonly made of a cross-fluted cellulose material somewhat similar in appearance to corrugated cardboard. Exhaust fans are placed on the opposite wall. Warm outside air is drawn in through the pad. Water in the pad, through the process of evaporation, absorbs heat from the air passing through the pad as well as from the surrounding pad and frame.

The fog evaporative cooling system, introduced in greenhouses in 1980, operates on the same cooling principle as the fan-and-pad system but uses quite a different arrangement. A high-pressure pumping apparatus generates fog containing water droplets with a mean size of less than 10 microns (one-tenth the thickness of a human hair). These droplets are sufficiently small to stay suspended in air while they are evaporating. Fog is dispersed throughout the greenhouse, cooling the air everywhere. People and plants stay dry throughout the process. This system is equally useful for seed germination and cutting propagation since it eliminates the need for a mist system.

Both types of summer evaporative cooling systems can reduce the greenhouse air temperature well below the outside temperature. The fan-and-pad system can lower the temperature of incoming air by about 80 percent of the difference between the dry and wet bulb temperatures, while the fog cooling system can lower the temperature by nearly the full difference. Thus, the drier the air, the greater the cooling that is possible (see Figure 4-2). Consider air at a dry bulb temperature of 90°F (32°C) but at relative humidities of 20 percent in Arizona and 60 percent in Florida. The wet bulb temperatures (lowest cooling points) would be 63 and 78°F (17 and 26°C), respectively. The fan-and-pad system could lower the temperature to approximately 68°F (20°C) in Arizona and 80°F (27°C) in Florida.

The fundamental difference between summer and winter cooling systems lies in the temperature of the air that is external to the greenhouse. It is desirable to cool the air during the summer before passing it over the plants. Large volumes of cooled air are introduced directly and uniformly into the plant canopy. During the winter, cold external air must be introduced indirectly above the plants and mixed with the undesirable warm air within the greenhouse prior to making contact with the plants in order to prevent cold spots at the plant level.

Originally, greenhouses were constructed with ventilators adjacent to the ridge and on the side walls. When cooling was required on winter days, the ridge ventilators were opened. Cold air, being more dense than the warm air inside, would drop to the floor beneath the ventilators, where it would spread laterally and increase in temperature as it mixed with the warm air. The result was a temperature gradient across the house at plant height. This led to uneven growth rates and subsequently to variation in maturation dates. The convection-tube and HAF ventilation systems used today for winter cooling correct the horizontal temperature gradient problem. Each circulates the air in the greenhouse.

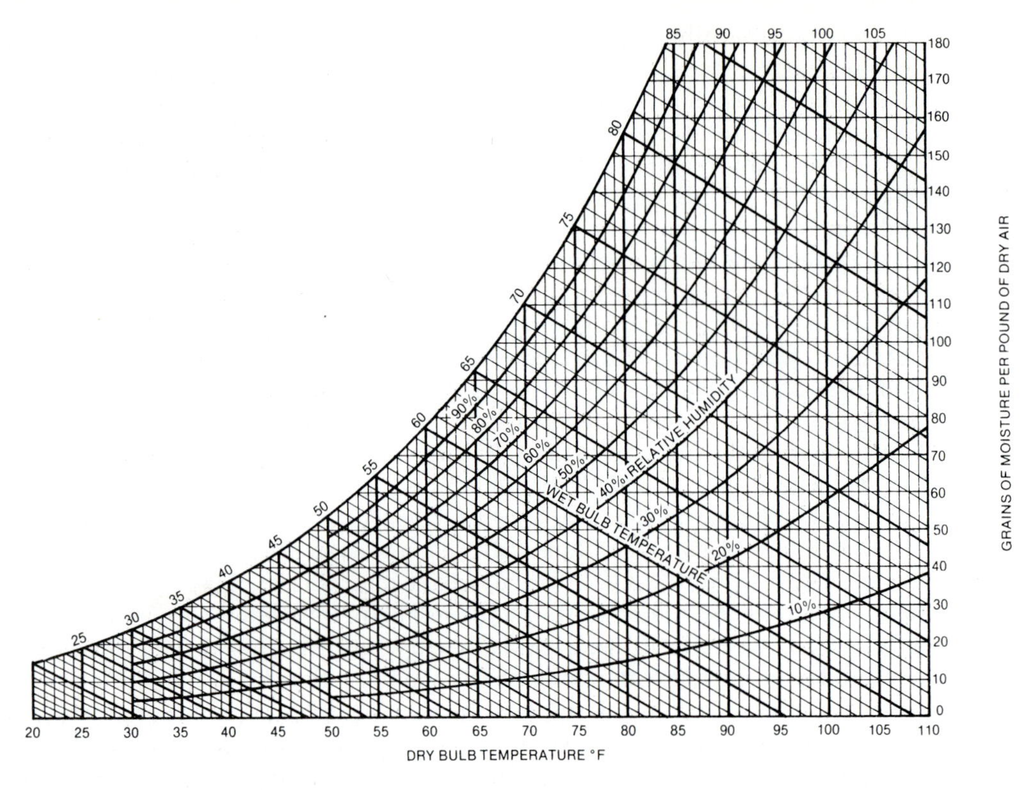

Figure 4-2 **The relationship of dry bulb temperature, wet bulb temperature, and relative humidity. To determine the wet bulb temperature (the lowest cooling temperature possible), start with the dry bulb temperature (read on a standard thermometer) along the lower horizontal axis, follow the vertical line above that temperature up to its intersection with the curve for the relative-humidity level in the outside air, move from this intersection along the diagonal line toward the upper left corner of the chart until the leftmost curve is reached, and read the wet bulb temperature from this leftmost curve.** ***(Graph courtesy of Acme Engineering & Manufacturing Corp., Muskogee, OK.)***

Passive Ventilator Cooling

Until the 1950s, all greenhouses were glass and were cooled by passive air movement through ventilators. Ventilators were located on both roof slopes adjacent to the ridge and on both sides of the greenhouse. The combined roof ventilators had an opening area about equal to 10 percent of the total roof area. The combined side ventilators on single greenhouses were also equal to about 10 percent of the roof area. During the winter cooling phase, the south roof ventilator was opened in stages to meet cooling needs. As greater cooling was required, the north ventilator was opened in addition to the south ventilator. In the summer cooling phase, the south-side ventilator was opened first followed by the north-side ventilator. Air entered the side ventilators. As the incoming air moved across the greenhouse, it was warmed by sunlight and by mixing with the warmer greenhouse air. With the increase in temperature, the incoming air became lighter and rose up and out the roof ventilators. This set up a chimney effect that, in turn, drew in more air from the side ventilators. This system did not adequately cool the greenhouse. To compensate, the interior walls and floor were frequently syringed with water on hot days.

Beginning in 1954, a series of active cooling systems were introduced to achieve greater and more uniform cooling. These included the summer fan-and-pad and fog cooling systems and the winter convection-tube and HAF systems. These systems are considerably more expensive to install and operate than a passive ventilation system. During the past decade, engineers have revisited the passive system and have improved it to where it is now functionally and economically effective for a number of applications.

Many standard low-profile greenhouses in cooler climates today have roof ventilators equal to approximately 20 percent of the roof area. Models of these greenhouses are also available with roof ventilators equal to 40 percent of the roof area and are preferred in warmer climates. One company in The Netherlands is currently constructing a low-profile greenhouse with nearly 100 percent of the roof occupied by ventilators. Each roof slope has a ventilator row attached to the ridge and a second row attached to the gutter.

Similar changes have been made in film plastic greenhouses. Roof ventilators are available on gutter-connected greenhouses, as shown in Figure 2-14. Roll-up side ventilators are also available on Quonset and gutter-connected greenhouses (see Figure 2-13). Film plastic greenhouses equipped with roll-up sides and roof ventilators have far more opening than the earlier glass greenhouses. Retractable-roof, film plastic greenhouses are also available (see Figure 2-15), in which nearly 100 percent of the roof opens. While the roof ventilators on film plastic greenhouses have a double layer of plastic, the roll-up side ventilators and the retractable roofs are constructed from a single layer of plastic.

Modern glass and film plastic greenhouses with their larger side and roof ventilators have adequate passive ventilation to meet cooling needs in temperate climates for crops in general. In warmer climates, these greenhouses are currently being used for heat-tolerant crops such as bedding plants, baskets, garden mums, and foliage plants. It is possible to install a fog cooling system in these greenhouses if passive cooling is found to be inadequate. Fog lines can be installed overhead to cool the air that is passively moving through the greenhouse. The retractable-roof greenhouses are very new and their utility is still being explored. Their initial use was for crops tolerant of low temperatures, such as bedding plants, garden mums, and green (foliage) plants. More currently they are being used for poinsettia. It appears that the only limitation will be the tolerance of the crop to the full light intensity it will receive when the roof is retracted. African violet and gloxinia could pose a problem in this respect.

One advantage of passive cooling is obvious. It is cheaper to operate than active cooling systems. Initial construction cost is not necessarily cheaper; it costs about \$1.25 extra per square foot (\$13.45/m^2) of greenhouse floor area to add ventilators to a polyethylene greenhouse. An alternative fan-and-pad system costs between \$0.75 and \$1.25/ft^2 (\$8.07 and \$13.45/m^2) of floor area. Both passively and actively cooled greenhouses still require a winter convection-tube or HAF system for air distribution during heating and between heating and cooling. Bedding-plant growers with retractable-roof greenhouses often state that plants produced in this system are compact and hardier. They feel this is due to the full light intensity that is possible during warm days in late winter and early spring and to the cool temperatures that are possible during the early morning. This latter topic will be revisited in Chapter 12 in the section on DIF. A disadvantage for these greenhouses is the single layer of plastic on the roll-up ventilators and the retractable roof and the possibility of air leaks where these coverings close. These factors lead to extra heat consumption and offset part of the gains in cooling energy.

Active Summer Cooling Systems

Fan-and-Pad Systems

The main considerations of the fan-and-pad system, in the order in which they will be discussed are (1) the rate at which warm air must be removed from the greenhouse to allow cool air to be drawn in, (2) the types of pads used for evaporating water and their specifications, (3) the placement of fans, and (4) the path of the airstream.

Rate of Air Exchange The rate of air exchange is measured in cubic feet of air per minute (cfm) or cubic meters per minute (cmm). The National Greenhouse Manufacturers' Association (NGMA) indicates in its 1993 standards for ventilating and cooling greenhouses that a rate of removal of 8 cfm/ft^2 (2.5 cmm/m^2) of greenhouse floor is sufficient. This applies to a greenhouse under 1,000 ft (305 m) in elevation, with an interior light intensity not in excess of 5,000 foot-candles (fc) (53.8 klux) and a temperature rise of 7°F (4°C) from the pad to the fans. In warmer climates, it is advisable to remove one full greenhouse volume per minute. Since this volume can be tedious to calculate, it is easier to use the recommendation of Willits (1993), which calls for removal of 11 to 17 cfm/ft^2 (3.4 to 5.2 cmm/m^2) of floor area. The NGMA recommendation of 8 cfm has an origin at the time when greenhouse eaves and gutters were typically 8 ft (2.4 m) high. Many greenhouses today have gutters 12 and 14 ft (3.7 and 4.3 m) high.

The rate of air removal from the greenhouse must increase as the elevation of the greenhouse site increases. Air decreases in density and becomes lighter with increasing elevation. The ability of air to remove solar heat from the greenhouse depends upon its weight and not its volume. Thus, a larger volume of air must be drawn through the greenhouse at high elevations than is drawn through at low elevations in order to have an equivalent cooling effect. Listed in Table 4-1 are factors (F_{elev}) used to correct the rate of air removal for elevation.

The rate of air removal is also dependent upon the light intensity in the greenhouse. As light intensity increases, the heat input from the sun increases,

Table 4-1 Factors Used to Correct the Rate of Air Removal for Elevation above Sea Level[1]

Elevation above Sea Level in ft (m)									
	Under 1,000 (300)	*1,000 (300)*	*2,000 (600)*	*3,000 (900)*	*4,000 (1,200)*	*5,000 (1,500)*	*6,000 (1,800)*	*7,000 (2,100)*	*8,000 (2,400)*
F_{elev}	1.00	1.04	1.08	1.12	1.16	1.20	1.25	1.30	1.36

[1]From National Greenhouse Manufacturers' Association (1993).

requiring a greater rate of air removal from the greenhouse. Factors (F_{light}) used to adjust the rate of air removal are listed in Table 4-2. An intensity of 5,000 fc (53.8 klux) is accepted as a desirable level for crops in general and is achieved with a coat of shading compound on the greenhouse covering or with a screen material above the plants in the greenhouse.

Solar energy warms the air as it passes from the pad to the exhaust fans. Usually, a 7°F (4°C) rise in temperature is tolerated across the greenhouse. If it becomes important to hold a more constant temperature across the greenhouse, that is, to reduce the rise in temperature, it will be necessary to raise the velocity of air movement through the greenhouse. Factors (F_{temp}) used for this adjustment are given in Table 4-3 for various permissible temperature rises.

Table 4-2 Factors Used to Correct the Rate of Air Removal for Maximum Light Intensity in the Greenhouse[1]

Light Intensity in fc (klux)									
	4,000 (43.1)	*4,500 (48.4)*	*5,000 (53.8)*	*5,500 (59.2)*	*6,000 (64.6)*	*6,500 (70.0)*	*7,000 (75.3)*	*7,500 (80.1)*	*8,000 (86.1)*
F_{light}	0.80	0.90	1.00	1.10	1.20	1.30	1.40	1.50	1.60

[1]From National Greenhouse Manufacturers' Association (1993).

Table 4-3 Factors Used to Correct the Rate of Air Removal for Given Pad-to-Fan Temperature Rises[1]

Temperature Rise in °F (°C)							
	10 (5.6)	*9 (5.0)*	*8 (4.4)*	*7 (3.9)*	*6 (3.3)*	*5 (2.8)*	*4 (2.2)*
F_{temp}	0.70	0.78	0.88	1.00	1.17	1.40	1.75

[1]From National Greenhouse Manufacturers' Association (1993).

Table 4-4 **Factors Used to Correct the Rate of Air Removal for Given Pad-to-Fan Distances**[1]

Pad-to-Fan Distance in ft (m)	20 (6.1)	25 (7.6)	30 (9.1)	35 (10.7)	40 (12.2)	45 (13.7)	50 (15.2)	55 (16.8)
F_{vel}	2.24	2.00	1.83	1.69	1.58	1.48	1.41	1.35

Pad-to-Fan Distance in ft (m)	60 (18.3)	65 (19.8)	70 (21.3)	75 (22.9)	80 (24.4)	85 (25.9)	90 (27.4)	95 (29.0)	100 (30.5) and over
F_{vel}	1.29	1.24	1.20	1.16	1.12	1.08	1.05	1.02	1.00

[1]From National Greenhouse Manufacturers' Association (1993).

The pad and fans should be placed on opposite walls. These walls may be the ends or the sides of the greenhouse. The distance between the pad and the fans is an important consideration in determining which walls to use. A distance of 100 to 200 ft (30 to 61 m) is best. Distances greater than 200 ft (61 m) can result in higher temperature rises across the greenhouse than are desired. When the distance is reduced below 100 ft (30 m), the cross-sectional velocity of air movement becomes lower and the air often develops a clammy feeling. This situation must be compensated for by increasing the size of the exhaust fans or, in other words, the velocity of air movement. This increases the cost of the system. Factors (F_{vel}) used to compensate for this point are listed in Table 4-4.

It is now possible to calculate the rate of air removal required for a specific greenhouse by using the factors given in Tables 4-1 through 4-4. First, determine the rate of air removal required for a greenhouse under standard conditions using the following equation, where L and W represent the greenhouse length and width, respectively. This equation calls for the removal of 8 cfm/ft^2 (2.5 cmm/m^2) of floor area:

$$\text{standard cfm} = L \times W \times 8 \tag{4-1}$$

or

$$\text{standard cmm} = L \times W \times 2.5 \tag{4-2}$$

Now, correct the standard rate of air removal by multiplying it by the larger of the following two factors, F_{house} or F_{vel}. F_{vel} is read directly from Table 4-4. F_{house} is calculated as follows:

$$F_{house} = F_{elev} \times F_{light} \times F_{temp}$$

Thus, the final capacity of the exhaust fans must be

$$\text{total cfm} = \text{standard cfm} \times (F_{\text{house}} \text{ or } F_{\text{vel}}) \tag{4-3}$$

or

$$\text{total cmm} = \text{standard cmm} \times (F_{\text{house}} \text{ or } F_{\text{vel}}) \tag{4-4}$$

Next, the size and number of exhaust fans must be selected. The collective capacity of the fans should be at least equal to the rate of air removal required and should be rated to do so at a static water pressure of 0.1 inch (30 Pa). If slant-wall-housing fans (with the fan outside the louvers) are used, the fans should be rated at a static water pressure of 0.05 inch (15 Pa). The static pressure figure takes into account the resistance the fans meet in drawing air through the pad and the fan itself. Air-delivery ratings for various sizes of fans are listed in Table 4-5. Fans

Table 4-5 **Air-Delivery Ratings and Required Pad Areas for Various Sizes of Steel Fans**[1]

			Pad Area per Fan (ft²)		
Fan Size (in)	*Horsepower (hp)*	*cfm at 0.1 Inch Static Pressure*	*Excelsior*	*Cellulose (4-inch)*	*Cellulose (6-inch)*
24	0.25	4,500	30	18	13
	0.33	5,700	38	23	16
	0.50	6,500	43	26	19
	0.75	7,600	51	30	22
30	0.33	7,400	49	30	21
	0.50	8,800	59	35	25
	0.75	10,200	68	41	29
36	0.33	8,800	59	35	25
	0.50	10,600	71	43	31
	0.75	12,700	85	51	37
	1.00	14,200	95	57	41
42	0.50	12,500	84	50	36
	0.75	15,000	100	60	43
	1.00	16,800	112	68	48
48	0.50	14,700	98	59	42
	0.75	17,800	119	72	51
	1.00	19,600	131	78	56
54	1.00	22,900	153	92	66
	1.50	25,800	172	104	74

[1]The data in the first three columns are from Acme Engineering and Manufacturing Corp., Muskogee, OK.

should not be spaced more than 25 ft (7.6 m) apart. If the end of the greenhouse is 60 ft (18.3 m) wide, a minimum of three fans will be necessary. The required capacity of each fan in this example can be determined by dividing 3 into the total cfm (or cmm) of air removal required. It is then a matter of finding fans in the table that are rated for this performance level. These fans should be evenly spaced along the end of the greenhouse, at plant height if possible, to guarantee a uniform flow of air through the plants.

Pad Types and Specifications Originally, excelsior (wood fiber) pads were used that were 1 to 1.5 in (2.5 to 4 cm) thick. They had to be replaced annually. Excelsior pads must be encased in a 1-inch-by-2-inch (2.5-cm-by-5-cm) mesh wire frame for support. Other types of pads, including aluminum fiber, glass fiber, and plastic fiber, are available today. Most cooling pads installed today are constructed of cross-fluted cellulose material (Figure 4-3a). These pads have the appearance of corrugated cardboard. They can last 10 years if properly handled. They should be protected from beating rain and heavy water streams and should be moved only if dry. These pads have sufficient strength to stand alone without a mesh wire support. The cellulose is impregnated with soluble antirot salts, rigidifying saturants, and wetting agents to give it lasting quality, strength, and wettability. Although cellulose pads are more expensive initially, they are cheaper over their 10-year useful life expectancy than are their predecessor, the excelsior pads.

Cross-fluted cellulose pads come in units of 1 foot (30 cm) wide and 2, 4, 6, or 12 in (5, 10, 15, or 30 cm) thick. Heights are available in 1-foot (30-cm) increments. Units are oriented vertically so that each adds 1 foot (30 cm) to the length of the overall greenhouse pad. Four inches is the most common thickness used today. Pads 6 in thick are useful in walls that are too small to accommodate the greater pad area required for a 4-inch-thick pad. A 4-inch-thick pad will accommodate an air intake of 250 cfm/ft^2 (75 cmm/m^2) of pad, while a 6-inch-thick pad will accommodate 350 cfm/ft^2 (105 cmm/m^2) of pad. Pads 12 in thick would be used in excessively hot and humid locations. The required area of 4- and 6-inch-thick cellulose pads is only 60 and 43 percent of the area required for excelsior pads, respectively. Excelsior pads support an airflow rate of 150 cfm/ft^2 (45 cmm/m^2).

The total area of pad required is determined by dividing the volume of air that must be removed from the greenhouse in one minute by the volume of air that can be moved through a square foot of pad in one minute. Alternatively, the pad area may be read directly from Table 4-5. The cooling pad should extend the entire length of the wall of the greenhouse in which it is installed, to ensure that all plants receive cooled air. The height of the pad is determined by dividing the total area of the pad by the length of the pad. Pads are most often placed immediately inside the side or end wall. The pad wall should be equipped with ventilators exterior to the pad to permit air entry during hot weather and for sealing off the outside air during cooler spring and fall nights. In this case, the ventilator arms and gears are located exterior to the greenhouse (Figure 4-4). Exhaust fans must be located in the wall opposite the pad to ensure that an even blanket

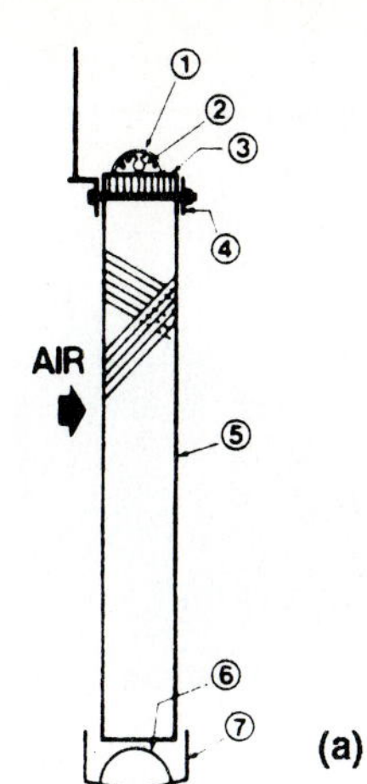

1. Impingement cover
2. Water distribution pipe
3. Water distribution pad
4. Support flashing and bolt
5. Cross-fluted cellulose pad
6. Spacer
7. Gutter

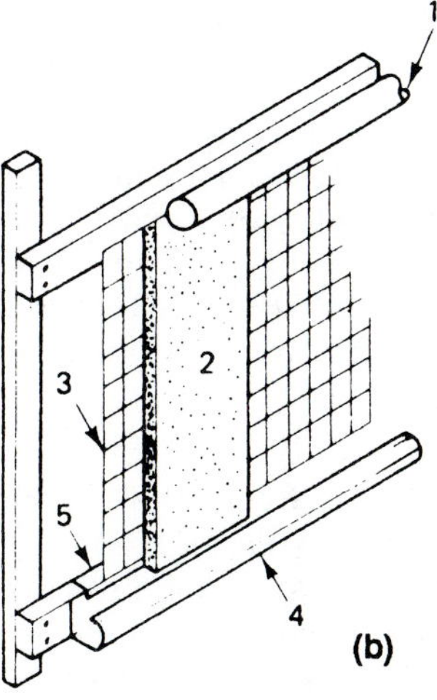

1. Water distribution pipe
2. Excelsior pad
3. Welded wire frame
4. Water return gutter
5. Galvanized flashing

6. Water distribution pipe
7. Excelsior pad
8. Water return gutter
9. Pump
10. Water inlet with float valve
11. Sump
12. Bleed-off valve

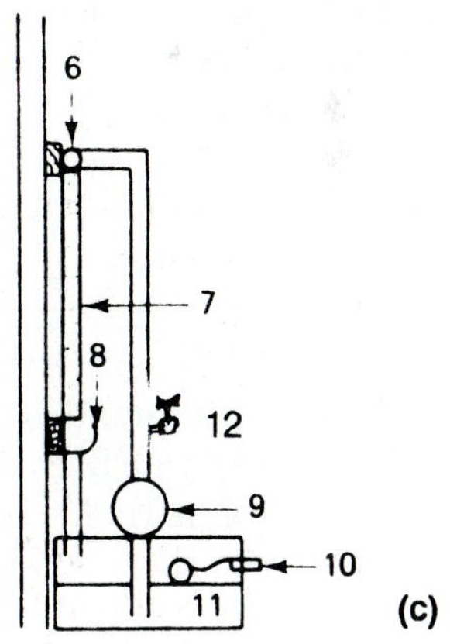

Figure 4-3 **A diagram of (a) the components of a cross-fluted cellulose pad system for evaporative cooling, (b) the components of an alternative excelsior pad system, and (c) the water-distribution system for either cooling pad system, including sump, float valve, and pump.**

Figure 4-4 **An evaporative cooling system arrangement with the pad located inside the greenhouse and the ventilator mechanism outside the greenhouse. This system permits cooling on warm autumn days when the evenings are too cold for the side wall to be left open.** ***(Photo courtesy of J. W. Love, Department of Horticultural Science, North Carolina State University, Raleigh, NC 27695-7609.)***

of cool air passes through all parts of the greenhouse. Pads and fans should be at plant height to keep the cooling air in the plants.

Water must be delivered to the top of a 4-inch (10-cm) -thick pad at the rate of 0.5 gpm per linear foot of pad (6.2 L/min/m of pad). For pad lengths of 30 to 50 ft (9.1 to 15.2 m), a 1.25-inch (32-mm) water-distribution pipe is required, while for lengths of 50 to 60 feet (15.2 to 18.3 m), a 1.5-inch (38-mm) pipe is needed. Sixty feet (18.3 m) is the longest recommended pipe length. A 120-foot (36.6-m) pad length could be serviced from a water supply at the midpoint supplying two 60-foot (18.3-m) distribution pipes. At every 3 in (7.6 cm), ⅛-inch (3-mm) holes should be made in the pipe.

The flow rate for a 6-inch (15-cm) pad is 0.75 gpm per linear foot of pad (9.3 L/min/m of pad). A 1.25-inch (32-mm) distribution pipe is used for pads 30 ft (9.1 m) and shorter, while a 1.5-inch (38-mm) pipe is used for 30- to 50-foot (9.1- to 15.2-m) pad lengths. The longest pipe length recommended is 50 ft (15.2 m). Again, ⅛-inch (3-mm) holes are spaced 3 in (7.6 cm) apart in these distribution pipes.

Holes in the distribution pipes for cross-fluted cellulose pads are oriented upward. An impingement cover is placed over the distribution pipe. Water squirting upward from holes in the distribution pipe strikes the inner side of the impingement cover and is dispersed. Half of a 4-inch (10-cm) plastic pipe provides a good impingement cover. Deflected water drips onto a distribution pad that is

2 in (5 cm) high and the thickness of the cellulose pad below it. This pad further disperses the water to more thoroughly wet the top of the cellulose pad. It is important that all of the pad be wet. There is less resistance to the flow of air through dry pad; thus, air will channel through dry areas and reduce the overall effectiveness of the pad. A gutter at the base of the pad collects water and permits it to flow to a sump, where it is pumped back to the top of the pad. Between the gutter and the base of the pad is a spacer. Half of a 4-inch (10-cm) plastic pipe provides a good spacer. The sump volume should be 0.75 gal/ft^2 (30.5 L/m^2) of 4-inch-thick pad and 1 gal/ft^2 (40.7 L/m^2) of 6-inch-thick pad. These sump volumes are designed for an operating water level at half the depth of the tank and will provide room to accommodate water returning from the pad when the system is turned off.

Water should be delivered to the top of an excelsior pad at the rate of 0.33 gpm per linear foot (4.1 L/min) of pad, regardless of the height of the pad. Since all water will return to the sump tank when the system is turned off, a sump capacity of 1.5 gal/ft (19 L/m) of pad is required.

As much as 1 gallon of water per minute can evaporate from 100 ft^2 of pad (0.4 L/min from 1 m^2 of pad) on a hot, dry day. Therefore, a water line with a float valve should be plumbed into the sump tank to automatically maintain the water level. As water evaporates from the pad surface, salts in the water are left behind. If this occurs for long, a white salt deposit will solidify on the pad whenever the system is turned off. Depending upon the salt content of the water used, it may be necessary to bleed off 1 to 2 percent of the recirculating water to avoid a salt buildup. A 3/8-inch (9.5-mm) bleed-off valve can be located on the pump discharge pipe. It should be adjusted to a flow rate that just eliminates signs of scale on the pad. Scale buildup in an excelsior pad is not as noticeable, since this type of pad is used for only one season and water is spilled from the pad as it passes downward.

The fan-and-pad system can be automated or operated manually. When cooling is demanded in the automated system, the ventilator over the pads is opened and the exhaust fans are turned on in stages. If this does not satisfy the cooling requirement, the pump providing water to the pads is turned on. When the cooling requirement is satisfied, the system turns off step-by-step in the reverse order.

Algae may build up in cross-fluted cellulose pads after two or three years. Algae buildup does not destroy the cellulose, but it can plug passages in the pad. A 1 percent solution of sodium hypochlorite (bleach) can be injected into the water supply line to the pad. This will provide the required 3 to 5 parts per million (ppm) free chlorine in the pad. As little as 30 gal (114 L) of solution per month can keep 100 linear feet (30 m) of 6-inch (15-cm) -thick pad free of algae. A problem with bleach is the rise in pH that it causes. The pH level should not go above 9.0 because it will soften the pad, nor should it drop below 6.0. Some growers inject hydrogen peroxide, which does not raise the pH level, into the water supply line. Chlorine and hydrogen peroxide break down rapidly and must periodically be applied. Small firms that cannot justify the injection equipment may spray the pads periodically with a chlorine solution. Biocides that can be added to the sump are also available for cleaning cooling water in various industrial applications. Oakite Biocide 20 (Oakite Products, Inc., 50 Valley Rd., Berkeley Heights, NJ 07922), when added once or twice per week to the sump, will control algae, bacteria, and fungi in greenhouse pads. It is used at an initial rate of 2.4 to 6.0 fl oz/1,000 gal (20 to 47 mL/1,000 L) and at a subsequent rate of 0.6 to 6.0 fl oz/1,000 gal (4 to 47 mL/1,000 L).

More recently, Agribrom tablets (Great Lakes Chemical Corp., P.O. Box 2200, West Lafayette, IN 47906), containing bromine and a lesser amount of chlorine, have been used. One injection system makes use of a homemade applicator constructed from 6-inch (15-cm) clear PVC pipe (Figure 4-5). A supply of tablets is maintained in the applicator. For removal of a heavy buildup of algae,

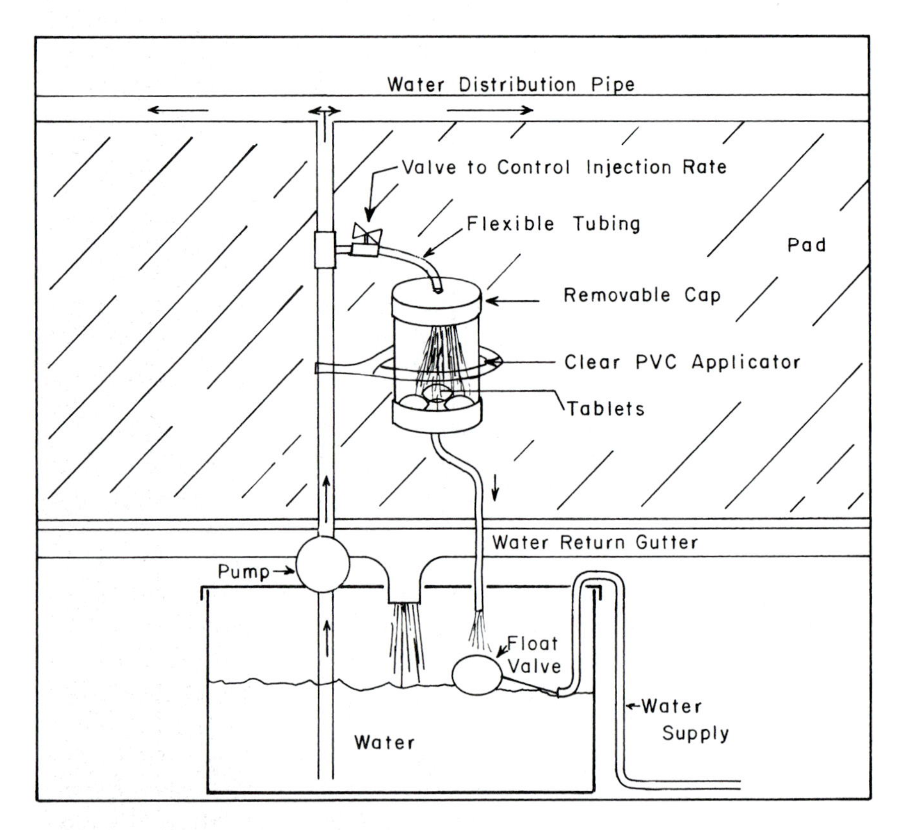

Figure 4-5 Design and installation of a bromine-and-chlorine applicator for control of algae in cooling pads. The main body of the applicator consists of a 1-foot (30-cm) long piece of clear PVC pipe to which one cap is cemented at the bottom and a second cap is pressed onto the top. Agribrom tablets containing bromine and chlorine can be added to the applicator as previously applied tablets dissolve by removing the top cap. The tablets rest on a wire screen. Each cap is drilled and fitted with a flexible tube. The upper tube is supplied with water from the main leading from the pump to the top of the pad and has a valve on it to control the flow rate. The lower tube permits water that has passed through the applicator and picked up bromine and chlorine to return to the sump tank or to the return gutter.

a concentration of 1 to 3 ppm of bromine should be used at the top of the pad. Following cleanup, a maintenance concentration of 0.1 to 1 ppm bromine is held. A bromine tester is available from Great Lakes Chemical Corporation for about $25. Concentrations can be controlled by adjusting the water inlet valve below the applicator.

Horizontal pads are also being tried by growers. A horizontal screen is constructed outward from the greenhouse. One of a variety of materials, including gravel, vermiculite, or excelsior, is placed on the screen to serve as an evaporative surface, yet permit air percolation. Mist nozzles keep the pad wet, and air is drawn from outside through the pad into the greenhouse. Several pads may be installed in a stack along one greenhouse wall, offering an economy of space. Another advantage is the long life expectancy of the pads, since more permanent types of materials can be used.

If the required pad area exceeds the area of the greenhouse wall, it is necessary to place it exterior to the greenhouse wall (Figure 4-6). The opening in the greenhouse should be at least half the area of the pad. The pad should be set back from the opening half of the distance by which the height of the pad exceeds that of the opening. Ideally, the extra height of the pad should be equally divided above and below the opening. It is important that the pad be connected to the greenhouse on the top and ends by a transparent covering material to ensure that any air drawn through the pad enters the greenhouse.

Figure 4-6 **An excelsior cooling pad exterior to the greenhouse. This arrangement can accommodate a pad larger than the wall of the greenhouse. The pad is set back from the wall at a distance equal to at least half the excess height of the pad over the wall and is connected to the greenhouse by a transparent covering to ensure that air entering the greenhouse comes through the pad.**

Fan Placement Whenever possible, it is best to place the fans on the leeward side of the greenhouse and the pads on the side toward the prevailing winds, so that the winds will assist rather than counteract the cooling system. If fans exhaust into the windward side, their capacity should be increased 10 percent or more. When two or more houses are located adjacent to one another, factors more important than wind direction dictate the placement. Fans from one greenhouse should not exhaust warm moist air toward the pads of an adjacent greenhouse unless it is located at least 50 ft (15.2 m) away.

When fans are located in adjacent walls of greenhouses located within 15 ft (4.6 m) of each other, they should be alternated so that they do not blow directly against each other. Adjacent service buildings can also present a problem. There must be a clearance of 1.5 fan diameters between the fan and adjacent obstacles. If this is not possible, special roof-mounted fans should be installed.

A waterproof housing should enclose the fan to protect it from the elements. Air-activated louvers give protection on one side. It is imperative that a screen or welded-wire guard be placed on the other side of the fan to protect workers and visitors from serious injury.

The Airstream The pads should be located at and slightly above plant height in order to bring the cool air in on the plants. Because of resistance of the foliage and plant supports, as well as the rising temperature, the airstream will rise at an angle of 7° (1 foot in every 8) and will soon pass over the plants, leaving a pocket of hot air below at the plant height. If air is drawn across ridge-and-furrow or gutter-connected greenhouses, the gutters will keep the flow of air down on the plants. If air is drawn longitudinally along the length of the greenhouse, it will rise. In this case, transparent film or rigid plastic vertical baffles should be installed in the gable of the greenhouse, perpendicular to the airstream, to direct the flow of air down to the plants. Baffles should be installed every 30 ft (9.1 m). The bottom of the baffle should be well above the plant height to permit passage of air.

If pads are located near the floor and the benches are tall, considerable air may pass beneath the benches, where little benefit occurs. In this case, baffles should be placed beneath the benches near the pads.

The situation encountered in a greenhouse more than 200 ft (61 m) long and less than 100 ft (30 m) wide can be remedied by placing pads at each end of the greenhouse and exhaust fans in the side walls or roof halfway between the ends. In this situation, the greenhouse is cooled by the equivalent of two systems, each half the length of the greenhouse. When roof fans are used, a transparent baffle should be installed about 5 ft (1.5 m) before the fan and just above plant height to force the cooling air down into the plants.

EXAMPLE PROBLEM

The following example illustrates the calculations involved in designing an evaporative cooling system. Consider a single greenhouse 50 ft (15 m) wide and 100 ft

(30 m) long located at an elevation of 3,000 ft (915 m). The greenhouse has a moderate coat of shading compound on it; thus, the maximum light intensity inside is 5,000 fc (53.8 klux). A 7°F (4°C) rise in temperature can be tolerated from pad to fans. Step-by-step calculations for developing a 4-inch (10-cm) -thick cross-fluted cellulose cooling system for this greenhouse follow.

1. Multiply the greenhouse floor width by the length and by 8 to determine the quantity of air to remove per minute under standard conditions:

$$\text{cfm}_{\text{standard}} = L \times W \times 8$$
$$= 50 \times 100 \times 8 = 40{,}000 \text{ cfm}$$

or

$$\text{cmm}_{\text{standard}} = L \times W \times 2.5$$
$$= 15 \times 30 \times 2.5 = 1{,}125 \text{ cmm}$$

2. Determine a factor for the house (F_{house}) by multiplying the three factors together: elevation, light intensity inside the greenhouse, and temperature rise from pad to fans. These factors are found in Tables 4-1 through 4-3, respectively:

$$F_{\text{house}} = F_{\text{elev}} \times F_{\text{light}} \times F_{\text{temp}}$$
$$= 1.12 \times 1.0 \times 1.0 = 1.12$$

3. Look up the factor for velocity (F_{vel}) in Table 4-4. Select two opposite walls that are 100 to 200 ft (30 to 61 m) apart or (2) as close to 100 ft (30 m) apart as possible, for installation of the pad and fans. The end walls, which are 100 ft (30 m) apart, should be used in this example:

$$F_{\text{vel}} = 1.00$$

4. Multiply the standard cfm value from step 1 by either F_{house} or F_{vel}, using whichever factor is larger—F_{house} in this case. This is the volume of air to be expelled from the greenhouse each minute:

$$\text{cfm}_{\text{adjusted}} = \text{std cfm} \times F_{\text{house}}$$
$$= 40{,}000 \text{ cfm} \times 1.12 = 44{,}800 \text{ cfm}$$

or

$$\text{cmm}_{\text{adjusted}} = 1{,}125 \text{ cmm} \times 1.12 = 1{,}260 \text{ cmm}$$

5. Determine the number of fans needed. Since they should not be over 25 ft (7.6 m) apart, divide the length of the wall housing the fans by 25 (7.6):

$$\frac{50 \text{ ft}}{25 \text{ ft}} = 2 \text{ fans}$$

or

$$\frac{15 \text{ m}}{7.6 \text{ m}} = 2 \text{ fans}$$

6. Determine the size of the fans needed by dividing the adjusted cfm of air to be removed (from step 4) by the number of fans needed:

$$\frac{\text{cfm}_{\text{adjusted}}}{\text{number of fans}} = \text{size of fan}$$

$$\frac{44{,}800 \text{ cfm}}{2} = 22{,}400 \text{ cfm per fan}$$

or

$$\frac{1{,}260 \text{ cmm}}{2} = 630 \text{ cmm per fan}$$

7. Purchase two fans of the size determined in step 6 and space them equidistantly on one end of the greenhouse. If the fans were to be purchased from the manufacturer of the equipment listed in Table 4-5, two 54-inch fans with 1-hp motors would be selected.
8. The pad area is determined next. One square foot of pad is required for each 250 cfm (1 m^2 per 75 cmm) of fan capacity. Divide the capacity of the required fan (22,400 cfm) by 250 cfm to arrive at a required pad area of 90 ft^2 per fan (630 cmm divided by 75 cmm = 8.4 m^2 of required pad area per fan). Since there are two fans, a total of 180 ft^2 (16.8 m^2) of pad is required. Approximately the same value could be read directly from Table 4-5.
9. The pad must cover the width of the wall in which it is to be installed—50 ft (15 m) in this example. The height of the pad is determined by dividing the total pad area by its width. A 4-foot (1.2-m) tall pad should be purchased:

$$\text{pad height} = \frac{\text{pad area}}{\text{pad width}}$$

$$= \frac{180 \text{ ft}^2}{50 \text{ ft}} = 3.6 \text{ ft}$$

or

$$\text{pad height} = \frac{16.8 \text{ m}^2}{15 \text{ m}} = 1.1 \text{ m}$$

10. The pump capacity is equal to 0.5 gpm multiplied by the length of the pad in feet (6.2 L/min/m of pad) and must be selected to have this flow rate for the given head under which it must operate. The head is the distance from the water surface in the sump to the top of the pads:

$$\text{pump capacity} = 0.5 \text{ gpm} \times 50 \text{ ft} = 25 \text{ gpm}$$

or

$$\text{pump capacity} = 6.2 \text{ L/min} \times 15 \text{ m} = 93 \text{ L/min}$$

11. The sump size is equal to 0.75 gal/ft^2 (30.5 L/m^2) of pad:

$$\text{sump volume} = 0.75 \text{ gal} \times 200 \text{ ft}^2 = 150 \text{ gal}$$

or

$$\text{sump volume} = 30.5 \text{ L} \times 18.6 \text{ m}^2 = 567 \text{ L}$$

Fog Cooling Systems

The second active summer cooling alternative is fog cooling. The cost of installation of a fog cooling system relative to a fan-and-pad system can range from less (when pure water is available) to more (when extensive filtering and chemical treatment of water is needed). However, the subsequent cost of electrical power is much less for the fog cooling system. Fog cooling involves dispersion of water particles in greenhouse air, where they extract heat from the air as they evaporate. The speed of evaporation of water and, consequently, the rate of cooling of air increase proportionately as water droplet size decreases. Mist droplets are in the range of 1,000 microns (0.040 inch) in diameter. If a cup of water were converted to mist, it would have 400 times as much surface area and would evaporate 400 times faster than the same water left in the cup. Mist droplets are large and will settle out of air, wetting surfaces of plants, soil, and people. In contrast, fog droplets are 40 microns or smaller (0.0016 inch). Their surface area and rate of evaporation is 10,000 times greater than the same volume of water in a cup. These droplets stay suspended in air while they evaporate to cool the air. This occurs without water condensing out on surfaces.

Greenhouse fog cooling systems are available that can convert 99.5 percent of the water in the system to 40 microns or smaller, with an average droplet size less than 10 microns (0.0004 inch) (Figure 4-7). These droplets evaporate at

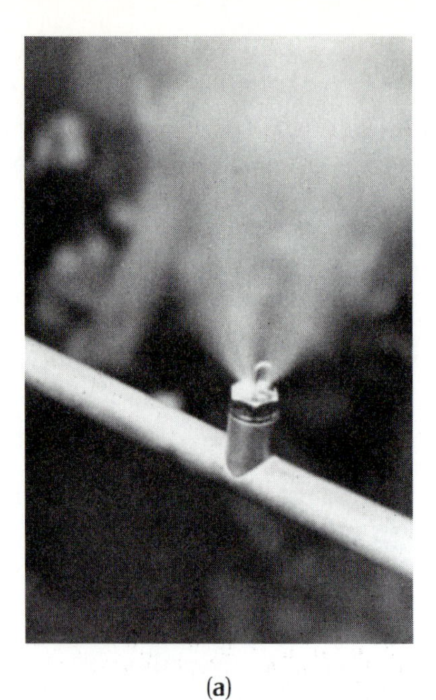

(a)

(b)

(c)

Figure 4-7 (a) A nozzle emitting 10-micron (.0004-inch) fog droplets in an evaporative greenhouse cooling system. (b) A fog system being used in a propagation greenhouse for cooling and for maintaining moisture in cuttings. (c) Water treatment and pumping apparatus for a fog cooling system. *(Photos courtesy of Mee Industries, Inc., 4443 N. Rowland Ave., El Monte, CA 91731.)*

40,000 times the speed with which water evaporates from a cup. With such a rapid evaporative response, air can be cooled at nearly 100 percent efficiency. The result is that wet bulb temperatures can essentially be obtained. Pumps used to provide fog droplets can typically operate at 1,000 psi (6.9 MPa) and possibly at pressures up to 1,500 psi (10.3 MPa).

The fog cooling system can be used in greenhouses built to be cooled by ventilators alone. Fog nozzles are spaced above plants throughout the greenhouse. Fog comes on intermittently to cool air that has entered the greenhouse through side wall ventilators. As the humid cooled air begins to warm and leave the greenhouse through roof ventilators, more outside air is drawn in and, in turn, is cooled by subsequent fog.

Greenhouses equipped with exhaust-fan cooling lend themselves well to fog cooling. A line of fog nozzles is installed just inside the inlet ventilators. Exhaust fans on the opposite wall draw outside air in through the open ventilators and then through the fog, where it is cooled. Only about half the exhaust-fan capacity, 4 to 5 cfm/ft^2 (1.2 to 1.5 cmm/m^2) of floor area, of fan-and-pad systems is used. If there were no more to the system, air would rise in temperature as it crossed the greenhouse, as happens in a fan-and-pad system. To prevent this, a second row of fog nozzles is installed further inside the greenhouse, parallel to the first row.

Water quality is extremely important. Particles of sand or clay can clog the fog nozzles. Multiple filters capable of screening down to 5-micron (0.0002-inch) particles are used. To prevent scale formation, carbonates and bicarbonates are removed. Sulfur and iron can support growth of slime organisms, which also plug nozzles. All of these factors are assessed, and appropriate filters and chemical treatment are provided by the firms that supply fog cooling systems.

Various control systems are used for fog cooling. Timers provide the simplest form of control. A 24-hour timer is used to select the time of day during which cooling is needed, which is usually the daylight hours. The circuit continues through a recycle timer that is typically set to apply fog from 30 seconds to 4 minutes out of each cycle of 1 to 20 minutes. More consistent control is achieved through the use of a humidistat. When temperature goes up in the greenhouse, humidity goes down. By holding a constant relative humidity, maximum cooling is achieved. The response time of the cooling system is much faster when a humidistat rather than a thermostat is used, since air can be cooled 20°F (11°C) or more in 30 seconds.

When fog cooling is used for growing crops to market stage, a relative-humidity level is set on the humidistat. This level is often in the range of 80 to 90 percent. However, growers fine-tune this setting by relating relative-humidity levels to responses of their crops over time. In this application, visible fog originating from the nozzles generally disappears for a few minutes before coming on again.

Fog is also used as a substitute for mist systems in cutting-propagation greenhouses. The object here is to stop water loss through transpiration by holding 100 percent relative humidity. In this case, fog is turned on just as the visible appearance of the previous fog pulse begins to disappear. Since fog cooling does not wet the foliage, less disease has been reported in greenhouses using this system.

Fog is used in a similar manner in seed-germination greenhouses. The goal here is to greatly reduce evaporation and transpiration, to the point where water is needed only at the frequency at which liquid fertilizer needs to be applied. In this way, the application of water between fertilizations is eliminated. The problem with watering, whether it occurs through mist or larger droplets, is the film of water that forms in the root substrate. This water film adversely reduces oxygen supply to the seeds and roots. A relative-humidity setting slightly lower than the 100 percent level used for cutting propagation is used in seed germination.

It is possible to fertilize seedlings, cuttings during rooting, and roots of plants in an aeroponic system by injecting nutrients into the water supplying the fog system. The system is then controlled to hold the air saturated so that moisture condenses on plant surfaces. Fog very effectively penetrates the plant canopy or root system, depositing a nutrient film on all upper as well as lower leaf and root surfaces for foliar uptake.

Advantages cited by greenhouse firms that have installed fog cooling include the following:

1. There is less electrical consumption, since the sum of the wattage of the fog pump and exhaust fans is less than that of the exhaust fans and pad water pumps in the fan-and-pad system.
2. Heat rise across the greenhouse is controlled.
3. Cooler average temperatures can be achieved across the greenhouse.
4. The system is a good substitute for the mist system in cutting-propagation greenhouses, where it uses less water and causes less disease.

Active Winter Cooling Systems

Convection-Tube Cooling

When the temperature set point for winter greenhouse cooling is reached, three events are activated simultaneously (Figures 4-8 and 4-9). An exhaust fan, located anywhere in the greenhouse, is turned on to create a vacuum. A louver is opened in a gable, through which cold air enters in response to the vacuum. A pressurizing fan in the end of the clear polyethylene convection tube turns on to pick up the cool air entering the louver, since the end of the convection tube is separated from the louvered inlet by 1 to 2 ft (0.3 to 0.6 m). Cold air under pressure in the convection tube shoots out of holes on either side of the convection tube in turbulent jets. The cold air mixes with the warm greenhouse air well above plant height. The cooled mixture of air, being heavier, gently falls to the floor, cooling the plant area.

The pressurizing fan directing incoming cold air into the convection tube must be capable of moving at least the same volume of air as the exhaust fan. If it moves less, excess incoming cold air will drop to the ground at the point of entry and cause a cold spot. When cooling is not required, the inlet louver closes, and

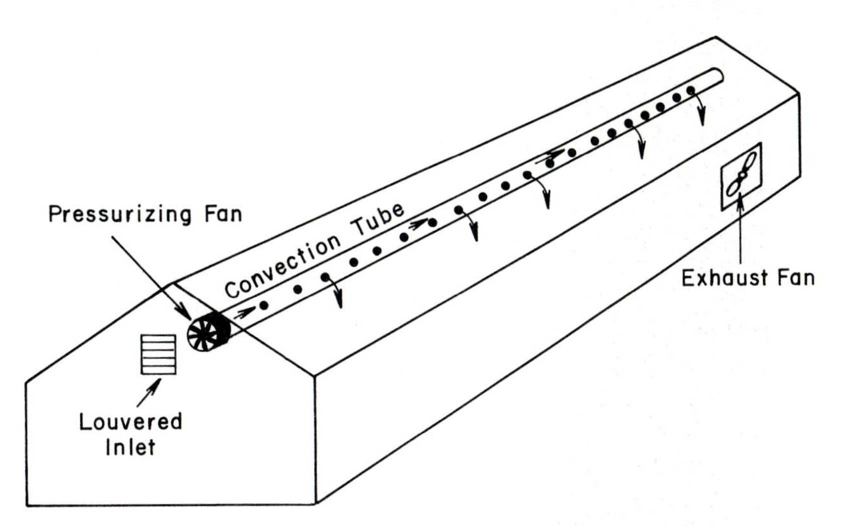

Figure 4-8 Diagram of a greenhouse showing the components of a convection-tube winter cooling system. When cooling is required, a thermostat activates an exhaust fan, opens the louvered inlet, and turns on the pressurizing fan. Cold air enters the louver and is directed down a transparent polyethylene convection tube by a pressurizing fan. Jets of cold air leave the tube through holes along both sides of the tube and thoroughly mix with warm greenhouse air before reaching the plants.

Figure 4-9 Air inlet components of a convection-tube winter cooling system, including the louvered inlet, a polyethylene convection tube located a short distance from the louvered inlet, and a pressurizing fan to direct air into the tube under pressure.

the pressurizing fan continues circulating air within the greenhouse. This step replaces the horizontal airflow (HAF) circulation system, but requires more power.

Under standard conditions, a volume of 2 cfm of air should be removed from the greenhouse for each square foot of floor area (0.61 cmm/m^2 of floor). (Remember that a minimum of 8 cfm of air for each square foot of floor (2.5 cmm/m^2) is required for summer cooling.) The volume obtained by multiplying the floor area by 2 would therefore define the capacity of the exhaust fan in terms of cubic feet of air movement per minute.

Various published specifications call for as little as 1.5 cfm and as much as 4 cfm of air to be exhausted for each square foot of floor area (0.46 and 1.22 cmm/m^2 of floor area). The high-capacity system costs more to set up but can be operated earlier in the fall and later in the spring to extend the winter cooling season. This can have an advantage, since frosts usually occur during these extension periods. A high-capacity convection-tube system eliminates the necessity of switching back and forth between the summer fan-and-pad and the winter fan-tube systems at these times.

When convection-tube ventilation is used, standard conditions specify a maximum inside temperature of 15°F (8°C) above the outside temperature. The temperature inside the greenhouse can become adversely high on a winter day when the sun is shining, even though the outside temperature is below the desired level. The convection-tube cooling system is designed to reduce the internal temperature to within 15°F (8°C) of the outside temperature.

If a lower inside temperature is desired, cold air must be introduced into the greenhouse at a greater rate. The compensating factors to be used in this case are given in Table 4-6. As in the case of the summer cooling system, standard conditions also specify an elevation under 1,000 ft (305 m) and a maximum interior light-intensity of 5,000 fc (53.8 klux). If other elevation or light intensity specifications are desired, factors must be selected from Tables 4-1 and 4-2 and used to correct the rate of air entry.

Convection tubes are conventionally oriented from end to end in the greenhouse. Each convection tube can be used to cool up to 30 feet (9.1 m) of greenhouse width, although it is desirable to use two tubes for greenhouses 30 ft (9.1 m) wide. One tube placed down the center of the house will cool houses up to 30 ft

Table 4-6 Factors (F_{winter}) for Adjusting the Standard Rate of Air Removal in a Winter Greenhouse Cooling System for the Temperature Difference between the Inside and Outside of the Greenhouse[1]

Greenhouse Temperature above Outdoor Temperature in °F (°C)										
	18 (10.0)	*17 (9.4)*	*16 (8.9)*	*15 (8.3)*	*14 (7.8)*	*13 (7.2)*	*12 (6.7)*	*11 (6.1)*	*10 (5.6)*	*9 (5.0)*
F_{winter}	0.83	0.88	0.94	1.00	1.07	1.15	1.25	1.37	1.50	1.67

[1]From National Greenhouse Manufacturers' Association (1993).

(9.1 m) in width. Greenhouses 30 to 60 ft (9.1 to 18.3 m) wide are cooled by two tubes placed equidistantly across the greenhouse. Holes along the tube exist in pairs on the opposite vertical sides. The holes vary in size according to the volume of greenhouse to be cooled. The number and diameter of tubes needed to cool a greenhouse can be determined from Table 4-7. If two or more tubes are needed, they should be of equal size and should be spaced evenly across the greenhouse. Recommendations in Table 4-7 are based on an airflow rate of approximately 1,700 cfm/ft^2 (518 cmm/m^2) of cross-sectional area in the tube. When the greenhouse is large and the required number of 30-inch (76-cm) -diameter tubes becomes cumbersome, tubes may be installed with air inlets in both ends. These inlets double the amount of cool air that can be brought in through a single tube.

The winter cooling system requirements should be taken into consideration when fans are ordered for the summer cooling system. In this way, one or more of the summer fans could be used for the winter exhaust fan requirement. Fans used for the summer system should all be of equal size or at least nearly equal. However, one fan could be purchased with a two-speed motor that provides half its capacity at the lower speed.

EXAMPLE PROBLEM

Determine the winter cooling specifications for a greenhouse measuring 50 ft (15 m) wide by 100 ft (30 m) long and situated at an elevation of 3,000 ft (915 m). The maximum interior light intensity anticipated is 5,000 fc (53.8 klux), and the desired interior-to-exterior temperature difference is 15°F (8°C).

1. The capacity of the exhaust fan is equal to 2 cfm (0.61 cmm) times the greenhouse floor area under standard conditions:

$$\begin{aligned} cfm_{standard} &= 2 \times \text{length} \times \text{width} \\ &= 2 \times 100 \times 50 = 10{,}000 \text{ cfm} \end{aligned}$$

   or

$$cmm_{standard} = 0.61 \times 30 \times 15 = 275 \text{ cmm}$$

2. Correct the exhaust-fan capacity just calculated for deviations from standard conditions. The only deviation in the sample problem is the elevation of 3,000 ft (915 m), which has an F_{elev} value of 1.12 (from Table 4-1). An exhaust fan with a capacity of 11,200 cfm (308 cmm) at a static water pressure of 0.1 inch (30 Pa) is needed:

$$\begin{aligned} cfm_{adjusted} &= cfm_{standard} \times F_{winter} \times F_{elev} \times F_{light} \\ &= 10{,}000 \times 1.0 \times 1.12 \times 1.0 = 11{,}200 \text{ cfm} \end{aligned}$$

Table 4-7 **Number (*N*) and Diameter (*D*) of Air-Distribution Tubes Required for Winter Cooling of Greenhouses of Various Widths and Lengths**[1]

Greenhouse Length in ft (m)

Greenhouse Width in	50 (15)			100 (30)			150 (46)			200 (61)						250 (76)					
													IBE						IBE		
		D			D			D			D			D			D			D	
ft (m)	N	in	cm	N	in	cm	N	in	cm	N	in	cm	N	in	cm	N	in	cm	N	in	cm
15 (4.6)	1	18	46	1	18	46	1	24	61	1	30	76	—	—	—	1	30	76	1	24	61
20 (6.1)	1	18	46	1	24	61	1	30	76	1	30	76	—	—	—	2	24	61	1	24	61
25 (7.6)	1	18	46	1	24	61	1	30	76	2	24	61	1	24	61	2	30	76	1	30	76
30 (9.1)	2	18	46	2	18	46	2	24	61	2	30	76	—	—	—	2	30	76	—	—	—
35 (10.7)	2	18	46	2	24	61	2	24	61	2	30	76	—	—	—	3	30	76	2	24	61
40 (12.2)	2	18	46	2	24	61	2	30	76	2	30	76	—	—	—	3	30	76	2	24	61
50 (15.2)	2	18	46	2	24	61	2	30	76	3	30	76	2	24	61	3	30	76	2	30	76

[1]Tubes run the length of the greenhouse and are spaced equidistantly across the greenhouse. Tubes derive cold air from a louvered air inlet on one end only, unless otherwise specified. Those open to louvered inlets on both ends are identified as IBE.

or

$$cmm_{adjusted} = 275 \times 1.0 \times 1.12 \times 1.0 = 308 \text{ cmm}$$

3. The number of air-distribution tubes can be determined from Table 4-7. Two 24-inch (61-cm) tubes are needed for this greenhouse with its 50-foot (15-m) width and 100-foot (30-m) length.

4. The diameter of individual holes along the side of distribution tubes and the distance between them must next be decided. The catalogs of greenhouse-supply companies include tables that specify the model of tube required for given tube diameters and greenhouse lengths. The model identification, unfortunately, does not indicate the size or distance between holes in these tubes.

   The hole specifications can be calculated if you wish to purchase unpunched tubing or purchase tubing from a company that punches holes to your specifications. Work in England by G. A. Carpenter specifies that the total area of all holes in a single tube should be 1.5 to 2 times the cross-sectional area of the tube. The cross-sectional area of a 24-inch (61-cm) tube is 3.14 ft^2 (890 cm^2). Thus, the combined area of all holes in a tube should be between 4.71 and 6.28 ft^2 (1,334 and 1,778 cm^2). As the required tube length increases, the distance between holes should increase to maintain a reasonable diameter hole. Distances of 2 to 4 ft (61 to 122 cm) are common.

5. The pressurizing fan in the inlet end of the convection tube should be equal to the exhaust fan in capacity. If this is not possible, then the pressurizing fan should be larger. The two pressurizing fans needed in the example greenhouse should have a combined capacity of 11,200 cfm (308 cmm), which is equal to the exhaust-fan capacity. Each pressurizing fan thus has half the capacity, or 5,600 cfm (154 cmm) at a static water pressure of 0.1 inch (30 Pa).

HAF Cooling

This system makes use of the same exhaust fans, inlet louvers, and controls as the convection-tube system. The only difference is the use of HAF fans in the place of convection tubes (see Chapter 3 for a description of the HAF system). Cold air entering through the louvers high in the gables of the greenhouse is picked up in the air-circulating pattern of the HAF fans and is distributed throughout the greenhouse. Just like the convection tubes, the HAF fans can be used to distribute heat in the greenhouse and can be used for air circulation when neither heating nor cooling is in operation.

Cooling Hobby Greenhouses

The principles are basically the same for cooling hobby greenhouses. A fan-and-pad system is used during the summer. To ensure proper vertical distribution of cool air, the pad should be at least 2 ft (61 cm) in height. Cooling problems inherent in these small greenhouses demand a higher-capacity system. A minimum of 12 cfm/ft^2 (3.66 cmm/m^2) of floor area should be exhausted from the greenhouse. If the greenhouse is attached to the east, west, or especially the south side of another building, then considerable solar heat will be collected inside the greenhouse by this wall. Half the area of the wall should be added to the floor area in calculating the ventilator requirement.

Package evaporative coolers are practical for small greenhouses. A package cooler, as pictured in Figure 4-10, consists of a cubical structure with evaporative pads on three sides. Water conduction and collection lines, as well as a pump, are built into the package. A fan is located inside the package to draw air in through the pads and expel the cool air to the greenhouse interior. A ventilator or automatic shutter must be open at the opposite end of the greenhouse to serve as an air exit. Package coolers should be sized to provide 15 cfm of air per square foot (4.5 cmm/m^2) of floor area. Package coolers can be less expensive for small greenhouses. They are easier to install and are more aesthetically pleasing than conventional fan-and-pad systems.

The winter cooling system for a hobby greenhouse again follows the principles of a larger greenhouse. Convection-tube ventilation works well, except that 4 cfm/ft^2 (1.2 cmm/m^2) of floor area should be exhausted under standard conditions. The HAF system may be used in lieu of the convection tube. The fan

Figure 4-10 An Arctic Air package evaporative cooler for a small or hobby-type greenhouse. Water is circulated through pads on three sides of the package. A fan is located within the unit to draw air in through the pads. *(Photo courtesy of J. W. Love, Department of Horticultural Science, North Carolina State University, Raleigh, NC 27695-7609.)*

within a heater in one corner of the greenhouse and an opposing fan in the diagonal corner will handle a greenhouse up to 50 ft (15 m) long.

An alternative arrangement of the convection-tube cooling system saves the price of the inlet louver and the pressurizing fan in the tube. In this system, the convection tube may be connected directly to a stovepipe elbow mounted in the wall of the greenhouse, with the end outside pointing down. The elbow serves as an air inlet. When this system is off, the polyethylene tube hanging from attachments along its upper side collapses and seals itself off from the outside. The tube inflates when an exhaust fan in the greenhouse turns on. An elbow inlet should be used rather than a straight pipe, to prevent wind from blowing into the system and bringing about cooling at times when it could not be tolerated. To further prevent wind entry, it is best to place the inlet for this type of cooling system on the leeward side of the greenhouse. Although heat cannot be distributed through this convection-tube arrangement, the fan in most heaters is generally powerful enough to circulate heat in small hobby greenhouses.

Integration of Heating and Cooling Systems

There are days in the spring and fall when it is necessary to use the heating, winter cooling, and summer cooling systems in order to maintain the desired temperature settings. All three systems, as pictured in Figure 4-11, can be integrated to operate in the following way. Suppose that the fan-and-pad evaporative cooling system is operating on a hot autumn afternoon. As the afternoon wears on and outside temperatures drop, the cooling requirement diminishes. A reduction in the inside temperature is translated into a signal that turns off the water-

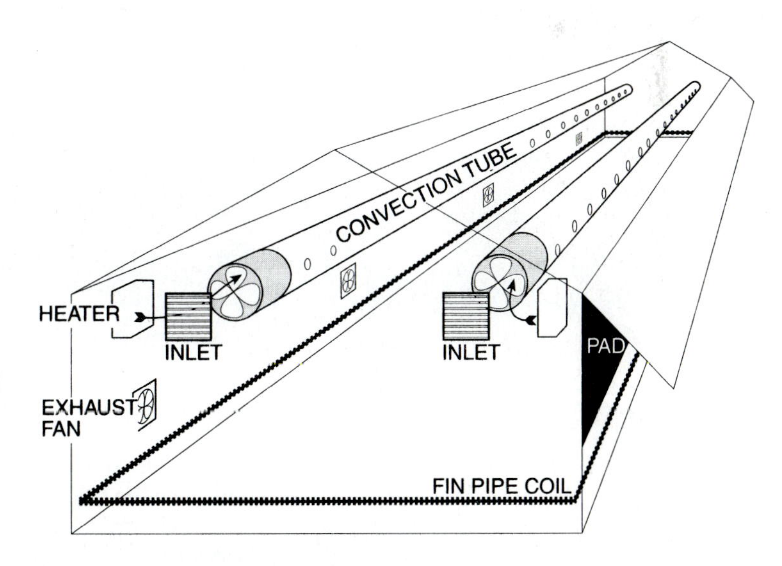

Figure 4-11 A completely integrated system for fan-and-pad evaporative cooling, convection-tube cooling, and heating.

circulation pump in the evaporative pad. The exhaust fans continue to operate, drawing air in through the dry pads. As the need for cooling diminishes further, half of the exhaust fans turn off. With further cooling, an additional quarter of the exhaust fans turn off, leaving only those needed for the winter fan-tube cooling system in operation. At this same time, the ventilators adjacent to the pads close. The winter cooling system is now in operation, with the pressurizing fan in the convection tubes running and the louvered inlet in the end wall open. As evening approaches, the temperature drops further and no cooling is necessary. The exterior louvered inlet closes. Air is now circulated within the greenhouse through the convection tubes. (If HAF fans were used as an alternative to the convection tubes, the HAF fans would have circulated the cold air entering through the inlet louver during winter cooling and would now circulate interior air in the greenhouse.) Temperatures continue to drop during the night. Heat is first supplied through the perimeter pipe coil. When this cannot hold the desired temperature, half of the overhead unit heaters are activated and later, if necessary, the remaining half. When the overhead unit heaters turn on, they blow warm air toward the convection-tube inlet, where it is picked up by the pressurizing fan in the tube inlet and is forced down the tube (Figure 4-12). In the morning, the reverse sequence of events begins to occur.

This complex system of heating and cooling devices might be handled by a

Figure 4-12 **General view of an alternative winter cooling-heating system with two unit heaters mounted apart from the tube inlet. Warm air is expelled from the heaters toward the tube inlet, where it is picked up by the pressurizing fan. The air-intake louver is located in the end wall opposite the inlet of the convection tube. Note the pressurizing fan in the inlet of the convection tube.**

very elaborate multistep thermostat. A dedicated microprocessor would probably handle the system more economically and certainly with more precision. If this firm had several similar production zones or several more types of equipment and sensors to coordinate, a computer would serve best. These controls will be discussed in Chapter 5.

Summary

1. Summer cooling requires that large volumes of air be cooled and brought into the greenhouse. The cool air must pass in a smooth pattern throughout the entire plant zone. A fan-and-pad system is one evaporative cooling alternative used for this purpose. It consists of pads on one wall, through which water is circulated, and exhaust fans on the opposite wall. Air entering through the pads is cooled and then drawn across the greenhouse to the exhaust fans. Air is drawn through the greenhouse at the minimum rate of 8 cfm/ft^2 (2.5 cmm/m^2) of floor area under standard conditions of an elevation under 1,000 ft (300 m), a maximum interior light intensity of 5,000 fc (53.8 klux), an air-temperature rise of 7°F (4°C) between the pads and the exhaust fans, and a distance of 100 ft (30 m) or more between the pad and the fans.
2. Fog cooling is an alternative evaporative cooling system to the fan-and-pad system. Water droplets of 40 microns or smaller (0.016 in) are generated under high pressure (1,000 psi, 6.9 MPa). Fog introduced into the incoming air, just inside the intake ventilators along one wall, cools the air as it evaporates. A second set of fog nozzles, arranged perpendicular to the path of air, counteracts any temperature rise as cooled air moves toward the exhaust fans opposite the intake ventilators. An air exhaust rate of 4 to 5 cfm/ft^2 (1.2 to 1.5 cmm/m^2) of floor area is used.
3. Winter cooling calls for the introduction of a small volume of already cold air from the outside. This air needs to be mixed with the warm greenhouse air above the plants before it is introduced into the plant zone. The convection-tube system is one option for winter cooling. The system consists of an exhaust fan used to develop a negative pressure in the greenhouse; a louvered air inlet in the gable; a clear polyethylene convection tube situated above plant height along the length of the greenhouse that has holes along opposite sides for turbulent air emission; and a pressurizing fan in the inlet end of the convection tube to pick up the air entering through the louver. A flow rate of 2 cfm of air for each square foot (0.61 cmm/m^2) of floor area is satisfactory for standard conditions, which include elevation under 1,000 ft (305 m), a maximum interior light intensity of 5,000 fc (53.8 klux), and a capacity to bring the inside temperature down to within 15°F (8°C) of the colder outside temperature.
4. An alternative for winter cooling is the horizontal airflow (HAF) system. The same exhaust fan and air-intake louver as in the convection tube system are used. The difference lies in the distribution of air throughout the greenhouse.

Small fans are placed above plant height at 50-foot (15-m) intervals down one half of the greenhouse and back up the other half. The fans are designed to set up a horizontal circular flow of air, which will conserve fuel by bringing hot air down from the gable and will also minimize temperature gradients at plant height.

5. Passive ventilator cooling has been greatly improved in recent years. Greenhouse designs are available in glass and polyethylene that permit the opening of 40 percent and in some cases nearly 100 percent of the roof. In cooler climates, these greenhouses can be used for growing most crops without additional cooling. In warmer climates, they are working well for crops tolerant of temperature extremes and are being tested with other, less tolerant crops to determine the latitude of their utility.
6. Even when greenhouses are cooled in the winter by ventilators, they are usually equipped with convection-tube or HAF air-distribution mechanisms for distribution of the heated air during heating. When neither cooling nor heating is required, the convection-tube or HAF fans are used to bring warm air down from the gable and to provide uniform temperatures in the plant zone.
7. The principles are the same for cooling hobby greenhouses. Somewhat simpler systems can be utilized. When the greenhouse is attached to an existing building on any side but the north, half of the attachment wall area is added to the floor area in calculating summer ventilation rates. For summer cooling, a minimum airflow of 12 cfm/ft^2 (3.66 cmm/m^2) of floor area is necessary. For winter cooling, an exhaust rate of 4 cfm/ft^2 (1.2 cmm/m^2) of floor area is used.

References

Various manufacturers of cooling and ventilation equipment offer valuable literature covering products, prices, and technical information.

1. Aldrich, R. A., W. A. Bailey, J. W. Bartok, Jr., W. J. Roberts, and D. S. Ross. 1976. *Hobby greenhouses and other gardening structures.* Pub. NRAES-2. Northeast Reg. Agr. Eng. Ser., Cornell Univ., 152 Riley-Robb Hall, Ithaca, NY 14853.
2. Aldrich, R. A., and J. W. Bartok, Jr. 1994. *Greenhouse engineering,* 3rd rev. Pub. NRAES-33. Northeast Reg. Agr. Eng. Ser., Cornell Univ., 152 Riley-Robb Hall, Ithaca, NY 14853.
3. American Society of Agricultural Engineers. 1995. Heating, ventilating, and cooling greenhouses: ASAE EP406.2 Feb. 95. In: *American Society of Agricultural Standards 1995,* 42nd ed., pp. 559–566. St. Joseph, MI: Amer. Soc. of Agr. Engineers.
4. Bartok, J. W., Jr. 1970. Fan tube greenhouse ventilation. *Connecticut Greenhouse Newsletter* 32:9–12.
5. Bohanon, H. R., C. E. Rahilly, and J. Stout. 1993. *The greenhouse climate control handbook.* Form C7T. Muskogee, OK: Acme Eng. and Manufacturing Corp.
6. Bucklin, R. A., R. W. Henley, and D. B. McConnell. 1993. *Fan and pad greenhouse evaporative cooling systems.* Inst. of Food and Agr. Sci., Cir. 1135. Univ. of Florida, Gainesville, FL.

7. Electricity Council. 1975. *Ventilation for greenhouses.* Farmelectric Centre, National Agr. Centre, Stoneleigh, Kenilworth, Warwickshire CV82LS, England.
8. Gray, H. E. 1956. *Greenhouse heating and construction.* Florists' Publishing Co., 343 S. Dearborn St., Chicago, IL.
9. Koths, J. S., and J. W. Bartok, Jr. 1985. Horizontal air flow. College of Agriculture and Natural Resources. Leaflet 85-14. Univ. of Connecticut, 1376 Storrs Rd., Storrs, CT 06268.
10. Ministry of Agriculture, Fisheries, and Food. 1974. *Glasshouse ventilation.* Ministry of Agr., Fisheries and Food, Mechanization Leaflet 5. London: Her Majesty's Stationery Office.
11. National Greenhouse Manufacturers' Association. 1994. *Standards for ventilating and cooling greenhouses.* Natl. Greenhouse Mfg. Assoc., 7800 S. Elati, Suite 113, Littleton, CO 80120.
12. Willits, D. H. 1993. Greenhouse cooling. *North Carolina Flower Growers' Bul.* 38(2):15–18.

CHAPTER 5

Environmental Control Systems

From the origin of greenhouses to the present, there has been a steady evolution of controls. Five stages in this evolution include manual controls, thermostats, step controllers, dedicated microprocessors, and computers. This chain of evolution has brought about a reduction in control labor and an improvement in the conformity of greenhouse environments to their set points. The achieved benefits from greenhouse environmental uniformity have been better timing of crops, higher quality of crops, disease control, and conservation of energy.

Types of Controls

Manual Controls

During the first half of the 20th century, it was common for greenhouse firms to hire a night watchperson to regulate temperature. This person made periodic trips through the greenhouses during the night, checking the temperature in each greenhouse and opening and closing valves on heating pipes as required. During the day, employees opened and closed ventilators by hand to maintain temperature (Figure 5-1). Temperatures had to be manually controlled on weekends and holidays as well. Obviously, there were large deviations above and below the desired temperatures.

Thermostats

A common range of prices for thermostats is $200 to $600. (See Chapter 3, "Greenhouse Temperature Sensing," for a description of thermostats.) Simple

Figure 5-1 **A manually operated greenhouse roof ventilator in the process of being opened.**

on/off thermostats were used at first. These have the advantages of being inexpensive and simple to install. Their disadvantages include poor accuracy, which can lead to low energy efficiency, and no coordination of equipment—each piece of equipment in the heating or cooling system requires its own thermostat. For this reason, heating and cooling systems are restricted to one or two steps each when using simple thermostats. Simple thermostats are often used in Quonset greenhouses where there is one stage of heating and cooling. One thermostat might turn the heater on at 60°F (16°C) and off at 62°F (17°C). The second thermostat might turn the exhaust fan on and open the air-intake louver for the winter cooling system when the temperature reaches 75°F (24°C) and then deactivate this equipment at 73°F (23°C). The set points of the thermostats are far enough apart that the inaccuracies that develop in each of them will not cause them to overlap so that heating and cooling occur simultaneously.

Step Controllers

When one desires to use a more complicated but accurate climate-control package, such as two stages of heating and three stages of cooling, it becomes cheaper and more effective to use a step controller. These sell for $800 to $1,800. The step controller, like the on/off thermostat, has one temperature sensor but several output connections. A step controller might be used to activate half of the unit heaters at a temperature of 58°F (14°C), the remaining half at 56°F (13°C), the winter cooling equipment at 75°F (24°C), the summer cooling system without water in the pads at 77°F (25°C), and finally turn the water on in the pads at 79°F (26°C). Where there is only one temperature sensor, there is no risk of the five functions falling out of sequence. By using two stages of heating and three stages of cooling, it is possible to reduce the energy input and wear on the equipment when less than full capacity is needed. With multiple stages, some equipment turns on only when maximum heating or cooling is required. In a single stage system, all equipment cycles on and off whenever heating or cooling is required.

Step controllers can receive signals from two or more sensors. As an example, temperature and humidity can be integrated. Step controllers are dedicated to the tasks for which they are purchased. They must be restructured or replaced when it is necessary to perform a different task.

Dedicated Microprocessors

The dedicated microprocessor can be thought of as a simple, limited computer (Figure 5-2). They typically cost from $800 to $2,500. Microprocessors currently used for greenhouse control typically have a keypad and a two- or three-line LCD display of perhaps 80-character length for programming. They generally do not have a floppy disk drive. They have far more output connections than the step

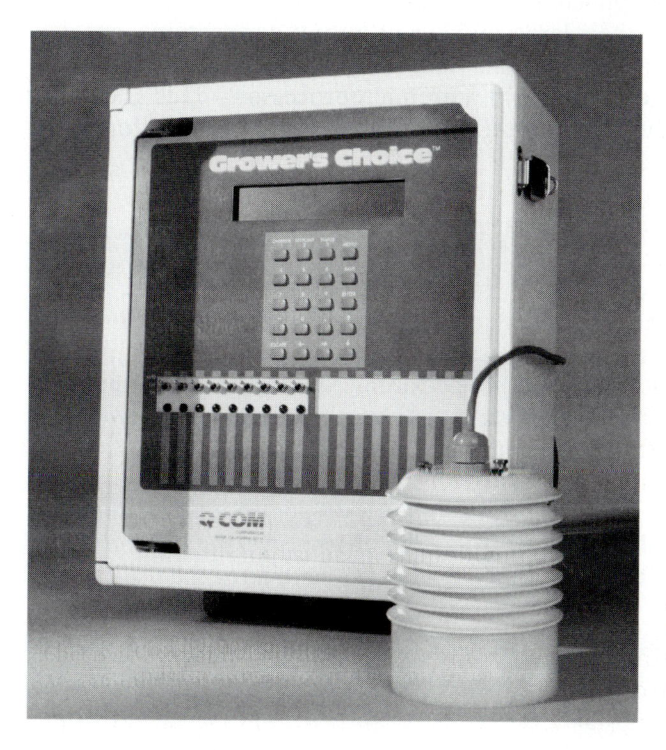

Figure 5-2 **A dedicated microprocessor used to control such operations as heating, cooling, and relative humidity.** ***(From Q COM Corp., 17782 Cowan Ave., Irvine, CA 92614.)***

controller; some can control up to 20 devices. With this number of devices, it is cheaper to use a microprocessor than several step controllers. Microprocessors permit integration of a diverse range of devices, which is not possible with a thermostat and is impractical with a step controller. The accuracy of a microprocessor for temperature control is quite good. Unlike a thermostat that is limited to a bimetallic-strip or metallic tube for temperature sensing, the microprocessor more often uses a thermistor. The bimetallic strip sensor has less reproducibility and a greater range between the on and off steps.

Open roof ventilators can be damaged by high-speed winds blowing directly into their openings. A wind-direction and -speed sensor can feed information to the microprocessor, which in turn sends a signal to close the ventilators on the windward side of the greenhouses while leaving the ventilators open on the leeward side. Likewise, a rain sensor could be set up to override the temperature sensor during a period of rain in order to lower the ventilators sufficiently to prevent a moisture sensitive crop inside from getting wet. Thermostats cannot integrate wind and rain signals into temperature control.

A microprocessor can be used to control carbon dioxide (CO_2)–generating equipment. Since CO_2 is of benefit to the crop only during daylight, this equipment needs to be turned on at dawn and off at dusk. If this is done with a time clock, an error occurs because the time clock does not take into account the daily shift in the time of sunrise and sunset. A microprocessor can be set to activate a CO_2 generator anytime the light intensity exceeds a given set point. The light sensor automatically adjusts to the changing day length, thereby saving CO_2 that would be wasted if it were applied during the dark period.

Microprocessors, with their large number of output connections, are well adapted to temperature control. Systems with four or five stages each of heating and cooling are handled effectively with microprocessors. This allows for greater efficiency of energy, less wear on the equipment, and more uniform temperature control.

Computers

The point of distinction between microprocessors and computers (Figure 5-3) is very arbitrary. There is a smooth progression of numerous models of equipment ranging from the simple models, just described as microprocessors, to computers, which are about to be described. A basic greenhouse computer, usually located in a central point such as the service building, costs about $5,000 to $7,000. A greenhouse range consisting of several discrete growing zones would require one microprocessor for each zone if a central computer was not used. The complexity of programming and collecting data from each microprocessor and the cost of the microprocessors would soon give an economic advantage to the use of a single central computer.

A central computer may operate under two different plans (Figure 5-4). In the first, set points such as heating and cooling temperatures, carbon dioxide level, and relative humidity are programmed into the computer by the crop pro-

Figure 5-3 **A greenhouse control computer located in the central service building. This computer can control several separate greenhouse zones. Sensor data are logged in the computer and can be plotted as needed for examination.** *(From Q COM Corp., 17782 Cowan Ave., Irvine, CA 92614.)*

duction manager. The computer then receives input signals from sensors in each greenhouse zone. The computer integrates the signals, deciding what equipment to activate to achieve the set points. Then the computer sends output signals to the equipment in each greenhouse zone.

In the second central-computer plan, the production manager again programs set points into the computer. Within each greenhouse zone there is a controller, which is actually a microprocessor. Controllers cost about $1,000 to $1,500. The computer sends set points to the controllers. Each controller then receives input signals from sensors within the greenhouse in which the controller is located. These signals are integrated in the controller and output signals are sent to the equipment.

Systems under the control of a central computer have several capabilities that are absent or very limited in dedicated microprocessor systems typically found in greenhouses. These advantages include record logging, data plotting, expandability, and PID control.

Record Logging and Data Plotting Records from each sensor, including indoor and outdoor temperature, light intensity, relative humidity, CO_2 level, and outside wind speed and direction, can be stored in the computer for future study.

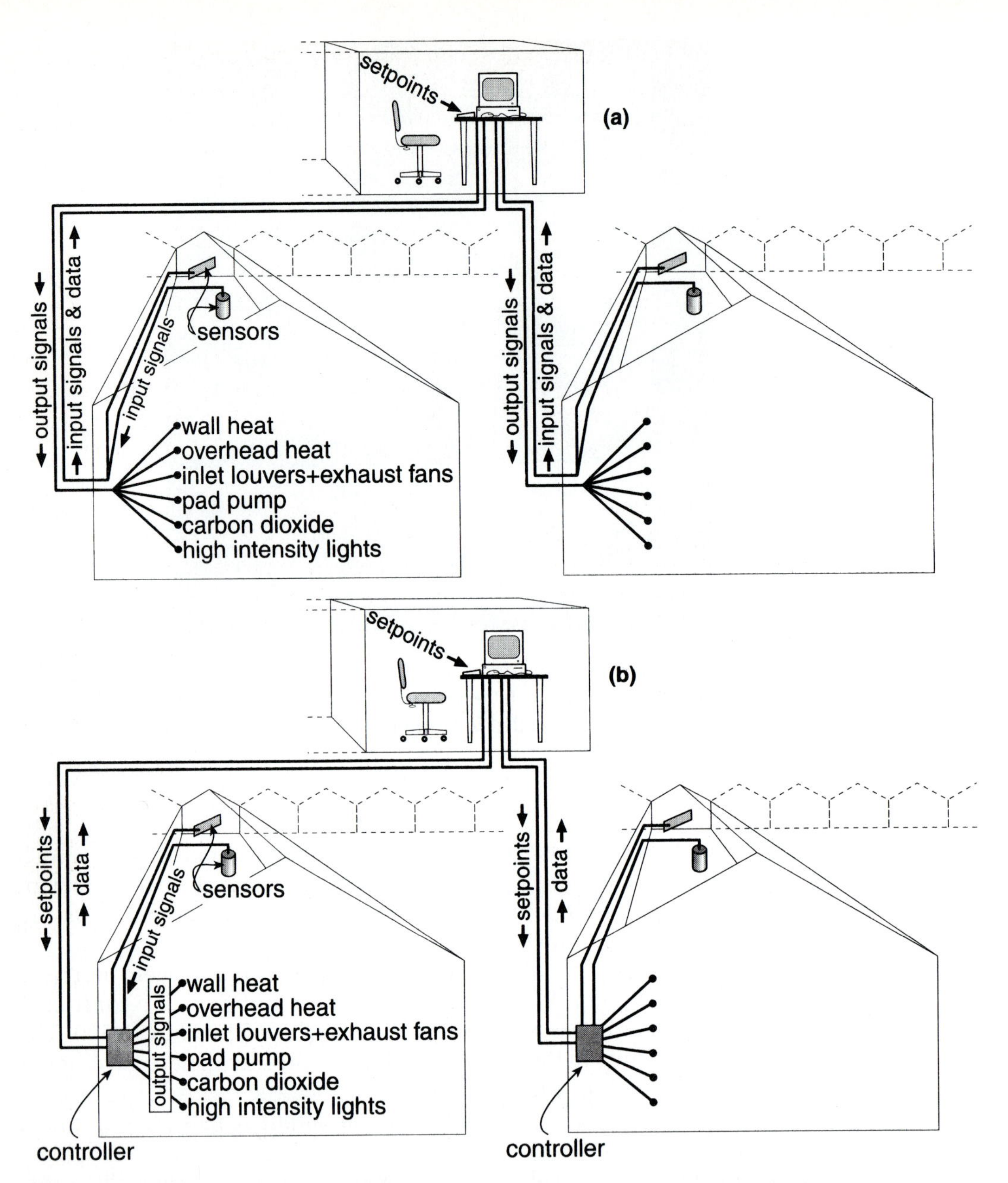

Figure 5-4 **Schematic plans showing two methods for integrating a computer into the control system for a multiple-zone greenhouse range. (a) The computer receives signals from sensors in the various greenhouse zones, processes them, and sends output signals to climate-control devices in the various greenhouses. (b) Greenhouse environment set points are transmitted from the computer to a microprocessor in each greenhouse zone. Signals from environmental sensors in a given greenhouse zone are received by the microprocessor within that zone, where they are processed. An output signal is then sent from the microprocessor to the environmental-control devices in that same greenhouse zone.**

These are handy when evaluating the efficacy of various pieces of equipment, or assessing the cause of crop quality or timing problems. The computer also has the capacity to tabulate or plot these records for easier historical analyses of the greenhouse control system. Current dedicated microprocessors typically used in greenhouses do not have these capabilities.

Expandability Dedicated microprocessors usually have a set number of output connections and may even be limited to specific control functions such as temperature or irrigation. When more climate-control equipment is installed in the greenhouse, more microprocessors must be purchased. Each microprocessor is independently managed. This situation can become difficult to manage and soon leads to a cost advantage for the computer system. In a computer system, additions of crop zones or new equipment in existing zones can be placed under the command of the original computer. In this way, the total greenhouse range is coordinated.

When new generations of microprocessor greenhouse controls are developed, growers usually must replace existing microprocessors to avail themselves of this. In contrast, the computer can be more easily updated as new control capabilities are developed, by simply changing the software.

Greenhouse technology is changing rapidly. A case in point is the trend away from high-volume pesticide sprayers toward low-volume (LV) pesticide applicators. LV applicators fill the greenhouse volume with a pesticide mist or fog for a period of several hours. This is best done during the night when no one is around. However, the greenhouse atmosphere needs to be exchanged with outside air before employees return to work. A firm with a computer can easily add these control functions into its program. Without a computer, the firm would have to purchase a separate time-clock controller. While the time clock is feasible, it further complicates the control system because it offers no coordination between temperature control, purging of greenhouse air during the night for humidity control, and pesticide application. The system must ensure that no ventilation occurs during the pesticide treatment period and that ventilation is allowed during the subsequent pesticide purge period, regardless of heating needs. This is a simple task for a computer but not for a microprocessor.

Consider an integrated heating and cooling system with four stages of cooling and three stages of heating. This system involves too many stages for many step controllers; even if it did, the precision and narrower range of a microprocessor would be more desirable. If additional functions are to be overlaid on the basic heating/cooling system, it becomes cheaper and more effective to use a computer.

One overlying function might be humidity control. A humidity sensor could be installed that would call for evacuation of warm moist air via the winter cooling system whenever the relative humidity reached a set point, such as 90 percent. The incoming cold air would cause the heaters to turn on, thereby warming and drying this air. The air purge would continue, regardless of the need for heating, until the relative humidity was lowered to a predetermined point. If the air purge were under the control of a time clock, which is often the case in greenhouses today, it would be necessary to purge for a fixed length of time, usually four minutes to achieve one full air exchange. However, the amount of air that must be purged is dependent on the relative humidity and temperature of the incoming air. Many times, less than one air exchange will suffice. A computer will exchange air only as long as necessary to bring the relative humidity down to the desired level. This would save heating fuel because less incoming air would have to be heated. An additional overlying function could be the application of LV pesticide mist and its ensuing purge phase. Humidity control would have to be deactivated during the pesticide exposure period so that the pesticide would remain in the greenhouse, in contact with the crop, and not be evacuated. Then, during the following pesticide purge phase, greenhouse air would have to be exchanged with the outside air even if heat was required. The combined complexity of temperature control, humidity control, and pesticide application favors the computer.

PID Control *PID* stands for proportional, integral, and derivative control. In a PID system, all three methods of control are integrated together. Initially we will use the example of a greenhouse heating system to illustrate each of these three forms of control.

In *proportional control,* the output of a piece of equipment such as a heater is set in proportion to the set-point error. For temperature, the set-point error is the deviation between the set point and the actual temperature in the greenhouse. A proportional temperature-control system could operate as follows in a greenhouse with a set point of 60°F (16°C). When the temperature falls to 59°F (15°C), a set point error of 1°F, a wall coil of pipes turns on. If this does not correct the temperature error and a set point error of 2°F occurs at a temperature of 58°F (14.5°C), half of the overhead unit heaters turn on in addition to the wall pipe coil. Finally, if the temperature continues to fall to 57°F (14°C), a set point error of 3°F, the remaining unit heaters are activated. The proportional-control system could also be applied to an in-floor heating system. Water temperature in the pipe coil inside the concrete floor would be set according to the differential between the 60°F (16°C) set point and the actual temperature in the greenhouse. In this case higher set point errors would result in higher water temperatures.

The *integral control* system brings into play a time factor. In this system, the output of the heating system is increased with increasing time that the actual greenhouse temperature remains below the set point. Consider a situation where the greenhouse temperature falls to 59°F (15°C) and the first stage of heating is activated. This heating stage is sufficient to stop any further decline in temperature but insufficient to raise the temperature back to the set point. Under the proportional system the error would remain 1°F (0.5°C), while under the integral system it would be driven up to the set point by activating a second phase of heating after a prescribed period of time passes.

Under *derivative control,* the rate of increase in error is used to determine the increase in output of the heating system. The faster the temperature drops from the set point in a given period of time, the greater the output of the heating system. The derivative-control system works well with fast-reacting heating systems. It permits a quicker reaction to a declining temperature than is possible in the integral-control system because less time is required to determine that a greater output is needed.

Up to this point we have considered the application of PID control only to heating systems. It can be applied equally well to greenhouse cooling, relative-humidity adjustment, carbon dioxide injection, and other environmental controls.

Future Computer Capabilities

Several new dimensions are under development for computer software. Many of the initial advances that occur in software are generic across a broad spectrum of industries, including the greenhouse industry. This is fortunate because it allows the relatively small greenhouse industry to be swept along with the mega-industries. It is the final stages of adaptation to greenhouse application that must occur within our industry. Toward this end, a sufficient number of companies are now well positioned to bring this about. Without question, the availability of new computer applications for control of greenhouse climate and operations will continue to escalate. Soon it will be unthinkable to operate a greenhouse without a computer.

In the immediate future we will experience considerable growth in the current integrated computer control systems. This growth will be driven by the development of physiological and economic data pertaining to crop production. To illustrate, let's consider the balance of temperature, carbon dioxide (CO_2), and ambient light intensity. The temperature at which a given crop will yield an optimum response increases as light intensity in the greenhouse increases from suboptimum to optimum intensity. The optimum concentration of CO_2 in the greenhouse atmosphere for supporting photosynthesis likewise increases with increasing light intensity, up to the optimum light intensity for photosynthesis. In order to realize an increased plant response from an increase in either temperature or CO_2, both of these factors must be increased in proportion. Otherwise, the factor in lowest supply will limit the plant response, rendering the increased level of the other factor ineffective. This three way relationship becomes much more complicated if we bring in the possibility of supplemental lighting in the greenhouse to bring light intensity closer to the optimal level during darker periods of the year. Now it becomes even more important to adjust temperature and CO_2 levels appropriately so that costly increases in supplemental light are not wasted. One day an algorithm, based on physiological response data, will be developed that will identify optimum levels of temperature and CO_2 for every possible light intensity level. Further, the cost of achieving each level of production will be simultaneously determined in this algorithm. This program will be capable of identifying for any given ambient light intensity the level of supplemental lighting that is profitable and the corresponding levels of temperature and CO_2 required to match the light level. This program will guard against achieving a yield increase where the cost of the inputs to achieve the increase are greater than the economic return of the yield increase.

Later, in Chapter 12, the practice of elevating night temperatures and lowering day temperatures (DIF) to yield compact plant shoots in lieu of expensive chemical growth regulators will be discussed. An integrated system can be supplied with a curve of optimum plant height for all points in time during its production. Such a system would be capable of making daily judgments as to whether the rate of plant height increase needs to be hastened or slowed. This would be accomplished by adjusting the day-to-night temperature ratio. While doing so, the daytime system would adjust levels of CO_2 and any supplemental light to augment the new temperature settings.

The possible realm of integrated systems spreads even further to encompass humidity control, fully automated pesticide application, watering, and fertilization. Obviously, the challenge before those developing such systems is the procurement of crop physiology and economic data. More intricate relationships must be understood between levels of environmental factors and plant response. For instance, what is the best rate of decline in temperature from the high day setting to the lower night setting? How long should the lower light intensity in the greenhouse, caused by an occasional cloud in the sky, exist before corrective action is taken in the form of lower temperature and CO_2 level? Would the quality of crops be better if all environmental stresses were eliminated? This is the direction in which control programs are currently moving. Or, are growers dependent on temperature, water, nutritional, or other stresses to achieve the quality levels that they have come to expect? If this latter question is answered in the affirmative, what are these stresses and their magnitudes? They will need to be programmed into the integrated systems.

Summary

1. Over the years, control of the greenhouse environment has undergone a steady evolution. Various stages have included manual controls, thermostats, step controllers, dedicated microprocessors, and computers.
2. Thermostats respond to temperature signals only. They are generally the simple on/off type. Several thermostats may be used together to offer two stages of heating along with one or two stages of cooling. However, these thermostats are not coordinated and are not as accurate as other control alternatives.
3. Step controllers can receive signals from several different types of sensors, such as temperature and relative humidity. They can likewise control several pieces of equipment in a coordinated method.
4. Dedicated microprocessors can be thought of as simple computers. They can receive signals of several types, such as temperature, light intensity, and wind speed. Also, the number of signals received can be greater, often 20 or more, than can be received by step controllers.
5. Computers are an extension of dedicated microprocessors. They can be used to record sensor data and to plot data for ease of interpretation. Computers are easily expandable and readily updated. Most important, computers have

the capacity to handle complex, cutting-edge programs such as the PID system.

6. Future integrated computer control of greenhouse production will undoubtedly include economic-optimization programs. These programs will determine and activate those sets of production inputs that maximize profits.

References

Much of the background information for this Chapter was provided by Dr. Michael Carnes, C-Quest, Inc., P.O. Box 5866, Cary, NC 27512, and by Dr. Dan Willits, Department of Biological and Agricultural Engineering, North Carolina State University, Raleigh, NC 27695-7625.

1. Aldrich, R. A., and J. W. Bartok, Jr. 1994. *Greenhouse engineering.* Pub. NRAES-33. Northeast Reg. Agr. Eng. Ser., Cornell Univ., 152 Riley-Robb Hall, Ithaca, NY 14853.
2. Hashimoto, Y., G. P. A. Bot, W. Day, H.-J. Tantau, and H. Nonomi. 1993. *The computerized greenhouse.* New York: Academic Press.
3. Kano, A., ed. 1994. Greenhouse environment control and automation. *Acta Hort.* No. 399.

 CHAPTER 6

Root Substrate

Once the greenhouses are constructed and the heating and cooling systems are set into operation, it is time for the first cultural consideration—that of selecting a root substrate. Taken at face value, this appears to be a monumental task. Some 15 or more components including field soil, sand, perlite, polystyrene, peats of many types, barks of various origins, coconut fiber, sawdust, and rock wool are to be found in a myriad of formulations used by growers, sold as commercial preparations, or recommended by research institutions. Many are well proven, while others are ineffective. There is a magical lure about concocting one's own root substrate, which often leads to poor combinations of components and the use of more components than are needed or can be justified economically. Selection of a root substrate, however, should be an easy matter once some fundamentals are understood.

Functions of Root Substrate

There are four functions that a root substrate must serve in order to support good plant growth:

1. It must serve as a reservoir for plant nutrients.
2. It must hold water in a way that makes it available to the plant.
3. At the same time, it must provide for the exchange of gases between roots and the atmosphere outside the root substrate.
4. It must provide an anchorage or support for the plant.

Some individual materials can provide all four functions, but not at the required level of each. Sand, for instance, provides excellent support and gas exchange but has insufficient water- and nutrient-supplying capacity. The coarse particles of sand have little surface area per unit of volume compared to the finer particles of soil or peat moss. Since water is held on the surfaces of particles, sand has a small water reserve. Plants grown in sand would need to be watered three or more times per day in the summer. Since most nutrients in a sand medium are held in the water films, there is likewise little nutrient reserve.

Clay, however, has a high nutrient- and water-holding capacity and provides excellent plant support, but the small particles of clay are close to one another. The water films of adjacent particles come into close contact, leaving little open space for gas exchange. Carbon dioxide produced by the roots and by microorganisms cannot adequately leave the clay. In high concentration, carbon dioxide suppresses respiration, which in turn slows growth. Oxygen, which is also needed to keep the processes of respiration going, cannot adequately diffuse into the clay. Consequently, clay is a poor medium for plant growth.

Water is sometimes used as a root substrate. It provides water and nutrients but lacks the ability for gas exchange and plant support. When plants are grown in water, air must be bubbled into it and the plants must be supported in some sort of frame. This cultural procedure is known as hydroponics. Aside from hydroponics, greenhouse root substrates typically contain two or more components to ensure that all four functions are met.

Adaptation of Field Soil to Containers

You might ask, why not use only field soil in greenhouse containers? Greenhouse crops often can be grown in the field without significant alteration of the soil, but when this soil is transferred to containers and the same crop is grown, failure ensues. While all four functions are provided by soil in the field, the function of aeration is usually not adequately provided by this soil in containers.

Water retention and aeration go hand in hand. Drainage is proportional to the depth of the soil above the water table (free water). The bottom of any container is equivalent to a water table. Most fresh (cut) -flower beds contain a 7-inch (18-cm) depth of soil, while pot plants range from 7 in down to 0.75 inch (1.9 cm) in various pots, flats, and plug seedling trays. The water content in a bedding-plant container shortly after watering would be similar to that in a soil situated 2 in (5 cm) above freestanding water—in other words, a swamp situation. The soil pores would be filled mostly with water, and little room would remain for gas exchange.

One dimension by which soil is classified is texture. Texture is the size distribution of particles in a soil. Field soil is composed of three mineral components (Figure 6-1). The finest particles, clay, extend up to a maximum diameter of 0.002 mm. Silt is composed of particles from 0.002 mm up to 0.05 mm, and the third component, sand, is everything larger. Moist clay feels sticky to the touch, silt is floury, and sand is gritty. Texture terms include *sandy loam, silt loam,* and *clay loam* for soils that are predominately composed of sand, silt, and clay, respectively. *Loam* refers to a reasonable balance of all three materials.

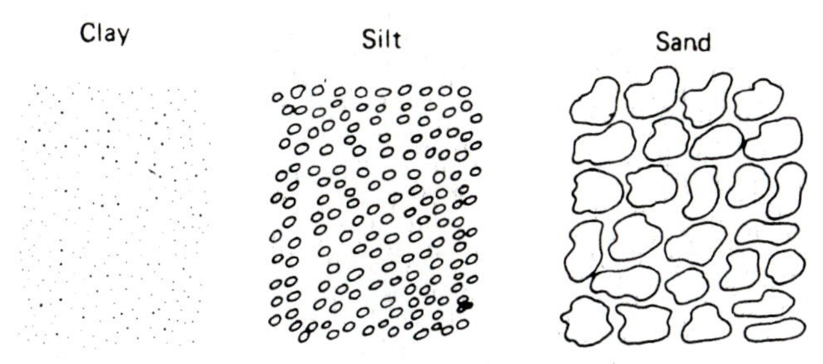

Figure 6-1 **Soil is composed of three mineral components, the relative sizes of which are shown above. Clay particles are the smallest and have a maximum diameter of 0.002 mm. Silt particles have a diameter extending from 0.002 mm to 0.05 mm. The largest particles are sand and are larger than 0.05 mm. The texture classification of a soil gives an indication of the proportion of these three mineral particles contained in it.**

Texture relates to water retention for a very simple reason. Water will remain in soil because it is attracted to the surface of soil particles. This is known as the *matrix force.* Water exists as a film or layer coating each soil particle. The thickness of the water layer depends upon the *gravitational force* attempting to pull the water out of the soil and down to the water table. The greater the distance from a given soil particle to the water table, the stronger the gravitational force. Within the water layer, water that is farthest from the soil particle surface is held the least tightly. This water will be pulled away first by the gravitational force. Thus, as the gravitational force, or the depth of the soil, increases, the thickness of the water layer on the soil particle surfaces decreases (Figure 6-2), and the air-filled center of the pore gets larger, permitting better gas exchange.

The logical solution to the shallow-container problem would appear to be a change toward coarser texture to increase the diameter of the pores. This does solve the problem of aeration, but it creates a new problem by reducing the water-holding capacity of the soil. When the diameter of particles making up a soil is increased, the total surface area of these particles in a given volume decreases. Since water is held on the surface of these particles, the total amount of water in the soil decreases as the particle diameter increases (as the texture becomes coarser).

There is another dimension of soil that can be altered to increase aeration without decreasing its water-holding capacity. *Structure* is the degree of combining of particles into aggregates. A soil with good structure is said to be *friable* (loose). The product of organic-matter degradation is humus, which, along with microbial secretions and hyphae, acts as a cement to bind particles together into aggregates. This is the greatest importance of organic matter in field soils.

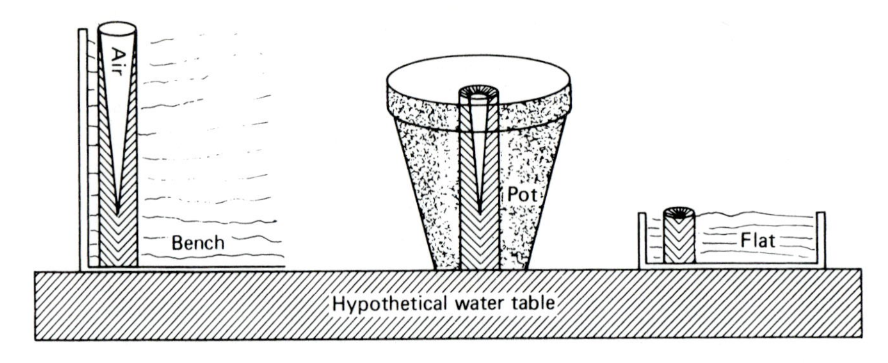

Figure 6-2 A greatly magnified soil pore is shown in various greenhouse containers filled with soil. All are shown perched on a water table, or reservoir of freestanding water, which is effectively the situation existing in the greenhouse shortly after watering. The pore depicts the moisture situation that exists within each container. Water is attracted to the walls of the pore, and at the base the entire pore fills. Higher in the pore, the downward gravitational pull on the water becomes greater, and the water farthest away from the pore wall is removed. The layer of water in the pore becomes thinner with increasing height. Spaces between soil particles in the container are interconnected and form pores running the depth of the soil. Pores in the bench are filled with water at the bottom and are mostly open at the top. Shortly after watering, roots can grow in the upper layers of soil in this bench but not in the lower layer, where there are no open pores for gas exchange. Pores in the pot do not rise as high above the water table; thus, in the upper layer of soil there is more water and less aeration than in the upper layer of the bench soil. The poorest situation for growth exists in the flat, where the pores are so short as to be completely filled with water.

Through the development of structure, a dimension is given to soil that cannot be achieved through alterations in texture. The high water retention of fine-textured soil can be combined with the excellent drainage of coarse-textured soil. This is accomplished by extensive retention of water in the small-diameter pores within each aggregate and rapid percolation of water out of—and, conversely, good gas exchange into—the large pores between the aggregates (Figure 6-3).

It should now be apparent that field soil must be prepared for use in containers by altering it to a coarser texture and by increasing its structure prior to planting. There is no luxury of time to permit the structure to form as a consequence of decomposing organic matter. A coarser texture can be achieved by mixing coarse sand into the soil. Structure can be improved instantly by incorporating large aggregate particles such as sphagnum peat moss and bark into the soil. Numerous materials can be added to the soil, but before a selection can be made, one more set of properties should be understood. These properties pertain specifically to greenhouse root substrates.

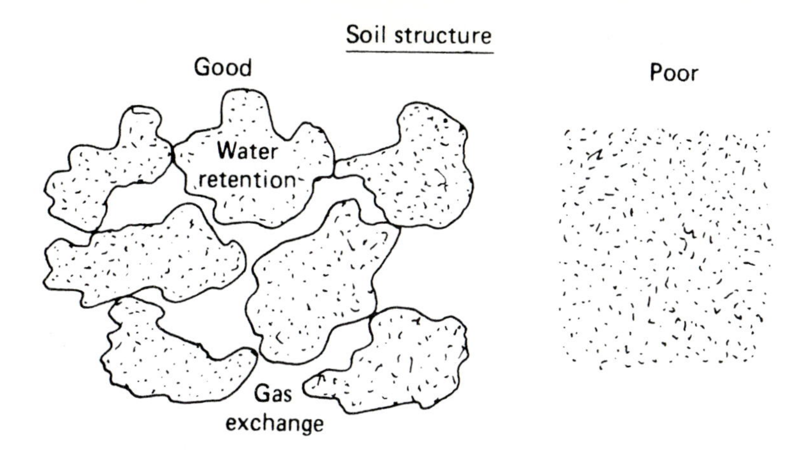

Figure 6-3 **An important property of soil is structure. The soil on the left has good structure because the particles composing it are cemented together into larger aggregate particles. Small-diameter pores still exist within the large aggregates, and because of the large surface area of these pores, a large volume of water is held in each aggregate. Between the aggregates are very large pores that do not fill with water, thus providing a channel for gas exchange. The soil on the right has very little structure. Only small pores exist. While water retention is good, aeration is poor.**

Desirable Properties of a Root Substrate

Stability of Organic Matter

Most greenhouse crops have a production period of one to four months. Good structure must exist when seed is sown or plants are potted. It is important that decomposition of organic matter in the root substrate be minimal. Decomposition of the organic aggregates will lead to finer texture and, consequently, poorer aeration. Also, since the volume of the root substrate available within the pot for root growth is small, any significant reduction in the volume during growth of the plant is detrimental. Straw and sawdust, except redwood sawdust, decompose rapidly and therefore are not desirable in a root substrate used for pot crops.

The situation is somewhat different for fresh-flower and vegetable crops in benches where the volume of the root substrate is sufficiently great to permit shrinkage. Substrate is used indefinitely for fresh-flower crops. With time, organic matter deteriorates and requires replacement, which is generally done on an annual basis. Direct addition of materials such as sphagnum peat moss or coarse composted bark is done in quantities sufficient to compensate for the lost volume.

Carbon-to-Nitrogen Ratio

The amount of nitrogen (N) relative to carbon (C) in a root substrate amendment is important. Decomposition of organic matter occurs largely through the action of living microorganisms. The largest component of organic matter (50 percent or more) is C, which is utilized by the microorganisms. N in the organic matter must be available to the microorganisms in the quantity of at least 1 pound for every 30 pounds of C; otherwise, decomposition slows down. Whenever this ratio of 30 C:1 N is exceeded (when more than 30 pounds of C exist for each pound of N), N already present in the root substrate or N added as fertilizer will be utilized by the microorganisms rather than by the crop plants. The crop will become deficient in N. If this situation occurs slowly and continuously, a grower can easily compensate for it by increasing the N fertilizer application. The decomposition of materials such as straw and sawdust occurs rapidly, however, and thereby creates a peak of N demand followed by a quickly diminishing demand for N as the organic matter available to the microorganisms runs out. Only the most experienced growers can compensate for this process.

The C:N ratio for sawdust is about 1,000:1. It has been reported that, in addition to the small amount of N already present in the sawdust, 24 pounds of N must be added to facilitate the decomposition of 1 ton of sawdust by microorganisms (12 kg N/ton). Bark has a C:N ratio of about 300:1 and requires an addition of 7 pounds of N to facilitate the decomposition of 1 ton (3.5 kg N/ton). It is not only the C:N ratio that determines the suitability of a root substrate component but also the rate of decomposition. While the bark has an undesirably wide C:N ratio of 300:1, its rate of decomposition is slow and steady, requiring as long as three years to decompose. The drain of 7 pounds of N per ton (3.5 kg N/ton) of bark, carried out over three years, presents a negligible N tax at each fertilization date. Bark is therefore a desirable root substrate component in spite of its wide C:N ratio. Sawdust, however, will decompose in a few months and has a wider C:N ratio of 1,000:1. The N tax in this case is great, and this material should be avoided by inexperienced growers. Redwood sawdust is an exception, since the waxes and similar compounds in it slow down its rate of decomposition. It is frequently used in greenhouses in the western United States.

Bulk Density

The bulk density of a root substrate relates to support of the plant. Nearly any solid substrate will provide for anchorage of the plant roots, but it is also important that the substrate be sufficiently heavy to prevent a pot plant from falling over due to the weight of the plant. A mixture of equal parts of sphagnum peat moss and perlite is sufficiently heavy just after watering. But, when the available water has been used, large plants in this substrate easily topple over when handled. On the other hand, a high bulk density is uneconomical because of the extra labor of handling such a substrate. An acceptable range for bulk density of potting substrate is 40 to 60 lb/ft^3 (640 to 960 g/dm^3) just after watering at container capacity (CC). *Container capacity* is the maximum amount of water that the root substrate in a container can hold against gravity.

As mentioned earlier, soil must be amended with coarse particles such as sand to provide aeration. Wet soil and sand at CC can weigh 100 pounds or more per cubic foot (1,600 g/dm^3) (see Table 6-1). Therefore, perlite, with a wet density of 32 lb/ft^3 (500 g/dm^3), and polystyrene, with a wet density of 7.5 lb/ft^3 (120 g/m^3), are often used as substitutes for sand in spite of their higher cost. The problems of bulk density are not nearly as important for substrates used in greenhouse benches.

Moisture Retention and Aeration

A wet root substrate is composed of (1) the solid particles of the substrate, (2) the liquid water coating the surfaces of the particles, and (3) the air occupying the center of the pores. To ensure a suitably long interval between waterings and to provide adequate aeration at all times, the balance of water and air in the root substrate pores must be controlled through selection of the particles composing the substrate. After watering, 10 to 20 percent of the volume of the root substrate should be occupied by air in a 6-inch (17-cm) azalea-type pot. The available water content should be as high as possible, provided that the air porosity and density of the total root substrate are adequate. A survey of Table 6-1 indicates that the property of aeration can be provided by components such as vermiculite, pine bark, perlite, polystyrene, and rock wool. The coarse concrete-grade sand (rather than the fine sand) listed in the table would also be an excellent additive for aeration. Excellent components for providing a high available-water content are sphagnum peat moss and rock wool.

When all three properties—sufficient density, adequate available water, and aeration—cannot be met by one component, a mixture is required. Four common mixes and their physical properties in a 6.5-inch (17-cm) azalea pot are listed at the bottom of Table 6-1. The first mix would ordinarily be satisfactory but in this case it is not, due to the use of a heavy clay soil. The fine clay particles result in a high available-water content (40.2 percent) at the expense of aeration (5.9 percent air at CC). A larger proportion of sand is needed in this mix. The sphagnum peat moss and vermiculite mix is excellent, with a wet bulk density of 53.3 lb/ft^3, an air content of 16.6 percent at CC, and an available water content of 46.2 percent of the volume of the pot. The pine-bark mix has good bulk density and aeration but is lower in available-water content. The water content is acceptable to most growers. But to those faced with long periods of shipping or poor maintenance in the marketplace, a mix with a higher available-water content would be better. The rock wool and sphagnum peat moss mix is excellent on all counts.

It should be recalled that water retention is related to the depth of the root substrate. As seen in Table 6-2, the water content of root substrate just after watering increases with decreasing depth of the container. The matrix force holding water to the surfaces of the root substrate is equal in each pot, but the gravitational force pulling water out of the pot becomes greater as the pot in-

Table 6-1 **Percentage of Total Volume in a 6.5-Inch (17-cm) Azalea-Type Pot Occupied by Solids, Water, and Air at Moisture Tensions of Container Capacity (CC) and 15 Bar for Various Root Substrate Components and Formulas[1]**

							Bulk Density			
	Solid	*Water (%)*		*Air (%)*		*Available Water[4]*	*CC*		*15 bar*	
Material/Mix	*(%)*	*CC[2]*	*15 bar[3]*	*CC*	*15 bar*	*(%)*	*lb/ft³*	*g/dm³*	*lb/ft³*	*g/dm³*
Soil (sandy clay)	53.3	39.8	6.4	6.9	40.3	33.4	106.0	1698	85.3	1364
Sand (concrete-grade)	59.3	35.4	4.4	5.3	36.3	31.0	107.1	1714	87.8	1404
Sphagnum peat moss[5]	15.4	76.5	25.8	8.1	58.8	50.7	53.7	859	22.0	352
Vermiculite (Progro No. 2)[5]	17.3	53.2	29.1	19.5	43.6	24.1	46.1	738	31.1	497
Pine bark (aged, < 3/8 inch, < 10 mm)[5]	20.7	58.9	30.3	20.4	49.0	28.6	50.6	809	32.7	523
Perlite (Krum, horticultural grade)[5]	36.9	38.3	20.2	24.8	42.9	18.1	32.1	514	20.8	333
Polystyrene beads[5]	64.6	10.5	1.0	24.9	34.4	9.5	7.5	120	1.6	25
Rock wool (Pargro, medium, granular)	8.9	65.0	4.4	26.1	86.7	60.6	54.4	870	16.5	264
1 soil:1 peat moss:1 sand	45.4	48.7	8.5	5.9	46.1	40.2	99.7	1595	74.6	1193
1 peat moss:1 vermiculite	13.1	70.3	24.1	16.6	62.8	46.2	53.3	853	24.4	391
3 pine bark:1 sand:1 peat moss	29.5	53.4	21.5	17.0	49.0	31.9	58.9	942	38.9	623
1 rock wool:1 peat moss	8.3	70.9	11.3	20.8	80.4	59.6	51.8	829	14.6	233

[1]Data provided by William C. Fonteno, Department of Horticultural Science, North Carolina State University, Raleigh, NC 27695-7609. Components used in the formulas include: heavy clay soil, sphagnum peat moss, concrete-grade sand, and aged < 3/8-inch pine bark.

[2]Container capacity = the amount of water a root substrate can hold against gravity just after watering.

[3]15 bar = the permanent wilting point at which the plant has essentially taken up all available water.

[4]Available water = the amount of water released between container capacity and 15-bar tension, unless otherwise stated; is expressed as a percentage of the total pot volume.

[5]Computation of percentage of water, percentage of air, and percentage of available water is based on a soil moisture tension of 300 cm rather than the stated 15 bar. Since water content for these five components differs very little between 300-cm and 15-bar tension, only small differences occur as a result.

Table 6-2 **Percentages of Total Container Volume Occupied by Water and Air at Container Capacity for Three Root Substrates in Five Different Sizes of Containers**[1]

	Standard Pots			*Flats*	
	8 in (20 cm)	*6 in (15 cm)*	*4 in (10 cm)*	*48 cells*	*512 plugs*
			1 soil:1 sand:1 peat moss		
Water (%)	45.0	47.2	51.2	52.9	54.3
Air (%)	9.5	7.4	3.4	1.7	0.3
			1 peat moss:1 vermiculite		
Water (%)	64.4	67.9	75.2	79.5	84.8
Air (%)	22.5	19.0	11.7	7.4	2.1
			3 pine bark:1 sand:1 peat moss		
Water (%)	48.7	51.5	57.6	61.4	66.9
Air (%)	21.8	18.9	12.9	9.1	3.6

[1]Data provided by William C. Fonteno, Department of Horticultural Science, North Carolina State University, Raleigh, NC 27695-7609. Components used in the formulas include: heavy clay soil, sphagnum peat moss, concrete-grade sand, and aged < 3/8-inch pine bark.

creases in depth. With increasing water content comes decreasing air content. Fortunately, the range in acceptable air- and water-content values is wide. Well-formulated root substrate with high air- and water-retention values is suitable for a wide range of pot sizes. Two cases for which special mixes are frequently formulated are shallow plug flats and larger (deep) green-plant containers.

One should not rely entirely on selection of substrate components for achieving proper aeration. There are three steps that can be taken after procurement of the substrate to ensure that a desirable level of aeration is achieved and maintained. When substrate is excessively compacted during planting, pores are compressed to smaller diameters. While this allows for a greater quantity of substrate in the pot, which desirably raises the water holding capacity, it also reduces aeration because of the smaller pores. Substrate should be compacted during planting only to the extent necessary to support the plant.

Shipping costs constitute a sizable proportion of the price of substrates or of the components used for formulating substrates. As a consequence, many materials are received in a drier state than desired. If water is added to a dry substrate in a mixer and the mix is allowed to sit for a period, it will expand to a coarser structure. Its facility for aeration increases and this facility persists after planting. A desirable substrate contains a weight of water equal to about 67 percent of the dry weight of the substrate.

The method of water application can also affect aeration. A strong force of water beating against substrate can break up aggregates. The finer particles that are released can lodge in the larger pore spaces below, thereby plugging them. This results in a decline in aeration. Automatic watering systems are desirable because of their gentle delivery of water to the substrate surface. When hand watering is necessary, a water breaker should be used on the end of the hose to reduce the pressure of emitted water.

Cation Exchange Capacity

Root substrate components such as clay, silt, organic matter, and vermiculite have fixed negative electrical charges. These charges will attract and hold positively charged nutrient ions (cations). Most fertilizer nutrients have electrical charges, some negative and others positive. Positively charged fertilizer ions are ammonium nitrogen, potassium, calcium, magnesium, iron, manganese, zinc, and copper. Field soil and greenhouse substrate electrically attract and hold these nutrients so that they are not washed away during a rain or heavy watering. At the same time, these electrically held nutrients are available to the plant.

Cation exchange capacity (CEC) is a measure of the ability of the fixed negative electrical charges in substrate to hold positively charged ions. It is generally expressed as milliequivalents per 100 cubic centimeters (me/100 cc) of dry substrate component. A level of 6 to 15 me/100 cc is considered desirable for greenhouse root substrate. Higher levels are not common but are very desirable. With lower levels, the substrate will not act as a suitable reservoir for nutrients, and frequent fertilizing will become necessary. Clay, peat moss, vermiculite, and most composted organic matter have a high CEC; sand, perlite, polystyrene, and noncomposted materials such as rice hulls and peanut hulls have an insignificant CEC. When preparing a root substrate, it is desirable to include a component with a high CEC.

pH

The importance of root substrate pH level will be developed further in Chapter 9. It is sufficient to say here that pH level controls the availability of nutrients to the plant. It is often said that half of crop nutritional problems can be averted by holding substrate pH in the desired range. Greenhouse crops fall into two categories. Most grow best in a slightly acid pH range of 6.2 to 6.8 in soil-based substrates and 5.4 to 6.0 in soilless substrates. A small number of crops are termed *acid-loving* because they grow best in a strongly acid pH range of 4.5 to 5.8. Sphagnum peat moss, pine bark, and many composts are acid. Peat moss can have a pH level below 4.0. Sand and perlite are neutral (pH 7.0). Vermiculite and some hardwood barks are alkaline (pH above 7.0). Field soil can range from acid (pH 3.5) to alkaline (pH 8.5). Rock wool can range from neutral to mildly alkaline. It is important to check the pH level of the substrate one has formulated and to adjust it to the proper level prior to planting. (Instructions are given in Chapter

9.) Commercial root substrates are usually adjusted to the proper pH level by the manufacturer.

Components of Root Substrate

The components of a root substrate can be selected from numerous materials. Listed in Table 6-3 are the more common components, the functions that each performs, and the cost of each. Alternative components exist for each of the four needed functions of a root substrate. Selection of components is based on the required function, cost, and availability.

Field Soil

Prior to the practice of soil pasteurization, which took hold in the early 1950s, it was customary to replace greenhouse substrate annually, usually during the summer. Much attention was paid to the type of field soil used. Soil with a high degree of structure and a loamy texture proved to be the most desirable.

Texture was ensured by locating the greenhouse range in a region of proper soil type. Structure was developed in the field soil by growing a mixed crop of grasses and clover in the soil for one to three years. These crops continually renew their root systems, leaving behind vast quantities of roots that decompose into humus and lead to good structure development. Crops commonly used were Kentucky bluegrass, timothy, redtop, red clover, alsike clover, and ladino clover. The crop was mowed twice per year and allowed to lie on the ground. During the fall prior to the summer when the soil was to be moved into the greenhouse, the crop was disked and the soil was placed in piles, where decomposition of the crop took place.

Table 6-3 **Root Substrate Components, Their Functions, and Cost Including Delivery**

Component	*Water Retention*	*Nutrient Retention*	*Aeration*	*Light Weight*	*\$/ft³*	*\$/m³*
Field soil	X	X			0.80	25.84
Sphagnum peat moss	X	X			0.85	27.46
Bark (0–3/8 in)	X	X			0.65	21.00
Sawdust (rotted)	X	X			0.65	21.00
Manure	X	X			0.65	21.00
Vermiculite	X	X		X	1.45	51.18
Calcined clay	X	X	X		2.45	86.50
Bark (3/8–3/4 in)	X	X	X		0.65	21.00
Sand (concrete grade)			X		0.77	24.87
Perlite			X	X	1.45	51.18
Polystyrene			X	X	0.50	17.66

Since fresh-flower and vegetable growers no longer replace their root substrate, it is important only at the time of establishing the greenhouse range that a proper field soil be developed. Pot-plant growers, however, must have a continuous supply of proper field soil if they utilize soil-based substrate. Many established greenhouse areas have been inundated by residential and commercial development, and newer ranges have located in regions of poor soil in order to take advantage of the availability of other factors such as transportation, labor supply, and utilities. Such businesses, lacking suitable field soil, have purchased soil with only sporadic success because of the cost or variation from one lot to another. These growers have found it expedient to use soilless substrates.

Peat Moss and Peats

There are different types of peat. Peat moss that is light tan to brown in color is the least decomposed, and is formed from sphagnum or hypnum moss (mostly the former). It has a nitrogen content of 0.6 to 1.4 percent and decomposes slowly; thus, nitrogen tie-up is not a problem. It has the highest water-holding capacity of all the peats, holding up to 60 percent of its volume in water. Sphagnum peat moss is the most acid of the peats, with a pH level of 3.0 to 4.0, and requires 14 to 35 lb of finely ground limestone per cubic yard (8 to 20 kg/m^3) to bring the pH up to the level that is best for most crops. In areas with hard water, containing calcium, the lower rate may be suitable. The CEC of sphagnum peat moss is in a desirable range, from 7 to 13 me/100 cc.

The fine structure of the moss can still be seen in peat moss. Large quantities of water are held on the extensive surface area of the moss, while good gas exchange occurs in the large pores between the aggregates (chunks) of peat moss. For this latter reason, peat moss should not be finely ground down to the level of fibers prior to use. Hypnum peat moss has a pH level in the range of 5.2 to 5.5. When it is used with vermiculite, no limestone is needed since vermiculite is mildly alkaline. One successful greenhouse pot-crop mixture calls for 50 volume parts hypnum peat moss to 40 parts perlite to 10 parts vermiculite.

Reed-sedge peat is brown to reddish-brown in color and is formed from swamp plants, including reeds, sedges, marsh grasses, and cattails. It occurs in varying degrees of decomposition but is generally more highly decomposed than peat moss. As a result, more fine particles are present, giving a poorer structure than that of peat moss. Also, the water-holding capacity of reed-sedge peat is lower than that of peat moss. Depending on the source, the pH level of reed-sedge peat can vary from 4.0 to 7.5. The salt level can also vary with the source. Although sphagnum peat moss is preferred for the general range of greenhouse applications, reed-sedge peat can be used in root substrate for pot and bench crops if pH is properly adjusted, high salt sources are avoided, and proper aeration is built into the final mix. There is little technical information available for using reed-sedge peat.

Peat humus is dark brown to black in color and is the most highly decomposed of the peats. It is usually derived from hypnum peat moss or reed-sedge peat. Original plant remains are not distinguishable, and water-holding capacity is less than that of other peats. The pH level can range from 5.0 to 7.5. Peat humus has a moderately high nitrogen content, which makes it undesirable in seed-flat substrate or substrate used for salt-sensitive plants. Ammonium nitrogen released from peat humus can build up to levels that are toxic to more sensitive plants such as young seedlings, African violet, azalea, and snapdragon. Ammonium nitrogen is released during microbial decomposition of peat humus because more than 1 pound of nitrogen is available per 30 pounds of carbon. Peat humus is rarely used in the greenhouse.

Bark

Redwood bark and fir bark have been used on the West Coast of the United States for many years as a component of nursery and greenhouse root substrate. Coarse fir bark provides an excellent substrate for orchids. Pine bark (Figure 6-4) is extensively used throughout the United States. Hardwood barks are used in many of the interior states. All are highly satisfactory. Most are inexpensive compared to the sphagnum peat moss that they replace.

Figure 6-4 **Barks of various origins are widely accepted today as a substitute for peat moss. Pine bark is pictured here. Typical processing calls for composting in a pile for three months or longer and then screening into different sizes for various markets. Particles passing through a 3/8-inch mesh screen are used in pot-plant substrate (left); those between 3/8 and 3/4 inch are used for fresh-flower substrate amendment (right); and larger particles are used for landscape mulches (center).**

Most bark is purchased by growers after it has been partially composted and screened. When bark is removed from logs, varying quantities of cambium and young wood are included. These materials decompose faster than bark and accentuate the nitrogen tie-up problem. The wood content tends to be highest in the spring when growth is more active. A period of composting rids the bark of these components and brings it to a stage where the rate of decomposition is slow and steady and nitrogen tie-up is not a problem.

Accounts have been given of fresh hardwood and softwood barks causing growth suppression and injury to plants. Unknown compounds are apparently destroyed during composting for a period of 30 days. One explanation is that the toxic material is acetic acid that is given off in the initial stages of composting and then quickly destroyed in subsequent stages of composting (Hoitink, H., personal communication). Composting has an additional beneficial effect for bark and sawdust as well. Fresh bark and sawdust do not hold fertilizer nutrients very well because of a low CEC of about 8 me/100 g (about 1.6 me/100 cc). After composting takes place, the CEC rises to a level of 60 me/100 g (about 12 me/100 cc) or higher, which imparts a strong nutrient-retention capacity to the bark and sawdust.

Composting is accomplished in two ways. Nitrogen is mixed in at the rate of 3 lb of actual nitrogen per cubic yard (1.8 kg/m^3), and the bark is set in piles in the field. Ammonium nitrate is a good source of nitrogen and is used at the rate of 9 lb/yd^3 (5.3 kg/m^3), since it contains 33 percent nitrogen. A period of four to six weeks is sufficient to complete the rapid phase of decomposition. In the second system, no nitrogen is used and a period of three months to a year is required. Both systems result in destruction of inhibitory compounds, degradation of wood, and fragmentation of larger particles into smaller ones. Because bark used for root substrates is not completely composted, it is often referred to as aged bark.

Compost piles must not be over 12 ft (3.7 m) deep because during the process of composting, heat is given off that—if permitted to become too intense—can set the pile on fire. The surface layer should be turned into the pile after one to two weeks of composting to ensure that all the bark has been processed. The heat given off by fermentation is sufficient to pasteurize the bark. Harmful disease organisms, insects, nematodes, and weed seeds are thus eliminated. It is important that subsequent handling be carried out in a way to maintain this cleanliness. The bark should not be piled where crops have been grown or where the runoff from crop lands has accumulated. Equipment used for moving bark should be sterilized first if it has been used on crops. If the bark is bagged, clean handling is almost ensured. Larger growers find economy in purchasing bark in bulk (unpackaged).

Prior to sale, bark is screened for various purposes. Particles 1/8 inch (3 mm) and smaller are used as soil conditioners in applications such as golf-course greens. Particles 3/8 inch (10 mm) and smaller are preferred for greenhouse pot-plant substrates, and those from 3/8 inch to 3/4 inch (10 to 19 mm) are used for organic-matter amendment of greenhouse fresh-flower substrate. Larger pieces are used for landscape mulching.

Since the largest part of the cost of bark often lies in the shipping expense—$2 or more per mile for a 60-yd^3 (46-m^3) truckload—it is beneficial to obtain bark from local sources. Consequently, numerous types of bark are used throughout the United States. In general, processed bark will cost from one-fourth to two-thirds the price of imported sphagnum peat moss.

Coir

Nutritive products, including coconut meat, milk, and oil, are derived from the inner seed of the coconut fruit. Three shells surround the white edible tissue (meat) of the seed. These are the hard inner *endocarp,* which lies immediately external to the meat; the softer, fibrous middle *mesocarp*; and the hard, thin outer *exocarp.* The fibrous mesocarp is known as *coir.* The mesocarp-exocarp husk also yields products. The husk is soaked in water and then shredded to release its long fibers. These fibers are used in the manufacture of brushes, automobile-seat and mattress stuffing, filters, twine, and other products. During the process of retrieving the longer construction fibers, a tan-brown material consisting of short fibers along with granular pith material is generated from the mesocarp. This material is referred to as coir dust, coir pith, or more simply in the horticultural trade as coir. Horticultural coir, used as a component in root substrate, has the consistency of coffee grounds.

At the point of production, wet coir dust is compressed into bricks, some the size of standard clay building bricks and others much larger. The bricks are then dried. Dry bricks are more easily shipped from the various tropical areas in the world where coconuts are grown. Before use, coir must be rehydrated, either by the supplier or by the grower. This is done by soaking bricks or ground bricks in sufficient water overnight to bring the final mix to approximately 70 percent moisture content by weight. The hydrated bricks yield a ready-to-use material that has a volume between 5 and 10 times that of the dry fiber, depending on the supplier (Figure 6-5).

The texture of coir, after hydration, is finer than that of sphagnum peat moss and aeration is not quite as high. Its water-holding capacity is similar to that of sphagnum peat moss. A decided advantage of coir over peat moss is its superior rewetting capacity. When allowed to dry beyond the desired point, peat moss repels water, while coir continues to absorb it. The pH range of coir is from 4.9 to 6.8. Its moist bulk density can range from 20 to 28 lb/ft^3 (321 to 449 g/L) and its dry bulk density ranges from 3 to 5 lb/ft^3 (48 to 80 g/L). Coir has a CEC of 60 to 130 me/100 g. Assuming an average bulk density of 4 lb/ft^3 (64 g/L) and an average CEC of 95 me/100 g, this equates to 6.1 me/100 cc. Coir has a carbon-to-nitrogen ratio of 80:1. In an alternative material, this could lead to nitrogen tie-up. However, this is not the case for coir. Coir is about two-thirds lignin, which suppresses microbial decomposition. An added advantage of the low decomposition rate is the stability of volume in the pot over crop production time.

Coir is sometimes used in mixes along with pine bark as a substitute for the peat moss component. To avoid a heavy bulk density of the resulting mix, a lightweight component such as vermiculite or perlite is also advisable in coir mixes. A good trial formula for a coir-based mix would be 30 percent coir plus 30 to 40

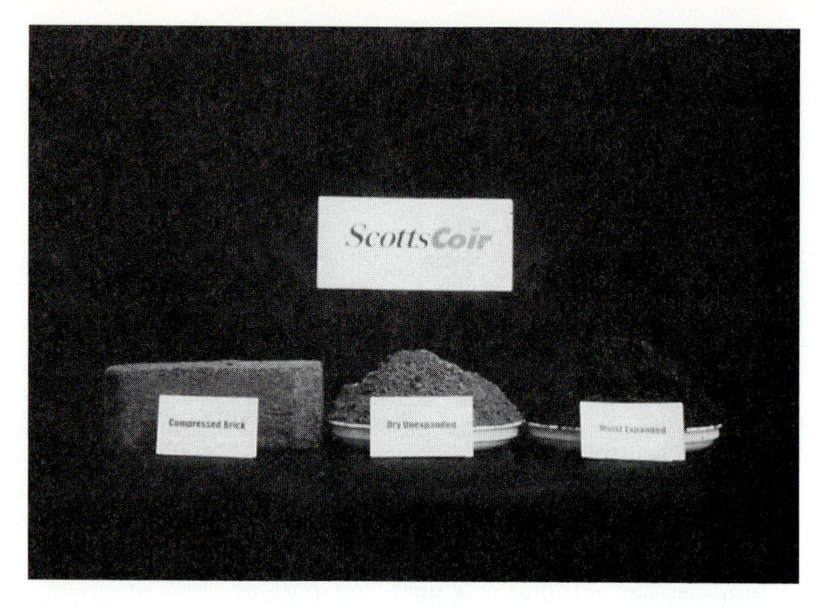

Figure 6-5 **Various stages of coir during preparation for use in root substrate. Coir is compressed into dry bricks (left) for economy of shipping to consumer countries. Firms in the destination country may sell the bricks directly to growers or may grind the dry bricks and sell the loose, bagged product (center). Growers add water to the bricks or loose coir in sufficient quantity to compose 70 percent of the final weight and allow the mix to stand overnight for hydration. The resulting hydrated coir can be 5 to 9 times greater in volume than the original bricks (right) and is ready for incorporation into root substrate.** ***(Photo Courtesy of Scott-Sierra Horticultural Products Co., Marysville, OH 43041)***

percent pine bark plus 30 to 40 percent vermiculite. At this time, the physical and chemical properties of coir can vary from one source to another. Salt content can be a problem in some sources. A major element contained in these salts is potassium, but sodium and chloride can also be present. Knowledge about the supplier is important. The price of coir is expected to be competitive with sphagnum peat moss. Coir-based mixes are on the market for the greenhouse industry.

Sawdust

Sawdust in many respects is similar to bark. It should be partially composted because in the fresh state its rate of decomposition and nitrogen tie-up is excessive and it may contain toxic substances such as resins, tannins, or turpentine. Even after composting, sawdust decomposes at a faster rate than bark, and, because of its wider C:N ratio (1,000:1), a greater amount of nitrogen is tied up in the root substrate. Whereas the problem is insignificant with bark, it must be taken into account in fertilizing a substrate containing composted sawdust.

Abandoned piles of sawdust are often available for the cost of transportation in forested areas. If a pile has existed for a year or more, the sawdust below the surface layer should be well composted. Care should be taken to avoid unleached areas deep in the pile, which are strongly acid and injurious to plants. These areas did not receive sufficient oxygen during fermentation, and, as a result, volatile organic acids were formed and trapped here. These problem areas can be identified by the exceptionally dark color of the sawdust and its pungent, acrid odor. This sawdust can be reclaimed by exposing it to the air and to leaching rains for a season, but it still will be more acid than properly composted sawdust.

Sawdust composted with additional nitrogen for one month, to the stage appropriate for use in root substrates, is itself acid and requires limestone to neutralize it. In this stage, it is granular and is medium-dark brown in color. It continues to decompose during use in the pot or greenhouse bench. Various types of pine and some types of hardwood sawdust require further additions of limestone as time passes. Sawdust, like other plant materials, ends up close to neutral in pH when thoroughly composted; however, this is well beyond the stage at which it is initially used in greenhouse root substrates.

Manure

Annual addition of manure, generally rotted cow manure, was a standard practice in fresh-flower beds and quite frequently was used in bedding- and pot-plant substrate until the middle of the twentieth century. When soil pasteurization became popular, ammonium toxicity problems arose as a result and discouraged the further use of manure. (Ways around this problem will be discussed in Chapter 7, "Root Substrate Pasteurization.") A few growers use manure today and realize good benefits from it.

Manure has a high CEC and thus serves as a good reservoir for nutrients. In addition, it is a good source of nutrients, and thus micronutrient deficiencies rarely occur when manure is used. As a matter of fact, micronutrient deficiencies were rare in the days when manure was used routinely. Today, such deficiencies present a serious problem. Manure also contains low levels of nitrogen, phosphorus, and potassium (see Table 6-4). Because large quantities of manure are used in root substrate, a significant part of the total requirement of these three nutrients is met. Manure has a high water-holding capacity, which is a basic requirement of greenhouse substrate. Peat moss perhaps comes the closest to manure in the functions that it serves in root substrate and, indeed, has been the component substituted for manure since the 1950s.

Rotted cow manure is the best type to use in the greenhouse. Other types are stronger and must be used cautiously and in smaller quantities. Often, as in the case of poultry manure, the ammonia content is too high and causes root and foliage injury. Cow manure is incorporated into a root substrate at the volume rate of 10 to 15 percent. The substrate is then pasteurized with steam or chemicals to get rid of harmful disease organisms, insects, nematodes, and weed seed. (Manure contains a sizable quantity of weed seed, which would otherwise become

Table 6-4 **Primary Nutrient Contents of Some Types of Fresh Animal Manure**

	Nutrient Content (% of fresh weight)		
Type of Manure	*Nitrogen (N)*	*Phosphorus (P_2O_5)*	*Potassium (K_2O)*
Cattle (cow)	0.5	0.3	0.5
Chicken	1.0	0.5	0.8
Horse	0.6	0.3	0.6
Sheep	0.9	0.5	0.8
Swine	0.6	0.5	1.0

troublesome.) Following pasteurization, it is very important that each time water is required, a sufficient quantity be applied to ensure leaching so that a buildup of ammoniacal nitrogen originating from the manure does not occur. Even if a crop is not planted in the substrate, it must be leached periodically. A buildup of ammoniacal nitrogen contributes to the total soluble-salt content of the root substrate and can be detected readily by a soluble-salt test. This test, which will be discussed in Chapter 9, can easily be performed by growers.

Manure has traditionally been used in a moist state, which renders it a difficult material to introduce into a mechanized system of root substrate preparation. Its after-pasteurization problems preclude its use in substrate to be stockpiled for later use. Its messy physical condition and heavy weight prevent its shipment more than a few miles from its origin. Until an economical process is devised for drying, grinding, and getting around the problem of ammoniacal nitrogen buildup, the use of manure will be limited to a few growers who have a local supply and the technical knowledge to handle it.

Crop By-Products

Straw is occasionally used as a root substrate amendment but must be chopped into pieces 3 in (8 cm) or less in length to permit uniform incorporation into the substrate. The labor input is expensive. Since straw decomposes rapidly, it must be added two to three times per year, which is also an expensive proposition. A variety of other organic amendments are occasionally used, including peanut hulls, bagasse (sugar cane fiber), and rice hulls. All of these can be used successfully, but require knowledge and careful handling. Materials such as straw, peanut hulls, and bagasse have a wide C:N ratio that causes nitrogen tie-up. If this is gauged and extra nitrogen is added, no problem arises.

Flower-crop stubble—the foliage, stems, and roots left in the benches after the harvesting of fresh-flower crops—has logically been looked upon as a source of organic matter. Growers have chopped the stubble into small pieces and rototilled it into the root substrate. Because this organic material is the very crop being grown, it is an excellent host for carrying diseases over from one crop to another. It should be pasteurized with the root substrate. Since many growers do not pasteurize after each crop, crop remains are generally removed from the greenhouse. Crop remains thoroughly composted outside the greenhouse can be used successfully as a root substrate amendment.

Composted Garbage

Many municipalities combine the collection and disposal of kitchen wastes and solid household trash. When a compost is produced from this waste, most metals, rags, and large items are first reclaimed, and then the remaining refuse is ground and set out in heaps to compost. The action of microorganisms breaking down the organic matter in these heaps generates heat, which destroys harmful organisms and results in a dark-brown, somewhat granular product. Glass is ground fine enough to prevent it from becoming a safety hazard. The pH level is about 8.5, and the salt content is moderately high but subject to removal by leaching. Processed garbage has worked well as a mulch in landscaping but has not been as satisfactory as a root-substrate component. The problem stems from the variation in refuse ingredients. When a high proportion of kitchen waste is present, a product rich in humus is produced that makes a good peat moss substitute in a traditional soil-based substrate. When high proportions of wood, paper, plastic, or other such materials are present, a product is produced that can tie up nitrogen in root substrate or simply act as an inert component that would be a better sand replacement. This variability within single batches of product has led to variable results within trials, ranging from excellent to poor. More work is needed before this product can be fully accepted as a component of greenhouse root substrate.

Vermiculite

Vermiculite ore is mined principally as a micalike, silicate mineral. The ore itself has a dry bulk density of 55 to 65 lb/ft^3 (880 to 1,040 g/dm^3), but when expanded to the state used in root substrates, the density drops to 7 to 10 lb/ft^3 (110 to 160 g/dm^3) (Figure 6-6). This lightweight property makes it very desirable in pot-plant substrate. Each particle of vermiculite ore contains numerous thin plates lying parallel to one another. Between the plates is moisture that expands into steam when heated to high temperatures, causing the plates to move apart into an open accordionlike structure. The expanded volume can be as much as 16 times the volume of the original ore. The water-holding capacity of expanded vermiculite is high because of the extensive surface area within each particle. Aeration and drainage properties are also good because of the large pores between particles. The common size is 6 to 10 mesh (USS).

Numerous negative electrical charges on the surface of each vermiculite platelet give rise to a CEC of 19 to 22.5 me/100 g (1.9 to 2.7 me/100 cc). The predominant fertilizer nutrients in vermiculite are potassium, magnesium, and calcium. The potassium content of U.S. vermiculite will provide part, but certainly

Figure 6-6 **Expanded vermiculite as it is used in greenhouse root substrate. The exceptional water- and nutrient-holding capacities of vermiculite make it an excellent component of soilless substrate.** *(Photo courtesy of Scotts-Sierra Horticultural Products Co., Marysville, OH 43041.)*

not all, of the total needs of a crop. The magnesium content of African Palabora vermiculite is high and has been known to provide the total needs of a greenhouse crop. Vermiculite varies in pH level. U.S. vermiculite is slightly alkaline, while African Palabora vermiculite tends to be very alkaline, with pH levels approaching 9 in some cases. The alkaline African vermiculite constitutes no problem when combined with an acidic substrate component such as peat moss or pine bark. If this vermiculite is used alone, in a propagation bed or in a hydroponic operation, its pH level should be adjusted downward. U.S. vermiculite can be used without alteration.

Vermiculite is a very desirable component of soilless root substrate because of its high nutrient and water retention, good aeration, and low bulk density. It is commonly included in soilless substrate. Expanded vermiculite can be compressed easily between the fingers. Under the weight of soil-based substrate, expanded vermiculite tends to compress, which greatly reduces aeration. Vermiculite is generally not used with soil.

Calcined Clay

Aggregates of clay particles are heated to high temperatures (calcined) to form hardened particles that resist breakdown in root substrate. These aggregates are large (mostly 8 to 45 mesh; 2.36 to 0.355 mm) and irregularly shaped. As a result, they fit together loosely in a root substrate, creating large pores for drainage and aeration. Within each calcined clay aggregate are numerous clay particles forming a myriad of small water-holding pores. One pound of calcined clay can contain over 13 acres of surface area within its structure. Calcined clay brings the property of structure to root substrate in the form of a hardened, buff-colored aggregate weighing about 30 to 40 lb/ft^3 (480 to 640 g/dm^3). The pH levels of different calcined clay products range from acid to alkaline (4.5 to 9.0), but they have only a small influence on the pH level of root media. Calcined clays have a sizable CEC, 6 to 21 me/100 g (3.4 to 11.8 me/100 cc), which gives them the property of good nutrient retention. The variation in properties of calcined clays stems back to the type of clay used. Examples are montmorillonite clay from the Mississippi Valley and attapulgite clay from Florida and Georgia. Lusoil, a brand of calcined clay made from attapulgite clay, has a pH of 7.5 to 9.0 and a CEC of 21 me/100 g (11.8 me/100 cc). Terragreen and Turface are derived from montmorillonite clay.

Calcined clays should be used in a quantity equal to 10 to 15 percent of the volume of fresh-flower substrate. For pot-plant substrate, they should constitute 25 to 33 percent of the total volume, the remainder being composed of either soil, peat moss, or a combination of the two.

Sand

Sand is used in soil-base root substrate for adding the coarser texture needed to induce proper drainage and aeration. For this reason, concrete-grade sand (a sharp, coarse sand) is used. Concrete-grade sand has the specifications listed in Table 6-5. Washed sand should be purchased since it is nearly free of clay, silt, and organic matter. In regions where there are snowfalls, caution should be exercised during the winter to avoid purchasing sand containing road salt (sodium chloride or calcium chloride). Road salt is added to batches of sand to be sold to highway departments because it melts road ice. The level used in sand is injurious to greenhouse crops.

Perlite

Perlite is a good substitute for sand for providing aeration in root substrate. Its main advantage over sand is its light weight of about 6 lb/ft^3 (95 g/L), as compared to 100 to 120 lb/ft^3 (1,600 to 1,920 g/L) for sand. Perlite is a siliceous volcanic rock that, when crushed and heated to 1,800°F (982°C), expands to form white particles with numerous closed, air-filled cells. Water will adhere to the surface of perlite, but it is not absorbed into the perlite aggregates. Perlite is sterile, chemically inert, has a negligible CEC (0.15 me/100 cc), and is nearly neutral with a pH value of 7.5. It does not appreciably affect the pH level of root substrates. Perlite costs considerably more than sand. As a result, it is used when low root-substrate density constitutes an economic advantage.

There have been reports that perlite releases quantities of fluoride that are injurious to lilies and certain green-plant crops. Research has shown this to be

Table 6-5 **ASTM (American Society for Testing and Materials) Specifications for Concrete-Grade Sand**

% of Total Passing the Screen	*Screen Size*	*Particle Size (mm)*
100	3/8 inch	9.5
95–100	No. 4[1]	6.4
80–100	No. 8	3.2
50–85	No. 16	1.6
25–60	No. 30	0.85
10–30	No. 50	0.51
2–10	No. 100	0.25

[1]These figures refer to the number of holes per inch. A No. 4 screen has holes slightly smaller than 1/4 inch due to the width of the wire between each hole.

unfounded. Quantities of perlite up to 50 percent of the volume of the substrate did not cause injury to sensitive crops including lilies, *Chlorophytum,* and Tahitian bridal veil.

Polystyrene Foam

This material is known more commonly by trade names such as Styrofoam, Styropor, and Styromull. Like perlite, it constitutes a good substitute for sand, bringing improved aeration and light weight to root substrate. It is a white synthetic product containing numerous closed cells filled with air. It is extremely light, weighing less than 1.5 lb/ft^3 (24 g/L). Like sand, it does not absorb water and has no appreciable CEC. It is neutral and thus does not affect root substrate pH levels.

Polystyrene can be obtained in beads or in flakes. Beads from 1/8 to 3/8 inch (3 to 10 mm) in diameter and flakes from 1/8 to 1/2 inch (3 to 13 mm) in diameter are satisfactory for pot-plant substrate. Larger particles may be used in bench substrate and for epiphytic plants such as orchids (Figure 6-7). Depending upon the source, the price of polystyrene can vary considerably. The edges cut from large blocks prior to cutting into sheets or the leftover pieces from shapes stamped from sheets can be ground to form an excellent substrate component. Polystyrene has been banned in some coastal regions due to its movement in wind and surface water to beaches, where it becomes an aesthetic problem. In other localities, it has been banned from landfills. The future of polystyrene as a root substrate component is questionable at this time.

Figure 6-7 Equal parts of polystyrene foam (Styromull) and sphagnum moss make a good root substrate for the orchid plant shown here. Polystyrene is an excellent lightweight substitute for sand in root substrate. *(Photo courtesy of BASF-Wyandotte Corporation, Wyandotte, MI 48192.)*

Rock Wool

A description of rock wool propagation cubes and slabs for growing fresh flowers and vegetables, as well as the method of manufacture, are presented in Chapter 10. Rock wool is also available in granular form for use as a component in formulating root substrate. As seen in Table 6-1, the granular form has very high available-water and aeration properties. Although slightly alkaline, it is not buffered. Mixing it with an acid component, such as pine bark or peat moss, will immediately lower the pH level. Rock wool has a negligible CEC. It neither contributes nor holds nutrients to any extent. This property should be provided by other components in the mix, such as sphagnum peat moss. Granular rock wool may be purchased by itself or in commercially formulated mixes. A blend of equal volume parts of rock wool and sphagnum peat moss makes an excellent mix.

Other Coarse-Textured Components

In the future, numerous substitutes for sand will appear—some derived from minerals and some perhaps by-products of industry. Their usefulness will be determined by their bulk density, size, shape, and cost. A few interesting and effective products have appeared on the market in recent years. One product consists of short lengths of plastic wire-coating stripped from the ends of electrical wire during the process of making electrical components. It serves as a lightweight sand substitute. A second product, Polytrol, consists of pellets and flakes, from 1/8 to 1/4 inch (3 to 6 mm) in size, made from plastics. Rejected plastic materials that cannot be used in prime products are combined with recycled municipal solid-waste plastics to make this product. Many other substitutes can be found for sand if one bears in mind the function of sand.

Soil-Based Substrate

The largest division in root substrate types falls between those containing soil and the soilless types. One type is not necessarily superior to the other. Plants of equal quality can be grown in each if cultural adjustments are made. Selection of a root substrate type is made on the basis of economics and the physical situation in which it must serve. During the 1960s, a discussion of soilless substrate would have been included in this book more as a curiosity or a prediction of things to come. Today, when one looks around the pot-plant industry, it seems conceivable that a discussion of soil-based substrate for container crops will be obsolete in a few years. This should not be the case, however, when a grower has an abundant source of good, uniform soil and has developed an efficient mixing procedure.

Formulation

A minor percentage of the pot plants in the United States are grown in soil-based substrates. By contrast, virtually all of the fresh-flower crops are grown in such substrates. Traditionally, a soil-based substrate has been composed of equal parts by volume of loamy field soil, concrete-grade sand, and sphagnum peat moss amended with phosphorus and adjusted to the proper pH level. Sandy field soil is compensated for by an increase in the proportion of peat moss and field soil and a decrease in sand, while clay soil calls for more sand.

Sand is used in soil-based substrate to develop large-diameter pores for good aeration. Two materials, perlite and polystyrene, have proven to be good substitutes. Like sand, both materials resist compaction and absorption of water. Unlike sand, they are very light in weight. A moist mixture of equal parts of soil, sand, and peat moss weighs about 100 lb/ft^3 (1,600 g/L), which is suitable for use in greenhouse benches but not for pot plants that must be handled in the greenhouse and shipped. Perlite can cost as much as three times the price of sand. Polystyrene is similar to sand in price.

Field soil provides reasonable nutrient (cation exchange) and water-holding capacities. When one-third of the soil is replaced by sand, these two properties are significantly reduced. To restore them, sphagnum peat moss, an amendment with a high CEC and water-holding capacity, has traditionally been added to the substrate at the expense of an additional one-third of the field soil.

Coarse peat moss should be obtained when possible. The structure of peat moss can vary from one peat bog to another and even within a single bog from year to year. Some peat moss is hydraulically mined, and the particles are so small that much of the effect of structure is lost. The importance of particle size is so great that some large growers obtain samples of peat moss from their suppliers, select the lot with the best structure, and then purchase a year's supply from that particular lot. Coarse peat moss has large pieces that fit together loosely to form wide pores for aeration. At the same time, the fine leaf structure within the large pieces forms copious narrow pores for holding water, up to 60 percent of its volume. Thus, sphagnum peat moss provides good water-holding capacity and a fair amount of aeration if it is coarse. Sand provides the balance of aeration.

Sphagnum peat moss is compressed into bales for shipment and sales purposes. If excess drying occurs, hard pieces can form in the bale. These pieces should be broken with a hoe. If this does not work, the peat moss can be passed through a soil-shredding machine. If passed through the machine more than once, however, the pieces are broken down nearly to individual moss filaments, and much of the desirable aeration property is lost.

Soil-based substrates generally require three nutrient amendments during formulation. First, the pH level should be adjusted to within the range of 6.2 to 6.8 with agricultural limestone. When neutral to alkaline soils are used, no upward adjustment is required. Acid soils may require as much as 10 lb of limestone per cubic yard (6 kg/m^3). The second amendment should consist of up to 1.5 lb of 0-45-0 superphosphate per cubic yard (0.9 kg/m^3) to provide phosphorus for up to one year. The third amendment is a complete micronutrient mixture, of which a number of commercial preparations are available. (For further details on fertilizer amendments, see Chapter 9.)

Maintenance

The structure of root substrate is sufficiently stable to persist until the time when the final purchaser of pot plants would ordinarily repot them. At that time, one to two years, some of the old root substrate can be removed and a new substrate prepared to fill the larger pot, thus restoring the original level of structure. In any event, loss of structure is not a problem for pot-plant growers. It is a problem for fresh-flower growers, since they maintain substrate permanently in their ground beds and benches.

The action of decomposition results in the loss of organic matter and the periodic need to add more. This is customarily done once each year at the time when the root substrate is pasteurized. The standard additive has been coarse sphagnum peat moss rototilled into the bench in a quantity equal to about 10 percent of the volume of the root substrate in the bench.

Coarse bark from 3/8 to 3/4 inch (10 to 19 mm) has proven to be a good alternative to peat moss in bench substrates. The decomposition rate of bark is slow, requiring up to three years for complete breakdown. In the first year, a quantity equal to 10 percent of the bench volume should be incorporated into satisfactorily drained substrate and 15 percent into poorly drained substrate. As a rule of thumb, each year thereafter a quantity equal to 5 percent of the bench volume is added. In any event, organic matter should be added in sufficient quantity to make up the volume loss in the bench.

Figure 6-8 A fresh-flower substrate containing too much clay. Note the cracks that occur upon drying. This substrate has inadequate gas exchange, as witnessed by symptoms of oxygen deficiency in the chrysanthemum plants. Growth is stunted, leaves are light green in color with veins lighter than the rest of the leaf blade, and the plants wilt on bright days.

Sometimes, the organic-matter level is adequate but the clay content is too high. Poor drainage and excessive cracking of the root substrate upon drying (Figure 6-8) are symptoms of this condition. This is particularly prevalent when clay soil is used. The problem is remedied by a single addition of concrete-grade sand to the substrate. Perlite is generally not used since weight is not a problem in benches. Calcined clay is sometimes used because, in addition to providing macropores for drainage and aeration, it contains numerous micropores within each particle to improve water-holding capacity and it has a high CEC, which improves nutrient retention. A quantity equal to 10 to 15 percent of the bench volume is incorporated into the substrate. It is expensive but need be applied only once, since it is resistant to breakdown.

Soilless Substrate

Growers who do not have their own sources of field soil have found it difficult to purchase soil of consistent texture from load to load. This demands considerable attention on the part of managers to compensate for soil changes in root substrate formulations. When such changes are overlooked, poor crops and loss of profits ensue. Soilless substrates are attractive to these growers.

Wholesale growers typically ship their pot plants to market by truck. The lighter weight of soilless substrates reduces the cost of handling labor. Still other growers look on soilless substrates as a form of automation since they can be purchased ready for use, thus eliminating the need for any labor input or mixing facilities. Such growers may be in a labor market of high wages or in a situation of limited labor availability.

Components of Soilless Substrates

So many materials are available for soilless substrates that growers often make the mistake of mixing too many or the wrong types together. The four functions of root substrate—plant support, aeration, nutrient retention, and moisture retention—should be considered in developing a formulation. Organic matter or clay is needed to provide CEC for nutrient retention. Unless the organic matter or clay is in coarse aggregates to facilitate aeration, coarse-textured particles such as sand, perlite, or polystyrene will be required. If the organic matter or clay selected has a high water-holding capacity, as does peat moss, no further components are necessary. However, if organic matter of insufficient water-holding capacity is used, such as coarse bark, it will be necessary to include a second organic material or clay component (such as peat moss or calcined clay) to increase the water-holding capacity. The desired density of the substrate can be attained by avoiding heavy coarse particles or clay components.

Good root substrates need not contain more than one to three components. The selection of components will generally depend upon their availability and cost. A grower who markets peat moss and thus can obtain it at wholesale cost, or who is located close to the point where it is dug so that transportation costs are minimal, should use peat moss for its superior water-holding capacity and CEC. If substrate weight is not a problem, the grower should mix it with the cheapest coarse-textured component, which is sand. If light weight is required, the considerably more expensive components perlite or vermiculite may have to be used. If light weight is required and the grower is fortunate enough to be located near a source of polystyrene flakes or beads, the lighter density can be achieved with less cost than perlite. A grower located in a timber area will probably find bark economical. Sand is added to bark because it nests between the bark particles, thus adding more surface area and, as a consequence, more water retention to a given volume of substrate. Remember that sand is added to soil for the opposite purpose, that of pushing soil particles apart to open up large pores for aeration. Often, sphagnum peat moss is also added to bark to further increase water-holding capacity as well as nutrient retention.

Formulations

Peat Moss–Based Formulations One of the earliest commercially prepared soilless substrates developed was Einheitserde (standardized soil), a mixture of half peat moss and half well-aggregated subsoil clay amended with nitrogen, phosphorus, and potassium and limed to a pH level between 5 and 6. It was introduced by Dr. A. Fruhstorfer in Hamburg, Germany, in 1948. Einheitserde is marketed by several companies in Europe and is used for a wide range of crops and applications from seed germination to plant finishing.

The University of California mixes were some of the earliest soilless substrates adopted in the United States during the 1950s. These are a series of five substrates ranging from 100 percent sphagnum or hypnum peat moss to 100 percent fine sand, with intermediate combinations of the two. These substrates are formulated by individual growers. The most popular greenhouse pot-plant substrate of this series is the half–peat moss, half–fine sand mixture. The designation of *fine sand* indicates sand between 0.5 and 0.05 mm in diameter, which is equivalent to 1/50 to 1/100 inch, or to sand that passes a 30-mesh screen but is retained on a 270 mesh screen.

The Peat-Lite mixes were introduced by Boodley and Sheldrake at Cornell University in the early 1960s. Mix A is composed of half sphagnum peat moss and half horticultural-grade vermiculite. Mix B contains horticultural perlite in place of the vermiculite. While some growers formulate Peat-Lite mixes, there are a number of commercial preparations of soilless substrate on the market similar to Peat-Lite Mix A.

It is significant that the substrates thus far discussed have been composed of only two components. This is possible because one is peat moss, which has the highest water-holding capacity of any of the components discussed, a significant CEC, and a modest degree of aeration if not too finely shredded. Peat moss comes very close to an ideal substrate by itself if it contains coarse aggregates. European growers have learned to grow top-quality crops in peat moss alone. If this system is used, it is important to guard against overwatering (watering too frequently). Because peat moss effectively retains nutrients, it is important that overdoses of fertilizer not be applied and that the peat moss is thoroughly watered each time water is needed to ensure that excess nutrient salts are leached.

Formulators of soilless substrates must remain competitive. A significant part of the expense of these substrates to the grower is the shipping cost. For this reason, substrates are in a rather dry state when shipped. Dry peat moss, particularly when it is finely ground, can be exceedingly difficult to wet because it repels water. Therefore, wetting agents are incorporated into peat moss–based soilless substrates. When a substrate does not contain a wetting agent and it is allowed to dry excessively, it becomes nearly impossible and economically unfeasible to wet the substrate once it is in a pot or flat. Such a substrate must be placed in a mixer, where water can be blended into it.

However, the wetting agent does not completely correct the problem. Substrates containing only peat moss and vermiculite can be difficult to re-wet when allowed to dry beyond the desired level, even though they contain a wetting agent. They may require one or two extra applications of water, which is a waste of labor.

This tedious wetting procedure can be avoided by adding coarse-textured particles such as sand or perlite to the substrate. These components provide large pores that allow quicker penetration of water throughout the substrate. In this way, lateral movement of water into the smaller pores of peat moss can occur more quickly, resulting in less labor. There are a number of commercially available substrates that contain sand and/or perlite in addition to peat moss and vermiculite.

Bark-Based Formulations Since bark is cheaper than peat moss, bark-based substrates are typically cheaper than peat moss–based substrates. This is one reason bark mixes are very commonly used for larger containers of green plants and nursery crops. While the water- and nutrient-holding capacities of bark are generally not as good as those of peat moss, the aeration and wettability of substrates containing bark are excellent. Concrete-grade sand is often added to bark to increase its water-holding capacity. This occurs because sand particles are smaller than many of the bark particles and, as such, will nest between the bark particles. A unit volume of the bark-sand mixture contains nearly as many bark particles as when there was only bark, but it now contains sand particles in addition. This combination increases the particle surface area per unit volume of substrate and, consequently, the water-holding capacity.

By contrast, the addition of concrete-grade sand particles to soil increases aeration and lowers water retention. This occurs because the sand particles are larger than the soil particles. The sand particles now occupy space formerly held by smaller soil particles. A unit volume of substrate therefore has fewer but larger particles. This reduces the particle surface area per unit volume of substrate and thereby reduces its water-holding capacity.

Vermiculite or peat moss is commonly used in addition to sand in commercial greenhouse preparations of bark substrates to further add moisture and also nutrient retention. These are not used in nursery substrates because of their price, but more important, they are not used because of their high water-retention property. During periods of rain, a nursery substrate requires higher aeration capacity than a substrate inside a greenhouse. If it is not properly aerated, the wetter nursery substrate will have too little gas exchange for adequate plant growth.

A ratio of 3 parts bark to 1 part sand to 1 part peat moss is favored by growers who mix their own substrate. Preferred bark sizes are $^1/_4$ inch (6 mm) and smaller for small pots, $^3/_8$ inch (10 mm) and smaller for intermediate pots, and $^1/_2$ inch (13 mm) and smaller for larger green plant and nursery pots. Pine, fir, and hardwood barks are commonly used. Many commercial substrates based on bark are available.

Formulation Summary Any given grower usually has access to 10 or more companies producing root substrates. Each substrate company typically offers three or more of five categories of greenhouse substrates. These categories include a germination mix, a general-crop/high-moisture mix, a general-crop/peat moss–based mix, a general-crop/bark-based mix, and a high-aeration mix.

Germination substrates (Category I) are usually composed of peat moss and vermiculite. The particle size of the peat moss and vermiculite is finer than in general-crop substrates to insure greater uniformity of texture. From the earlier discussion of container depth versus aeration, it would seem logical to use a very coarse-textured substrate in germination containers because of their shallow depth. However, this is not the practice because of the need for a higher degree of texture uniformity in germination containers than in larger finish-plant pots. The greater uniformity is achieved through the use of smaller component particles. Texture needs to be uniform to ensure that each of the tiny cells in a plug tray has similar water retention and aeration and to ensure that seeds lodge at approximately the same depth during sowing so that they will germinate at the same time. Texture also needs to be finer to ensure adequate water as the seedling grows larger in a very small volume of root substrate. The adverse effect of fine texture on aeration is compensated for by the use of peat moss of high aggregate structure. Germination substrates tend to cost more due to the price of their components.

General-crop/high-moisture substrates (Category II) usually contain peat moss and vermiculite. The particle size of each is larger than in the germination substrates because the need for texture uniformity is not as great as the need for aeration. In the larger pots in which these substrates are used, there is a greater need for aeration, probably due to the greater distances from the substrate-air interfaces to the inner zone of the root ball. Growers need to be very precise in their watering habits with these substrates. If they water too frequently, insufficient aeration will become a major problem. If watering is not frequent enough, it may become very difficult to re-wet the substrate.

General-crop/peat moss-based substrates (Category III) contain perlite and possibly sand in addition to peat moss and vermiculite. Some contain peat moss and perlite only. These substrates are more forgiving in situations of over- or underwatering. Thus, they are very popular.

General crop/bark-based substrates (Category IV) have bark as a partial or full substitute for peat moss. Because of the bark, they absorb water freely and have good aeration. In some cases, these substrates can be cheaper than the peat moss–based substrates. The physical properties of these bark-based substrates depend on the particle size of the bark and the other components used with the bark. With more extensive composting and finer particle size, bark takes on more of the higher water- and nutrient-retention properties of peat moss. When well composted, fine bark is used with vermiculite, or vermiculite and some peat moss, the substrate functions similarly to the Category II substrates. When a substrate contains bark that is coarser, less composted, or used in combination with perlite or polystyrene, it has less water retention and greater aeration. Such substrates can be similar to Category III substrates or even more extreme. One needs to test several lines of bark-based substrates before assessing their efficacy in a given situation.

High-aeration substrates (Category V) contain larger bark particles. They are often cheaper than the other substrates. This is a decided advantage for plants in large containers. These substrates also have the advantage of very high aeration. Crops grown outdoors, where they are subject to rain, benefit from this property.

There is no single substrate that is best for each crop. The best substrate depends on the cultural habits of the grower. A grower who waters infrequently or is faced with a market channel where plants are not watered sufficiently will do best with a high-moisture-retention Category II substrate. Conversely, a grower who waters frequently will find a well-drained Category III or IV substrate best. Each grower needs to test a range of substrates before deciding which is best. Alternatively, growers need to be aware that they can alter their watering practices from crop to crop so that a minimum of substrates can be used for a wide range of crops.

These commercially available substrate categories can also be formulated by growers. Ten root substrate formulas are presented in Table 6-6. These are the more common mixes produced by growers themselves and are representative of many of the commercially prepared mixes. The first mix is the classical soil-based mix containing equal volumes of loamy soil, sphagnum peat moss, and concrete-grade sand. If one considers vermiculite a substitute for soil, then the remaining soilless mixes in the table emerge. This is reasonable, since vermiculite has high nutrient- and water-retention properties. Note that the peat moss of the classical mix can be retained or can be partially or completely replaced by bark. The formulas and components used with pine bark and hardwood bark are the same. When peat moss is combined with bark, it is not necessary to use vermiculite. Other organic matter, including the various composted plant materials, should be looked upon as a partial or complete replacement for peat moss. Perlite, polystyrene, or any other coarse particle capable of imparting drainage and aeration properties to the substrate can be substituted for the sand of the classical soil-based mix. Firms producing large quantities of substrate often avoid sand because it rapidly wears out the mixing equipment and can increase shipping costs.

Future Formulations Root substrates discussed thus far include the more common components. Numerous other components exist, and many new ones will be developed as the trend away from soil-based substrate continues. You should now be able to determine, with only a minimum of testing, whether these components are useful to you and how they should be used. First, select a com-

Table 6-6 **Several Currently Popular Greenhouse Root Substrate Formulas and Their Functions**

Substrate Components	*Function*
1 soil:1 peat moss:1 sand	Pot and bench mix
1 vermiculite:1 peat moss	Germination mix
1 vermiculite:2 peat moss:1 perlite	Pot-plant mix
3 peat moss:1 perlite	Pot-plant mix
1 vermiculite:1 pine bark	Pot-plant mix
2 vermiculite:2 pine bark:1 perlite	Pot-plant mix
2 vermiculite:1 peat moss:1 pine bark:1 perlite	Pot-plant mix
1 peat moss:3 pine bark:1 sand	Pot-plant mix
1 peat moss:3 hardwood bark:1 sand	Pot-plant mix
1 rock wool:1 peat moss	Pot-plant mix
3 rock wool:7 peat moss	Pot-plant mix

ponent that provides adequate moisture and nutrient retention. If one component does not provide both functions adequately, two components may be required. Seek components that have aggregate structure so that optimum aeration is provided. If this is not possible, a coarse-textured component will be needed to provide aeration. The fewer components, the better, because of the cost of mixing. Be sure that none of the components provides an excessive quantity of nutrients or salt, such as excessive ammonium released by rapidly decomposing peats or from chicken manure.

Fertilizer Amendments As in the case of soil-based substrates, soilless substrates require three nutrient amendments. If needed, dolomitic limestone should be added to bring the pH level into the range of 5.4 to 6.0. Most often, 10 lb of agricultural dolomitic limestone per cubic yard (6 kg/m^3) is used for this purpose. Phosphorus is added as superphosphate (0-45-0) at a rate up to 2.25 lb/yd^3 (1.3 kg/m^3). The third nutrient additive is a micronutrient mix in a quantity sufficient to last at least one crop time (three to four months). In addition to these three amendments, a wetting agent is almost always included. Quite often, but not always, nitrogen and potassium, in amounts sufficient to last about two weeks, are likewise included. (Specific instructions for nutrient amendment of soil-based and soilless substrates are presented in Chapter 9 under "Pre-Plant Fertilization.") Most commercial formulations contain all of these amendments.

Economics of Substrate

The greenhouse grower who elects to use a soilless substrate must decide whether to purchase it ready for use or to formulate it. This decision must be made individually and is based on economics. The grower should calculate the cost of the substrate he or she formulates and compare it with the price of commercial substrate, including shipment. In calculating the formulation cost, be sure to include management time, office expenses, depreciation cost of the mixer, any conveyor belts and front-end loaders used to fill the mixer, buildings used for holding components of the substrate, cost of pasteurization if it is necessary, and all labor costs. Whether you formulate a soil-based or a soilless substrate, you may be startled by the true cost.

Commercial substrates, while expensive at face value, are not very different in cost from individually formulated substrates and can actually be cheaper if a steady source of relatively inexpensive components is not available for mixing your own substrate. Several of the widely available brands of commercial substrate cost \$1.25 to \$2.50/ft^3 (\$44 to \$88/m^3) delivered to the grower. The lower price reflects the lower-priced bark category of mixes, packaged in compressed bales rather than in loosely filled bags, a short shipping distance, and a competitive market zone. Shipping itself costs between \$1.50 and \$2.00 per loaded mile (\$0.94 to \$1.25/km) for a truckload of 90 to 100 yd^3 (69 to 77 m^3) of peat moss–based substrate or 80 to 90 yd^3 (61 to 69 m^3) of bark-based substrate. Many local formulators sell at even lower prices, particularly when the substrate is pur-

Table 6-7 **Number of Pots and Flats That Can Be Filled from 1 ft^3 of Root Substrate**

Pot Size (in)	*Number of Pots or Flats/ft^3*	*Pot Size (in)*	*Number of Pots or Flats/ft^3*
Standard Round Pot		Azalea Pot *(cont.)*	
2 1/4	296	6	18
2 1/2	176	6 1/2	15
3	120	7	11
3 1/2	80	7 1/2	9
4	44	8	6.7
5	24	10	4
5 1/2	19	12	2
6	14		
7	9	Low Pan	
8	5.6	5	40
10	3	6	31
12	1.6	7	14
Standard Square Pot		Hanging Basket	
2 1/4	283	6	30
2 1/2	234	8	14
3	151	10	7.4
4	57	Bedding-Plant Flat (11.25 in × 21.25 in × 2.5 in)	
4 1/2	50		
Azalea Pot		32-cell	5.4
4	64	48-cell	6
5	32	72-cell	7
5 1/2	23		

chased unpackaged by the truckload. The figure of \$2.50/ft^3 (\$88/m^3) appears to be high but is not necessarily intolerable. Fifteen 6-inch (17-cm) azalea-type pots used for pot mum culture can be filled from 1 ft^3 of substrate (485 pots/m^3) (Table 6-7) at a cost of \$0.17 per pot. If each finished pot wholesales for \$4.50, the substrate cost will be less than 4 percent of the total costs of production.

Preparation and Handling of Substrate

You have now made several important decisions. First, you have decided whether to use a soil-based or a soilless substrate. Next, you have decided whether to purchase or prepare your own substrate. If you have decided to formulate your own substrate, you have determined the minimum number of components necessary to ensure reasonable nutrient and moisture retention without sacrificing aeration. You have further studied all possible component substitutes to reduce costs by investigating materials ranging from bark, sawdust, and coconut fiber to polystyrene and calcined clay. Now, you have worked out a formulation that is tailored to your conditions and needs. It meets the required functions of a root substrate, has the proper weight for your format of handling and shipping, and incorporates the most economical combination of various components available in your locality. You must complement this plan with an efficient system for mixing and handling.

Small-Batch Handling

Very small batches of a substrate (up to 5 or 6 ft^3, 0.14 to 0.17 m^3) may be mixed by hand shovel on a potting bench or on any hard surface (Figure 6-9a). Components are piled on one another and the nutrient amendments, including limestone, superphosphate, and micronutrient mix, are broadcast over the pile. The pile is thoroughly mixed in three or four shifts. The pile is methodically removed by shovel from its base in the front. As material is removed, other material higher up tumbles downward, mixing as it falls. The new pile is built in front of the original pile by continually dropping material on the top point of this conically shaped pile. As material is added, it tumbles down all sides of the pile, mixing as it goes. This procedure is repeated two to three more times by moving the pile to the side and then to the back.

Intermediate-Volume Handling

Preparation of larger batches requires motorized equipment. The simplest system calls for a concrete pad and a tractor with a bucket. Components are piled on the pad and then are mixed by tractor in a similar fashion to that described for hand mixing. Although this is the most common mixing system, one questions the uniformity of the product.

A more sophisticated system makes use of a mixer 2 to 10 yd^3 in capacity (1.50 to 7.5 m^3) (Figures 6-9b, 6-9c). Growers sometimes purchase old concrete trucks. The mixer is removed, reconditioned, and set up for greenhouse operation. The mixer is located near piles of the root substrate components, which are fed into it by either a conveyor belt or a tractor-mounted front-end loader. Upon mixing, the substrate is automatically discharged from the mixer into a potting trailer. Then the trailer is moved to a convenient location for potting plants, and the sides are lowered to a horizontal position to serve as a potting bench (Figure 6-10).

Soilless substrates are not pasteurized, whereas soil-based substrates generally are. To accommodate pasteurization, the bed of one commercial-design trailer has a perforated plate with a chamber below it. For homemade trailers, a series of 1¼-inch (3-cm) pipes spaced 1 foot (30 cm) apart are fixed to the bottom of the trailer. The pipes are connected to a manifold pipe that has a single steam inlet. The other ends of all pipes are capped. Holes from ⅛ to ¼ inch (3 to 6 mm) in diameter are drilled in pairs every 6 in (15 cm) on opposite sides of each pipe to permit the escape of steam.

When the trailer is filled with the substrate, a tarp is fastened over it and

(a)

(b)

(c)

Figure 6-9 (a) A hand-shovel procedure for mixing small batches of a root substrate. (b) A small-scale root substrate mixing operation. The soil shredder on the left is used to break up clods in field soil. Components of the substrate, including fertilizer amendments, are mixed in the 2-ft^3 (60-L) cement mixers. The freshly prepared substrate is placed in the pasteurizing wagons in the background and is pasteurized. (c) An intermediate-sized root substrate mixing operation making use of the mixer from a concrete truck.

Figure 6-10 **A potting trailer with sides lowered to the horizontal position to serve as a potting bench.**

steam is injected into the chamber below the false bottom or into the pipe distribution system. Steam rises through the perforations and percolates up through the root substrate, thereby destroying harmful disease organisms, insects, and weed seeds. This pasteurization process requires about two hours. The tarp is then removed, and when the substrate is sufficiently cool for handling, it is used.

Large, Fully Automated Systems

Large growers are in the best position to automate. Substrate handling systems of two general types can be purchased ready-built or can be designed and assembled by the grower. Where an automated system is justified, it is generally in daily use and therefore is placed under a roof to permit its use regardless of weather.

The first system (Figure 6-11) begins with storage bins for the root substrate components, which can be filled directly by trucks. Components are then moved by tractor to a hopper mounted over a conveyor belt that feeds them into a mixer. Chemical amendments are added directly to the mixer. If pasteurization is required, steam is injected into the mixer to pasteurize the root substrate. The substrate is then expelled into a storage bin by reversing the mixer. Later, it can be moved from the storage bin to an automatic pot- or flat-filling machine by conveyor belt.

The second system is a continuous-belt mixing system. Each of the substrate components is loaded by tractor into a hopper mounted over a belt. A gate at the bottom of each hopper can be adjusted to control the ratio of components in the mix. Nutrient amendments are placed in a smaller hopper that meters these out

(a)

(b)

(c)

Figure 6-11 A soil-based substrate mixing system for large operations. (a) Components of the substrate are placed in the hopper, through which they drop into a shredder and then pass up an elevator into a mixer. Chemical amendments are added directly into the mixer from above. The root substrate is steam pasteurized in the mixer, with steam produced in the portable steam generator at the right. After pasteurization, the rotation of the mixer is reversed to expel the substrate into the 6-yd^3 (4.6 m^3) storage hopper on the left. The duct over the elevator is used to blow cool air into the mixer when it is emptying, in order to reduce the time before the root substrate can be handled. (b) Further along the system, the root substrate is automatically brought by conveyor belt to a pot-filling machine as needed. In this scene, 157 3-inch (7.6-cm) pots are being filled per minute. Pots leave the filling machine on a belt and can be planted directly or can be removed and planted elsewhere. (c) A close view of the pot-filling machine. Excess substrate is recycled back to the hopper on the filling machine. This machine can fill flats or pots of any size, including 3-gallon (11-L) cans. *(Photos courtesy of Soil Systems, Inc., Apopka, FL 32703.*

over the other components, such as peat moss, already on the belt. Further along, the belt passes through a box in which spinning tines are located for mixing the various substrate components and nutrient amendments together. The mixed substrate continues along the belt to either a storage bin or the hopper on a flat- or pot-filling machine. When the substrate runs low in the hopper of the potting machine, a button is pushed to turn on the belt mixing system, which results in the hopper being refilled. All that is necessary is to keep the raw-ingredient hoppers in the belt system supplied.

Summary

1. Root substrates must serve four functions: to provide water, to supply nutrients, to permit gas exchange to and from roots, and to provide support for plants.
2. Desirable properties of greenhouse root substrates include the following:
   - **a.** For pot-plant substrates, a stable organic-matter content that will not diminish significantly in volume during growth of a crop.
   - **b.** Organic matter with a reasonable carbon-to-nitrogen ratio and rate of decomposition, so that nitrogen tie-up is not troublesome.
   - **c.** For pot-plant substrates, a bulk density light enough to enhance handling and shipping but sufficiently heavy to prevent toppling of plants—40 to 60 lb/ft^3 (640 to 960 g/dm^3) when wet at container capacity.
   - **d.** At least 10 to 20 percent air by volume at container capacity (CC) in a 6-inch (17-cm) azalea-type pot with as high an available-water content as possible without sacrificing bulk density or aeration needs.
   - **e.** A high cation exchange capacity (CEC) for nutrient reserve (6 to 15 me/100 cc).
   - **f.** A pH level of 6.2 to 6.8 in soil-based substrate and 5.4 to 6.0 in soilless substrate for crops in general but lower for acid-requiring plants.
   - **g.** A sufficient level of all nutrients other than nitrogen and potassium to prevent a deficiency for the duration of at least one crop.
3. A long list of potential components exists for use in greenhouse substrates. One should select substrate components on the basis of meeting the four functions of substrate and of economics, steady availability, and use of a minimal number of components.
4. Soil-based substrates have traditionally been used in greenhouses. Soil provides water and nutrient retention. Concrete-grade sand is added to increase aeration, and peat moss is used to restore moisture and nutrient retention lost by the addition of sand. A standard formulation of 1 part loamy soil to 1 part sand to 1 part peat moss can be altered to accommodate various soil textures.
5. Soilless substrates are an asset where soil procurement or weight is a problem. Peat moss alone or combined in equal volume amounts with vermiculite or

sand constitutes an effective root substrate. Composted bark, of species ranging from pine to hardwoods, also provides a good base for soilless substrates. Two successful formulations include either equal volumes of bark and vermiculite or 3 parts bark to 1 part peat moss to 1 part concrete-grade sand.

6. It is desirable to amend root substrates with three nutrient packages: agricultural dolomitic limestone to achieve desired pH levels, superphosphate, and a micronutrient mixture. In addition, soilless substrates should be amended with a wetting agent. Although not necessary, some growers and most producers of commercial formulations include sufficient nitrogen and potassium to last about two weeks. (Specific recommendations are given in Chapter 9.)

7. The preparation and handling of root substrates poses an important economic consideration for growers. A root substrate may be purchased already mixed and chemically amended, thus circumventing considerable labor, or it may be formulated by the grower. Various degrees of automation are available for formulating and handling substrates and should be considered.

References

1. Baker, K. F., ed. 1957. *The U.C. system for producing healthy container-grown plants.* Univ. of California Agr. Exp. Sta. and Ext. Ser. Manual 23. Univ. of California, Berkeley, CA.
2. Boodley, J. W., and R. Sheldrake, Jr. 1982. *Cornell peat-lite mixes for commercial plant growing.* New York State College of Agr. and Life Sci. Ext. Info. Bul. 43.
3. Bunt, A. C. 1988. *Media and mixes for container-grown plants.* London: Unwin Hyman.
4. DeBoodt, M., ed. 1974. First symposium on artificial media in horticulture. *Acta Hort.* No. 37.
5. Johnson, P. 1968. *Horticultural and agricultural uses of sawdust and soil amendments.* P. Johnson, 3106 Simbar Rd., Bonita, CA 92002.
6. Kelly, J. C., ed. 1978. Symposium on production of protected crops in peat and other media. *Acta Hort.* No. 82.
7. Lemaire, F., ed. 1982. Symposium on substrates in horticulture other than soils in situ. *Acta Hort.* No. 126.
8. Maulopa, E., and D. Gerasopoulo, eds. Symposium on updating the soilless cultivation technology for protected crops in mild climates. *Acta Hort.* No. 408.
9. Penningsfeld, F. 1972. Third symposium on peat in horticulture. *Acta Hort.* No. 26.
10. Poincelot, R. P. 1975. *The biochemistry and methodology of composting.* Connecticut Agr. Exp. Sta. Bul. 754.
11. Reed, D. W., ed. 1996. *Water, media and nutrition for greenhouse crops.* Batavia, IL: Ball Publishing.
12. Robinson, D. W., and J. G. D. Lamb, eds. 1975. *Peat in horticulture.* New York: Academic Press.
13. Rynk, R., ed. 1992. *On-farm composting handbook.* Northeast Regional Agr. Eng. Ser., Cornell Univ., 153 Riley-Robb Hall, Ithaca, NY 14853-5701.
14. Sonneveld, C., ed. 1994. International symposium on growing media and plant nutrition in horticulture. Acta Hort. No. 401.

15. Tattini, M., ed. 1992. Symposium on horticultural substrates other than soil in situ. *Acta Hort.* No. 342.
16. Van der Borg, H. H., ed. 1975. Symposium on peat in horticulture. *Acta Hort.* No. 50.
17. Wilson, G. C. S., ed. 1980. Symposium on substrates in horticulture other than soils in situ. *Acta Hort.* No. 99.

 CHAPTER 7

Root Substrate Pasteurization

Subtropical conditions exist in the greenhouse that are conducive to the development of plant disease organisms. The environment never freezes, the atmosphere is continually moist, and temperatures are always warm. The continuous culture of one or, at best, a few crops accentuates the disease problem by providing a continuous host on which disease organisms can build.

Prior to 1950, vegetable and fresh-flower firms avoided soil-borne disease problems by removing root substrate from greenhouses annually. New substrate was brought in that had been carefully prepared by a proper succession of crops in the field and by composting, as described in Chapter 6. During the 1950s, this cumbersome labor-consuming system became less prevalent as root substrate pasteurization was adopted.

Root substrate pasteurization is a standard practice today for virtually all fresh-flower greenhouse ranges and those vegetable ranges that grow in ground beds. It generally is done on an annual basis, although a number of growers are pasteurizing their root substrate before every crop. The need for such an increase in frequency is occasionally dictated by the buildup of disease in the greenhouse. For a relatively short crop, such as chrysanthemums, substrate pasteurization could be required every 12 to 16 weeks. The summer is a preferred time for pasteurization because crop production is usually at a low point, student labor is more available, root substrate is warmer, and, in the case of steam pasteurization, all or much of the boiler capacity is available.

Root substrate pasteurization, in addition to eliminating disease organisms, is used to control nematodes, insects, and weeds. Field operators have been known to pasteurize soil for the single benefit of weed control.

Firms that grow containerized plants that are sold with the substrate fall into two groups. The first group uses soil-based substrates and, because of the presence of soil, must practice substrate pasteurization. The second group uses soilless substrate and generally does not pasteurize the substrate.

Pasteurization may be accomplished by injecting steam into the substrate or by injecting one of several chemicals, such as methyl bromide and chloropicrin. These two methods will be discussed separately.

Steam Pasteurization

Temperature Requirements

A number of organisms are injurious to plants, and each organism has its own conditions under which it is destroyed, as set forth in Figure 7-1. It has been customary to apply steam for 30 minutes beyond the time when the coldest spot

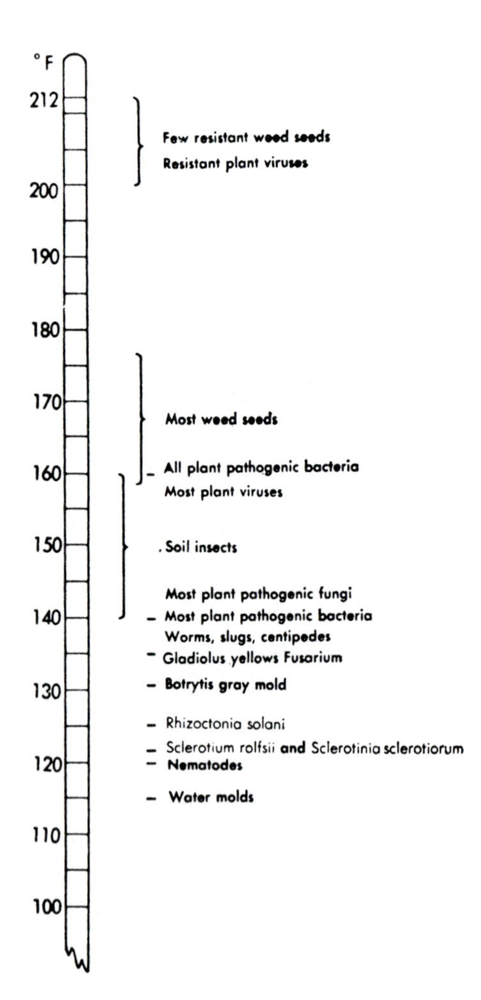

Figure 7-1 **Temperatures necessary to kill pathogens and other organisms harmful to plants. Most of the temperatures indicated here are for exposures of 30 minutes under moist conditions.** ***(From Baker [1957].)***

in the batch of root substrate being pasteurized reaches 160°F (71°C). While this practice guarantees 30 minutes at a minimum temperature of 160°F (71°C), the root substrate temperature usually rises to 212°F (100°C), the temperature of steam.

This chapter refers to pasteurization rather than to sterilization because sterilization would imply destruction of all organisms in root substrate, whereas pasteurization indicates that only selected organisms are killed. Root substrates contain, in addition to the harmful disease organisms, many beneficial organisms. A root substrate heavily occupied by beneficial organisms is not readily infected by disease organisms. By virtue of their strong foothold, the beneficial organisms compete successfully for oxygen, space, and nutrients, and present resistance to the establishment of disease organisms. Pasteurization at a temperature of 212°F (100°C) results in considerable destruction of beneficial organisms. This situation is not harmful if beneficial organisms are the first to reinoculate a root substrate. However, if disease organisms are first, they will develop rapidly without resistance or competition.

Figure 7-2 A steam aerator that permits pasteurization of root substrate at a temperature of 140 to 160°F (60 to 71°C) rather than at 212°F (100°C), the temperature of steam. Steam from a boiler enters the aerator at the right while a blower introduces air. The two gases mix in the tank at the top and are then conducted through the hose on the left to a covered potting wagon containing the root substrate to be pasteurized. A hand valve is used to regulate the amount of air introduced, which, in turn, controls the temperature of the mixture. The temperature is indicated on the temperature gauge shown.

Equipment is available today that will mix air with steam (Figure 7-2). The temperature of the mixture can be adjusted to a desired level below 212°F (100°C) and injected into the root substrate. The root substrate will rise to this desired temperature and no higher. This system is known as aerated steam pasteurization. Several recommendations call for a temperature of 140°F (60°C) for 30 minutes, although many growers pasteurize at a temperature of 160°F (71°C) for 30 minutes. Most harmful organisms are destroyed at either temperature, while only a minimum of beneficial organisms are killed.

Root Substrate Preparation

Root substrate should be loosened before pasteurizing. If it is in a bench, it should be rototilled. Heat moves more rapidly within root substrate by convection through the pores than it does by conduction from particle to particle. The large pores in loose root substrates facilitate the movement of steam and thereby cut down the length of time required to pasteurize bench or container root substrates.

Root substrate should not be dry. Dry root substrate acts very much as an insulator, resisting the conduction of heat and causing the substrate to warm up slowly. The addition of water speeds up the rate of pasteurization, but there is an optimum level of water beyond which further additions again slow down the speed of pasteurization. Water requires five times as much heat to raise its temperature as does an equal weight of soil. Since all the excess water in the root substrate must also be raised to the desired temperature of pasteurization, the process becomes very slow and consequently expensive. As a general rule, the root substrate should be at the moisture level one would desire at the time of planting a crop.

A few types of weed seed can survive temperatures approaching 212°F (100°C). Some of these are bur clover, buttonweed, Klamath weed, morning glory, and shepherd's purse. This problem can be circumvented by moistening the root substrate a week or two prior to pasteurization. As soon as seed begins the germination process by taking up moisture, it is easily killed at lower temperatures typically used for pasteurization.

Since root substrate must be mixed prior to pasteurization, it is desirable to add the various chemical and physical amendments at that time. Superphosphate, limestone, fritted micronutrients, inorganic complete fertilizers, and the slow-release fertilizer MagAmp can undergo the process of pasteurization without adverse effect. The slow-release fertilizers Osmocote and Nutricote should not be steam pasteurized. The high temperature can damage the coating, resulting in an increased rate of release. This leads to a high soluble-salt injury.

Fresh-flower and vegetable substrates used for a succession of crops in ground beds and benches require periodic additions of organic matter such as peat moss or bark. These are most easily incorporated at the time a root substrate is rototilled prior to pasteurization. It is also a good practice to carry these amendments through the pasteurization process to destroy any harmful organisms that might be in them.

Steam Sources

The temperature of 1 cubic foot of a greenhouse root substrate, on the average, can be raised 1°F by the addition of 24 Btu of heat. (One cubic meter of a root substrate can be raised 1°C by the addition of 1.6 MJ or 381 kcal of heat.) The lower the initial temperature of the substrate, the greater the quantity of heat that must be applied to pasteurize it. Table 7-1 lists the heat required to raise soil-based substrate to 180°F (82°C) from various starting temperatures.

Steam pasteurization efficiency may be as low as 50 percent. Half of the heat generated in the boiler may be lost from the boiler itself, the lines leading to the root substrate, the walls of the bench, and the cover over it. It is therefore necessary to double the figures in Table 7-1 for determining the size of boiler needed. Since 1 boiler horsepower (hp) is equal to 33,475 Btu per hour, a total of about 6 ft^3 (0.17 m^3) of substrate at 65°F (18°C) can be pasteurized in 1 hour with 1 boiler hp of heat at 50 percent efficiency (5,580 Btu/ft^3). This would be equivalent to about 12 ft^2 of bench area. (One cubic meter of a root substrate requires 208 MJ or 50,000 kcal during pasteurization.)

Boilers can also be rated in terms of pounds of steam generated. In this case, we are referring to 1 lb of water heated to the state of steam. When 1 lb of steam at 212°F changes state to 1 lb of water at 212°F, it releases 970 Btu of heat (1 g steam releases 540 cal). One more Btu is released for each degree the water drops below this point (1 cal released from 1 g water for each decrease of 1°C). If the root substrate is pasteurized at 180°F, the water will drop 32°F, releasing an additional 32 Btu beyond the 970 Btu released when it changed states. Thus, 1 lb of steam contributes 1,002 Btu to the job of pasteurization. About 6 lb of steam are required to pasteurize 1 ft^3 (96 kg steam/m^3) of a root substrate.

A steam boiler used for heating a greenhouse can be used for pasteurization. A tee and valve should be installed in the main steam line at a convenient point in each greenhouse from which steam can be obtained.

Table 7-1 **Heat Required to Raise 1 ft^3 or 1 m^3 of Greenhouse Root Substrate Containing 15 Percent Moisture from Various Starting Temperatures to 180°F (82°C)[1]**

Start (°F)	*Heat (Btu/ft^3)*	*Start (°C)*	*Heat (kcal/m^3)*
70	2,640	20	21,824
60	2,880	15	23,584
50	3,120	10	25,344
40	3,340	5	27,104
30	3,600	0	28,864

[1]Adapted from Gray (1960).

Steam does not have to be generated under high pressure for pasteurization purposes. Once it is released in the root substrate, it is under very low pressure—considerably less than 1 psi (6.9 kPa). Pressure at the boiler serves the purpose of driving the steam through the lines to the root substrate. For this purpose, a pressure at the boiler of 10 to 15 psi (70 to 100 kPa) is practical. It is true that the heat content of steam rises as it is put under pressure. However, the increase in heat content is small, and a high-pressure system must be justified on other grounds, such as heat distribution in a large greenhouse range. When steam pressure is increased to 50 psi (345 kPa), the temperature rises to about 297°F (147°C), and the additional heat content of 1 lb of water increases by only 29 Btu over steam at zero pressure (the heat content of 1 kg water increases by 67 kJ or 16 kcal).

Steam Distribution

Steam should be conducted from the portable steam generator or main steam line in the greenhouse through a low-pressure steam hose of at least 1.25 inches (32 mm) in diameter. Couplings on the hose should be full flow. If steam is provided from a central boiler, there should be a valve in each greenhouse section from which steam can be obtained (Figure 7-3).

Steam is distributed in fresh-flower and vegetable ground beds or benches through buried perforated pipes. For beds 3 ft (0.9 m) wide, one row is buried;

Figure 7-3 **When steam is provided by a central boiler for root substrate pasteurization, it is best to have a permanent steam line in each greenhouse from which steam can be obtained for this purpose. A subsurface steam line with periodic risers is used in the situation here to minimize the length of steam hose and the amount of labor required.**

Figure 7-4 **An easily constructed steam-line manifold. The 4-foot (1.2-m) -wide bench pictured here is best pasteurized with two perforated steam-conduction pipes buried in the root substrate.**

for 4-foot (1.2-m) beds, two rows are used. Used rain gutters, used boiler flue tubes, irrigation pipe, and other materials can be used for this purpose. A pair of holes from 1/8 to 1/4 inch (3 to 6 mm) in diameter should be drilled on opposite sides every 6 in (15 cm) to distribute steam. The end of each pipe is plugged with a cap. A simple pipe manifold can be assembled to distribute steam from the inlet hose to each pipe (Figure 7-4).

Many older ground beds, particularly in rose ranges, were constructed with a concrete V-shaped bottom. At the lowest point in the V, drainage tile was installed along the length of the bed. Steam can be very effectively applied through this tile, minimizing the equipment and labor of setup needed. Ground beds without bottoms can present another problem. Disease organisms and nematodes can exist below the point to which the soil has been loosened. Steam does not penetrate rapidly into this hard area. Harmful organisms below this point can return after pasteurization to the upper levels where roots grow. It is best to bury the steam-conduction pipes at the bottom of the rototilled root substrate. This results in deeper penetration of steam and also prevents nematodes and symphilids from escaping by burrowing deeper, ahead of the steam.

Raised benches filled with root substrate may be pasteurized with or without buried steam-conduction pipes. If pipes are used, they are buried at half the depth of the substrate. This is the best system. Some growers inject steam between the cover and the root substrate through 5-inch (13-cm) -diameter canvas hoses. Once the cover is inflated, steam readily penetrates the loosened root substrate. Although this system is easier to set up than the buried steam-pipe system, it can require a longer time for steam to penetrate the root substrate.

Empty raised benches also can be pasteurized with steam distributed through a 5-inch (13-cm) -diameter canvas hose. The hose is slipped over the end of the steam hose and tied in place. It is then placed on the root substrate, and the distant end is tied closed with a piece of wire. The hose should be wet before pasteurizing to speed up the initial release of steam.

Potting substrates are best pasteurized in a wagon equipped with perforated steam pipes at the bottom or a perforated false bottom with a steam chamber below. Such wagons have already been described in Chapter 6. Ideally, the sides of such wagons can be lowered to a horizontal position to serve as potting benches.

Fields of soil also can be steam pasteurized, which is commonly done in chrysanthemum production areas such as Florida and California. It would be quicker to inject methyl bromide into the soil by tractor, but this practice would not completely kill verticillium wilt—a very devastating and prevalent disease of chrysanthemums in production fields. Steam is effective against this disease. The boiler may be in a fixed central location or may be mounted on a truck so that it may be moved from field to field. Steam is conducted from the boiler by a steam hose across the field to a steam rake (Figure 7-5). The rake consists of a 4-inch (10-cm) pipe header 12 ft (3.65 m) long, drawn perpendicular to a cable that pulls it across the field. Projecting down into the soil from the header are 16- to 18-inch (40- to 46-cm) blades spaced 9 in (23 cm) apart. Behind each blade is a ½-inch (1.3-cm) pipe carrying steam from the header to the soil at the lower rear side of each blade. A winch is often used to draw the rake across the field at a rate of 10 to 20 in (25 to 50 cm) per minute. One acre (0.4 ha) of soil can be pasteurized by a single rake in 40 to 70 hours of operating time. A sterilizing cover is attached to the back side of the header and is thus dragged across the field. The cover should be sufficiently long to require 30 minutes to pass over any given point in the field. The cover should be 50 ft (15 m) long for a rake moving at 20 in (50 cm) per minute. The cover serves to hold the steam in the soil so that the soil temperature will be maintained at or above 160°F (71°C) for 30 minutes.

The coldest spot during pasteurization is at the end of the bench or trailer where the steam enters and usually near the outer wall at this end. A thermometer should be placed in the coldest spot. Pasteurization should not be stopped until the coldest spot reaches the temperature and time conditions desired. If the thermometer were placed in a warm spot, pasteurization would stop before harmful organisms were killed in the colder areas. These areas would become a source of inoculation for the remainder of the soil. Because of the lack of competition, the harmful organisms would spread rapidly. It would be better not to pasteurize the soil than to do an incomplete job such as this.

The cold spot in a greenhouse bench can be corrected by applying an extra quantity of steam at that point. Figure 7-6 shows a system for doing this. A short piece of pipe is connected to and run parallel to the steam-conduction pipe at the point where it enters the root substrate being pasteurized. The extra piece of pipe runs about a third of the length of the bed and has numerous perforations on opposite sides spaced about 2 in (5 cm) apart.

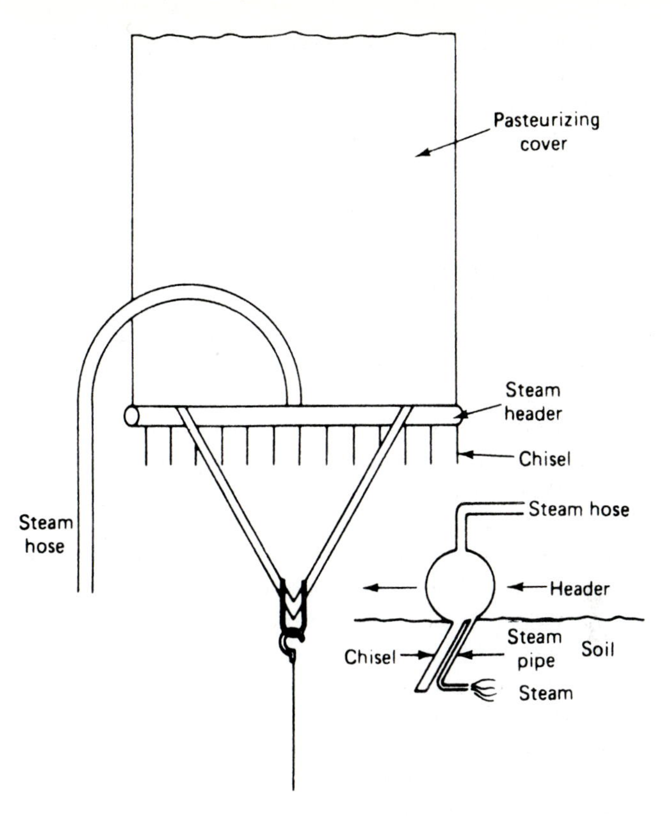

Figure 7-5 A steam rake used for pasteurizing soil in the field. Steam is conducted via a hose to a 12-foot (3.65-m) -long header. Chisels about 9 in (23 cm) apart project into the soil at a distance of 16 to 18 in (40 to 46 cm). Small pipes behind each chisel carry steam into the soil to the depth of the chisels. A pasteurizing cover is drawn behind the rake to maintain a high soil temperature for 30 minutes. The rake itself often is drawn across the field by a cable and winch.

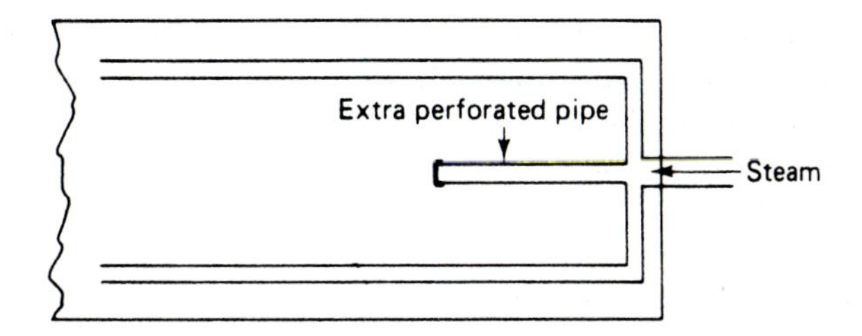

Figure 7-6 An extra perforated steam-conduction pipe in the center of the bed used to deliver additional steam to the cold end of a bed to prevent excessive pasteurization time.

Covers

Without a cover, steam will quickly rise through the root substrate and be lost, further reducing the efficiency of an already inefficient use of steam. A cover is placed over the root substrate during pasteurization to catch and hold steam in close contact with the root substrate so it can be of further value in raising the temperature.

Basically, there are two types of covers: polyethylene and vinyl. Polyethylene film has the shortest life expectancy but is the cheapest. An inexpensive, 4-mil (0.1-mm) construction grade costs about 2.5¢/ft^2 (27¢/m^2). It may be used several times during one season of pasteurization but cannot be stored from season to season. Vinyl covers are the most popular and are usually purchased in 8-mil (0.2-mm) thicknesses. They are advertised to last for up to 25 uses. Actually, these covers last much longer with proper handling and storage away from sunlight. Ultraviolet light breaks down vinyl plastic. These vinyl covers cost about 33¢/ft^2 ($3.55/m^2).

Covers used on benches with smooth outer side walls do not have to be fastened to benches. They should overhang each side by 1 foot (30 cm) or more. As steam contacts the inner side of the cover, it condenses and moistens that side. The film of water that forms between the outer side of the bench and the inner side of the cover causes the two to stick together, preventing the cover from blowing off as steam builds up under it. Covers used on benches with outside posts or rough side boards must be fastened to the bench. The simplest method is to lay a chain or other heavy object over the cover against the inner side wall of the bench (Figure 7-7). A reusable plastic tube is commercially available that can be inflated with water for this same purpose. Some growers squeeze the cover between the top of the bench side and a lath strip with a clamp. Other growers with wooden benches place a lath strip over the cover at the top outer sides of the bench and nail the lath to the bench. This puts holes in the cover, which is undesirable, particularly in the case of a higher-quality cover that must be used many times.

Thirty minutes after 160°F (71°C) has been achieved, the steam should be shut off. The cover will fall back to the root substrate in a few minutes, and then it can be cautiously removed. When the substrate has cooled to a comfortable working temperature, seeds and young plants may be planted. This cooling can require from four to eight hours, depending upon depth and moisture content. Substrate pasteurized with aerated steam can be cooled much faster by using the aerator to pass cool air through the root substrate for 30 minutes after the cover has been removed.

After-Steaming Problems

Two toxicity problems can occur as a result of steam pasteurization: manganese toxicity and ammonium toxicity. Large quantities of manganese exist in many soils. Fortunately, a small but adequate amount is available for plant use, while the majority is in an unavailable form. Steam pasteurization results in further

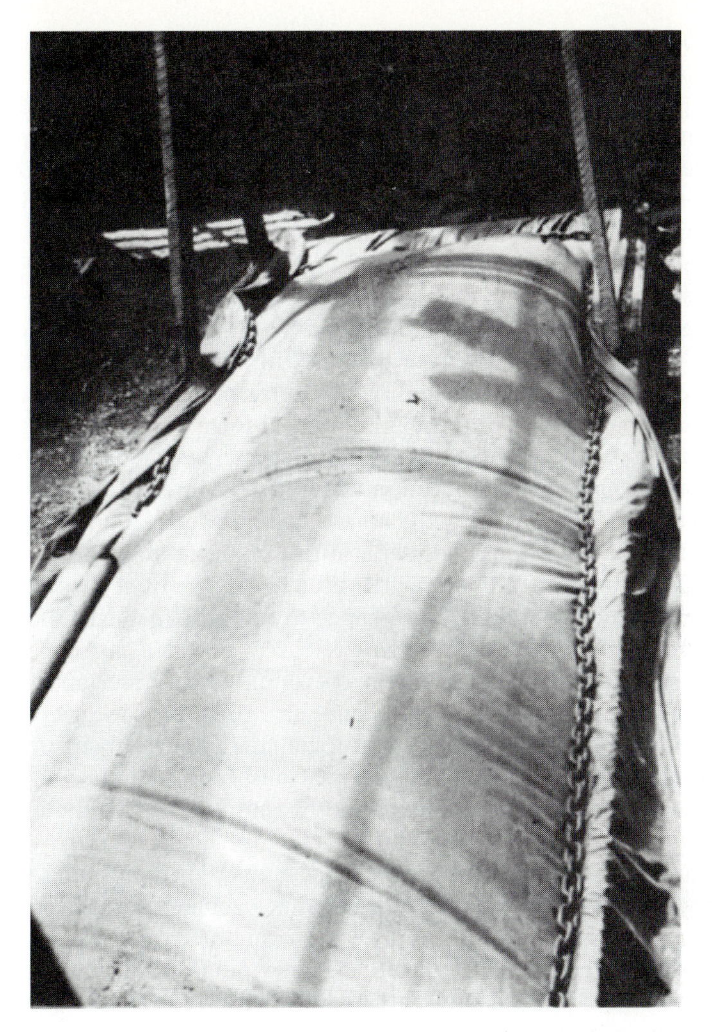

Figure 7-7 **When the outer side of a bed or bench wall is uneven, rough, or short, the pasteurizing cover must be fastened to prevent lifting by steam beneath. The simplest method is to weight it down with heavy objects such as chain or pipe.**

conversion of unavailable to available manganese. The longer the soil is steamed, the greater is the buildup of available manganese and, hence, the greater the risk of manganese toxicity. It is important that substrates containing field soil be pasteurized at the recommended temperature for only the length of time necessary—30 minutes. Soilless substrates usually present no problems, since the components contain little or no manganese.

A high level of manganese in the plant is toxic in itself, causing tip burn of older leaves. A high level of manganese in the root substrate also interferes with root uptake of iron. In fact, iron deficiency is commonly caused by high available-manganese levels.

Root substrates that contain organic matter rich in nitrogen can release toxic levels of ammoniacal nitrogen after pasteurization. Manure, highly decomposed peats, leaf mold, and composts are examples of such materials. Microorganisms feed upon the organic matter for the carbon, nitrogen, and other elements contained in it. When an overabundance of nitrogen is contained in it, much will be released for plant use. As illustrated in Figure 7-8, ammonifying microorganisms convert nitrogen in organic matter to ammoniacal nitrogen, and then nitrifying bacteria convert the ammoniacal nitrogen to nitrate nitrogen.

Most plants grow best in a mixture of ammonium and nitrate forms of nitrogen. Some plants, including poinsettia and gloxinia, are sensitive to a high proportion of nitrogen in the ammoniacal form. Normally, ammoniacal nitrogen is continuously converted to nitrate nitrogen by soil bacteria, so there is a mixture. During pasteurization, ammonifying and nitrifying bacteria are nearly eliminated. In a few weeks, the ammonifying bacterial population builds back to an effective level, and sizable quantities of ammoniacal nitrogen are released from organic matter. It is not until three to six weeks after pasteurization that nitrifying bacteria generally build back to a population size where they can cope with the ammoniacal nitrogen being released. In the meantime, two to six weeks after pasteurization, toxic quantities of ammoniacal nitrogen may develop. This may burn the roots of plants and cause stunting of the entire plant as well as wilting of the tops. Then, any type of nutrient deficiency can ensue as a result of the root injury. Once the nitrifying population becomes large, the high levels of ammoniacal nitrogen are converted to nitrate nitrogen, which is less toxic to plants and is more readily leached from the root substrate during watering. Because of these lower levels and the fact that many plants can tolerate higher levels of nitrate than ammoniacal nitrogen, the problem usually ends at this time. It is mainly for this reason that the use of manure gave way to peat moss during the 1950s when pasteurization became popular. Peat moss, because of its low nitrogen content and slow rate of decomposition, does not support a toxic buildup of ammoniacal nitrogen.

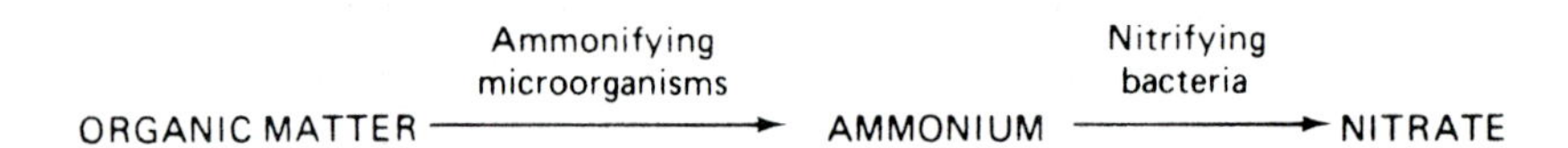

Figure 7-8 **Ammonium toxicity can be a problem when organic materials rich in nitrogen are pasteurized with either steam or chemicals. Nitrogen contained in organic matter is released as ammonium when ammonifying microorganisms, including bacteria, fungi, and actinomycetes, break down the organic matter in the process of utilizing its carbon content. Nitrifying bacteria, in turn, convert ammonium to nitrate nitrogen. During pasteurization, both populations of microorganisms are reduced to low levels. The ammonifying organisms build back to an effective level before the nitrifying organisms do. During the interim period, plants are prone to ammonium injury.**

A third event that can occur from oversteaming of substrate attracts considerable attention but is not harmful. The fungus *Peziza ostrachoderma* will build into a large conspicuous population when competition from other microorganisms is reduced by overpasteurization (too high a temperature or excessive time). The fungus forms spores at the substrate surface that are at first white, then yellow, and finally brown. The fungus *Pyronema sp.* forms pink spores. These fungi do not attack plants. They disappear in a week or two as competitive organisms develop in the substrate. However, the appearance of these fungi serve to illustrate the ease with which a disease organism can get a foothold in overpasteurized substrates where competition has been suppressed.

Chemical Pasteurization

Chemicals offer an alternative to steam for growers who do not heat with a steam boiler and who are not in a position to afford a portable steam generator. Field growers of crops other than chrysanthemum would also see a value in chemical pasteurization because it can be set up for less cost and applied much more rapidly than steam.

Counterbalancing these advantages of chemicals are three disadvantages. First, chemically treated substrate cannot be used for young plants for 10 days after treatment. For fresh-flower crops, costly overhead continues during this time. Second, chemicals are injurious to humans, and stringent safety precautions must be taken. Third, while steam and methyl bromide may be used in a greenhouse that contains plants, chloropicrin may not.

Methyl Bromide

Methyl bromide is available under various trade names and in different combinations with chloropicrin. Methyl bromide is extremely hazardous to humans, and for this reason a small quantity, usually 2 percent, of tear gas (chloropicrin) is added as an irritant to warn against exposure. It is available in 1- and 1.5-lb (454- and 680-g) cans, or larger cylinders for tractor mounting. It is a liquid under pressure and turns to gas when released. Methyl bromide is effective against disease organisms, insects, nematodes, and weed seeds.

Due to toxic residues of bromide in the environment, plans are underway to halt the sales of methyl bromide. Various termination dates have been set between now and 2002 depending on the country and the end use of this fumigant. At this time, it appears that use in developing countries will be permitted well beyond the termination dates in the remainder of the world. Termination of methyl bromide use in the United States for container substrate pasteurization will most likely be extended until a suitable substitute is found. In the meantime, methyl bromide remains the fumigant of preference for greenhouse substrate. Basamid (also Mylone, Microfume, and Crag), known by the common name DMTT or dasomet, while commonly used in several countries, is not used to any extent in the

United States because of the long (three- to four-week) period between application and planting.

Root substrate should be worked up to a loose state for rapid penetration of methyl bromide gas and should be at the moisture content desired for planting. Root substrate at 40°F (4°C) or lower should not be treated. It is best if the substrate is at 50°F (10°C) or higher. A potting substrate should be placed in a container or on a hard, flat surface preferably not over 1 foot (30 cm) deep. Then cans of methyl bromide are placed adjacent to the pile or to the bench and are used at the rate of 1 lb/yd^3 (0.6 kg/m^3) of root substrate. Each can is placed in an applicator, as pictured in Figure 7-9. A tube extends from the applicator to the top of the root substrate, where it is placed in an open saucer or can to collect any liquid that might come out with the gas. From here, the liquid can quickly evaporate. If it entered the root substrate as a liquid, it might take several days longer to evaporate than anticipated and in the meantime would be injurious to plants. A polyethylene cover is placed over the bench or pile. The bench cover can be weighted down by chain along the edges, and the pile cover can be weighted down along the edges with sand. Clay pots or wooden blocks should be placed on the root substrate to hold the cover up so that the gas can contact all of the surface.

When the substrate is ready for pasteurization, the handle on the applicator is closed. As this is done, a hollow spike is driven into the can, allowing methyl bromide to escape through the tube to the space between the substrate and the cover. An alternative method of release consists of a plastic tray with hollow spikes in the bottom. Cans are placed over the spikes. Then the tray is placed on the root substrate and a plastic sheet is placed over the substrate and the tray. Cans are released by pushing down on them from outside the plastic sheet.

The substrate should be exposed to the gas with the cover on for at least 24 hours at temperatures of 60°F (16°C) and higher. At the cooler temperature of

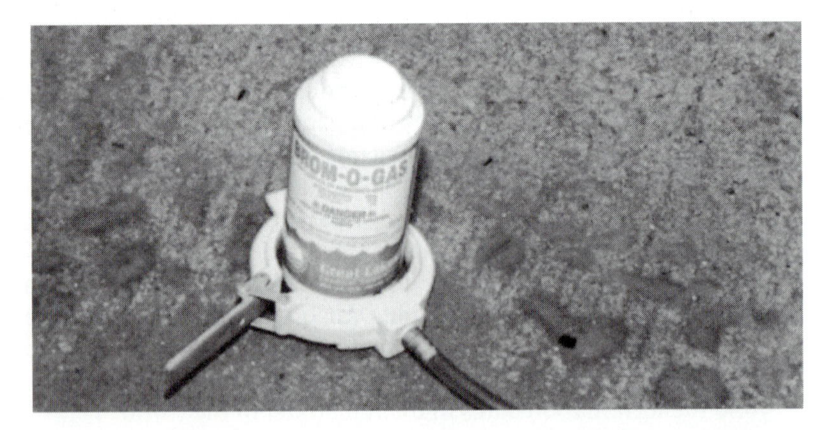

Figure 7-9 One type of applicator for small cans of methyl bromide. The can is placed in a ring. As the ring is clamped tight, a hollow spike punctures the can. Methyl bromide liquid under pressure expands to gas as it exits from the spike into a plastic tube that conducts it to the root substrate pile. It is used at the rate of 1 lb/yd^3 (0.6 kg/m^3) of substrate.

50 °F (10 °C), the exposure time should be 48 hours. The cover is then removed, and the root substrate is left undisturbed to aerate for 24 hours (48 hours at 50 °F, 10 °C). After this time, the substrate may be handled. Seeds can be safely planted after 3 days of aeration, but plants including cuttings and seedlings should not be planted for 7 to 10 days.

Methyl bromide should not be used in a greenhouse containing plants unless full ventilation can be provided during the process. Carnations are susceptible to even slight concentrations of methyl bromide, so this chemical should not be used for this crop. Cauliflower, salvia, and snapdragon may sustain a moderate degree of distorted growth from substrate not thoroughly aerated.

Methyl bromide can be applied by tractor in the field. Methylbromide, and more often a mixture of methylbromide and chloropicrin, gas from a pressurized cylinder is conducted through plastic tubes to a row of chisels mounted behind the tractor 6 to 8 inches (15 to 20 cm) apart (Figure 7-10). The gas is released at the bottom of the chisels 4 to 6 inches (10 to 15 cm) deep. Also mounted to the back of the tractor is a roll of polyethylene film. The end is buried at the beginning of the row to anchor it. As the tractor moves across the field, the plastic sheet is unrolled. The leading edge of the first sheet is buried with soil by a disk mounted on the tractor. Subsequent sheets are glued to the preceding sheet to form one continuous covering over the field. Rigs such as the one described cost between $3,600 and $4,400 and can be depreciated over 10 years. Three workers with one tractor rig can treat up to 5 acres (2 ha) in one day.

A less expensive rig buries both sides of the plastic sheet. No gluing is involved. A 5- to 12-foot (1.5- to 3.7-m) strip of the field is treated; the next strip is skipped. This procedure is repeated across the field. In a few days, when sufficient exposure time has lapsed, the plastic is removed and the untreated strips are treated. These rigs cost from $2,000 to $3,600 and the labor and time inputs are about equal to the former rig.

Other materials for field fumigation include 1.7 rolls of 10.5-foot-by-3,000-foot (3.2-m-by-914-m), 1-mil (0.025-mm) polyethylene ($230) and 325 lb (150 kg) of actual methyl bromide ($400). If the glue system is used, 5 gal of glue are required per acre (47.25 L/ha) at a total cost of $68. Custom applicators can be contracted for a 3 to 5 acre plot at the rate of $1,200 to $1,500 per acre ($3,000 to $3,750/ha) by those who are not set up to apply the chemical themselves. Everything is included in the custom applicator price except two laborers to bury the ends of each plastic strip applied and the labor of removing the plastic after a few days of exposure.

Methyl bromide, like steam, greatly reduces the populations of ammonifying and nitrifying bacteria. The same toxic buildup of ammoniacal nitrogen can occur if organic matter rich in nitrogen and capable of rapid breakdown is used.

Chloropicrin

This fumigant, also known as tear gas, is a popular choice for carnation crops because of their sensitivity to methyl bromide. Chloropicrin, however, cannot be

Figure 7-10 Tractor-mounted rig for injecting gaseous chemicals into field soil for pasteurization. (a) Chisels extend into the soil. Behind each is a tube that delivers the gas into the soil. A rake follows to seal the holes made by the chisels. (b) A roll of polyethylene film is attached behind the rake. At the beginning of the row, the end of the polyethylene is anchored by burying it in the soil. As the tractor moves across the field, the film unrolls. (c) One side of the film is buried by a disk while the other side is glued to the previous sheet of plastic. Up to 5 acres (2 ha) of field can be treated by three people in one day with a single tractor. *(Photos courtesy of W. A. Skroch, Department of Horticultural Science, North Carolina State University, Raleigh, NC 27695-7609.*

used in a greenhouse where there are plants. Another disadvantage of chloropicrin is its poor penetration into plant tissue in root substrate.

Chloropicrin is used at the rate of 3 cc/ft^2 (32 cc/m^2) of bench or field surface and at the rate of 3 to 5 cc/ft^3 (106 to 177 cc/m^3) for bulk substrate. Chloropicrin is injected into substrate in the greenhouse by a hand injector. A spike at the base of the injector is pushed by one's foot into the substrate. As this occurs, 3 cc of liquid chloropicrin is released from the end of the spike. The injector is inserted on 12-inch (30-cm) centers. After application, the bench or pile of substrate is covered with a polyethylene cover. Chloropicrin may be injected through tractor-drawn chisels in the field. The treated field is covered with polyethylene film, as in the case of methyl bromide.

Chloropicrin should not be used at substrate temperatures below 60°F (16°C); 70°F (21°C) is best. An exposure time of one to three days is needed, the longer time being needed at 60°F. Substrate should be aerated for 7 to 10 days before planting in it.

Residue Test

Wet, heavy, or cold substrates are slow to release chemical fumigants upon aeration. A grower must be certain that the residue is below an injurious level before planting a crop. A simple lettuce test can determine this. The procedure is as follows:

1. Fill a few jars three-quarters full with treated substrate. The substrate should be at a moisture content for planting. Wet it if necessary.
2. Place wet, 1-inch (2.5-cm) squares of absorbent cotton on the substrate, and on each square of cotton place 10 to 15 lettuce seeds that have been presoaked in water for 30 minutes.
3. Seal the jars tightly as soon as possible and place at room temperature in an area where they receive daylight. (Some types of lettuce seed will not germinate in the dark.)
4. Prepare some other jars in the same manner, using untreated substrate as a control.

After two days, the seeds in the control jars should have germinated as well as those in the treated-substrate jars if the residue is down to a safe level. If the seeds fail to germinate in the treated-substrate jars, aerate the substrate longer. It will help to mix or rototill the substrate.

Reinoculation

Root substrate pasteurization serves the purpose of eradicating biological pests. It does not provide a resistance to these pests. Growers must think through all of the operations in which they might introduce contaminated substrate to clean substrate.

Pots and flats are of first concern. If they have been used before, they contain substrate and possibly plant tissue that could be contaminated. They should be cleaned. Clay and wooden containers can be steamed along with root substrate. Because plastic pots will distort at high temperature, they (as well as clay and wooden containers) can be treated with chemical fumigants.

Tools are another source of disease inoculum. They should be periodically disinfected to prevent the spread of any disease that might be present in the range and certainly before using a recently pasteurized substrate. Many florist-supply firms sell a hospital-type disinfectant that can be maintained in a bucket conveniently located in the headhouse and in the greenhouses so that tools can be easily dipped in it. Household bleach diluted 1:9 with water serves well also, but it will break down in sunlight in a day or so. These disinfectants also work well for plastic pots.

Another common source of contamination is the soil on the soles and heels of shoes. Placing one's foot on the side of a bench in the greenhouse is an efficient way to transfer inoculum. Visitors with an interest in plants have probably been in another greenhouse range or a garden, and the probability of their carrying contaminated soil is great. It should be a standard rule around the greenhouse that feet are to be kept off the benches. Some growers place a fiber mat in a shallow tray of a disinfectant solution at the entry to their ranges so that everyone steps through it, disinfecting his or her shoes before entering. This is a particularly wise practice for a propagation greenhouse, where disease prevention is an even more serious matter.

Plastic watering systems become distorted when left under the cover during steam pasteurization. They are customarily removed or raised above the bench during pasteurization. They should be syringed with household bleach diluted 1:9 with water or with a disinfectant before they are placed back on the root substrate. In a small operation, these pipes may be wiped with a rag saturated with disinfectant. The thin tubes and weights used in automatic pot-watering systems should be dipped in a container of disinfectant. Wire and string supports for fresh flowers should likewise be sterilized before they are reused on a recently pasteurized bench.

For pasteurization to be effective, the grower must think through all operations to identify and correct those that can cause reinoculation of the root substrate. Means exist to ensure a clean range. Where failure occurs, it is due to a lack of foresight.

Economics

Steam pasteurization is particularly desirable when the steam can be obtained from an already existent heating system. An average of 5,600 Btu of boiler capacity is required to supply the 2,800 Btu needed to pasteurize 1 ft^3 (50,000 kcal or 208 MJ of boiler capacity/m^3) of root substrate. One gallon of oil with an approximate heat content of 139,000 Btu and a thermal efficiency of 85 percent provides sufficient heat to pasteurize 21 ft^3 of root substrate. The fuel cost is therefore \$0.029/ft^3 (\$1.02/m^3) based on a price of \$0.60/gal (\$.16/L).

Methyl bromide in 1-lb (454-g) cans can be purchased for \$2.55 and is sufficient to treat 27 ft^3 (0.76 m^3) of soil. The unit price here is \$0.094/ft^3 (\$3.33/m^3).

There are other economic considerations. A large operator without a central steam source could easily depreciate the cost of a steam generator and still end up pasteurizing root substrate for an acceptable cost. Small growers, or those who operate only during one season per year, often find the cost of steam pasteurization unacceptable if they must purchase a steam generator.

For the purpose of illustration, a 1-acre greenhouse range could be expected to have 29,000 ft^2 of bench area containing 16,900 ft^3 of root substrate (1 ha has 6735 m^2 of bench containing 3925 m^3 of substrate). A \$6,000 steam generator financed for 10 years at 10 percent interest would add \$0.053 (\$1.87/m^3) to each cubic foot of substrate for a total fuel-plus-equipment cost of \$0.082/ft^3 (\$2.89/m^3).

One last factor must be considered, and that is the cost of labor required to carry out each method of pasteurization. Preparation of substrate is the same in each alternative, calling for a loose, moderately moist condition. Covers are needed for each alternative also. The differences lie principally in the steam-conduction hose or pipes to and in the bench. This is a very small consideration and would not likely influence the decision as to which system to use.

Summary

1. All greenhouse root substrates used for successive crops, such as fresh flowers and vegetables, should be pasteurized at least once per year, and more often as required, to rid them of harmful disease organisms, nematodes, insects, and weed seed. All substrates containing soil need to be pasteurized before initial use. Soilless substrates used for a single crop are generally not pasteurized before initial use.
2. Numerous microorganisms develop in root substrate that are not harmful. These can be beneficial by providing competition for harmful microorganisms, which might otherwise proliferate. For this reason, root substrates are pasteurized and not sterilized; that is, only some organisms are killed.
3. A root substrate may be pasteurized with steam by raising it to a temperature of 160°F (71°C) for 30 minutes.
4. Volatile chemicals also are used for pasteurizing root substrates. Methyl bromide is most popular, although chloropicrin is used as well, especially for carnation crops, which are injured by methyl bromide residues for a few months after application. Chemical pasteurization precludes the need for a steam boiler—an advantage for field and small-greenhouse growers.
5. Both steam and chemical pasteurization require that the root substrate be loose and of a moisture content suitable for planting. Amendments such as

peat moss, manure, and bark should be incorporated prior to pasteurization to prevent introduction of diseases or pests.

6. Pasteurization can result in ammonium and manganese toxicities in certain situations. If the root substrate contains organic matter rich in nitrogen, such as manure, steam and chemical pasteurization can result in an excessive release of ammonium, particularly in the period of two to six weeks after pasteurization. Either these materials should be avoided, or an adjustment should be made in the watering practice to ensure adequate leaching of ammonium. Many soils contain large levels of manganese, most of which is unavailable. Steam pasteurization causes a conversion of unavailable manganese to an available form. A toxic level is sometimes reached. This is another reason for pasteurizing root substrate at a low temperature (160°F, 71°C) and for only the necessary length of time (30 minutes).
7. Pasteurization of root substrate is designed to eliminate harmful organisms. It does not protect against future infestation. Good sanitation practices must be employed to maintain clean conditions. Some considerations include disease-free seeds and plants, sterilization of containers and tools, a pesticide program, foot baths, a clean working area, sanitation outside the greenhouse, and proper control of temperature and humidity.

References

1. Baker, K. F., ed. 1957. *The U.C. system for producing healthy container-grown plants.* Univ. of California Agr. Exp. Sta. and Ext. Ser. Manual 23. Univ. of California, Berkeley, CA.
2. Ball, V. 1975. Soil sterilizing—steam. In Ball, V., ed. *The Ball Red Book,* 13th ed., pp. 91–107. West Chicago, IL: Geo. J. Ball.
3. Bunt, A. C. 1988. *Media and mixes for container-grown plants.* London: Unwin Hyman.
4. Gray, H. E. 1960. Steam sterilization. *Florists' Review* 127 (3292):13–14, 77–79.
5. Horst, K. 1985. Chemicals for sterilizing. In Ball, V., ed. *The Ball Red Book,* 14th ed., pp. 133–145. Englewood Cliffs, NJ: Reston Publishing.

CHAPTER 8

Watering

Watering is the greenhouse operation that most frequently accounts for loss in crop quality. Taken at face value, it would appear to be the simplest operation. When performed correctly, it is simple and perhaps a bit boring. For this reason, the task is often mistakenly assigned to a less experienced employee. If this employee waters at the wrong times or uses an incorrect amount of water, the crop will be injured. The original quality cannot be regained.

The decision of when to water should be made by the greenhouse manager. He or she should inspect every bench of plants at least twice daily and should supervise the watering operation when it is carried out by another employee. Actually, with the wide variety of inexpensive automatic watering systems available today, the range should be equipped with a system simple enough to permit the manager to do the actual watering while inspecting the range. The few minutes it takes for each section to be watered affords the manager a chance to further inspect plants in each section for insects, disease, nutritional disorders, and any other problems. Success or failure is due not so much to the quantity of labor expended as to the correct timing of the various labor operations. It is of utmost importance that the greenhouse range be inspected daily by the most knowledgeable person and that work plans be altered according to his or her findings.

Effects of Watering on Plants

Underwatering

When water is not applied frequently enough, plants wilt, thus retarding photosynthesis and slowing growth. The elongation of young developing cells is reduced, resulting in smaller leaves, shorter stem internodes (the length of stem between leaves), and, in general, a hardened appearance to the plants. In more extreme cases, burns may begin on the margins of leaves and spread inward, affecting whole leaves. On plant species capable of leaf abscission, the leaves will drop off. Before the days of chemical height retardants, it was customary to control height of some crops by allowing plants to wilt between waterings. Today, this practice is restricted primarily to bedding plants. The availability of chemical height retardants and the risk of foliar injury from drying has brought about this change.

Overwatering

When water is applied a little too frequently, new growth may become large but soft as a result of high water content, and, as a whole, plants tend to be taller. This situation is undesirable because some of these plants wilt easily under bright light or dry conditions and do not ship or last well. If water is supplied even more frequently, the oxygen content of the root substrate is reduced by the higher average water content in the pores, resulting in damage to the roots. A damaged root system cannot readily take up water or nutrients. This condition causes wilting, hardened growth, an overall stunting of the plants, and several nutrient-deficiency symptoms.

Rules of Watering

Rule 1: Use a Well-Drained Substrate

The importance of texture and structure was brought out in Chapter 6. If the root substrate is not well drained and aerated, proper watering cannot be achieved. Either you will underwater to achieve aeration, or you will provide the required water at the expense of aeration. In either case, poor plant quality will result. A well-drained substrate of high water-holding capacity is required for use in containers. This calls for coarse texture and a high degree of stable structure—in short, a formulated substrate and not field soil alone.

Rule 2: Water Thoroughly Each Time

Because substrates cannot be partially wetted, it is important to water all of the substrate in a container each time water is applied (Figure 8-1). Water applied to the root surface of the substrate enters the pores at the top and adheres to the particle surfaces making up the pore walls. Additional water causes the layer on the particle surfaces to become thicker. Eventually, the layer of water becomes thick enough that any additional water is too far away from the particle surface to be held, and gravity pulls this additional water down to the next particle below. There, it is attracted to the particle surface, and as more water enters, the water

Figure 8-1 **Only half the amount of water that the soil in the beaker is capable of holding was applied. Instead of all of the particles being partially wetted, those at the top are thoroughly wetted, while those at the bottom remain completely dry. This points out the fallacy of trying to partially water a root substrate.**

layer on this particle grows thicker. This process keeps repeating itself until water finally reaches the particles at the bottom of the container. Additional water then flows through the substrate and out the bottom of the container.

If 6 oz of water are required to water the root substrate in one pot and only 3 oz of water are applied, the substrate in the top half of the pot will be thoroughly wetted while the substrate in the lower half will remain dry. Late in the afternoon or on Saturdays, there is always a temptation to water a crop partially to carry it over until more time is available for watering. From the preceding discussion, one can readily see the fallacy of doing this. The root substrate in the lower part of the pot or bench will not receive water, and, as it continues to dry, roots will die.

In open cultural systems where water is applied to the top of the pot or bench and excess water drains from the bottom of the pot or bench, it is important to apply more water or fertilizer solution than the substrate can hold. Ten to 15 percent of the water applied to the top of the container should run out of the bottom. This is done to leach excessive fertilizers and nonfertilizer elements that might otherwise build up to toxic levels. Some fertilizers contain elements that are not used in large quantities by plants. As fertilizer is repeatedly applied to provide the elements needed in large quantity, these other elements accumulate.

As a general rule of thumb for soil-based substrates, $^1/_{15}$ gallon of water should be applied to each square foot of bench for each inch of root substrate depth. A typical bench 8 in deep containing 7 in of substrate should receive $^7/_{15}$ gallon of water, or in practical terms, $^1/_2$ gallon of water per square foot. (Apply 1.1 L/m^2/cm of depth, or 20 L/m^2 to an 18-cm-deep bed.) A 6-inch (15-cm) azalea pot requires about 10 to 12 oz (300 to 350 mL) of water. This rule applies to the application of nutrient solution as well as to water. These quantities of water should be checked out for each individual situation, since substrates vary in water-holding capacity. Larger or smaller quantities may be justified for various soilless substrates.

Rule 3: Water Just before Moisture Stress Occurs

It is apparent from Rule 2 that overwatering does not refer to the amount of water applied during a single application. Overwatering indicates that water is applied too frequently. When this is done, too much of the lifetime of the root is spent under conditions of minimum aeration, and as a result root development is suppressed.

Water should be applied just before the plant enters the early symptoms of water stress. For each plant, these signs are different. Some plants, such as chrysanthemum, take on a darker leaf color; others, such as begonia, exhibit a gray-green leaf color. By observing a crop, one can quickly learn the early warning signs of moisture stress. There are always individual plants that will run short of water ahead of the other plants and serve as an indication of impending water stress. It is also important to learn the color and feel of the root substrate associated with early moisture stress. Some crops, such as azalea, do not show signs of moisture stress until permanent damage occurs to the roots. Judgment of when to water rests in this case entirely on substrate appearance, feel, and weight.

Water Quality

It is very important to know the chemical content of water to be used in the greenhouse. Irrigation water should be tested anytime a new source is established, whether it be from a well, river, pond, or municipal system. During the first two years it is advisable to test the water source at least twice a year. It should be tested during a wet period and during a dry period. Introduction of large amounts of water into a surface water source or an aquifer servicing a well can result in a reduction of the chemical content of this water. Conversely, water impurity concentrations can increase during a drought. Once the water-quality pattern is established, water needs to be tested only every few years. Governmental and private laboratories that run substrate tests can usually analyze water.

Soluble Salt (EC)

An all-encompassing test in virtually all water analyses is the soluble-salt (EC) test (Table 8-1). Typically, this test is conducted with an electrical conductivity (EC) meter. This test measures all electrically charged ions dissolved in water. The meter probe has two electrodes that are placed in the liquid sample to measure the conductance of electricity between them. The higher the salt content, the

Table 8-1 **Guidelines for Greenhouse Irrigation Water**

Electrical conductivity (EC)			
0.75 dS/m	Seedlings; maximum tolerable level		
1.5 dS/m	General crops; maximum tolerable level		
Alkalinity			
1.5 me (75 ppm $CaCO_3$ equivalent)	Caution		
3.0 me (150 ppm $CaCO_3$ equivalent)	Troublesome		
Hardness			
3.0 me (150 ppm $CaCO_3$ equivalent)	Caution—check the calcium-to-magnesium balance		
Calcium and magnesium			
3 to 5 ppm calcium per 1 ppm magnesium	Tolerable		
Specific elements (maximum tolerable levels)			
sodium 50 ppm		zinc	0.3 ppm
chloride 70 ppm		copper	0.2 ppm
chlorine 2 ppm (less in hydroponics)		borate-B	0.5 ppm
iron 4 ppm		fluoride	0.5 ppm
manganese 1 ppm		lithium	0.5 ppm

greater the flow of electrical current through the sample. EC is measured in terms of mho/cm; the mho is a unit of electrical conductivity and is the reciprocal of the ohm, a unit of electrical resistance. Since the conductivity is very low, it is recorded in fractions of a mho—thousandths of a mho in most cases (mho × 10^{-3}/cm), also called a millimho. Today, another term for EC measurement is becoming popular: the dS/m, which has the same value as a mmho/cm (1 dS/m = 1 mmho/cm). These terms will be of little concern to the grower. Each laboratory will use only one test and will make available an interpretation chart. The upper acceptable EC level for water used for irrigating seedlings is 0.75 dS/m and for older crops in general it is 1.5 dS/m.

Occasionally, a well is drilled that yields water of low quality (it contains quantities of impurities). The impurities may be sulfate in areas of old coal-mine shafts, sodium chloride (table salt) or sodium bicarbonate (baking soda) along coastal areas, calcium bicarbonate in areas of limestone deposits, and sodium in the alkaline areas found in arid parts of the world. Surface waters can contain fertilizer salts that have leached or drained from adjacent farms or landscapes.

Salts in water are transferred to the substrate during irrigation. There, these salts add to other salts derived from fertilizers and possibly from the breakdown of nitrogen containing components of the substrate. High salt levels in the substrate are injurious to plants. Salt concentrations in the substrate solution and in the root cells determine to a large extent the flow of water between the two. Water flows in the direction of the higher salt concentration, which generally exists in the root cells. Thus, when the salt level in the substrate solution equals or exceeds the solute concentration in the root cells, water no longer moves into the root. Following transpirational loss of water from the foliage, cells begin to desiccate (lose adverse amounts of water).

An excessive level of soluble salts in the root substrate is first seen as wilting of plants during bright times of the day, even though the root substrate is moist. Overall growth slows down. Roots die from the tips back, particularly in the drier zones of the root substrate. Leaves become necrotic, in some cases along the margin and in others as circular spots scattered across the leaf blade. Ultimately, deficiency symptoms of many nutrients will occur as a result of acutely impaired nutrient uptake by the injured root system.

Alkalinity

Note that pH guidelines are not presented in Table 8-1. Too much importance has been placed on water pH. It is a misconception that substrate pH is mainly under the control of water pH. The primary impact of water on substrate pH stems from water alkalinity. Two water sources can exist, each with a pH of 7.4. If the first has an alkalinity level of 2 me and the second has a level of 6 me, the first will raise substrate pH very little while the rise from the second will be unacceptably high. Judgment of the quality level of these two water sources would have been 50 percent accurate on the basis of pH and 100 percent accurate based on alkalinity. A high water pH, that is, 7.2 or higher, should be merely a warning to look at the alkalinity level.

The alkalinity test has far-reaching implications. Alkalinity is a measure of the facility of water to raise pH. It is primarily a measure of the amount of carbonate plus bicarbonate in water. These are the active ingredients in liming materials used to raise substrate pH. Thus, applying alkaline water is equivalent to applying limestone.

Alkalinity is generally reported as milliequivalents (me) or as parts per million (ppm) of equivalent calcium carbonate. One me of alkalinity is equal to 50 ppm of equivalent calcium carbonate. Precise upper critical water-alkalinity levels cannot be assigned. An excessive alkalinity level is one that causes the pH of substrate to rise to an unacceptable height by the end of a crop. There are three crop situations that dictate the upper critical alkalinity level: the length of the crop period, the plant-to-substrate ratio, and the upper substrate pH level tolerated by the crop.

Water alkalinity causes substrate pH to rise gradually over time, as the cumulative quantity of bicarbonate plus carbonate increases with additional applications of water. A 3-week crop of marigolds could possibly tolerate an alkalinity level of 6 me, while a 12-week crop of chrysanthemums would not. During the short 3-week period of marigold growth, the substrate pH might rise only to a safe level of 6.8. On the other hand, the 12-week cultural period of the chrysanthemums could provide the time necessary for an unacceptable pH rise to 7.5.

The effect of shoot mass to root substrate mass on the critical alkalinity level is best seen in plug seedling production. The roots of a plug seedling can be confined to a cell with as little as 0.1 in^3 (2 cc) of substrate. Yet the shoot of this seedling will be grown over six weeks to a large size greatly out of proportion to the substrate volume. The great cumulative volume of shoot transpiration throughout this period will demand an unusually large amount of water application per unit volume of substrate. If the irrigation water is alkaline, substrate pH will rise surprisingly fast because there is little substrate present to neutralize the bicarbonates and carbonates supplied in the water.

Finally, crops that need to be grown at low-substrate pH levels will be less tolerant of water alkalinity. Some examples are pansy, petunia, vinca, and snapdragon. These crops perform best at a pH level of 6.0 of lower; otherwise, iron deficiency becomes a problem.

Alkalinity guidelines can be found in Figure 8-2. Alkalinity levels up to 1.5 will probably not require any preventive action. Levels in the range of 1.5 to 3.0 might call for action. The cheapest action includes use of less limestone in the substrate and/or a shift to acid-reacting fertilizers. Most greenhouse substrates require limestone in their formulation. Therefore, less limestone can be used to compensate for the lime that will be applied in the alkaline water. The amount of limestone to use is that amount that achieves the minimum substrate-pH re-

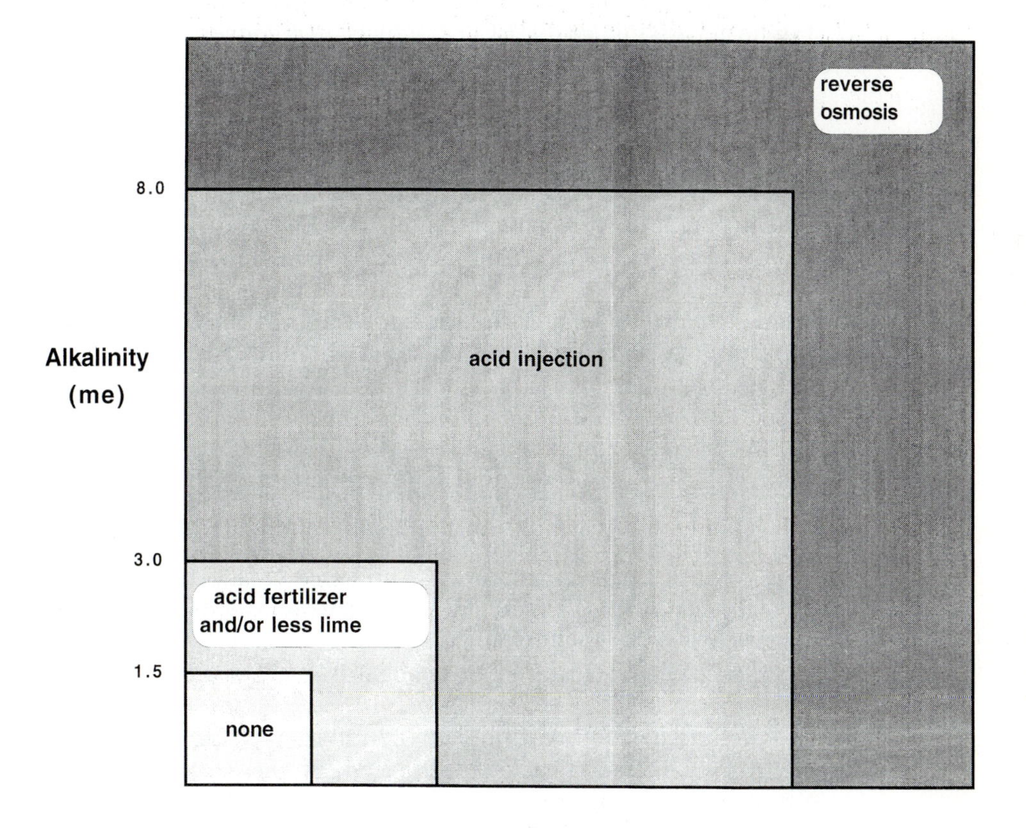

Figure 8-2 A graphic presentation of the steps to be taken in sequence for correcting high-water-alkalinity problems as waters of increasing alkalinity are encountered.

quirement of the crop at the start of the crop, yet does not result in an excessive pH level at the end of the crop. Some plug seedling growers in areas of extremely alkaline water apply no limestone to their germination substrate, leaving the initial pH at about 4.5. By the time seedlings have germinated and developed a bit, bicarbonate in the irrigation water has raised the substrate pH to a safe level of 5.2 or higher. If the pH does not rise above 6.8 by the end of the seedling crop, then no further costly measures are needed to combat the water alkalinity. As will be discussed in the next chapter, some fertilizers are basic (alkaline) in that they raise the substrate pH, while others are acidic. Fertilizer acidity/basicity can be used to manipulate moderate pH shifts in substrate. Use of an acidic fertilizer may be all that is required to counteract alkalinity in moderately alkaline water.

Often when the alkalinity level of water is 3.0 me or higher, it becomes necessary to take stronger action to prevent the substrate pH level from rising too high by the end of the crop. The next level of action in effectiveness and cost is injection of acid into the water (Figure 8-2). Bicarbonate ($HCO3_3^{-3}$) and carbonate (CO_3^{-2}) in water cause the substrate pH to rise because bicarbonate (Equation 8-1) and carbonate (Equation 8-2) combine with substrate acidity (H^+) to form carbonic acid (H_2CO_3), which then converts to water (H_2O) and carbon dioxide (CO_2):

$$HCO_3^- + H^+ \rightarrow H_2CO_3 \rightarrow H_2O + CO_2 \qquad (8\text{-}1)$$

$$CO_3^{-2} + 2H^+ \rightarrow H_2CO_3 \rightarrow H_2O + CO_2 \qquad (8\text{-}2)$$

In these processes the acidity and the bicarbonate are consumed. Since pH is a measure of H^+ ions, the loss of H^+ ions in substrate results in a higher pH level. When acid is added to water it dissociates to yield H^+, as in the case of nitric acid in Equation 8-3. These H^+ ions consume much of the bicarbonate before it reaches the substrate through the same reaction seen in Equation 8-1.

$$HNO_3 \rightarrow H^+ + NO_3^- \qquad (8\text{-}3)$$

Water below pH 8.3 contains only bicarbonate, while water at or above pH 8.3 can contain bicarbonate plus carbonate.

Four types of acids are used for neutralizing alkalinity in irrigation water. The first three, purchased as liquids, are nitric, phosphoric, and sulfuric acids; the fourth is citric acid, which is purchased as a solid. Nitric acid, available in concentrations of 61.4% and 67%, and sulfuric acid in a concentration of 93% are dangerous to handle, but are often used in greenhouses. Phosphoric acid is available in 75% and 85% concentrations and is much safer to handle and more commonly used. Recently, 35% sulfuric acid has become popular among greenhouse growers due to its greater safety in handling and low cost. This latter acid can be purchased at automotive-supply companies as battery acid.

The quantity of acid required depends on the initial alkalinity and pH of the water to be treated and the final desired pH or alkalinity level of the treated water. An alkalinity calculator is available on the World Wide Web for calculating the quantity of each of the four listed acids for any given situation. The Web address **http://www2.ncsu.edu/unity/lockers/project/floriculture/www/programs/index.html** opens up the NCSU floriculture software page. There one can select the alkalinity calculator or the plant growth regulator calculator that will be described in Chapter 13. These calculators require an Excel 5.0 software program and will run on Macintosh or IBM-compatible computers. The alkalinity calculator was used to determine the quantities of 35% sulfuric acid required to bring water of various initial alkalinity and pH levels to a final pH level of 5.8 (Table 8-2). Although other endpoints are sometimes desired, a final pH level of 5.8 works well for most crops. Equivalent quantities of other acids can easily be determined by multiplying values in Table 8-2 by factors presented in the footnote to Table 8-2.

Acid is introduced into the water stream in various ways. Growers with low-volume wells pump water into a large holding tank. Water is then pumped at a much higher flow rate from this tank to the crops during the few hours of the day when irrigation is required. In this situation, acid may be pumped directly from the drum in which it was purchased to the holding tank any time that water is being supplied to the holding tank. The two pumps that separately deliver water and acid to the holding tank are activated by the same water level sensor in the holding tank. The proportion of acid delivered to the holding tank is controlled by adjusting the flow rate of the acid pump. Growers who pump directly from the water source (well, river, pond) to the irrigation lines can use a proportioner to meter acid into the water line. As water passes through the proportioner under the pressure of the pump, acid is drawn into the proportioner and mixed with water. Often the proportion of acid to water is lower than the range of the proportioner. In this case, diluted acid is used in the supply tank. Remember, it is imperative that acid be added to water during the dilution process and that it be done slowly. If water is added to acid, there is a great risk of acid splattering out of the solution onto the worker due to the evolution of heat. Be sure that acid resistant pumps and proportioners are used.

It is important to determine the quantities of essential nutrients that are supplied along with various acids so that fertilizer rates can be reduced by these amounts. Nitrate nitrogen is supplied by nitric acid, phosphate phosphorus by phosphoric acid and sulfate sulfur by sulfuric acid. The concentrations of each of these three nutrients that result from the use of 1 fl oz of acid in 100 gal of water and from 1 mL in 1 L of water are presented in Table 8-3. Consider a grower who fertilizes a crop of bedding plants with fertilizer containing 100 ppm nitrate-N, 50 ppm phosphate-P_2O_5, and 100 ppm potassium-K_2O, and has water with an alkalinity level of 10 me and pH of 7.8. If this water is treated with nitric acid, it will contain 109 ppm nitrate-N which is more than can be tolerated in the fertilizer program. If phosphoric acid is used, the contribution of phosphate-P_2O_5 will be 540 ppm, which is also more than is desired in the fertilization program. This leaves only sulfuric acid that can be used. Although it will supply 125 ppm sulfate-S, this is tolerable. Whenever it is necessary to supply more of one nutrient element than is desired in a fertilization program, it is generally best to supply sulfate or calcium. These nutrients have less solubility in the

Table 8-2 **Quantities of 35% Sulfuric Acid Required to Treat Alkalinity in Water of Various Initial pH and Alkalinity Levels When an Endpoint of pH 5.8 Is Desired**[1]

	Initial Water Alkalinity (me/L)				
Initial pH	*2*	*4*	*6*	*8*	*10*
	fl oz Acid/100 gal				
7.2	2.15	4.30	6.45	8.60	10.75
7.4	2.18	4.37	6.55	8.73	10.91
7.6	2.20	4.41	6.61	8.82	11.02
7.8	2.22	4.44	6.65	8.87	11.09
8.0	2.23	4.45	6.68	8.91	11.13
8.2	2.23	4.47	6.70	8.96	11.17
8.4	2.24	4.48	6.72	8.96	11.20
8.6	2.25	4.49	6.74	8.98	11.23
8.8	2.25	4.51	6.76	9.01	11.27
9.0	2.26	4.53	6.79	9.05	11.32
9.2	2.28	4.55	6.83	9.11	11.39
9.4	2.30	4.59	6.89	9.18	11.48
9.6	2.32	4.64	6.96	9.29	11.61
	mL Acid/L				
7.2	0.168	0.336	0.504	0.672	0.840
7.4	0.171	0.341	0.512	0.682	0.853
7.6	0.172	0.344	0.517	0.689	0.861
7.8	0.173	0.346	0.520	0.693	0.866
8.0	0.174	0.348	0.522	0.693	0.870
8.2	0.175	0.349	0.524	0.698	0.873
8.4	0.175	0.350	0.525	0.700	0.875
8.6	0.175	0.351	0.526	0.702	0.877
8.8	0.176	0.352	0.528	0.704	0.880
9.0	0.177	0.354	0.530	0.707	0.884
9.2	0.178	0.356	0.534	0.712	0.889
9.4	0.179	0.359	0.538	0.718	0.897
9.6	0.181	0.363	0.544	0.725	0.907

[1]Equivalent quantities of other types of acid can be determined by multiplying values in the table by the following factors: 0.25 for 93% sulfuric acid, 0.67 for 61.4% nitric acid, 0.60 for 67% nitric acid, 0.72 for 75% phosphoric acid, and 0.59 for 85% phosphoric acid. Always add acid to water and not vice versa.

substrate, and the plant is able to restrict their uptake more than most other nutrients. This point, coupled with price and safety of handling, renders 35 % sulfuric acid the preferred material for many growers.

Citric acid is a much more expensive acid to use than the previous three. However, it has a decided advantage in that it does not supply an essential nu-

Table 8-3 **Concentrations (ppm) of Essential Nutrients Resulting from the Addition of 1 fl oz of Various Acids in 100 gal of Water and from 1 mL in 1 L of Water**

Acid	*Nutrient*	*From 1 fl oz/100 gal (ppm)*	*From 1 mL/L (ppm)*
Nitric acid (61.4%)	nitrate-N	14.6	187
Nitric acid (67%)	nitrate-N	16.3	208
Phosphoric acid (75%)	phosphate-P_2O_5	67.1	857
Phosphoric acid (85%)	phosphate-P_2O_5	81.8	1045
Sulfuric acid (35%)	sulfate-S	11.3	145
Sulfuric acid (93%)	sulfate-S	45.1	578

trient to interfere with the fertilization program. Citric acid is often used for acidifying water used in fertilizer concentrates. In greenhouses, most fertilizer is dissolved into a concentrate and then injected into the water line through a proportioner. The concentrate is typically 50 to 200 times more concentrated than the fertilizer solution applied to the plants. Solubility of the concentrate diminishes with increasing pH. Phosphoric and sulfuric acids will acidify water and thereby increase fertilizer solubility. However, part of the increased fertilizer solubility is lost due to the contribution of phosphate or sulfate, which reduce fertilizer solubility.

Conceivably, enough acid could be injected into water to combat any level of alkalinity. This is not true for greenhouse irrigation purposes. The carbonates and bicarbonates that constitute much of the alkalinity in water are negatively charged ions. Such ions must be accompanied by positively charged ions, that is, cations such as calcium, magnesium, or sodium. Thus, you might think of alkalinity in terms of calcium bicarbonate rather than just bicarbonate. Acids are added to alkaline water to supply hydrogen, H^+, which removes the bicarbonate as seen in Equation 8-1. However, each acid contains anions such as nitrate (NO_3^-) in nitric acid or sulfate (SO_4^{-2}) in sulfuric acid. When bicarbonate is removed, a new salt forms to replace it. Just such a scenario is seen in Equation 8-4, where alkalinity in water is assumed to be mainly calcium bicarbonate [$Ca(HCO_3)_2$] and is neutralized with nitric acid (HNO_3), thereby forming calcium nitrate [$Ca(NO_3)_2$] in the place of calcium bicarbonate.

$$Ca(HCO_3)_2 + 2HNO_3 \rightarrow 2H_2CO_3 + Ca(NO_3)_2 \quad (8\text{-}4)$$

$$2H_2CO_3 \rightarrow 2H_2O + 2CO_2 \quad (8\text{-}5)$$

As we have just seen, the active ingredients in alkalinity are salts. Highly alkaline water usually has an undesirable high EC level. Since none of the measures listed thus far for combating alkalinity remove this salt, there is a limit to using acid to treat alkalinity.

When alkalinity is around 8 me (Figure 8-2), reverse osmosis (RO) might be necessary. This is a last resort because it is very expensive. The exact alkalinity level at which RO is required depends on the salt tolerance of the crop. In reverse osmosis, water is forced through sufficiently fine filters to remove ions such as sodium, calcium, and bicarbonate. This system effectively removes alkalinity and salts. Since most greenhouses do not need RO, this is an expense that could considerably reduce a firm's competitiveness. It is very expedient to evaluate the quality of water at a prospective greenhouse site prior to establishing a business.

Hardness

The presence of high alkalinity in water immediately dictates the need for a hardness test. Hardness is a measure of the combined content of calcium and magnesium in water. Hardness is expressed as milliequivalents or parts per million of equivalent calcium carbonate (limestone). As in the case of alkalinity, 1 me equals 50 ppm of equivalent calcium carbonate. When hardness exceeds 3 me, it is important to check the calcium and magnesium concentrations in irrigation water to determine if they are in proper balance. For each part per million of magnesium, there should be 3 to 5 ppm of calcium. If there is more calcium than this, it can block uptake of magnesium in the plant, causing a magnesium deficiency. This can occur even though the concentration of magnesium in the substrate would otherwise have been adequate. If the calcium level is lower than the 3–5 Ca: 1 Mg ratio, the relatively high proportion of magnesium will block calcium uptake, causing a calcium deficiency.

Specific Elements

It is difficult to state upper critical levels for nitrate and ammoniacal nitrogen, phosphate, and potassium. Water concentrations up to the fertilizer concentrations needed for application with every irrigation could be acceptable. It is important to reduce the fertilizer concentration by the amount that is found in the irrigation water to prevent overapplication. Requirements for these nutrients will be covered in Chapter 9. Excesses of nitrogen and potassium are generally expressed as high-soluble-salt damage. These two nutrients contribute more to the salt level of fertilizer solutions than the remaining nutrients combined.

Upper critical water levels for sodium and chloride are commonly stated to be 50 and 70 ppm, respectively (Table 8-1). These values are highly questionable. Sodium is not an essential nutrient, although a small concentration of a few parts per million has been found to be beneficial for carnation. Chloride is an essential micronutrient. The requirement for chloride is so low that the contaminant amount of chloride in water, substrate, and fertilizers is sufficient to meet plant needs. It would be best to apply no sodium or chloride in order to keep the EC level of water and fertilizer solutions as low as possible. However, for situations where these ions cannot be avoided, sensitive crops will tolerate only the stated limits, while others can tolerate much higher levels. High levels of sodium block uptake of potassium, ammonium, calcium, and magnesium in plants. When the sodium concentration in water is high, that is, 50 to 100 ppm, one should increase the application rates of potassium, calcium, and magnesium to bring these into balance with sodium so that their uptake is not impeded. Whenever sodium and chloride levels are high in water, one should observe the EC level because these ions contribute in large part to the total salt level in water.

Chlorine is quite different from chloride. Low concentrations of chlorine are injurious to plants. A chlorine concentration of 0.4 ppm can cause root-tip injury to some crops, including chrysanthemum and rose, when grown in hydroponic systems. Crops grown in solid substrates tolerate much higher chlorine levels in water than crops in hydroponic systems, because far less of the chlorine gets to the roots in substrate. Chlorine converts rapidly to chloride as it reacts with organic matter in substrate. Since the concentration of chlorine in water rarely exceeds 10 ppm, the low level of resulting chloride is insignificant. The quantity of chlorine added to municipal water varies. Also, the longer water remains in the water mains, the lower the chlorine content. Water drawn from a position near the end of a municipal water main, or water drawn on a rainy day when customers are using little water, will have been in the pipe system sufficiently long for the chlorine level to drop significantly. Typical levels of chlorine in municipal water are in the range of 1 to 2 ppm but can be as high as 10 ppm. A study of 23 species of foliage, flowering pot, and vegetable crops by Frink and Bugbee (1987) revealed that growth was inhibited by 2 ppm chlorine in irrigation water in begonia and geranium, 8 ppm in pepper and tomato, 18 ppm in kalanchoe, lettuce, and tradescantia, 37 ppm in broccoli, marigold, and petunia, and 77 ppm in English ivy, Madagascar palm, and Swedish ivy. Bridgen (1986) reported that zinnia seedlings began to react to chlorine in irrigation water when it reached 7.6 ppm. Symptoms included shorter growth, leaf curl, and veinal chlorosis on all leaves. Pot mums were not adversely affected until a concentration of 15.2 ppm chlorine was reached. The symptom was veinal chlorosis of leaves. It appears that less than 2 ppm chlorine is generally safe for crops in solid substrate.

It is difficult to state an upper critical concentration for the essential micronutrients in irrigation water. If the water source is applied to high-pH substrate, a high proportion of these nutrients will precipitate and not bother the plant. Tolerable concentrations of iron, manganese, zinc, copper, and borate in water become lower with lower substrate pH. A good way to set upper limits for these nutrients is to identify the higher concentrations used in hydroponic solutions. Roots in hydroponic solutions are fully exposed to nutrients 24 hours a day throughout their entire crop period. Typical upper concentrations for these nutrients in hydroponic solutions are 4 ppm iron, 1 ppm manganese, 0.3 ppm zinc, 0.2 ppm copper and 0.5 ppm borate-B (Table 8-1). These levels should be safe in irrigation water. How much higher they can be is not clearly known except in the case of boron. A concentration of 1 ppm borate-B is generally toxic.

Fluoride causes injury to some crops. The 0.5 to 1 ppm concentration of fluoride that is added to many water supplies to reduce the incidence of tooth decay is injurious to some green plants. Plants that are highly sensitive to fluoride are as follows: *Chlorophytum* (spider plant), *Cordyline terminalis* (mainly the cultivar 'Baby Doll'), and *Dracaena deremensis* (mainly the cultivars 'Janet Craig' and "Warneckii'). Sensitive plants include: *Dracaena fragens* (corn plant), *Maranta leuconeura erythroneura* (red nerve plant), *Maranta leuconeura kerchoviana* (prayer plant), *Spathiphyllum* (most species), *Yucca elephantipes* (spineless yucca), *Ctenanthe oppenheimiana, Ctenanthe amabilis,* and *Chamaedorea elegans.* Plants that are probably sensitive are *Chamaedorea sigfritzii, Aspidistra eleatior, Calathea insignis, Calathea makoyana, Dracaena marginata, Dracaena sanderana,* and *Pleomele thalioides.* It is interesting to note that most of the plants belong to the families Liliaceae and Marantaceae. For these crops, it is best to avoid fluoridated water. When fluoride is a problem, the pH level of root substrate can be raised to reduce the solubility of fluoride.

Lithium appears in some water supplies. At concentrations over 0.5 ppm some plants can be injured. The symptoms of toxicity are chlorosis of the margins of older leaves, followed by necrosis.

Watering Systems

Hand Watering

Hand watering today is uneconomical. Growers can afford hand watering only where a crop is still at a high density, such as in seed flats, or when they are "spot watering"—that is, watering a few select pots or areas that have dried sooner than others. We will first consider the price of hand watering a bench 4 ft (1.22 m) wide by 100 ft (30.5 m) long and then compare this price to those of automatic systems for watering a bench of equal size. The price of materials in each system is based on minimum order rates, which maximizes the cost. The water main to each bench is not calculated into the cost of each system. The labor of installation is not included either, but is quite minimal. A fair estimation of time required to install an automatic system on one bench would be four hours. Thus, the labor bill would be only about $45 at an hourly rate of $9 plus all benefits. (Benefits include two 15-minute coffee breaks per day, six holidays, five vacation days, five sick-leave days, unemployment insurance, worker's compensation, and social security, for a total of 23.5 percent of the hourly rate.) In all cases, the labor saved will pay for the automatic system in less than one year.

A bench area of 400 ft^2 (37 m^2) with a fresh-flower crop requires 200 gal (750 L) of water at each watering. The frequency of watering can range from once a week in the dark part of the winter to more than three times a week during the summer. Taking a conservative average of two times per week, the bench is watered 104 times in a year. At a water flow rate of 8 gpm (30 L/min), which is common for 3/4-inch (19-mm) hose, 25 minutes is required to apply 200 gal (750 L). For the whole year, 43.3 hours are spent watering one 4-foot-by-100-foot (1.22-m-by-30-m) bench. At an hourly rate of $6 plus 23.5 percent in benefits, the cost is $320.85.

It soon will become apparent that this cost is too high. In addition to this deterrent to hand watering, there is a great risk of applying too little water or of waiting too long between waterings. Hand watering requires considerable time and is very boring. It is usually performed by inexperienced (lower-paid) employees, who may be tempted to speed up the job or put it off to another time. Automatic watering is rapid and easy and is performed by a manager, who is less tempted to submit to error. Where hand watering is practiced, a water breaker should be used on the end of the hose (Figure 8-3). Such a device breaks the force of the water, permitting a higher flow rate without washing the root substrate out of the bench or pot. It also lessens the risk of disrupting the structure of the substrate surface.

Fresh-Flower Watering Systems

The following fresh-flower watering systems are *open systems.* In an open system, water is applied to the upper surface of substrate in a bench or pot and any excess water applied is allowed to drain from the bottom of the container. By contrast, a *closed system* is one in which water is applied either to the bottom or top of the substrate in a manner that does not permit any excess water to drain from the container. Closed systems are becoming more popular as one means for combating environmental contamination. Closed systems for fresh-flower and vegetable production are discussed in Chapter 10.

Figure 8-3 **The device on the end of the hose is a water breaker. It reduces the force of water striking the root substrate by increasing the cross-sectional area through which it flows. Reduced water pressure minimizes the breakdown of the root substrate structure and the loss of substrate from containers.**

Perimeter Watering A perimeter watering system consists of a plastic pipe around the perimeter of a bench with nozzles that spray water over the substrate surface below the foliage (Figure 8-4). Either polyethylene or PVC pipe can be used. While PVC pipe has the advantage of being very stationary, polyethylene pipe tends to roll if it is not anchored firmly to the side of the bench. This causes nozzles to rise or fall from proper orientation to the substrate surface.

Nozzles are made of nylon or a hard plastic and are available to put out a spray arc of 180°, 90°, or 45°. For fresh flowers other than roses in benches up to 42 in (107 cm) wide and for rose benches up to 48 in (122 cm) wide, the 180° nozzles are used and are spaced 30 in (76 cm) apart. For fresh flowers other than roses in benches 48 in (122 cm) wide, 180° and 90° or 45° nozzles are alternated 20 in (51 cm) apart. The 90° and 45° nozzles project water farther into a bed than the 180° nozzles do. Regardless of the types of nozzles used, they are staggered across the benches so that each nozzle projects out between two other nozzles on the opposite side. A hole is punched in the polyethylene pipe or drilled in the PVC pipe, and the threaded nozzle is then turned in with a wrench.

Perimeter watering systems with 180° nozzles require one water valve for benches up to 100 ft (30.5 m) in length. For benches over 100 ft (30.5 m) and up to 200 ft (61.0 m), a water main should be brought to the middle of a bench and $^3/_4$-inch (19-mm) water valves should be installed on either side, one to service each half of the bench. This system applies $^1/_{10}$ gpm of water per foot (1.25 L/min/m) of pipe. Where 180° and 90° or 45° nozzles are alternated, the length of a bench serviced by one water valve should not exceed 75 ft (23 m).

Figure 8-4 A perimeter watering system for fresh-flower production in benches or beds. A polyethylene or PVC pipe carries water around the perimeter of the bed. Plastic or nylon nozzles screwed into the perimeter pipe spray water into the bed below the foliage.

The cost of this perimeter watering system for a 4-foot-by-100-foot (1.22-m-by-30-m) bench with alternating 180° and 90° nozzles is $79.00. This includes PVC pipe and two water valves. The cost breakdown is as follows:

2	3/4-inch valves	$21.00
15	3/4-inch PVC pipe fittings	7.30
210 ft	3/4-inch PVC pipe	31.50
120	nozzles	19.20
		$79.00

Turbulent Twin-Wall The Turbulent Twin-Wall hose system (Figure 8-5) is popular because long lengths of bench can be handled from a single header (over 200 ft, 61 m) and because this hose equalizes water pressure along the length of sloping benches (slopes up to 2 percent). Turbulent Twin-Wall hose, when flat in the roll, is 1 inch (2.5 cm) wide and is constructed from 10- or 15-mil (0.025- or 0.038-mm) black polyethylene. While the 10-mil hose is most popular for cut-flower application, the 15-mil hose will last longer. Water outlet spacings are available at 2- or 4-inch (5- or 10-cm) intervals. Water flow rates for the 2- and 4-inch-spacing tubes are 1.5 and 1.0 gpm per 100 ft of tube (5.7 and 3.8 L/min per 30.5 m), respectively. The recommended pressure for this system is 10 psi (70 kPa).

Turbulent Twin-Wall tubes are placed on the surface of the substrate from end to end in the bench. Individual hoses are spaced 8 in (20 cm) across the bench. The distant end of the hose is folded over double and clamped with an end closer, supplied by the manufacturer. The inlet end of the hose is slipped over a special pipe fitting, which is an integral part of a manifold designed to deliver water from a 3/4-inch (19-mm) supply line to each of the hoses in a given bench.

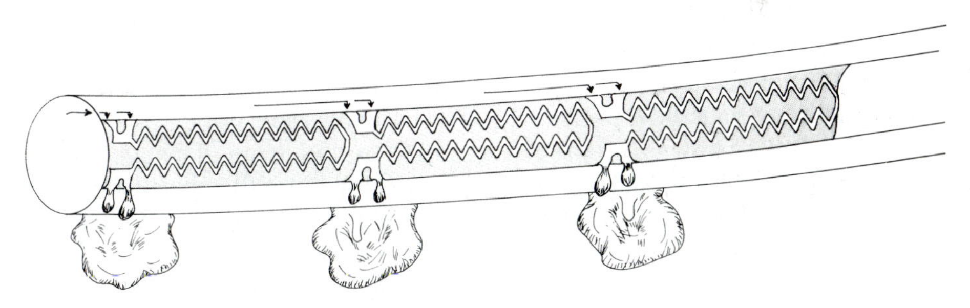

Figure 8-5 **A cutaway view of a Turbulent Twin-Wall hose. Water enters the large tube and quickly runs the length of the bed. Water then moves through pores in the upper wall of the large tube into the small turbulent channel. Water in the turbulent channel moves along a serrated path that produces a turbulent flow. After flowing a short distance, water turns 180° and flows back to a point opposite to where it entered the turbulent channel. At this point, water drips out onto the root substrate. The turbulent flow helps prevent blockage from debris.**

A Turbulent Twin-Wall hose system for a 4-foot-by-100 foot (1.22-m-by-30-m) bench making use of six lengths of 10-mil (0.025-mm) hose with outlets every 2 in (5 cm) and a 3/4-inch (19-mm) plastic header with a valve costs $38.96. The cost breakdown is as follows:

600 ft	10-mil Turbulent Twin-Wall hose	$18.00
1	manifold	7.58
6	end closers	0.48
3	3/4-inch PVC pipe fittings	0.90
10 ft	3/4-inch PVC pipe	1.50
1	3/4-inch valve	10.50
		$38.96

Containerized-Plant Watering Systems

Tube Watering Tube watering has been the standard for automatic watering of pot plants. It is an open system. Water is carried to each pot by a thin polyethylene microtube (Figure 8-6). The tube is available in various inside diameters from different sources, including 0.036, 0.045, 0.050, 0.060, 0.075, and 0.076 inch (0.9, 1.1, 1.3, 1.5, 1.9, and 1.9 mm). The number of pots that can be watered from a single 3/4-inch (19-mm) water main depends upon this inside diameter. The 0.036-, 0.050-, 0.060-, and 0.076-inch microtubes can handle 1,600, 900, 700, and 400 pots, respectively. The narrower-diameter microtubes are used for small pots, where the density is high in the bench and the water requirement per pot is low. This minimizes the expense of laying larger water lines. The 0.060-inch (1.5-mm) size is popular for 6-inch (15-cm) pots, and the 0.075- or 0.076-inch (1.9-mm) size is used for 2- to 5-gallon (7.5- to 18L) containers for such items as poinsettia stock plants and green plants.

The microtube must have an emitter at its end in the pot. The weight of the emitter prevents it from being thrown from the pot when the water is turned on. The emitter further serves the purpose of breaking the force of water so that it does not dig a hole in the substrate. Usually, there is a baffle in the emitter that breaks the flow of water and permits it to trickle out. The emitter also prevents light substrate components from being drawn into the microtube and plugging it. When the water is turned off, suction often occurs, which can draw particles into the microtube. Various types of emitters are sold. One consists of a plastic cylinder with the tube entering through the side (Figure 8-6). Other emitters are conical in shape and are made of noncorrosive metal. Some emitters are designed to be turned off when the pot is removed. This is accomplished in various ways, such as inserting a rod in the opening or simply pushing the emitter back against the microtube. The on-off-type emitters are particularly handy for overhead hanging baskets where a free stream of water could land on personnel or other plants.

When a coarse substrate is used in large pots, 2 gal (7.5 L) and larger, it becomes difficult to wet all of the substrate in the pot. Water slowly flowing from

Figure 8-6 **The tube system is used for automatic watering of pot plants. Water is carried the length of the bench in a plastic pipe generally located down the center. Each pot is connected to the central pipe by a separate small polyethylene tube. A weight is attached to the end of the tube in each pot to anchor the tube and to break the force of the water before it reaches the root substrate.**

a tube tends to channel down through the substrate in a conical shape, leaving much of the surface dry. In such situations, a spray-type water emitter can be used (Figure 8-7). Water is delivered at a greater flow rate and is sprayed out over the surface of the pot. This works well for plants that have little foliage at their base. Drip-ring emitters also serve the purpose of wetting multiple locations on the surface of a large pot (Figure 8-8). In this product, a polyethylene tubular ring is attached to the end of a microtube. Rings can vary in diameter from 4 to 10 in (10 to 25 cm) to meet the requirements of various sizes of pots serviced. The tube comprising the ring has several pin-sized holes in it for water to exit.

Generally, water is provided along the bench by a 3/4-inch (19-mm) polyethylene or PVC supply pipe. The latter lies straighter. Microtubes to each pot may be connected directly into holes punched or drilled into the supply pipe, which is usually run down the center of the bench from end to end. The microtube can be pushed into the hole directly, or a brass insert can be pressed into the hole first and the microtube inserted into it. The brass insert facilitates subsequent removal and replacement of microtubes. Plugs are available to fill holes left by removed microtubes. Microtube watering can be used for hanging baskets as well (Figure 8-9). A plastic water line is run along the length of a row of baskets. A separate thin polyethylene microtube connects each pot to the water line. Some growers install a galvanized water pipe and hang pots directly from it.

Figure 8-7 A Spray Tube water distributor. This method of water distribution is used in larger pots and in root substrate that is very porous, such as that used for orchids. Broad distribution of water over the surface of the pot reduces the problem of water channeling down through the root substrate.

Figure 8-8 A Dribble Ring water emitter for delivering water to large pots. Water is emitted from a series of small holes on the underside of the ring. This wets more of the substrate surface in a large pot than would be possible from a single emitting point.

Figure 8-9 **An automatic tube watering system for hanging baskets. Each pot is connected to the plastic water line above by a thin polyethylene tube.**

Each microtube must be the same length because the flow rate depends upon the length of the microtube. To cut down on the quantity of tubing, particularly in benches with numerous small pots, manifold systems are sold. Either the manifold is attached directly to the water-supply pipe on the bench or it is supplied water through a large microtube from the water pipe. The manifold then supplies water to 8, 10, 20, or more microtubes, each running to an individual pot.

Benches or beds that have a slope along their length create special problems. The water flow rate from tubes at the lower end of the bench is greater than at the high end. When the pot at the low end is watered properly, the one at the high end is underwatered. Conversely, if the pot at the high end is watered properly, the pot at the low end is overwatered and leached. A second problem occurs when water is turned off and the residual water in the system drains out at the lower end of the bed. These problems can be solved with pressure-compensating emitters. These emitters contain a device such as a diaphragm that requires a minimum pressure to open, commonly 5 to 10 psi (35 to 70 kPa). Once opened, the flow rate remains constant, even at pressures up to 60 psi (414 kPa) in some models.

A microtube system of pot-plant watering for a 4-foot-by-100-foot (1.22-m-by-30-m) bench costs $162.22. This includes a manual valve, a 3/4-inch (19-mm) PVC water main along the length of the bench in the center, and 400 24-inch (61-cm) tubes with weights and brass inserts where each is connected directly to the water main. The cost breakdown is as follows:

100 ft	3/4-inch PVC pipe	$15.00
400	24 inch 0.060 ID microtubes with weights	126.80
400	brass inserts	4.40
1	3/4-inch valve	10.50
6	3/4-inch PVC pipe fittings	5.52
		$162.22

Overhead Sprinklers While the foliage on the majority of crops should be kept dry for disease-control purposes, a few crops do tolerate wet foliage. These few crops can most easily and cheaply be irrigated from overhead. Bedding plants, azalea liners, and some green plants are crops commonly watered from overhead.

A pipe is installed along the middle of a bed. Riser pipes are installed periodically to a height well above the final height of the crop. A total height of 2 ft (0.6 m) is sufficient for bedding plant flats and 6 ft (1.8 m) for fresh flowers. A nozzle is installed at the top of each riser. Nozzles vary from those that throw a 360° pattern continuously to types that rotate around a 360° circle. Nozzles with a 180° arc can be obtained for the ends of beds. The spray diameter of various nozzles can range up to 36 ft (11 m) or more. Dripless overhead sprinkler systems are also available.

Trays are sometimes placed under pots to collect water that would otherwise land on the ground between pots and be wasted (Figure 8-10). Each tray is square and meets the adjacent tray. In this way nearly all water is intercepted. Each tray has a depression to accommodate the pot and is then angled upward from the pot toward the tray perimeter. Drain holes are located in the tray, a short distance out from the pot, in the event that excessive water is inadvertently applied. Without the holes, the trays could hold enough unwanted water to keep the base of the soil ball excessively wet for a day or more. Water held in the reservoir below the holes is generally absorbed by substrate in the pot within an hour or two. The object is to turn off the sprinklers before water rises above the drain holes.

Figure 8-10 A plastic tray designed for collecting water applied by overhead sprinklers that would otherwise land between pots.

Boom Watering Boom waterers can function either as an open or a closed system, and are used most often for the production of seedlings grown in plug trays. Plug trays are plastic trays that have width and length dimensions of approximately 12 by 24 in (30 by 61 cm), a depth of 0.5 to 1.5 in (13 to 38 mm), and contain from about 100 to 800 cells. Each seedling grows in its own individual cell. Watering precision is extremely important over the two- to eight-week production time of plug seedlings.

A boom watering system generally consists of a water-pipe boom that extends from one side of a greenhouse bay to the other (Figure 8-11). The pipe is fitted with nozzles that can spray either water or fertilizer solution down onto the crop. The boom is attached at its center point to a carriage that rides along rails, often suspended above the center walk of the greenhouse bay. In this way, the boom can pass from one end of the bay to the other. The boom is propelled by an electric motor. The quantity of water delivered per unit area of plants is adjusted by the speed at which the boom travels.

Often, each side of the water boom, right or left of its center point, is under separate control. In this way, one side of the greenhouse bay may be watered while the other side is not. In addition, the water boom can be turned on and off automatically as it passes over different blocks of plants on its way down the bay.

Most growers water when the crop approaches stress. They apply a sufficient quantity of water or fertilizer to cause 10 to 15 percent of the applied fluid to pass out the bottom of the trays. This is an open system. A few growers practice what

Figure 8-11 **A boom watering machine applying water to a crop of plug seedlings. The machine is contained within a carriage that travels along two overhead rails. Water may be automatically turned on or off in the right and left booms independently. Thus, individual crops may be skipped or watered as the boom travels the length of the greenhouse.**

is commonly called *zero-leach watering.* They water much more frequently during the day than the former group. Each time, a small quantity of water is applied by a fast-moving boom in order to replace only the water that has been utilized by the plants or has evaporated. No water leaches from the trays. This is a closed system. In the closed system the concentration of fertilizer must be reduced to approximately half of the concentration of the open system. While the fate of fertilizer in the open system can be leaching and plant uptake, only uptake occurs in the closed system.

Mat Watering Mat (capillary) watering offers a very good alternative for pot-plant growers who have different pot sizes in a given bench during the year (Figure 8-12). The tube system would require constant removal and addition of tubes to suit the changing pot densities from crop to crop. Mat watering is also beneficial to growers of small pot plants, where the number of tubes in a tube system would be cumbersome and it is best not to wet the foliage, as in a sprinkler system.

The mat watering system uses a mat from 3/16 to 1/2 inch (5 to 13 mm) thick, which is kept constantly moist with water or fertilizer solution. Pots on the mat take up water or nutrient solution by capillarity through holes in the bottom of the pots. Pots of any size can be placed on the mat at one time. No adjustment is needed for shifting pot sizes. Mat watering provides another form of a closed watering system.

Figure 8-12 A mat (capillary) watering system for pot plant crops. Water is applied to the mat several times a day through tubes, such as Turbulent Twin-Wall, that are spaced 2 ft (61 cm) apart. Water moves by capillarity from the mat into the root substrate in the pot and maintains a constant moisture content in the pot at all times. Various types of mats are available.

Mat watering is an old system based on subirrigation. Years ago, sand was placed in a bench and kept moist. Pots were set in the sand, and water continuously rose by capillarity into the root substrate in the pots. This system maintains a constant moisture content in the pot and greatly reduces the labor of watering.

A number of mats are available, but to use them, benches should be level. A polyethylene sheet is placed on the level bottom of each bench. It is preferable, but not necessary, that it be black to reduce light in the mat and, thus, algal growth. A sheet 2 mil (0.05 mm) thick is sufficient since its only role is to serve as a water barrier. The mat is placed on the polyethylene sheet. Care should be taken to keep the edges of the mat level with the plane of the mat. If they are lower, they act as a wick, drawing water from the mat and dripping it to the ground. Mats can be cut with scissors, and more than one piece can be used to line a bench by butting the edges together. Various types of mats are used. Some are composed of reprocessed cloth, while others are composed of virgin synthetic fiber.

Algal buildup is a problem on all mats, but particularly those supplied with fertilizer solution. Algae is unsightly on the mat and on pots, it harbors insects, and it emits a foul odor upon drying. To prevent most algal growth, black perforated polyethylene (up to 14,000 perforations/ft^2, 150,000 perforations/m^2) is available alone or as a component of some mats. The perforated polyethylene lies on the mat, and the pots rest on the plastic. This film restricts light from the mat, which blocks algal growth, and is easy to wash. Roots do not penetrate the polyethylene film. The perforated film also reduces humidity within the plant canopy because its surface is dry except during the time when water is delivered to it.

Watering tubes such as Turbulent Twin-Wall are used to deliver water to the mat. These tubes run the length of the bench and are placed 2 ft (61 cm) apart. The mat should be kept moist at all times. Often, water or fertilizer application is required several times per day. A time clock can be set to activate a solenoid water valve. Overwatering is not a problem because excess water simply drips from the edge of the mat.

Disease organisms can build up on mats. Following any crop in which disease appears, the mat should be sterilized with a disinfectant such as bleach (5.25% active ingredient). A 1:9 bleach-to-water solution can be sprinkled on, allowed to sit for 5 to 10 minutes, and then hosed off. Some mats will withstand steaming, which kills algae and pathogens.

Nutrient solutions applied to mats must be lower in concentration than those that would be applied to the top of the pot in a tube watering system. A good starting point is half the concentration used in a tube watering system. The more common problem encountered is a buildup of salts in the substrate. Water utilization generally exceeds nutrient uptake; thus, fertilizer salts concentrate, particularly at the top of the pot. This situation should be monitored through periodic soluble-salt tests (discussed in Chapter 9). When it occurs, pots should be heavily watered one time from the top to leach the substrate.

The cost of materials in a mat watering system can range from about $100 to $175. The variation is due to the type of mat used. The example system includes a 3/4-inch (19-mm) manual valve, a 3/4-inch (19-mm) header system, two lengths of Turbulent Twin Wall hose for water distribution, and a 4-mil (0.05-mm) black polyethylene underliner. The cost breakdown is as follows:

400 ft^2	mat	\$64.00–\$140.00
200 ft	Turbulent Twin Wall tube plus manifold	12.00
400 ft^2	black, 4-mil polyethylene	10.00
1	3/4-inch valve	10.50
5	3/4-inch plastic header fittings	2.40
		\$98.90–\$174.90

Ebb-and-Flood Systems The ebb-and-flood system is basically a closed subirrigation system for pot plants and bedding plants (Figure 8-13). Pots or flats are grown in a level, watertight bench. Nutrient solution is pumped into the bench during a period of 10 minutes or less to a depth of about 0.75 to 1 inch (1.9 to 2.5 cm) and is held there for 10 to 15 minutes, or long enough for the solution to

Figure 8-13 An ebb-and-flood bench for pot- and bedding-plant production. Nutrient solution enters through the plastic pipe in the foreground. Channels molded into the bottom of the watertight bench ensure rapid and even delivery of nutrient solution to the base of each pot. Solution is held at a depth of 0.75 to 1 inch (1.9 to 2.5 cm) for 10 to 15 minutes to allow time for it to move up through the substrate by capillarity. Then the dark-colored valve at bottom right in the picture is opened and the solution drains back to a holding tank to be used again the next time the plants require water.

rise to the top of the root substrate in each pot by capillarity. The solution is then allowed to drain back to a storage tank, over a period of 10 minutes or less, where it is held until the next bench requires watering. Fertilizer is applied at each watering. The nutrient solution is tested, altered as required, and recycled for months. This closed system is for container-grown plants, and it eliminates nearly all effluent from the greenhouse. It is not a nutriculture (hydroponic) system in that conventional substrates and containers are used. Also, it is not necessary to provide all of the essential nutrients in the fertilizer solution as in the case of nutriculture systems such as NFT and rock wool, which are discussed in Chapter 10. Calcium and magnesium can be provided as dolomitic limestone, phosphorus as superphosphate, and micronutrients as a commercial mixture—all of which are mixed into the substrate prior to planting. Estimates indicate that about 80 percent of Dutch and Danish pot- and bedding-plant crops are produced in ebb-and-flood systems. Adoption of this system has been slower in the United States but should gain momentum as the need to meet antipollution standards intensifies.

Benches are prefabricated primarily from plastic or fiberglass components in various widths of 4 to 6.5 ft (1.2 to 2.0 m) and a customary length of 39 in (1 m). Bench components are cemented together to form the desired bench length. Ebb-and-flood benches can work well as movable (rolling) benches. The floor of the bench has channels molded into it to conduct nutrient solution to all parts of the bench before reaching the bottom of the pots and also to aid in drainage from the bench. The bench must be absolutely level for uniformity of watering and for complete emptying of the bench after watering. Many designs have leveling screws on the bench legs. Below the bench is located a solution holding tank. The tank is covered to exclude dust and light and thus prevent algal growth. During watering, nutrient solution is pumped from this tank into the bench through a hole in the floor of the bench. After watering, solution returns to the tank through the same hole by way of a pressure-sensitive tee valve. A filter is built into the return line to remove debris such as plant tissue and substrate.

Pots with holes that extend up the sides work well. Pots with holes on the bottom work best only if there is a ridge on the bottom of the pot to elevate it off the bench floor. Situation of the pot drainage hole off the floor of the bench permits thorough drainage of water from the pot when the bench is emptied. It also separates the root substrate in the pots from slight puddles that could remain in the bottom of the bench after emptying.

Generally, ebb-and-flood fertilizer concentrations are half the level used for the same crop when it is fertilized from the top of the pot, as in a tube watering system. The optimum concentration can vary. When a substrate is used that does not fully absorb water during subirrigation, higher concentrations of fertilizer are needed to compensate. Incomplete water absorption can be caused by using a substrate that is too coarse or by allowing it to dry too much between watering, so that it repels water. The use of a fine-textured substrate reduces the former problem, while the use of a wetting agent in the substrate reduces the latter problem.

During subirrigation, water and fertilizer move up through the pot to the top of the substrate, where water evaporates and leaves fertilizer behind to accumulate as salt. The accumulating salt levels are generally not a problem when the proper substrate is selected and thus permits the use of low fertilizer concentrations. If salts build up at the top of the pot, it is necessary to leach them down through the profile of the pot and out the bottom. This must be done thoroughly; otherwise, salts will be moved down into the root zone, where injury can occur. Leaching is accomplished by applying a large volume of water to the top of the pots. The resulting effluent is not collected in the holding tank but is discarded.

Several zones of ebb-and-flood benches will often be fertilized in succession with the same solution. After a series of zones are fertilized, or at least once per week, the nutrient solution should be tested for pH and salt (EC) levels. If these levels have deviated, adjustments should be made. Generally, the fertilizer solution does not change. After entering the pots by capillarity, little (if any) fertilizer solution returns to the bench; thus, changes usually do not occur. Losses in volume of fertilizer solution returning to the holding tank are generally made up by adding more of the original fertilizer formulation through a fertilizer proportioner system. Some evaporation of water can occur from the fertilizer solution when it is in the benches. This will raise the salt concentration, which is determined by the EC meter. When this happens, additional water is added to lower the concentration to the desired level.

In more sophisticated systems, the solution is automatically tested in the holding tank after each zone is watered. Equipment commercially available to growers will automatically sample the fertilizer solution, analyze it, and make additions including an acid if the pH is too high, a base such as potassium hydroxide if the pH is too low, fertilizer concentrates if the salt level is too low, and water if the salt level is too high.

Disease pathogens would appear to be a very formidable threat to an ebb-and-flood system, where solution contacting one pot will ultimately contact all other pots over and over for weeks to come. Years of experience have demonstrated that disease is not a problem with this system. However, one would be well advised to take preventive measures. Benches should be cleaned and sterilized between crops. Chlorine bleaches, bromine from Agribrom, or hospital disinfectants work well. The disinfectant should be rinsed from the bench, plumbing, and tank after use. Care should be taken to use disease-free plants. During the crop production time, plant debris and any diseased plants should be removed from benches.

Fungicidal drenches can be applied to the top of pots in ebb-and-flood benches in the conventional manner. Growers will generally not allow the drench effluent from the bottom of the pots to return to the fertilizer holding tank, but rather will reroute it to waste. Although the feasibility of applying drench fungicides through subirrigation in the fertilizer solution has been demonstrated, this practice does not have label clearance. For this reason, it cannot be done. More research should be directed toward this objective because it would appear that lesser amounts of fungicide could ultimately be used. Also, this would be an excellent way of nearly eliminating the flow of fungicide to the environment. It would be contained in a closed system, where extensive degradation could occur. Should disease become a problem in an ebb-and-flood system, a grower could use one of the six in-line pasteurization systems described for the NFT system in Chapter 10.

The ebb-and-flood system is expensive, costing about $4.50/ft^2 ($48.44/m^2) of bench. This cost includes the holding tank, pump, and plumbing. Nevertheless, growers who have adopted this system find that it pays for itself in the following four ways:

1. Fertilizer effluent can nearly be eliminated, since the ebb-and-flood system is closed.
2. Less water and fertilizer are used. The exact amount saved depends on the amount that would normally be lost through leaching. For most greenhouses, this would probably be 30 to 40 percent.
3. The labor input is reduced because watering is automatically handled, and there is no need to install or remove microtubes before and after each crop.
4. It is easy to change from one pot size to another between crops. In a tube watering system, it would be necessary to change the density of tubes. However, it is important that all pots in a zone require the same watering frequency. Thus, the same species of plant, age of plant, and pot size are situated in a zone at the same time.

The disadvantage that goes along with an ebb-and-flood system is the high relative humidity that can build up in the canopy of plants. This situation can lead to condensation and ultimately to foliar diseases. Humidity rises because air is unable to circulate through the floor of the bench. To combat this problem, growers frequently place heating beneath their benches, which helps to dry the air in the plant canopy and sets up air currents during the heating season. It is likewise important to have an effective air movement system that will replace the moist air in the plant canopy resulting from plant transpiration with drier air. A final measure that can be taken is to apply fertilizer solution early in the day when drying can occur long before the cooler hours of the evening.

Flood Floor System Flood floors work well for pot crops that have a low labor input once they are placed in the greenhouse. Bedding plants are grown to a lesser degree on flood floors. These crops respond well to a modification of the closed ebb-and-flood recirculating system. The floor is paved with a lip around the edge and a drain in the center so that it can be flooded and drained in the same fashion as an ebb-and-flood bench (Figures 8-14, 8-15). The advantages of any floor system of production are the possibility of utilizing a greater portion of the floor for production and the ease of moving materials in and out by motorized equipment. The disadvantage of this system is the difficulty laborers have bending over to work on crops. The flood floor system costs between $4.50 and $5.50/ft^2 of floor ($48.44 and $59.20/m^2) for labor and all materials including the floor, plumbing, tanks, and controls.

Since the floor is much larger than a bench and it is difficult to construct an intricate pattern of channels in the floor for drainage, the floor is usually sloped from either side to a drain channel in the center. Slopes of 0.25 to 0.75 inch per 10 ft (2 to 6 mm/m) have been used. Individual zones should not be made larger

Figure 8-14 Poinsettias growing in a flood floor system. Nutrient solution can be seen bubbling up from a line of holes in the floor along the center of the greenhouse bay. Solution enters the bottom of the pots and is drawn throughout the substrate by capillarity. Then the excess fluid returns through these same holes in the floor to a holding tank from where it can be reused. *(Photo courtesy of Greenlink, 22 S. Pack Square, Suite 404, Asheville, NC 28801.)*

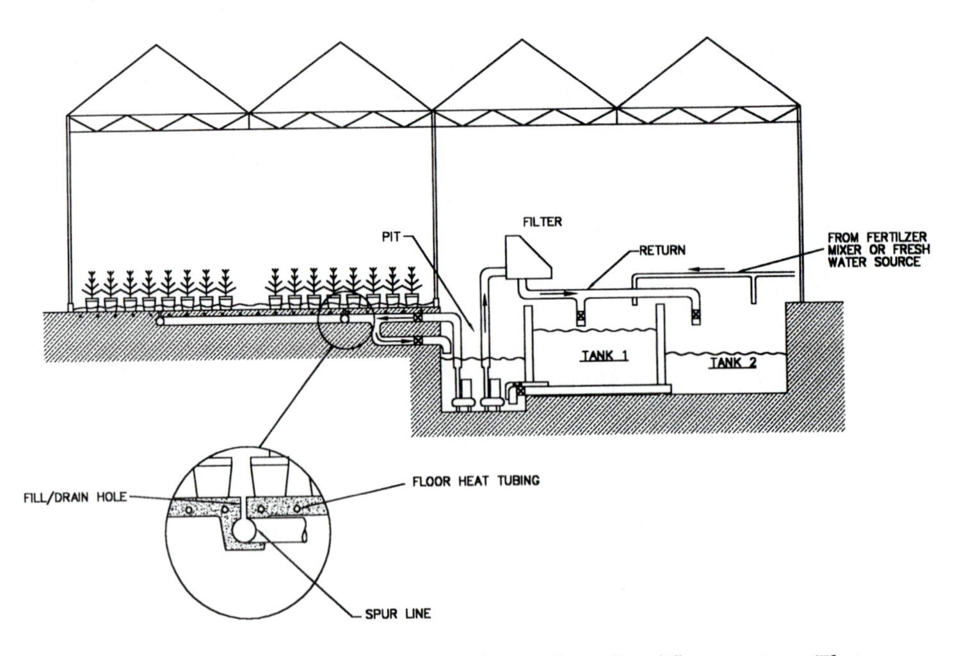

Figure 8-15 A schematic design of the mechanics for a flood floor system. *(Photo courtesy of Greenlink, 22 S. Pack Square, Suite 404, Asheville, NC 28801.)*

than can be flooded in 10 minutes, or preferably in 5 minutes. Without channels in the floor, the pots at the highest point of the floor are the last to begin taking up fertilizer solution and the first to lose contact with the solution. The width of the greenhouse bay constitutes the width of the ebb-and-flood zone. The drain is at the center point of the bay and runs along the length of the greenhouse.

Hot-water heat pipes are installed in the floor of most ebb-and-flood systems to dry the floor rapidly after watering. This lowers humidity in the plant canopy and prevents condensation on the foliage. Some precautions must be taken for bedding plants. A heated concrete floor is slow to cool. Heat application in the concrete may have to be restricted after the first few weeks of growth to ensure that the floor cools before the heat of day occurs. If not, plants will grow too tall.

Trough Culture Trough culture is a variation of ebb-and-flood culture that is used for growing pot plants (Figure 8-16). Single rows of plants are grown in watertight troughs. Several troughs parallel to one another, with spaces between the troughs, constitute the equivalent of a bench. Troughs are sloped from end to end

Figure 8-16 A trough culture system for pot-plant production. Each trough is inclined so that nutrient solution pumped to the top flows down around the base of each pot in the trough and finally spills from the lower end into a collection gutter. The gutter returns the solution to a holding tank. The nutrient solution is recirculated for a sufficiently long time to allow some of it to move throughout the substrate in each pot by capillarity. The unused solution is stored in the holding tank to be used the next time the plants require water. *(Photo courtesy of D. A. Bailey, Department of Horticultural Science, North Carolina State University, Raleigh, North Carolina 27695-7609.)*

at a decline of 0.25 to 0.5 inch per 10 ft (2 to 4 mm/m). Nutrient solution is pumped to the high end, where it slowly trickles down by gravity to supply each pot along the way. At the low end of each trough, the solution drops into a single gutter, perpendicular to the troughs, which returns it to the holding tank. The length of time the solution is left recirculating is determined by how long it takes the solution to rise by capillarity to the top of the substrate in the pot. The same fertilizer solutions and methods of handling are used in trough culture as are used in ebb-and-flood systems.

The advantage of the trough over the ebb-and-flood system is the air circulation that occurs through the plant canopy. The space between troughs allows natural convection currents to move up through the bench platform and the plants. This results in drier plants and consequently less foliar disease.

Float System In the United States, 4.6 billion tobacco seedlings are produced to support 700,000 acres of field production. Sixty percent of these seedlings (2.76 billion) are produced in greenhouses. Nearly all of the greenhouse seedlings are produced in the float system (Figure 8-17).

Seeds are sown in polystyrene trays that contain from 200 to 392 cells. The most common trays for tobacco contain 200 cells. The trays are floated on pools of nutrient solution. Nutrient solution is absorbed by the substrate in each cell through a hole at the base of the cell. A constant moisture content is thereby established in the substrate. Trays are approximately 13 in (33 cm) wide and 26 in (66 cm) long. The pools are on the floor of the greenhouse and can consist of ei-

Figure 8-17 **A float system for growing tobacco seedlings. Note the fertilizer solution ponds on the floor of either side of the greenhouse and the polystyrene trays of tobacco plants being removed for field planting.** ***(From W.D. Smith, Dept. of Crop Science, North Carolina State University 27695-7620.)***

ther a concrete floor with a concrete curb around its perimeter, or a level earthen floor surrounded by a 2-inch-by-6-inch (5-cm-by-15-cm) wooden frame, all of which is lined with 6-mil (0.15-mm) black polyethylene.

The pool is filled with water to a depth of about 4 in (10 cm), and sufficient fertilizer, typically 20-10-20, is added to the water to establish a nitrogen concentration of 150 ppm. Trays are filled with soilless substrate, preferably without nitrogen, phosphorus, or potassium. Seeds are sown in the trays, then floated on the pool of solution. Some growers start the system with water only and add the fertilizer at seven days, when seeds have germinated. About four weeks after sowing, when the solution level has fallen to the point where an addition is required, water is added to the existing solution to restore the original level. At this time, sufficient fertilizer is added to increase the concentration of the total volume of solution by 100 ppm nitrogen. After this date, no further fertilizer additions are made. However, about once per week water is added to maintain the original level. The total cropping time is typically 60 to 65 days.

Pulse Watering Construction of closed systems can be very costly. Modification of open systems, such as perimeter watering of fresh flowers or tube watering of pot crops, is an attractive alternative because many components of the traditional system can be retained. Pulse watering is a modification of an open system and is designed to reduce consumption and effluent of water and nutrients. Benefits of pulse watering include lower usage of fertilizer and water and reduced runoff into the environment of nutrients and any pesticides that might be in the substrate. We have already had a description of one pulse watering system, that is, zero runoff, in the "Boom Watering" section.

There are two main reasons why excesses of fertilizer and water are applied in many greenhouses. In times prior to governmental regulation of nutrient runoff from greenhouses, it was expedient to apply higher concentrations of fertilizer than were necessary to guard against deficiencies. The excess fertilizer was removed by applying 10 to 15 percent excess fluid at each application. Although wasteful, this system is in common use today because it is easier to administer than a more precise system where the exact requirement of nutrients are met. The excess fluid applied is actually much greater in many greenhouses where a time clock is not used for exact control. In these greenhouses, an extra minute of watering can easily occur and result in a doubling or tripling of the amount of excess water applied. It is probably more common to see 40 to 50 percent excess fluid and, consequently, fertilizer applied in greenhouses. The second reason for excess application relates to the porosity of the root substrate used. If there are excessively large drainage pores in the substrate, water will quickly pass through and out the bottom without moving laterally to the dry, fine-textured substrate. To compensate, growers leave the water or fertilizer solution running longer to allow sufficient time for lateral movement of fluid to occur. Excess fertilizer solution is lost during this time. Of course, partial solutions to these problems are to precisely time the application of water or fertilizer solution and to avoid substrates with too high a percolation rate.

Pulse watering goes a step further in reducing water or fertilizer application. Fertilizer solution is applied several times during a drying cycle rather than just once at the end of the drying cycle. However, less solution is applied each time, so that close to zero leaching occurs. The concentration of fertilizer must be lower, approximately half of the concentration used in an open system. The exact concentration of fertilizer required can be gauged by EC tests of the substrate. When salt level in the substrate becomes too high, it can be lowered by applying only water for a period to allow for plant uptake of the fertilizer. This concept is new to greenhouse firms. With time, various ways of implementing it will emerge.

Greenhouse Water Lines

A greenhouse area of 20,000 ft^2 (1,860 m^2) requires a 2-inch (51-mm) water main that can accommodate a 50-gpm (190-L/min) flow rate. An area of 50,000 ft^2 (4,645 m^2) needs 3-inch (76-mm) mains for a flow rate of 125 gpm (473 L/min). Plastic (PVC) pipes are commonly used because they are cheaper and have less pressure drop due to friction than iron pipes. Mains can be installed underground or overhead. More commonly they are placed overhead, which greatly reduces the cost of the system and facilitates subsequent repairs and alterations. Water mains are used for delivering fertilizer solutions as well as water to the crop. It will occasionally be necessary to switch from one fluid to the other.

Consider a greenhouse firm of 20,000 ft^2 (1,860 m^2) laid out in one block measuring 144 ft (44.0 m) wide and 139 ft (42.4 m) long. A roadway runs through the middle along the length of the greenhouse. Benches 6 ft (1.8 m) wide by 68 ft (21.0 m) long run out from either side of the roadway. Pot mums are grown in 6.5-inch (16.5-cm) azalea-type pots at a bench-space allotment of 1.25 ft^2 each (13.5 pots/m^2). There are 326 pots on each bench. Each requires 12 fl oz (350 mL) of water at each watering. Benches are grouped in sets of three, and each set is supplied water through a single valve and manifold. Water or fertilizer is applied to 1,008 plants simultaneously. The best arrangement of water mains calls for 2-inch (51-mm) mains running the length of the greenhouse, each perpendicular to the benches and running over the midpoint of each bench, for a total of about 350 ft (107 m) of 2-inch (51-mm) main pipe. Each linear foot of pipe holds 17.4 fl oz of water (1.7 L/m of pipe). The total main system holds 47.6 gal (180 L). Assume that one day after a watering, an application of fertilizer must be applied. The fertilizer proportioner is turned on at the beginning of the water main. At the opposite end of the water main, the valve on a three-bench station is opened. Before fertilizer reaches plants in that station, 47.6 gal (180 L) of water must be flushed from the lines onto the 1,008 pots. This provides 4.8 fl oz (142 mL) of water to each pot. Also, 1 fl oz (30 mL) of water is pushed out into each pot from the 0.75-inch (19-mm) supply pipes (2.44 fl oz/ft of pipe or 236 mL/m of pipe) located on the bench, for a total of 5.8 fl oz (171 mL). Of the 12 oz (354 mL) of fluid supplied, only 6.2 oz (183 mL) are fertilizer solution. After this station is fertilized, another station will be opened. This time 12 fl oz (354 mL) of fertilizer solution are applied.

A much worse situation occurs in a greenhouse area of 50,000 ft^2 (4,645 m^2) where 800 ft (244 m) of 3-inch (76-mm) main pipe are required. Each linear foot of pipe holds 39.1 fl oz (3.8 L/m) for a total of 244.4 gal (924 L) in the whole system. The combined water in the main system and in the 0.75-inch (19-mm) pipes on the bench provide 12 fl oz (354 mL) of water to the first 2,840 pots supposedly fertilized. Therefore, 10 percent of the plants in the firm receive water rather than fertilizer.

The problem can be rectified by installing double mains, one for water and one for fertilizer (Figure 8-18a). The appropriate valve is opened at each station, depending upon the need for water or fertilizer. A second answer calls for installing a solenoid water valve at the end of a single main system. In order to change from water to fertilizer, a fertilizer proportioner is turned on at the beginning of the water main, and the solenoid at the other end of the main is opened by a switch in the fertilizer room. When the fertilizer solution reaches the end of the greenhouse main, the solenoid is closed. The time required to reach this point is predetermined.

The number of benches in one water station is determined by the size of a single planting of one crop species. It can be safely assumed that all plants in this unit will be watered and fertilized on a single schedule. Generally, from one to three of the benches described earlier would constitute a station.

Water is distributed along the bench from a microtube watering system through two 0.75-inch (19-mm) plastic pipes running the length of the bench.

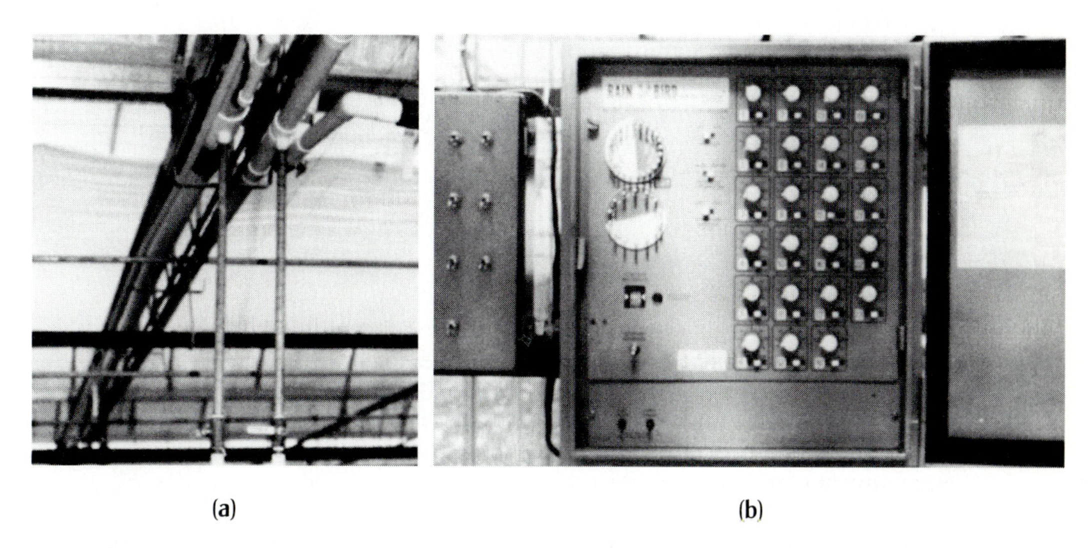

(a) (b)

Figure 8-18 **(a) A dual-main system suspended overhead at the midpoint and perpendicular to benches. One pipe carries water while the other carries fertilizer solution. Each main is connected to the water-distribution system on the bench below through a manual valve. The manual valves could be replaced with solenoid valves, and these, in turn, could be controlled by two stations on a sequential timer. (b) A sequential timer capable of watering 23 separate zones in a greenhouse individually.**

Water is supplied to the midpoint of each 0.75-inch (19-mm) pipe, rather than to the end, to minimize the pressure drop along the bench. A 0.75-inch (19-mm) pipe 70 ft (21.3 m) long supplied with water at one end can distribute water to 70 pots. The same length of pipe supplied with water at its midpoint can distribute water to 280 pots without an adverse differential in the amount of water delivered to each pot. These figures are based on 0.06-inch (1.5-mm) -inside-diameter microtubes 2 ft (61 cm) long. For further details on greenhouse water main systems, see Brumfield et al. (1981).

Further Considerations

Life Expectancy

It is difficult to assess a life expectancy for each watering system, and yet this must be done in order to give proper economic consideration to this form of automation. In general, it can be assumed that the more delicate parts of these systems, such as 8-mil (0.2-mm) polyethylene tubes and plastic nozzles, will last five to six years if properly maintained. The pipes, valves, and overhead metal nozzles can last considerably longer, at least 10 years, particularly if PVC is used.

Problems arise when particles are not strained from well or pond water. These particles accumulate in the smaller tubes and nozzles. A 150-mesh strainer should be used in all systems, even when city water is used. Metal fittings should be avoided after the filter to prevent clogging from rust. When river water is used, it is best to strain the water through a sand filter prior to the 150-mesh filter.

Sufficient light enters thin-walled white PVC pipe to permit algal growth, which can cause plugging of tubes and nozzles. To prevent this problem, pipes can be painted. The best color is aluminum since it restricts light and at the same time keeps the pipe cool. Water in black pipes that are not used continuously on summer days can become sufficiently hot to burn plants.

Sterilization

The plastic components of automatic watering systems should not be steam sterilized. This process tends to reduce the life expectancy of the polyethylene components and to distort PVC pipe and plastic nozzles. Prior to steaming, the flexible tubes should be rolled up, and the plastic pipe mains should be lifted above the bench and secured to the superstructure above.

Once these components are removed, the watering system must be sterilized; otherwise, there is the risk of recontaminating the bench substrate with particles adhering to the watering system. A sponge or rag dipped in a pail of disinfectant such as bleach can be used to wipe pipes and nozzles. Disinfectants could also be proportioned into the water line and applied by hose. The flexible water tubes can be removed and soaked in a barrel of disinfectant.

The problem is not as great for pot plant benches because there is no substrate to be pasteurized. These benches and the watering system may be hosed with a disinfectant.

Automation

It is perhaps best for small growers to use manual valves on the watering system. In this way, the owner or manager can check each bench daily, and the expense of further automation is avoided. Larger firms tend to have a heavier investment in management, which better guarantees careful daily monitoring of all growing areas. Because of the extensive area to be watered, these firms should install automatic valves.

A fixed interval cannot be set between waterings. Bright and warm conditions increase the frequency of drying, while cold or overcast conditions reduce it. Commonly in Europe and occasionally in the United States, a solar-control switch is used to determine the time to water. This instrument has a remote light-sensing mechanism that measures solar energy at the point in the greenhouse where the sensor is installed. A given level can be set on the instrument, and when it is reached, any electrical system plugged into it, such as a solenoid switch on the watering system, can be turned on.

Invariably, a firm investing in automatic water valves will have numerous water zones, each with a solenoid valve. To cut down on the size of the water main and pump, one area is watered at a time. A sequential-control instrument is used to coordinate the watering of a number of areas (Figure 8-18b). The sequential-control instrument may be turned on manually, or it may be activated by a solar-control instrument. Once activated, it will open and close any number of solenoid valves, one at a time, for any preset time from 15 seconds to 30 minutes. The price of sequential controllers begins around $200 for eight-station systems and increases according to the features included and the number of stations activated. Up to 40 stations can be handled by one controller.

Sequential timers can result in considerable labor savings in large firms. Consider the time required for a manager to open valves, wait three to five minutes for water to be applied, and close valves on each of 40 water zones. With a sequential timer, the manager walks through the 40 zones, making a list of those that need watering. Then, he or she programs the sequential timer to apply water to those zones and is thus free to perform other tasks. The cost savings are considerable since the person capable of making watering decisions should be one of the more experienced (higher-paid) employees.

Economics

As one manufacturer states, "Automatic watering doesn't cost, it pays." This statement applies more to automatic watering than to most other systems of automation in the greenhouse. The automatic systems discussed range in materials-plus-installation costs from $39 to $175 as compared to a labor cost of $321 for hand watering a 4-foot-by-100-foot (1.2-m-by-30.5-m) bench for one year. The labor of operation throughout the year is negligible, since it simply entails opening and closing valves or programming a timer. It can be done by the manager during the rounds he or she would ordinarily make. In a large range, a sequential timer could ensure that this time is minimized. Taking all materials and labor into consideration, automatic systems cost less the first year than hand watering does.

Other factors make automatic watering a necessity for greenhouse operators. First, as already mentioned, the ease of watering better ensures that water will be applied when needed and in the quantity required. Second, automatic watering provides a means of applying water without wetting the foliage, which is very important for the control of disease, particularly in such crops as African violet, cyclamen, gloxinia, primula, Rieger begonia, and the lower foliage of fresh-flower crops. Third, automatic watering systems provide the means through which liquid fertilizer can be automatically applied.

Summary

1. Watering would appear at face value to be a boring, unimportant operation, but poor watering practice is probably the most common cause of poor greenhouse crops. Underwatering can have as deleterious an effect on crops as overwatering.
2. Proper watering depends upon three rules:
   a. Use a well-drained substrate with good structure. This will allow for ample moisture retention along with good aeration, even immediately after application of water.
   b. Water thoroughly each time. Substrates cannot be partially wetted. Water should be applied until it flows from the bottom of the container. As a rule, a 10 to 15 percent excess of water is applied. In general, for soil-based substrates, water is applied at the rate of $^{1}/_{2}$ gal/ft^2 (20 L/m^2) of bench, or 10 to 12 fl oz (300 to 350 mL) per 6.5-inch (16.5-cm) azalea-type pot.
   c. Water just before initial moisture stress occurs. This can be determined in most crops by the occurrence of subtle foliar symptoms such as texture, color, and turgidity changes. Some crops, such as azalea, do not show symptoms until root damage has occurred. Color, feel, and weight of the substrate are the cues for these crops.
3. Water quality is very important and is often overlooked. Total salt-content levels, alkalinity levels, the balance of calcium to magnesium, and levels of individual ions such as boron and fluoride can all have a serious bearing on crop success. The water source should be tested before a greenhouse is established. See Table 8-1 for water-quality guidelines.
4. Hand watering is too expensive in today's labor market. Numerous automatic watering systems exist for both fresh-flower and pot-plant production. These

systems can pay for themselves within a year. In addition to having an economic advantage over hand watering, automatic watering systems better guarantee that sufficient water will be applied on time because of the ease of application they offer. Automatic watering systems help to foster disease control by keeping foliage drier. These systems are also used for automated fertilizer application.

5. Water systems can be of the open type, where water or nutrient solution is applied to the top of the root substrate and the excess drains from the bottom of the pot or bench to the environment; or they can be closed, where no excess is released to the environment. Some typical open systems are perimeter watering and Turbulent Twin-Wall for fresh flowers and tube watering for pot plants. Closed systems include mat watering, ebb-and-flood, and trough culture, all used for containerized plants. Fertilizer concentrations used in closed systems are usually about half of those used in open systems.

6. It is advisable to equip the greenhouse with dual water lines, one for water and the other for fertilizer. In this way one can immediately switch from watering to fertilization or vice versa at any location in the greenhouse. It is also beneficial to control application of water or fertilizer solution by a sequential timer or a computer to eliminate the need for one to wait for each station of plants to be watered.

References

Florist-supply company catalogs are available annually to greenhouse growers and are a primary source of information. The manufacturers themselves are another good source of literature.

1. Ball, V. 1991. New irrigation concepts. In Ball, V., ed. *The Ball Red Book,* 15th ed., pp. 97–115. West Chicago, IL: Geo. J Ball.
2. Biernbaum, J., M. Yelanich, W. Carlson, and R. Heins. 1991. Irrigation and fertilization go hand in hand to reduce runoff. In Ball, V., ed. *The Ball Red Book,* 15th ed., pp. 116–122. West Chicago, IL: Geo. J Ball.
3. Bridgen, M. P. 1986. Good for you, bad for your crops. *Greenhouse Grower* 4(1):58–59.
4. Brumfield, R. G., P. V. Nelson, A. J. Coutu, D. H. Willits, and R. S. Sowell. 1981. *Overhead costs of greenhouse firms differentiated by size of firm and market channel.* North Carolina Agr. Res. Ser. Tech. Bul. 269.
5. Farnham, D. S., R. F. Hasek, and J. L. Paul. 1985. *Water quality: its effects on ornamental plants.* Univ. of California, Coop. Ext., Div. of Agr. Natural Resources, Leaflet 2995.
6. Frink, C. R., and G. J. Bugbee. 1987. Response of potted plants and vegetable seedlings to chlorinated water. *HortScience* 22(4):581–583.
7. Matkin, O. A., and F. H. Petersen. 1971. Why and how to acidify irrigation water. *Amer. Nurseryman* 133:14, 73.
8. Reed, D. W., ed. 1996. *Water, media and nutrition for greenhouse crops.* Batavia, IL: Ball Publishing.

9. Weiler, T. C., and M. Sailus., eds. 1996. *Water and nutrient management for greenhouses.* Pub. NRAES-56. Northeast Reg. Agr. Eng. Ser., Cornell Univ., 152 Riley-Robb Hall, Ithaca, NY 14853-5701.
10. Whipker, B. E., D. A. Bailey, P. V. Nelson, W. C. Fonteno, and P. A. Hammer. 1996. A novel approach to calculate acid additions for alkalinity control in greenhouse irrigation water. *Comm. Soil Science and Plant Analysis* 27(5–8):959–976.
11. Wilcox, L. V. 1948. *The quality of water for irrigation use.* USDA Tech. Bul. 962.

CHAPTER 9

Fertilization

Greenhouse fertilization has no equal in agriculture. Heavy plant growth is forced year round under subtropical conditions. Root substrate volume is minimal by field standards and has no lower horizon, as in the field. In the field, the lower horizon catches leaching nutrients for subsequent plant uptake. As a result, nitrogen applications of 4,000 lb are commonly applied to an acre of chrysanthemums in a year (3,600 kg/ha). Excessive levels and imbalances of fertilizer nutrients more often account for difficulties than deficiencies. Micronutrient deficiencies are a constant threat because soils are held in continuous production, often under conditions of heavy leaching.

A typical plant is composed of about 90 percent water. The solid materials in the plant, commonly referred to as *dry weight,* are composed of 17 essential nutrient elements (Table 9-1) plus any of a number of nonessential elements that happen to be available in the root environment. Nearly 90 percent of the dry weight can be attributed to carbon, hydrogen, and oxygen—three essential elements that are not provided in a fertilization program but are obtained pursuant to other cultural procedures.

Carbon and oxygen are derived from carbon dioxide (CO_2) in the air, while oxygen and hydrogen are derived from water. Carbon deficiency is common in the greenhouse and is covered in Chapter 11. Oxygen deficiency is usually a result of slow diffusion of air into the root substrate due to excessively high water content. Hydrogen deficiency is essentially nonexistent. Since only a small quantity of water is needed to provide hydrogen, water-stress injuries are usually related to other factors, such as reduction in photosynthesis caused by closing of stomates or by desiccation of cells.

The remaining 10 percent of the dry weight includes 14 essential elements. Two of these, chloride and nickel, are available in sufficient quantities in root substrate components or as contaminants in fertilizers. Thus, twelve elements must

Table 9-1 Essential Plant Nutrients, Related Chemical Symbols, Classification, and Typical Foliage Composition for Greenhouse Crops Expressed as a Percentage of the Leaf Dry Weight

Nutrient Element	*Chemical Symbol*	*Classification*	*Typical Plant Content (% of dry weight)*
Carbon	C	Nonfertilizer	41.0
Hydrogen	H	Nonfertilizer	6.0
Oxygen	O	Nonfertilizer	42.0
Nitrogen	N	Macronutrient, primary	4.0
Phosphorus	P	Macronutrient, primary	0.5
Potassium	K	Macronutrient, primary	4.0
Calcium	Ca	Macronutrient, secondary	1.0
Magnesium	Mg	Macronutrient, secondary	0.5
Sulfur	S	Macronutrient, secondary	0.5
Iron	Fe	Micronutrient	0.02
Manganese	Mn	Micronutrient	0.02
Zinc	Zn	Micronutrient	0.003
Copper	Cu	Micronutrient	0.001
Boron	B	Micronutrient	0.006
Molybdenum	Mo	Micronutrient	0.0002
Chloride	Cl	Micronutrient	0.1
Nickel	Ni	Micronutrient	0.0005

be applied in a fertilization program. These elements fall into two categories: (1) six macronutrients, which are present in the plant in large (macro) quantities, and (2) six micronutrients, which are present in small (micro) quantities.

Fertilization Programs

Until the middle of the 20th century, dry fertilizers such as dried blood, Milorganite, ammonium nitrate, muriate of potash, and superphosphate were applied monthly to the substrate surface in accordance with monthly soil tests. With the advent of automatic watering systems during the 1950s, it made sense to dissolve water soluble fertilizer into the water during irrigation. In this way, one investment provided automation for both water and fertilizer application.

Today, the standard practice is to dissolve high-analysis fertilizers into concentrated solutions. The concentrate is then proportioned, as needed, by means of a fertilizer injector into the greenhouse water line at the final concentration desired for crop application. This alleviates the need for a large tank to hold single-strength solution. Automatic watering systems connected to the greenhouse water line deliver the fertilizer solution to individual pots or to the soil surface in cut-flower or vegetable beds. Fertilizer is most commonly applied either with each watering (this practice is known as *fertigation*) or on a seven-day basis.

Pre-Plant Fertilization

It is a complicated task to provide all of the 12 essential fertilizer nutrients on a continuous basis. Fortunately, several nutrients may be mixed into the root substrate prior to planting without further application. The results of a soil test provide the basis for pre-plant nutrient additions. Only four categories of nutrients need to be considered: calcium and magnesium, phosphorus, sulfur, and micronutrients.

Calcium and Magnesium The generally desired pH ranges for most crops are 6.2 to 6.8 for soil-based substrate, and 5.4 to 6.0 for soilless substrate. Some crops do well at lower levels, such as azalea and rhododendron, but very few require higher levels. Nutrient availability is controlled by the root substrate pH level, as illustrated in Figure 9-1. Low pH levels can result in excessive availability of iron, manganese, zinc, and copper; low availability of calcium and magnesium; increased leaching of phosphorus; and a higher sensitivity to high substrate levels of ammoniacal nitrogen (Figure 9-2). High pH levels, on the other hand, result in the tie-up of phosphorus, iron, manganese, zinc, copper, and boron. It is readily apparent in Figure 9-1 that the best compromise of nutrient availability lies

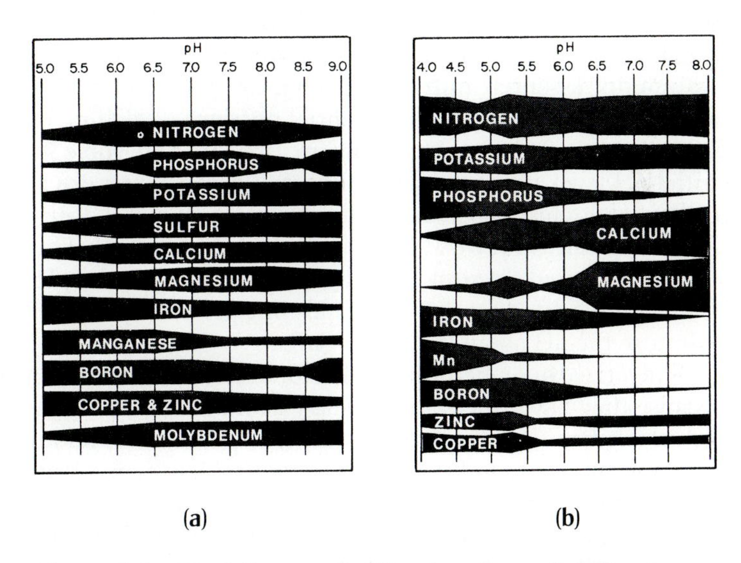

Figure 9-1 **The influence of pH level on the availability of essential nutrients in (a) a mineral soil (*from Truog [1948]*) and (b) a soilless root substrate containing sphagnum peat moss, composted pine bark, vermiculite, perlite, and sand (*from Peterson [1982].*)**

Low	<-- pH -->	High
Toxic: Fe, Mn, Zn, Cu Deficient: Ca, Mg Sensitive: NH4 Leached: PO4		Deficient: Fe, Mn, Zn, Cu, B

Figure 9-2 Nutrient disorders that occur as a consequence of excessively low or high substrate pH levels.

in the pH range of 6.2 to 6.8 for soil-based substrate, and in the range of 5.4 to 6.0 for soilless substrate. Soilless substrate pH levels up to 6.8 do not pose a great difficulty and may be necessary for growers with alkaline water.

A lower pH range is allowed in organic (soilless) substrate for two reasons. First, in organic substrate, there is less native iron, manganese, and aluminum to convert at low pH to a soluble form and thus become toxic or cause a phosphorus tie-up. Second, higher quantities of calcium and/or magnesium are required in organic substrate to attain a given pH level; thus, a sufficient level of calcium and magnesium can be attained at a lower pH level.

Most greenhouse root substrates tend to be acidic (low pH) due to the use of acidic amendments such as peat moss and pine bark. Agricultural limestone is used to raise the pH level. Unless the soil test indicates a high magnesium level, which is rare, dolomitic limestone should be used. This material contains magnesium, an essential nutrient, in addition to calcium. Regular (calcitic) limestone contains primarily one essential nutrient—calcium. Rates of addition vary for each type of root substrate and are presented in Table 9-2. Depending on the initial pH level and the clay content, soil-based substrates can require from 0 to 10 lb of dolomitic limestone per cubic yard (0 to 6 kg/m^3), while soilless substrates typically require 10 lb per cubic yard (6 kg/m^3). This is sufficient to provide the required calcium and magnesium as long as the pH level remains in the desired range. In general, 3 lb of limestone per cubic yard (1.8 kg/m^3) of substrate will raise the pH about 0.3 to 0.5 units.

Pulverized and micronized limestone should not be used at the rates just given because they have finer particle sizes than agricultural limestone. When used at the rates given for agricultural limestone, pulverized and micronized limestones can raise the pH to levels of 7.5 and higher. Lower rates must be used for these finer grades of limestone. Unfortunately, no national standard exists for the particle size of limestone. Each state sets its own definition of agricultural limestone. North Carolina defines agricultural limestone by indicating that 90 percent will pass through a 20-mesh screen ($^1/_{20}$ or 0.05 inch, 1.27 mm) and 35 percent will pass through a 100-mesh screen (0.01 inch, 0.25 mm). Particles larger than 20 mesh dissolve slowly and are of little value during a typical greenhouse crop length. If 90 percent of the limestone particles were smaller than 100 mesh, they

Table 9-2 **Nutrient Sources Commonly Added to Root Substrates during Formulation**[1]

Nutrient Source	Rate per yd^3 (m^3)	
	Soil-Based Substrates	Soilless Substrates
To Provide Calcium and Magnesium		
When a pH rise is desired:		
Dolomitic limestone	0–10 lb (0–6 kg)	10 lb (6 kg)
When no pH shift is desired:		
Gypsum for calcium	0–5 lbs (0–3 kg)	0–5 lbs (0–3 kg)
Epsom salt for magnesium	0–1 lb (0–0.6 kg)	0–1 lb (0–0.6 kg)
To Provide Phosphorus[2]		
Superphosphate (0-45-0)	1.5 lb (0.9 kg)	2.25 lb (1.3 kg)
To Provide Sulfur		
Gypsum (calcium sulfate)	1.5 lb (0.9 kg)	1.5 lb (0.9 kg)
To Provide Micronutrients: Iron, Manganese, Zinc, Copper, Boron, Molybdenum		
Esmigran	3–6 lb (1.8–3.6 kg)	3–6 lb (1.8–3.6 kg)
Micromax	1–1.5 lb (0.6–0.9 kg)	1–1.5 lb (0.6–0.9 kg)
Pro-Max	1–1.5 lb (0.6–0.9 kg)	1–1.5 lb (0.6–0.9 kg)
F-555HF	3 oz (112 g)	3 oz (112 g)
F-111HF	1 lb (0.6 kg)	1 lb (0.6 kg)
To Provide Nitrogen and Potassium (Optional)		
Calcium nitrate	1 lb (0.6 kg)	1 lb (0.6 kg)
Potassium nitrate	1 lb (0.6 kg)	1 lb (0.6 kg)

[1]Rates in this table are for crops other than seedlings. Only limestone is necessary in seedling substrates. Optional nutrient sources for seedling substrate include up to 1 lb (0.6 kg) each of superphosphate, gypsum, and calcium nitrate; no potassium nitrate; and the low end of the rate range for micronutrients.

[2]These are maximum rates designed to supply phosphorus for three to four months if pH is maintained in a desirable range for the crop and the leaching percentage is at or below 20 percent.

would dissolve more rapidly and result in an adversely high pH level unless a lower rate was used. Each bag usually has the screen size indicated on it.

Neutral substrates can result when alkaline (high-pH) components such as hardwood bark are used along with acidic components such as peat moss or pine bark. Limestone is not used in these substrates due to the threat of excessively high pH. Such substrates should be tested. If the calcium or magnesium levels are low, these nutrients can be supplied by incorporating calcium sulfate (gypsum) or magnesium sulfate (Epsom salt) into the substrate during its formulation at rates up to 5 lb/yd^3 and 1 lb/yd^3 (3 kg and 0.6 kg/m^3), respectively. These two salts do not affect the pH of substrates. The availability of these nutrients will generally remain sufficient as long as the pH stays in the desired range.

Phosphorus A second very important reading in the soil test is that of phosphorus. Unless it is high due to application to a previous crop, an application of 1.5 lb of superphosphate (0-45-0) per cubic yard (0.9 kg/m^3) should be made to soil-based substrates and 2.25 lb (1.35 kg/m^3) to soilless substrates (Table 9-2). These are the upper rates to be used to provide all of the phosphorus for the entire crop length through the pre-plant application.

Phosphorus will last this long only when the *leaching percentage* is low, 20 percent or less, and substrate pH is held in the earlier recommended ranges. Leaching percentage refers to the percentage of the water or fertilizer applied to the top of the root substrate that passes out through the bottom of the substrate. If 10 fl oz are applied and 2 fl oz drain from the bottom of the pot, the leaching percentage is 20 percent. Phosphorus is held in substrates mainly on clay particles. For this reason, a single pre-plant phosphorus application can last a year or more in a soil-based substrate. Since soilless substrates rarely contain clay, phosphorus rapidly leaches from these. Leaching increases with increasing leaching percentage and with decreasing substrate pH. One should never depend on pre-plant phosphorus to last for the entire crop length. Soil tests are necessary to determine if and when phosphorus runs out. Subsequent post-plant applications of phosphorus may be required.

When one intends to incorporate phosphorus into the continuous liquid, post-plant fertilizer program, it is not necessary to incorporate pre-plant phosphorus into the substrate. Post-plant phosphorus is sufficient. However, some phosphorus, less than the upper quantities previously recommended, is almost always incorporated into substrates as an insurance policy. The phosphorus source is most often superphosphate, but it need not be. Commercial root substrate formulators more commonly spray water-soluble salts such as monoammonium phosphate onto the substrate prior to mixing to better assure uniform distribution. Any of a number of calcium, potassium, or ammonium phosphate salts can be used dry or dissolved for the pre-plant phosphorus source. The phosphorus source in superphosphate is monobasic calcium phosphate and is water soluble.

Sulfur In the past, sulfur was incorporated into nearly all substrates through the application of single superphosphate (0-20-0). Although superphosphate was applied for its phosphorus source, it supplied sulfur as well. Single superphosphate is approximately 50 percent gypsum (calcium sulfate). Single superphosphate is rarely available today and has been replaced by triple superphosphate (0-45-0), which does not contain gypsum.

When sulfur is desired, it can be supplied by incorporation of 1.5 lb of gypsum into each cubic yard (0.9 kg/m^3) during substrate formulation (Table 9-2). This is a moderately soluble salt that will generally leach from the substrate within a month. Yet such an application can be effective because more sulfur than is required can accumulate in the early stages of plant growth. Later, the extra sulfur is translocated and utilized by newly forming plant tissue when substrate sulfur is low. Sulfur is a partially mobile nutrient in the plant.

Micronutrients The preceding amendments—limestone, superphosphate, and gypsum—provide 4 of the 12 essential fertilizer nutrients. Six more—the micronutrients iron, manganese, zinc, copper, boron, and molybdenum—can be applied by three different strategies. A single application can be made with one of several commercial pre-plant micronutrient mixtures available for incorporation during root substrate formulation (Table 9-2). Some examples follow. Pro-Max (ProSol Inc., Ozark, AL 36361) and Micromax (Scotts-Sierra Horticultural Products Co., Marysville, OH 43041) are mixtures of micronutrient salts. Esmigran (Scotts-Sierra Horticultural Products Co., Marysville, OH 43041) consists of micronutrients impregnated into clay granules, which increases the particle size and total volume for ease of blending during root substrate formulation. Frit-F-555HF trace element mix (Frit Industries, Inc., Ozark, AL 36361) is a mixture of slowly soluble nutrients in the fritted and oxide forms. This product is also available on a clay filler as Frit-F-111HF to increase its particle size and the amount added to a volume of root substrate for ease and accuracy of blending. While these two formulations are generally used for floricultural purposes, other Frit formulations are available for specialty purposes.

There are other pre-plant products in which micronutrients have been incorporated into a mix containing macronutrients, leaving only the possible need for limestone during formulation. Greencare 4-3-14 Soil Mix Charge (Blackmore Co., Belleville, MI 48111) contains N, P, K, Ca, and Mg, and Uni-Mix Plus III (Scotts-Sierra Horticultural Products Co., Marysville, OH 43041) contains all six macronutrients. Barring adversely high pH levels, all of these micronutrient applications can last up to one year.

A second alternative strategy for micronutrient application involves a single liquid application after planting. CHEMEC (Plant Marvel Laboratories, Inc., Chicago Heights, IL 60411) and Microplex (Miller Chemical and Fertilizer Corp., Hanover, PA 17331) contains chelated micronutrients. STEM (Scotts-Sierra Horticultural Products Co., Marysville, OH 43041) and Sol-Trace (Plant Marvel Laboratories, Inc., Chicago Heights, IL 60411) contains micronutrient salts. These products generally provide micronutrients for three to four months.

The third alternative micronutrient application system calls for a low concentration of micronutrients to be applied along with each application of complete liquid fertilizer. Most commercially prepared complete fertilizers contain the six micronutrients. Growers who make their own fertilizers can purchase micronutrient packages to add into their complete fertilizer. Compound 111 (Scotts-Sierra Horticultural Products Co., Marysville, OH 43041) contains chelated micronutrients and a dye for tracing the fertilizer solution, while Masterblend Formula 222 (Vaughan's Seed Co, Masterblend Div., Chicago, IL 60629) and Microplex (Miller Chemical and Fertilizer Corp., Hanover, PA 17331) contain chelated micronutrients. Other products listed earlier under the second strategy can be used at low rates as indicated on their labels toward this purpose.

The quantity of micronutrients required for a crop depends in greatest measure on substrate pH. The requirement goes up with increasing pH. When pH is maintained in the recommended ranges, that is, 6.2 to 6.8 for soil-based and 5.4 to 6.0 for soilless substrates, any one of the three micronutrient strategies can be used alone. When the upper end of the desired pH range is encountered, a combination of two sources becomes necessary; for example, micronutrients may be added as a pre-plant addition in the substrate and as a component of the complete fertilizer. As pH increases further, it may become necessary to use a third micronutrient source such as a single soluble application. Micronutrient increases can even go a step further. Complete fertilizer formulations are available with standard or super-high levels of micronutrients. Those with super-high levels carry names such as "peat-lite special" and "plus" and are designed for soilless substrates. Actually, they can be looked upon as a fourth alternative means for increasing the level of micronutrient addition to a crop.

If maintaining substrate pH in the recommended range makes it so easy to supply micronutrients, why do so many firms operate at higher substrate pH levels? Some erroneously feel that soilless substrate pH should be the same as soil-based substrate, that is, above 6.0. Others agree with the lower recommended pH ranges but can't keep the levels down because of alkalinity in their water. It is more economical to allow the substrate pH to rise and then fight the high pH problem with increased micronutrient applications. Regardless of the micronutrient addition strategy used, the crop should be carefully monitored through foliar analysis to determine when upward or downward adjustments in micronutrient application are needed.

Nitrogen and Potassium The previous four categories of pre-plant nutrient addition are practiced by most growers. A fifth category is optional. A two-week supply of nitrogen and potassium can be provided by 1 lb each of calcium nitrate and potassium nitrate in 1 yd^3 of substrate (0.6 kg each/m^3) (Table 9-2). When nitrogen and potassium are left out of the root substrate, seedlings or cuttings can be fertilized with a complete fertilizer on the day of planting. This practice will immediately establish sufficient levels of these nutrients. The problem of uniform mixing of potentially damaging fertilizers during root substrate formulation is thereby eliminated. Commercial substrates may be purchased with or without nitrogen, phosphorus, and potassium. Most contain these nutrients. More often, when these nutrients are included, growers will use water at the first irrigation after planting and begin their fertilizer program at the second irrigation. However, other growers begin fertilization with the first irrigation, and generally do not experience toxicity problems.

Post-Plant Fertilization

Ten of the 12 fertilizer nutrients, all except nitrogen and potassium, have been applied at this point in quantities sufficient to last for all or most of the crop. These two nutrients are not retained long enough in the root substrate to warrant pre-plant application exclusively unless a slow-release fertilizer is used. They are most commonly applied as a post-plant solution continually throughout crop production.

Complete fertilizers potentially contain the three primary macronutrients—nitrogen, phosphorus, and potassium—and are labeled with a numerical grade such as 10-5-10. The first number indicates the percentage of elemental nitrogen (N), the second indicates the percentage of phosphate expressed in the oxide form (P_2O_5), and the third indicates the percentage of potassium, also expressed in the oxide form (K_2O). A very common fertilizer grade used in the greenhouse is 20-10-20. Fifty or more complete greenhouse fertilizers are on the market, and a similar number of formulas are available for growers who make their own fertilizers. One can easily be intimidated when trying to decide which of these formulas to use. This need not be the case if seven simple questions are answered in proper sequence. The first two pertain to the concentration of fertilizer required and the final five provide the criteria for selecting the best fertilizer formula. These follow.

Frequency of Application Two frequencies of post-plant application are most common. The first is fertigation, in which fertilizer solution is applied at a dilute concentration during each irrigation. The second is a weekly application of a more concentrated fertilizer solution. In this latter program, fertilizer solution should not be applied precisely on every seventh day but rather should average a seven-day frequency over time. There will be times when the substrate has not sufficiently dried on the seventh day. To fertilize on that day would lead to excessively wet substrate and, consequently, insufficient aeration for the roots.

Equal-quality crops can be grown at either frequency. The decision is based on labor considerations. In a small firm where a fertilizer injector and tank of fertilizer concentrate must be moved throughout the greenhouses and connected to one bench at a time, it is more convenient to do this weekly. In large firms where the fertilizer injector is plumbed into the main water line, it is easiest to fertilize at every irrigation. To do otherwise would require continuous switching of the system between fertilizer and water, purging of lines, and tedious record keeping.

Rate of Nitrogen Once the frequency of application is known, the concentration of nitrogen to apply should be established. This rate is based on the crop to be grown, as presented in Table 9-3. Note that the nitrogen concentration ranges are narrow. For fertigation it ranges from 90 ppm nitrogen for cut snapdragon to 255 ppm for poinsettia, and for weekly application it ranges from 240 ppm to 600 ppm. Aside from the bulb crops listed in Table 9-3, all crops fit into one of four nitrogen-concentration categories: very light, light, moderate, and heavy. When a new crop is encountered, one simply needs to gather sufficient information to place it into one of the four nitrogen-requirement categories and then apply the appropriate rate. For a weekly program of application and a requirement category of very light, light, moderate, or heavy, this would be 240, 360, 480, or 600 ppm nitrogen, respectively. For a fertigation program, a rate of 200 ppm nitrogen would be a good starting point for any crop unless more spe-

Table 9-3 **Quantities of a 20 Percent Nitrogen Fertilizer to Dissolve in 100 gal of Water and Resulting Nitrogen Concentrations Required for Fertigation and Weekly Application for Several Greenhouse Crops**[1]

		Fertigation		*Weekly*	
Crop	*Concentration Category*	*oz/100 gal*	*ppm N (mg N/L)*	*oz/100 gal*	*ppm N (mg N/L)*
Daffodil	None	—	—	—	—
Iris	None	—	—	—	—
Hyacinth	None	—	—	—	—
Tulip[2]	Very light	—	—	—	—
Snapdragon	Very light	6	90	16	240
Bedding plants	Very light	13.5	200	16	240
Elatior begonia	Very light	8.5	130	17	255
Azalea	Light	—	—	20	300
Gloxinia	Light	13.5	200	24	360
Rose	Moderate	10	150	32	480
Carnation	Moderate	13.5	200	32	480
Geranium	Moderate	13.5	200	32	480
Easter lily	Moderate	13.5	200	32	480
Chrysanthemum	Heavy	13.5	200	40	600
Poinsettia	Heavy	17	255	40	600

[1]Rates are for any fertilizer containing 20 percent nitrogen. Use one-third more if the fertilizer contains 15 percent nitrogen and one-fifth less if it contains 25 percent nitrogen. (1 oz 20% nitrogen fertilizer/100 gal = 15 ppm nitrogen.)

[2]As an insurance against nitrogen and calcium deficiencies, calcium nitrate should be applied at the rate of 32 oz/100 gal (2.4 g/L) at the start and at the midpoint of the growth-room stages and at the start of greenhouse forcing.

cific information is available. It is not surprising that 200 ppm nitrogen is best for most crops on a fertigation program. The more constant the nutrient level is held in the substrate, the closer most crops come to a single optimum nutrient level. All rates listed thus far pertain to crops that are watered and fertilized from the top of the pot, so that leaching occurs from the bottom. Alterations for subirrigated crops are presented in the "Ebb-and-Flood" section of Chapter 10.

Although common fertilizers in use today contain 20 percent nitrogen, such as 20-10-20, fertilizers with other nitrogen concentrations can be used equally well. When using fertilizers with 15 percent nitrogen, such as 15-0-15, 15-16-17, and 15-5-25, use one-third more to achieve the same nitrogen concentration. If fertilizers with 25 percent nitrogen are used, use one-fifth less.

Proportion of Potassium (K_2O) The first criterion for selecting a fertilizer formula is the amount of potassium needed relative to nitrogen. Most crops develop best on a fertilizer equally balanced in nitrogen and potassium-K_2O. Thus, fertilizers such as 20-10-20, 20-9-20, 20-0-20, 13-2-13, and 15-0-15 are among the most commonly used. There are a few exceptions (Table 9-4). Azalea develops best when the ratio is 3 nitrogen to 1 potassium; 21-7-7 fertilizer is almost universally used for azalea. Elatior begonia grows faster and develops more side shoots when it is fertilized with a ratio of 2 parts nitrogen to 1 part potassium. The preference for nitrogen is slightly greater than potassium for foliage plants in general. The carnation requirement is quite different; a ratio of 1 part nitrogen to 1.5 parts potassium is favored. The greatest requirement for potassium is seen in cyclamen, where a ratio of 1 part nitrogen to 2 parts potassium is best. A fertilizer such as 15-10-30 serves this crop well.

As time passes, the balance of nitrogen and potassium in the root substrate can change. Many fertilizer ratios are commercially available that can be used temporarily to reestablish the desired ratio.

Proportion of Phosphate-P_2O_5 The second criterion for selecting a fertilizer formula is the amount of phosphate required relative to nitrogen. When there is no pre-plant phosphate in the substrate, post-plant fertilizers containing half as much phosphate as nitrogen will supply ample phosphate to meet greenhouse crop needs. This level is well met by fertilizers such as 20-10-20 and 20-9-20. In all probability, lower levels of phosphate will be used in the future, as greenhouse nutrient effluent is further reduced. The lowest level of fertilizer phosphate permissible is not clearly known. A light phosphorus deficiency is sometimes used to produce deep-green, compact plants, especially for bedding plants. This is achieved by leaving phosphate out of the pre-plant addition and supplying 10 to 15 percent as much phosphate as nitrogen in the post-plant fertilizer.

When pre-plant phosphate is supplied to soil-based substrates, post-plant phosphate can usually be omitted. However, when pre-plant phosphate is supplied to soilless substrates, it is important to conduct periodic soil tests to determine if and when leaching becomes sufficient to necessitate phosphate addition in the post-plant program.

Form of Nitrogen At this point it would appear that the universal greenhouse fertilizer has been defined as a 2:1:2 ratio or a 20-10-20 formula. This is true, except that the form of nitrogen has not been determined. There are three common fertilizer forms of nitrogen: ammonium, urea, and nitrate. Selection of which to

Table 9-4 **The Best Ratios of Nitrogen-N to Potassium-K_2O for Greenhouse Crops**

N : K_2O					
3 : 1 Azalea	2 : 1 Begonia	1.5 : 1 Foliage plants	1 : 1 General crops	1 : 1.5 Carnation	1 : 2 Cyclamen

use is the third criterion for selecting a fertilizer formula. Choice of nitrogen form is based on two effects of these forms: threat of ammonium toxicity and influence on root substrate pH. Before discussing each, it is important to realize that ammonium and urea have similar effects in each of these two considerations, while nitrate acts very differently. Urea is similar to ammonium because urea is split into ammonium and carbon dioxide in the root substrate by urease enzyme from microorganisms or within the plant by plant urease. In either event, it is ultimately ammonium that is utilized in the plant.

When plants are supplied more nitrogen than the minimum required, luxuriant quantities are taken up into the plant to be used subsequently if the need arises. Luxuriant fertilization is the common situation in greenhouse crops. Plants can safely store vast quantities of nitrate, but not ammonium. To reduce the probability of ammonium toxicity, it is important to keep the proportion of total nitrogen in ammonium-plus-urea forms below a critical level. Unfortunately, the critical level varies. However, 40 percent of total nitrogen in ammonium-plus-urea forms would appear to be safe in nearly all situations. Symptoms of ammonium toxicity can be seen in Figure 9-7.

Actually, all nitrogen could be in ammonium form if no extra nitrogen is supplied to the crop. In this case all nitrogen would be converted into organic form and little or none stored as ammonium. The second factor controlling the proportion of nitrogen permissible in ammonium-plus-urea forms in fertilizer is the activity of nitrifying bacteria in the substrate. These bacteria convert ammonium to nitrate. The higher their activity, the higher the proportion of permissible fertilizer ammonium. Nitrifying bacteria are suppressed by cooler winter substrate temperatures, anaerobic conditions in overly wet substrate, and by low substrate pH. The best pH for nitrifying bacteria is just above 7.0. Activity is acceptable at pH levels down to 6.0. At levels below 6.0, as often occur in soilless substrates, crops become very susceptible to ammonium toxicity.

During the early decades of liquid fertilizer injection in greenhouses, the most common fertilizer was 20-20-20. After 1970, when soilless substrates became well accepted, this fertilizer was replaced with 20-10-20. The reason was undoubtedly the shift from 70 percent of nitrogen in ammonium-plus-urea form in 20-20-20 to 40 percent in 20-10-20. Under the higher pH levels in soil-based substrates, ample quantities of ammonium were converted to nitrate, but this was not the case at the frequently lower pH levels in soilless substrates.

The second effect of nitrogen to consider when deciding upon the form to use in a fertilizer is its effect on root substrate pH. Generally speaking, ammonium and urea tend to lower the substrate pH, while nitrate tends to raise pH. The commercial fertilizers listed in Table 9-5 are arranged from the most acid at the top to the most basic (alkaline) at the bottom. Note that the percentage of nitrogen in the ammonium-plus-urea forms tends to diminish from the top to the bottom of the table. Eleven nutrient sources used for making fertilizers are listed in Table 9-6 along with the potential acidity or basicity of each.

Plants that grow well in highly acid root media, such as azalea and rhododendron, develop best on a high proportion of ammonium nitrogen. It is interesting to note that when acid-tolerant plants such as azalea are grown at an

Table 9-5 **A List of Several Commercially Available Fertilizers Along with the Percentage of Total Nitrogen in Ammonium-Plus-Urea Form, the Potential Acidity or Basicity of Each, and the Percentage of Calcium (Ca), Magnesium (Mg), and Sulfur (S) Wherever These Are Greater than 0.2 Percent**

Fertilizer	*NH_4[1] (%)*	*Potential Acidity[2]*	*Potential Basicity[3]*	*Ca (%)*	*Mg (%)*	*S (%)*
21-7-7 (acid)	90	1,700				
21-7-7 (acid)	100	1,560		—	—	10.0
24-9-9	50	822		—	1.0	2.2
20-2-20	69	800				
20-18-18	73	710		—	—	1.4
24-7-15	58	612		—	1.0	1.3
20-18-20	69	610		—	—	1.0
20-20-20	69	583				
20-9-20	42	510		—	—	1.4
20-20-20	69	474				
16-17-17	44	440		—	0.9	1.3
20-10-20	40	422				
21-5-20	40	418				
20-10-20	38	393				
20-8-20	39	379		—	0.9	1.2
21-7-7 (neutral)	100	369				
15-15-15	52	261				
17-17-17	51	218				
15-16-17	47	215				
15-16-17	30	165				
20-5-30	56	153				
17-5-24	31	125		—	2.0	2.6
20-5-30	54	118		—	0.5	—
17-4-28	31	105		—	1.0	2.2
20-5-30	54	100				
15-11-29	43	91				
15-5-25	28	76		—	1.3	—
15-10-30	39	76				
20-0-20	25	40		5.0	—	—
21-0-20	48	15		6.0	—	—
20-0-20	69	0	0	6.7	0.2	—
16-4-12	38		73			
17-0-17	20		75	4.0	2.0	—
15-5-15	28		135	5.0	2.0	—
13-2-13	11		200	6.0	3.0	—
14-0-14	8		220	6.0	3.0	—
15-0-15	13		319	10.5	0.3	—
15.5-0-0 (calcium nitrate)	6		400	22.0	—	—
15-0-15	13		420	11.0	—	—
13-0-44 (potassium nitrate)	0		460			

[1]NH_4(%) refers to the percentage of total nitrogen in ammonium-plus-urea form; the remaining nitrogen is nitrate.

[2]Pounds of calcium carbonate limestone required to neutralize the acidity caused by using one ton of the specified fertilizer (divide potential acidity or potential basicity values by 2 to convert to kg limestone per metric ton).

[3]Application of 1 ton of fertilizer has the effect of this many pounds of calcium carbonate limestone.

Table 9-6 **Effect of Various Nutrient Sources on the pH Level of Root Substrate**[1]

Source	*Potential Acidity*[2]	*Potential Basicity*[3]
Ammonium sulfate	2,200	
Urea	1,680	
Diammonium phosphate	1,400	
Ammonium nitrate	1,220	
Monoammonium phosphate	1,120	
Superphosphate	0	0
Potassium chloride	0	0
Potassium sulfate	0	0
Calcium nitrate		400
Potassium nitrate		520
Sodium nitrate		580

[1]From Mortvedt and Sine. (1994).

[2]Pounds of calcium carbonate limestone required to neutralize the acidity caused by using 1 ton of the specified fertilizer (divide potential acidity or potential basicity values by 2 to convert to kg limestone per metric ton).

[3]Application of 1 ton of fertilizer has the effect of this many pounds of calcium carbonate limestone.

adversely low pH for them, nitrate becomes the preferred nitrogen form. This is fortuitous, since the use of nitrate raises the pH.

When positively charged ions (*cations*) such as ammonium, potassium, sodium, and magnesium are taken up by plants, positive hydrogen ions (H^+) are released. This release lowers the substrate pH. Conversely, when negative ions (*anions*) such as nitrate, phosphate, sulfate, or chloride are taken up, hydroxide (OH^-) is often released by the plant. Hydroxide release causes a rise in substrate pH by eliminating H^+, as seen in Equation 9-1. Alternatively, anion uptake can be associated with the simultaneous uptake of a H^+ ion, in which case substrate pH rises due to the loss of H^+. Anion uptake may also be associated with release of a bicarbonate ion (HCO_3^-) from the root. Bicarbonate will also tie up a H^+ ion, causing a rise in substrate pH, as shown in Equation 9-2. In all three responses to root uptake of an anion, the net effect is a loss of hydrogen ions (acidity) in the substrate.

$$OH^- + H^+ \rightarrow H_2O \qquad (9\text{-}1)$$

$$HCO_3^- + H^+ \rightarrow H_2CO_3 \rightarrow H_2O + CO_2 \qquad (9\text{-}2)$$

Since it is the relative uptake of all cations to all anions that determines substrate pH, nitrogen ions alone do not control pH. However, more nitrogen ions are taken up than all other ions combined. Thus, the form of nitrogen supplied in a fertilization program tends to predict the pH effect. This is the reason that the transition from ammonium-plus-urea at the top of Table 9-6 to nitrate at the bottom is not completely smooth. Urea has a similar effect to ammonium because urea is converted to ammonium in the substrate or in the plant prior to its assimilation.

When a rapid decline in substrate pH is desired, a fertilizer such as 20-18-20 with a high potential acidity of 610 lb of calcium carbonate equivalent (Table 9-5) can be used. Application of 1 ton of this fertilizer theoretically causes acidification that requires application of 610 lb of calcium carbonate limestone to counteract (305 kg/metric ton). Substrate pH decline would be much slower with a 15-16-17 fertilizer having a potential acidity of 165 lb of calcium carbonate equivalent. The 15-0-15 fertilizer with a potential basicity of 420 lb of calcium carbonate equivalency will rapidly raise substrate pH. Application of 1 ton of this fertilizer is equivalent to the application of 420 lb of calcium carbonate limestone (210 kg limestone/metric ton of fertilizer). These pH shifts are the result of plant uptake of fertilizer and not of the mere presence of fertilizer in the substrate.

In summary, one must consider the following factors when selecting the form of nitrogen to use in a fertilizer. If a decline in substrate pH is needed, acidic fertilizers should be used. Most, but not all, of these will have more than 20 percent of nitrogen in the ammonium-plus-urea forms. Second, the total amount of nitrogen in the ammonium-plus-urea forms must not be excessive. Growers in northern climates will have to use a smaller proportion of ammoniacal nitrogen in the winter, due to their cooler soil temperatures, than those in warmer climates.

Secondary Macronutrients The fourth criterion for selecting a fertilizer formula is its secondary macronutrient content. Many greenhouse fertilizers contain one or more of the secondary macronutrients (calcium, magnesium, and sulfur) (Table 9-5). If one or more of these nutrients becomes deficient during production of a crop, an easy way to restore it is through a switch to a complete fertilizer that contains it. This points out once again the importance of monitoring the nutritional status of the crop through either soil testing or foliar analysis. As pointed out earlier, a decline in root substrate pH can result in loss of calcium and magnesium from the substrate. Fertilizers such as 15-5-15 (5% Ca, 2% Mg) and 13-2-13 (6% Ca, 3% Mg) contain both of these nutrients in balance. Use of either of these would raise the substrate level of both nutrients in proper proportion while continuing to provide the three primary macronutrients. If calcium is found to be low relative to magnesium, a 20-0-20 (5% Ca) fertilizer could be used to supply calcium without significant magnesium. Once the proper root substrate calcium-to-magnesium balance is restored, the 20-0-20 fertilizer could be discontinued and the previous fertilizer resumed. A shortage of sulfur could be corrected by switching to 20-9-20 (1.4% S) fertilizer. In all of these suggested corrective procedures the 1:1 N-to-K_2O ratio and the nitrogen concentration remains constant.

Micronutrients The fifth criterion for selecting a fertilizer formula is its micronutrient content. Most greenhouse fertilizers contain the six fertilizer micronutrients at "standard" concentrations. A few fertilizers are offered with no micronutrients, while several others are offered with higher-than-standard concentrations. This points out the need to read fertilizer labels. Availability of micronutrients is very high in substrates with a low pH level. In this situation, it might be best to have no micronutrients in the fertilizer if the root substrate has a micronutrient charge in it. At a very high substrate pH, micronutrients are tied up. In this situation it might be best to have a fertilizer with a higher-than-standard concentration of micronutrients. The best test for micronutrient status of a crop is foliar analysis. This further supports the necessity for having a crop nutrient monitoring system.

Finish Fertilization

Nitrogen status at the end of the crop determines to a large degree the plants' post-production longevity and resistance to handling. Plants stand up better to handling when nitrogen is low at the end of the crop. Nitrogen is generally reduced by reducing the total fertilization program. The specific steps to take depend on a soil analysis about two weeks before the end of the crop. When substrate analysis indicates that the available levels of nutrients, particularly nitrogen, are at or above the high end of the optimum range, fertilization can be discontinued for the last two weeks of production. If substrate nutrient levels are in the middle of the optimum range, fertilization can be cut in half for the last two weeks. This may be done by reducing the fertilizer concentration by half, or by applying full-strength fertilizer at half the frequency. If nutrient levels in the substrate are at the low end of optimum, it may be sufficient to stop fertilization for the last week only. In the unlikely event that the crop is receiving insufficient nutrition when the last two weeks start, it would be best to continue fertilization up to the market date.

Formulating Fertilizers

A greenhouse crop manager has to know the math for converting between parts per million (ppm) of a nutrient and its equivalent in ounces or pounds per 100 gal of water for two reasons. First, many fertilizer recommendations are given in parts per million of a nutrient, such as 240 ppm nitrogen. This gives no information relative to how much fertilizer to dissolve in a given volume of water. The solution requires math. If the recommendation were given as 2 lb of 20-10-20 fertilizer per 100 gal, there would be no problem.

The second need for fertilizer conversion math serves growers (fewer than half, and fewer as the years pass) who formulate their own fertilizer. These growers have a wider range of formulas available and save half or more of the purchase price of fertilizer. However, one should be aware that commercially prepared fertilizer constitutes only 1 percent or less of greenhouse income. Formulating fertilizers is simple in concept. Most fertilizers are formulated from combinations of two or more of nine nutrient sources. For example, 1 lb of potassium nitrate added to 1 lb of ammonium nitrate yields 2 lb of 23-0-22-grade fertilizer.

There are three procedures available to determine the amounts of fertilizer sources to dissolve in a given volume of water to achieve the nutrient concentrations in parts per million in a given recommendation. These follow.

Rule of 75 The first method calls for determining the amount of fertilizer carrier needed by using Equation 9-3:

$$\frac{\text{desired ppm}/75}{\text{decimal fraction of desired nutrient in nutrient source}} = \text{oz of nutrient source per 100 gal} \qquad (9\text{-}3)$$

Let us assume that a recommendation calls for 200 ppm nitrogen and that we have a 20-10-20-grade fertilizer available. Using Equation 9-3, we divide 200 ppm by 75, which results in a value of 2.66. Then, we divide this number by 0.20, which is the decimal fraction of nitrogen in the 20-10-20 fertilizer, to obtain a final answer of 13.33 oz of 20-10-20 fertilizer per 100 gal of water:

$$\frac{200/75}{0.20} = 13.33 \text{ oz}/100 \text{ gal}$$

Since this fertilizer also contains 10 percent phosphate-P_2O_5 and 20 percent potassium-K_2O, we end up with a final solution containing 200 ppm N + 100 ppm P_2O_5 + 200 ppm K_2O.

Assume now that we have potassium nitrate available and want 200 ppm potassium. This fertilizer source contains 13 percent nitrogen and 44 percent potassium-K_2O. Applying Equation 9-3, we find that 6.1 oz must be dissolved into each 100 gal of water to yield a final concentration of 200 ppm potassium-K_2O:

$$\frac{200/75}{0.44} = 6.1 \text{ oz}/100 \text{ gal}$$

We also obtain nitrogen from this fertilizer source, and it is important to know what quantity. Equation 9-4 is used to generate this information:

$$\begin{bmatrix}\text{oz of nutrient}\\ \text{source per 100 gal}\end{bmatrix} \times 75 \times \begin{bmatrix}\text{decimal fraction of}\\ \text{desired nutrient in}\\ \text{nutrient source}\end{bmatrix} = \begin{bmatrix}\text{ppm of desired}\\ \text{nutrient}\end{bmatrix} \qquad (9\text{-}4)$$

Applying this equation to our problem, we find that the concentration of nitrogen in the final solution is 59.5 ppm:

$$6.1 \times 75 \times 0.13 = 59.5 \text{ ppm}$$

Conversion Table The "rule of 75" equations can be cumbersome to use. A simplified alternative is to use Table 9-7. Consider formulating a fertilizer containing 200 ppm nitrogen (N), 100 ppm phosphate-P_2O_5, and 200 ppm potas-

Table 9-7 Conversion Table for ppm of Desired Nutrient to oz of Nutrient Source in 100 gal of Water (or g in 1 L) and Vice Versa[1]

Oz Nutrient Source/100 Gal	Percentage of Desired Nutrient in Nutrient Source																
	10	12	13	14	15	16	17	20	21	23	24	33	44	45	53	60	62
	ppm																
1	7.5	9	9.7	10.5	11.2	12.0	12.7	15.0	15.7	17.2	18.0	24.7	32.9	33.7	39.7	44.9	46.4
2	15.0	18	19.5	21.0	22.5	24.0	25.4	29.9	31.4	34.4	36.0	49.4	65.9	67.4	79.3	89.8	92.0
3	22.5	27	29.3	31.4	33.7	35.9	38.2	44.9	47.2	51.6	53.9	74.1	98.8	101.0	117.0	134.7	139.2
4	29.9	36	38.9	41.9	44.9	47.9	50.9	59.9	62.9	68.8	71.8	98.8	131.7	134.7	158.7	179.6	185.6
6	44.9	54	58.4	62.9	67.4	71.9	76.3	89.8	94.3	103.3	107.8	148.2	197.6	202.1	238.0	269.4	278.4
8	59.9	72	77.8	83.8	89.8	95.8	101.8	119.7	125.7	137.7	143.7	197.6	263.4	269.4	317.3	359.2	371.2
16	119.8	144	155.7	167.7	179.6	191.7	203.6	239.5	251.5	275.4	287.4	395.2	526.9	538.9	634.6	718.5	742.4
24	179.6	216	233.5	251.5	269.4	287.5	305.3	359.2	377.2	413.1	431.1	592.7	790.3	808.3	952.0	1,077.7	1,113.6
32	239.5	288	311.4	335.4	359.2	383.4	407.1	479.0	502.9	550.8	574.8	790.3	1,053.7	1,077.7	1,269.3	1,436.9	1,484.8
40	299.4	359	389.2	419.2	449.1	479.2	508.9	598.7	628.6	688.5	718.5	987.9	1,317.2	1,347.1	1,586.6	1,796.2	1,856.1
48	359.2	431	467.0	503.0	538.9	575.0	610.7	718.5	754.4	826.2	862.2	1,185.5	1,580.6	1,616.5	1,903.9	2,155.4	2,227.2
56	419.1	503	544.9	586.9	628.7	670.9	712.5	838.2	880.1	963.9	1,005.8	1,383.0	1,844.0	1,886.0	2,221.2	2,514.6	2,598.4
64	479.0	575	622.7	670.7	718.5	766.7	814.3	958.0	1,005.8	1,101.7	1,149.6	1,580.6	2,107.5	2,155.4	2,538.6	2,873.9	2,969.7
g Nutrient Source/L	ppm																
0.1	10	12	13	14	15	16	17	20	21	23	24	33	44	45	53	60	62
0.2	20	24	26	28	30	32	34	40	42	46	48	66	88	90	106	120	124
0.3	30	36	39	42	45	48	51	60	63	69	72	99	132	135	159	180	186
0.4	40	48	52	56	60	64	68	80	84	92	96	132	176	180	212	240	248
0.6	60	72	78	84	90	96	102	120	126	138	144	198	264	270	318	360	372
0.8	80	96	104	112	120	128	136	160	168	184	192	264	352	360	424	480	496
1.0	100	120	130	140	150	160	170	200	210	230	240	330	440	450	530	600	620
1.5	150	180	195	210	225	240	255	300	315	345	360	495	660	675	795	900	930
2.0	200	240	260	280	300	320	340	400	420	460	480	660	880	900	1,060	1,200	1,240
2.5	250	300	325	350	375	400	425	500	525	575	600	825	1,100	1,125	1,325	1,500	1,550
3.0	300	360	390	420	450	480	510	600	630	690	720	990	1,320	1,350	1,590	1,800	1,860
3.5	350	420	455	490	525	560	595	700	735	805	840	1,155	1,540	1,575	1,855	2,100	2,170
4.0	400	480	520	560	600	640	680	800	840	920	960	1,320	1,760	1,800	2,120	2,400	2,480

[1]Adapted from J. W. Love, Department of Horticultural Science, North Carolina State University, Raleigh, NC 27695-7609.

sium-K_2O from ammonium nitrate, potassium nitrate, and monoammonium phosphate. To use Table 9-7, we proceed as follows.

1. List the fertilizer sources to be used and the percentages of N, P_2O_5, and K_2O contained in each. (These percentages can be found in the first ten entries in Table 9-8).

ammonium nitrate (AN)	33-0-0
potassium nitrate (PN)	13-0-44
monoammonium phosphate (MAP)	12-62-0

2. Sketch a balance sheet as follows.

	oz/100 gal	*N*	*P_2O_5*	*K_2O*
Desired levels (ppm)		*200*	*100*	*200*
potassium nitrate				
monoammonium phosphate				
ammonium nitrate				
Total (ppm)				

3. Select a fertilizer source that contains two desired nutrients. In this case, select potassium nitrate (PN). Begin with the nutrient that is contained in this fertilizer source at the highest concentration. Since nitrogen is 13 percent and potassium is 44 percent, we will start with potassium. In Table 9-7, go to the column headed by 44. Read down the table until you come to 200 ppm. The value 197.6 is close enough. Stop at that point and read across that row to the extreme left entry, which is 6. This is the number of ounces of PN to dissolve in 100 gal of water to achieve a fertilizer solution that contains 197.6 ppm potassium-K_2O. Now determine how much nitrogen you will receive from the use of this rate of PN. Start in the leftmost column with the value 6 oz/100 gal and read across to the right until you reach the column under 13 (the percentage of nitrogen in PN). The value you find is 58.4, which is the ppm concentration of nitrogen provided by this rate of PN. Enter the two concentration values 58.4 ppm N and 197.6 ppm K_2O and the quantity value 6 oz/100 gal into the balance sheet as follows.

	oz/100 gal	*N*	*P_2O_5*	*K_2O*
Desired levels (ppm)		*200*	*100*	*200*
potassium nitrate	6	58.4		197.6
monoammonium phosphate				
ammonium nitrate				
Total (ppm)				

4. Select the next fertilizer source that contains two desired nutrients and the nutrient in it that is in highest quantity. The source is monoammonium phosphate (MAP) and the nutrient is phosphate-P_2O_5. Read down the column in Table 9-7 headed by 62 (the percentage of P_2O_5 in MAP) until you come to the closest value to 100 ppm. The value is 92 ppm. Now read across that row to the extreme left entry, which is 2 oz/100 gal. Two ounces of MAP in 100 gal will yield 92 ppm P_2O_5. However, we need 8 ppm more, or roughly 10 percent more; thus we need 2.2 oz/100 gal. This will give us 101.2 ppm P_2O_5. Determine the amount of nitrogen that is supplied by this concentration of MAP by starting in the leftmost column at the value 2 and reading across the row to the column headed by 12 (the percentage of nitrogen in MAP). The value you find is 18 ppm (nitrogen). Since you are actually dissolving 2.2 oz of MAP, you need to expand this nitrogen concentration value by 10 percent to 19.8 ppm. Enter these values into the balance sheet as follows.

	oz/100 gal	*N*	*P_2O_5*	*K_2O*
Desired levels (ppm)		*200*	*100*	*200*
potassium nitrate	6	58.4		197.6
monoammonium phosphate	2.2	19.8	101.2	
ammonium nitrate				
Total (ppm)				

5. We have already supplied 78.2 ppm nitrogen (58.4 from PN and 19.8 from MAP); thus, the concentration we still need to achieve our goal of 200 ppm is 121.8. This will be supplied by ammonium nitrate (AN) (33-0-0). Go to the column in Table 9-7 headed by 33 (the percentage of nitrogen in AN). The closest value to our desired 121.8 is 98.8. The reading at the left of that row tells us that 4 oz of AN/100 gal will yield a nitrogen concentration of 98.8 ppm. Note that the first entry in the column headed by 33 is 24.7 ppm and that this concentration is achieved by dissolving 1 oz AN in 100 gal. Therefore, we can use 5 oz of AN/100 gal to achieve a concentration of 123.5 ppm nitrogen (98.9 ppm from 4 oz AN + 24.7 ppm from 1 oz AN). This value is close enough to the 121.8 ppm we sought and should be added into the balance sheet. At this time the concentrations in the balance sheet are summed. The resulting fertilizer solution contains nutrient concentrations close enough to the desired levels and can be achieved by dissolving 6 oz of PN, 2.2 oz of MAP, and 5 oz of AN in 100 gal of water.

	oz/100 gal	*N*	*P_2O_5*	*K_2O*
Desired levels (ppm)		*200*	*100*	*200*
potassium nitrate	6	58.4		197.6
monoammonium phosphate	2.2	19.8	101.2	
ammonium nitrate	5	123.5		
Total (ppm)		201.7	101.2	197.6

This table can be used for determining the quantity of a commercially prepared fertilizer to dissolve in 100 gal of water. A recommendation might call for supplying 240 ppm nitrogen from 20-10-20 fertilizer. One would find 239.5 ppm in the column in Table 9-7 headed by 20, the percentage of nitrogen in this fertilizer. The number in the extreme left column opposite this concentration indicates that 16 oz of 20-10-20 fertilizer must be dissolved in 100 gal of water.

Table of Fertilizer Formulas The formulas for 20 fertilizers have already been calculated and are presented in Table 9-8. The first nine fertilizer entries are the individual nutrient sources from which the 20 subsequent fertilizer formulas are derived. For example, an 18-0-22 formula fertilizer can be formulated by blending together 1 lb of ammonium nitrate plus 2 lb of potassium nitrate plus 1 lb of ammonium sulfate. This formulation was determined by locating the 18-0-22 formula in the Analysis column. Then, the three numbers 1, 2, and 1 were located in the row after this formula. Each of these three numbers was traced to the *X* above it and then to the nutrient source to the left of the *X*.

Different fertilizers are recommended in Table 9-8 for blue and pink hydrangea crops. Aluminum serves to regulate flower color in hydrangea. Copious quantities of aluminum exist in most soils. When the pH is low, much of the aluminum is available to the plant and flowers are blue. When the pH is high, aluminum is rendered unavailable and flowers are pink. High levels of phosphate also render aluminum unavailable. You will note that fertilizer formulations used for blue flowers are devoid of phosphate and are very acid, while those used for pink flowers contain large quantities of phosphate and are not very acid.

Additional formulations can be found in Table 9-9. Three commercially formulated complete fertilizers are listed first, followed by three "make-your-own" formulations. The percentage of total nitrogen in each fertilizer that is in ammonium (NH_4)-plus-urea form is given. Quantities to dissolve in 100 gal of water to yield concentrations of 50 to 600 ppm each of nitrogen and potassium are also given.

While the make-your-own fertilizers do not contain micronutrients or dye, commercial preparations are available. They were discussed in the "Pre-Plant Fertilization" section, under "Micronutrients." One example is Compound 111 (Scotts-Sierra Horticultural Products Co., Marysville, OH 46041), which is added at the rate of 1 lb per 40 lb of macronutrient formulation.

Automated Fertilizer Application

The most expedient method for applying fertilizer is the automatic watering system present in most greenhouses. The fertilizer must be dissolved into a concentrated solution in order to conserve space in the mixing and holding tanks. This necessitates the use of a *fertilizer injector* (also known as a *proportioner*). This device mixes precise volumes of concentrated fertilizer solution and water together. By plumbing the proportioner into the main water line that serves the whole greenhouse range, all lines will carry a single-strength fertilizer solution. The pro-

Table 9-8 **Amounts of Nutrient Sources to Combine in Making Various Fertilizer Formulas**[1]

Fertilizer		Nutrient Sources[2]										
Name	*Analysis*	*33-0-0*	*13-0-44*	*15.5-0-0*	*16-0-0*	*21-0-0*	*45-0-0*	*0-0-60*	*12-62-0*	*21-53-0*	*% of N as NH_4 + urea*	*Reaction in Substrate*[4]
Ammonium nitrate	33-0-0	x									50	A
Potassium nitrate	13-0-44		x								0	N
Calcium nitrate	15.5-0-0			x							6	B
Sodium nitrate	16-0-0				x						0	B
Ammonium sulfate	21-0-0					x					100	A
Urea	45-0-0						x				100	SA
Potassium chloride	0-0-60							x			—	N
Monoammonium phosphate	12-62-0								x		100	A
Diammonium phosphate[3]	21-53-0									x	100	SA
Magnesium nitrate	10-0-0										0	B
Chrysanthemum green	18-0-22	1	2			1					47	A
General summer	20-10-24	1					1	2		1	83	A
General low phosphate	21-4-20	7						4		1	55	A
General summer	21-17-20	1					2	3		3	90	A
General	17-6-27	4						4		1	57	A
UConn Mix	19-5-24		6	2			2		1		49	N
Editor's favorite	20-5-30		13				4			2	57	SA
20-20-20 substitute	20-20-22		4				1			3	67	SA
Starter and pink hydrangea	12-41-15		1						2		65	SA
Starter and pink hydrangea	17-35-16						1	4		10	100	SA
N-K only	16-0-24	2			1			2			40	SA
N-K only	20-0-30	1	2								28	SA
Blue hydrangea	13-0-22					2		1			100	VA
Blue hydrangea	15-0-15					3		1			100	VA
Acid	21-9-9	3	1			7		1		2	79	VA
Spring carnation	11-0-17				5			2			0	B
Winter nitrate	15-0-15		1	2							5	B
Winter potash	15-0-22		1	1							4	B
Lily substitute	16-4-12	1	4	6						1	22	N
High K	15-10-30		7	1						2	28	N

[1]Adapted from Koths et al. (1980).

[2]For names of nutrient sources, see the first nine entries in the Name column.

[3]Diammonium phosphate may be pelletized and coated. To dissolve, use very hot water and stir vigorously. Do not worry about sediment. Use crystalline potassium chloride if possible.

[4]B = basic; N = neutral; SA = slightly acid; A = acid; VA = very acid.

Table 9-9 **Quantities of Fertilizers or Nutrient Sources to Dissolve in 100 gal of Water to Make Solutions Containing Concentrations of 50–600 ppm Each of Nitrogen (N) and Potassium (K_2O)**

		Concentration of N and K_2O						
Fertilizer	*% NH_4 + urea*	*50*	*100*	*200*	*300*	*400*	*500*	*600*
		oz/100 gal						
20-20-20[1]	70	3.3	6.7	13.3	20.0	26.7	33.4	40.0
15-15-15[1]	52	4.5	8.9	17.8	26.7	35.6	44.5	53.4
20-10-20[1]	40	3.3	6.7	13.3	20.0	26.7	33.4	40.0
Ammonium nitrate	36	1.4	2.9	5.7	8.6	11.4	14.3	17.1
+ potassium nitrate (23-0-23)		1.5	3.0	6.1	9.1	12.1	15.2	18.2
Calcium nitrate	4	3.0	6.0	12.0	18.0	24.0	30.0	36.0
+ potassium nitrate (15-0-15)		1.5	3.0	6.0	9.0	12.0	15.0	18.0
Ammonium nitrate	40	1.2	2.5	4.9	7.4	9.9	12.3	14.8
+ potassium nitrate		1.5	3.0	6.0	9.0	12.0	15.0	18.0
+ monoammonium phosphate (20-10-20)[1]		0.5	1.1	2.2	3.2	4.3	5.4	6.5

[1]These formulations also contain phosphate-P_2O_5 at equal or half the concentration of nitrogen.

portioner is located either (1) on a bypass line so that either water or fertilizer solution can be obtained from the lines (Figure 9-3), or (2) on a second water main leading to the greenhouse such that one main in the greenhouse supplies water and the second main supplies fertilizer solution, as described in Chapter 8.

It is advisable, and in most states mandatory, that in a potable water system a backflow preventor be installed on any water-supply fixture that has an outlet that may be submerged (Figure 9-4). Such fixtures include fertilizer proportioners and hoses used to fill spray tanks or equipment washtubs. The backflow preventor stops back-siphoning of contaminated water into the water system in the event that a negative pressure (suction) develops. Nitrate, commonly supplied in fertilizers, is harmful to humans. Babies are particularly susceptible to low levels of nitrate. The World Health Organization standards for drinking water set the maximum acceptable concentration of nitrate at 23 ppm in Europe and at 45 ppm in the United States. Backflow preventors are not required when there is a gap equal to twice the diameter of the supply line between the water-supply line and the highest possible level of water in a mixing tank receiving the water. In such a setup, it is not possible for contaminated water to enter the water-supply line.

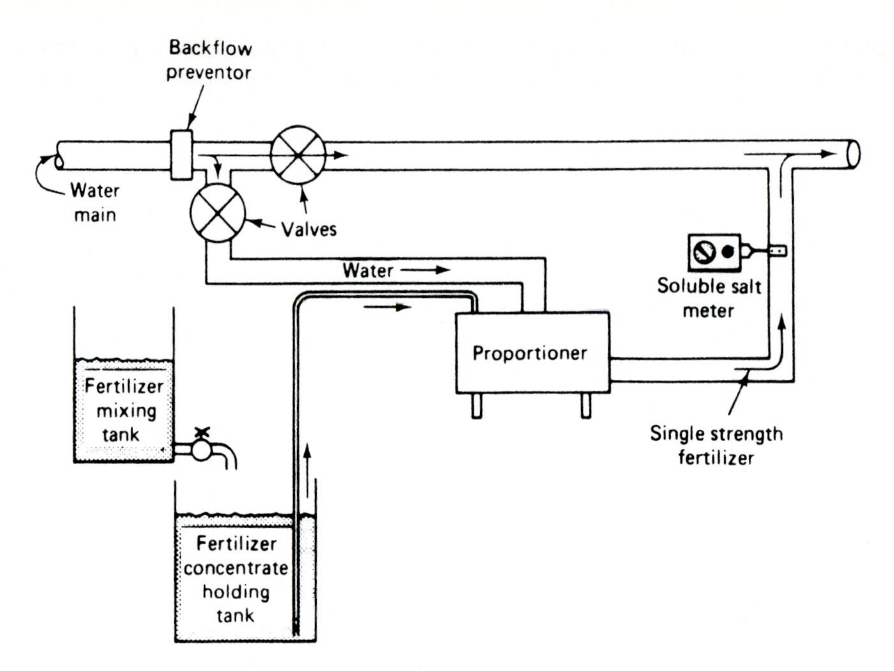

Figure 9-3 A typical arrangement of a fertilizer mixing tank, a holding tank, a fertilizer proportioner, and a soluble-salt (EC) meter along the main water line in a greenhouse range.

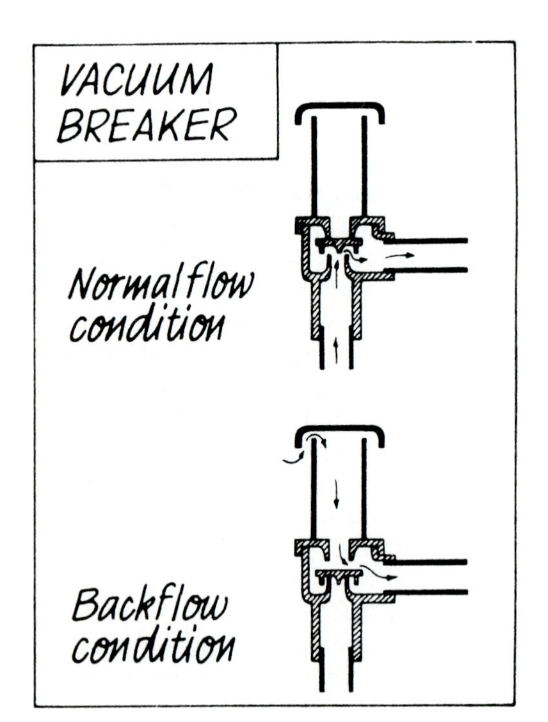

Figure 9-4 A backflow preventor in the open position permitting normal flow of water (*top*). When water pressure drops to a predetermined level, the check valve closes and shuts off the flow of water (*bottom*). At this point, air can enter the device, thus eliminating the negative pressure and any backflow of water. (*From Aldrich and Bartok [1994].*)

Most greenhouse fertilizer is purchased in solid form. It is dissolved in a mixing tank (Figure 9-3), where it is allowed to stand for the better part of a day to permit settling of solids, which occasionally occurs. A tap is located an inch or two from the bottom of the mixing tank for transferring the clear liquid to a holding tank. The proportioner draws fertilizer concentrate from the holding tank. This sequence of tanks prevents unwanted solids from getting into and plugging the proportioner and automatic watering system.

Several types of proportioners (Figure 9-5) are used, depending upon the application. The five criteria to use in selecting the proportioner best suited for a given greenhouse application are as follows:

1. The ratio of fertilizer concentrate to water should be sufficiently wide to keep the concentrate tank volume needed for one complete fertilizer application down to a manageable size. For a small firm, a 1:16 ratio would be sufficient; for a large firm, it would not. The solubility of greenhouse fertilizers allows them to be concentrated up to about 200-fold. Ratios higher than 1:200 are rarely used.
2. An adjustable ratio is highly desirable. In this way, a single fertilizer concentrate can be applied at different concentrations to each of several plant species.
3. The flow rate of the proportioner determines the area of plants that can be fertilized at one time. An 8-gpm (30-L/min) flow rate will service only one ¾-inch (1.9-cm) pipe and, thus, only one bench of plants at a time. In a very large firm, it may be necessary to fertilize 5 to 10 benches at a time in order to finish the task in half a day.
4. The concentrate tank should have sufficient volume to allow the entire fertilization job to be completed with one batch of fertilizer concentrate. Larger firms purchase proportioners without a built-in concentrate tank. In this way, they provide their own tank of required size.
5. As soon as a firm is large enough to afford a dual-head proportioner, it should obtain one. Such a proportioner siphons simultaneously from two separate fertilizer concentrates. Calcium or magnesium can be supplied in one concentrate, while sulfate or phosphate may be supplied in the other. These nutrients are compatible in this case because they do not mix until after they are diluted to single strength. When either calcium or magnesium is added to phosphate, or calcium is added to sulfate in the concentrate tank, they are subject to precipitation.

Proportioners such as the Hozon and the Syfonex operate on the Venturi principle, whereby water passing at a high velocity through an orifice sets up a suction in a line entering from the side. Fertilizer concentrate is drawn in through the side line. This type of proportioner is inexpensive and serves well in a small greenhouse, but is limited in application because of its fixed, narrow proportioning ratio of approximately 1:16. Where large volumes of fertilizer solution are needed, the concentrate tank would become prohibitively large. The ¾-inch

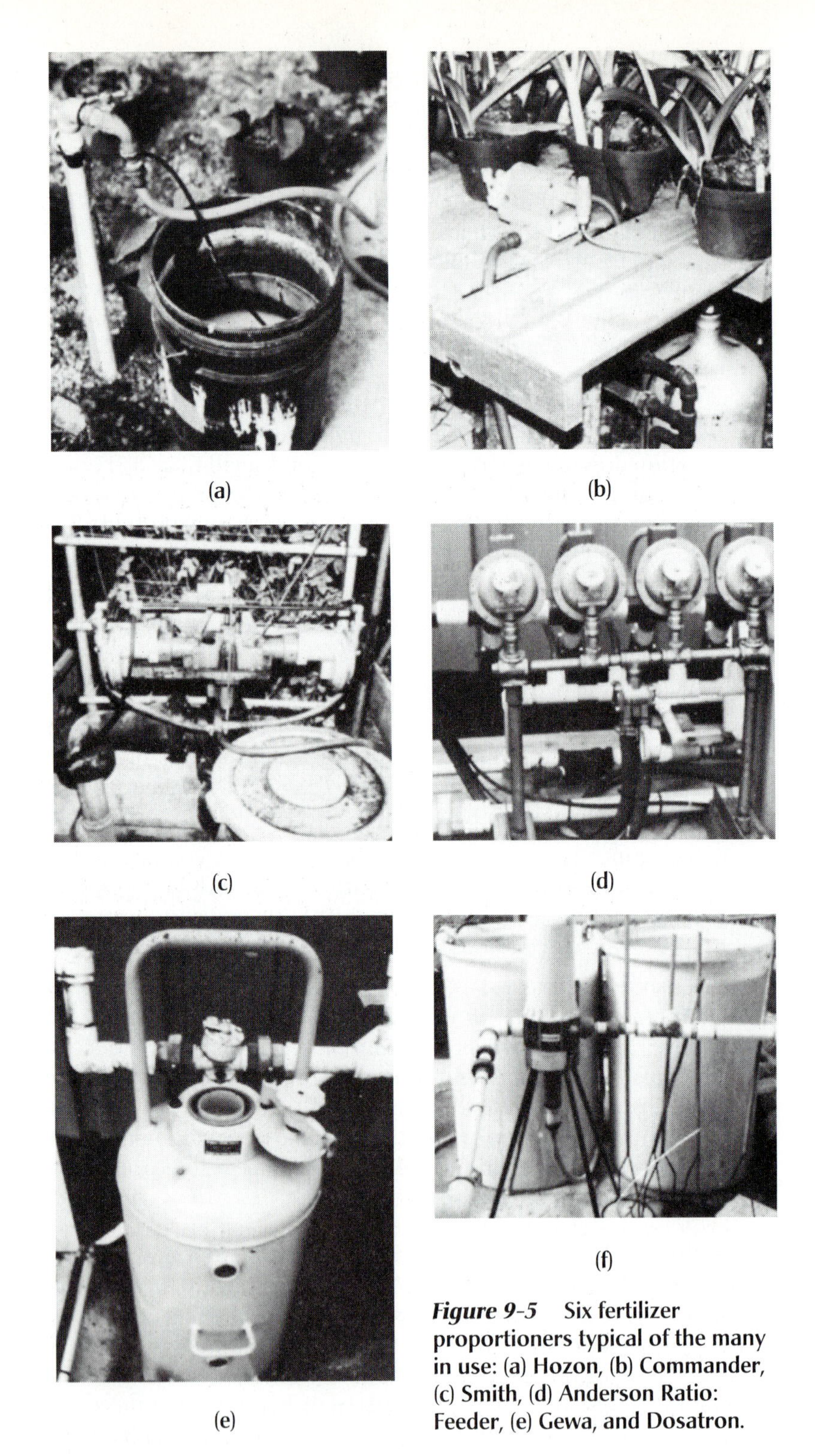

Figure 9-5 Six fertilizer proportioners typical of the many in use: (a) Hozon, (b) Commander, (c) Smith, (d) Anderson Ratio: Feeder, (e) Gewa, and Dosatron.

(1.9-cm) thread size and 3-gpm (8-L/min) flow rate limit the area that can be fertilized at one time.

The Commander proportioner draws in fertilizer concentrate by means of a pump that is driven by water passing through it. It requires a much smaller concentrate volume with its fixed ratio of 1:128 (1 fl oz/gal), but it is limited to a 3/4-inch (1.9 cm) thread size and 6.6-gpm (25-L/min) flow rate.

Fert-O-Ject and Smith Measuremix proportioners have factory-fixed ratios, including 1:100 or 1:200 for the Fert-O-Ject and 1:20 to greatly in excess of 1:200 for the Smith. They can be purchased to fit pipe sizes up to 6 in (15 cm). Their flow rates, which go up to 560 gpm (2,100 L/min) at a ratio of 1:200, permit simultaneous fertilization of many benches. These proportioners use the water-pumping principle to pick up fertilizer concentrate and do not have built-in concentrate tanks.

The Gewa proportioner adds versatility since its ratio is adjustable on site from 1:15 to 1:350. This proportioner includes a concentrate tank available in 4-, 6-, 15-, and 26-gallon (15-, 20-, 57-, and 100-L) sizes and fits up to a 2-inch (5-cm) water line. The concentrate is contained in a rubber bag positioned inside an iron tank. Entering water builds up pressure between the rubber bag and the iron tank wall, thereby pressing concentrate out into the line. The ratio is set by moving a lever to the appropriate numbered setting. The number is determined by dividing 300 by the dilution factor desired. For a ratio of 1:100, 300 would be divided by 100 to achieve a setting of 3.

Anderson Ratio:Feeder proportioners are available in several models, each with on-site adjustable step ratios ranging from 1:80 to greater than 1:200. A flow rate of 450 gpm (1,700 L/min) is possible at a ratio of 1:200. Separate concentrate tanks are supplied by the grower.

Dosatron proportioners may be adjusted on site from a ratio of 1:50 to greater than 1:200. Flow rates extend up to 100 gpm (380 L/min) in pipe sizes up to 2 in (5 cm). The concentrate tank is not a part of these proportioners.

Fertilizer proportioners may shift over time resulting in higher- or lower-than-desired fertilizer concentrations. The fertilizer output concentration should be checked routinely. This may be done by measuring the volumes of concentrate taken up (input) and single-strength solution emitted (output) for a given length of time. The dilution factor of the proportioner is then calculated by dividing the output by the input.

The alternative procedure calls for measuring the electrical conductivity (EC) of the single-strength output solution. Most fertilizer sources are salts that conduct electricity. Thus, EC indicates the fertilizer concentration for any given fertilizer. Each fertilizer company supplies on the fertilizer bag or in its technical literature a table of EC values for each of its fertilizers (Table 9-10). One simply collects fertilizer solution from emitters at individual pots or from spray nozzles on fresh-flower beds and measures the EC level with a meter available from most greenhouse-supply companies. The EC level should equal the one listed in the manufacturer's table for the concentration of fertilizer being applied. The EC values for a given fertilizer formula, such as 20-0-20, may vary from company to company. This is reasonable since each company may use different nu-

Table 9-10 **Soluble-Salt Conductivity (EC) Values in mmho/cm (dS/m) for Various Commercial Fertilizers and for Five Single-Salt Nutrient Sources[1]**

Nitrogen (ppm)	20-19-18 20-20-20	24-8-16 20-5-30	20-0-20	20-10-20	Ammonium Nitrate (34% N)	Calcium Nitrate (15.5% N)	Ammonium Sulfate (21% N)	potassium Nitrate (14% N)	Epsom Salt (10% Mg)
50	.20	.22	.25	.33	.23	.37	.45	.48	.38
100	.40	.45	.50	.65	.46	.74	.90	.95	.75
150	.60	.67	.75	.98	.69	1.11	1.35	1.42	1.13
200	.80	.90	1.00	1.30	.92	1.48	1.80	1.90	1.50
250	1.00	1.12	1.25	1.63	1.15	1.85	2.25	2.37	1.88
300	1.20	1.35	1.50	1.95	1.38	2.22	2.70	2.85	2.25
350	1.40	1.57	1.75	2.28	1.61	2.59	3.15	3.32	2.63
400	1.60	1.80	2.00	2.60	1.84	2.96	3.60	3.80	3.00
450	1.80	2.02	2.25	2.93	2.07	3.33	4.05	4.27	3.38
500	2.00	2.25	2.50	3.25	2.30	3.70	4.50	4.75	3.75
550	2.20	2.47	2.75	3.90	2.53	4.07	4.95	5.22	4.13
600	2.40	2.70	3.00	3.90	2.76	4.44	5.40	5.70	4.50
650	2.60	2.92	3.25	4.23	2.99	4.81	5.85	6.17	4.88
700	2.80	3.15	3.50	4.55	3.22	5.18	6.30	6.65	5.25
800	3.20	3.60	4.00	5.20	3.68	5.92	7.20	7.60	6.00
900	3.60	4.05	4.50	5.85	4.14	6.66	8.10	8.55	6.75
1000	4.00	4.50	5.00	6.50	4.60	7.40	9.00	9.50	7.50

[1]Values are from Scotts-Sierra Horticultural Products Co., Marysville, OH 43041.

trient sources to formulate this fertilizer. Also included in Table 9-10 are the EC values associated with five nutrient source salts commonly used by growers for formulating their own fertilizers.

To facilitate the measurement of fertilizer concentration, an EC meter should be installed in the plumbing downstream from the proportioner. The same meter used for testing root substrate can be used, but the probe containing the electrodes is different. The probe is contained in a pipe fitting that is permanently plumbed into the water line. Each time the proportioner is turned on, the EC level of the solution coming from the proportioner should be checked.

pH, Calcium, and Magnesium Control in Substrate

If any component in a fertilization program could be singled out as the most important, it would be root substrate pH. When pH is set and maintained at the proper level, more than half of all nutritional problems are prevented. The best pH range for soil-based substrates is 6.2 to 6.8; for soilless substrates it is 5.4 to 6.0. The soilless substrate range can extend up to 6.8 with moderate effort to supply more micronutrients.

Substrate pH cannot be treated in a singular fashion. Calcium and magnesium are intimately related to pH. Each factor that governs pH in greenhouse root substrates affects calcium and/or magnesium levels. The complicating issue here is that calcium and magnesium must be in balance. The ideal ratio of calcium to magnesium has not been precisely determined. Based on soil-test standards, a desirable range of ratios would appear to extend from 3 to 5 ppm calcium for each 1 ppm magnesium in irrigation water as well as in the soil solution when determined by the saturated-paste procedure, which is described later.

There are three factors that naturally set, and may be used to manipulate, the pH, calcium, and magnesium levels in root substrates. These are irrigation water, the liming materials used to adjust substrate pH, and fertilizers. They will be discussed in this order.

Irrigation Water

As discussed in the "Water Quality" section in Chapter 8, alkaline water contains bicarbonate and possibly carbonate. Most often these alkaline agents are accompanied by calcium and magnesium. Application of alkaline water is equivalent to the addition of limestone to substrate. This will cause a rise in substrate pH. The rise needs to be counteracted with a lower addition of limestone during substrate formulation and/or use of acid fertilizers. Since calcium and magnesium are also often associated with alkalinity, it is important to determine their ratio in water. A ratio of 3 to 5 ppm calcium for each 1 ppm magnesium is desired in water. If there is more calcium than this, it can block uptake of magnesium in the plant, causing a magnesium deficiency. This can occur even though the concentration of magnesium in the substrate would otherwise have been adequate. If calcium is lower than the 3–5:1 calcium-to-magnesium ratio, the relatively high proportion of magnesium will block calcium uptake, causing a calcium deficiency. When calcium and magnesium are out of balance, compensations must be made through the form of limestone used in the substrate or in subsequent applications of these two nutrients in the fertilization program.

Liming Materials

Pre-Plant There is little one can do within economic reason to change the calcium and magnesium content and ratio in water. Instead, it is easier to adjust the liming materials incorporated into the substrate during its formulation. When calcium and magnesium are in balance in water, dolomitic limestone is the best material to use in the formulation of the substrate. The calcium-plus-magnesium content in the dolomitic limestone and in the irrigation water will insure a proper balance throughout the crop. When magnesium is high relative to calcium in water, it is best to use calcitic limestone in the substrate. This will provide a large initial supply of calcium that will help balance the excessive magnesium supplied throughout the crop in the irrigation water. If calcium is high relative to magnesium, one can use dolomitic limestone plus magnesium sulfate (Epsom salt) up to 1 lb/yd^3 (0.6 g/L) in the substrate. The extra magnesium supplied by Epsom salt will help balance subsequent excessive calcium derived from irrigation water. In all situations, it is important to monitor the nutrient status of the substrate with soil tests to know when later adjustments in the calcium-to-magnesium ratio are required. It should be noted that injection of acid into irrigation water to reduce alkalinity does not lower the concentration of calcium and magnesium in that water.

Post-Plant When pre-plant limestone additions fail to maintain the proper pH later in the crop, post-plant lime additions are possible.

If the pH needs to be adjusted up and calcium and magnesium are in proper balance, flowable dolomitic limestone (Limestone F from W. A. Cleary Chemical Corp., Somerset, NJ 08875) may be used. This product is purchased as a liquid suspension of finely ground limestone. It can be injected into water through a fertilizer proportioner. For each pound of regular limestone required to accomplish the pH rise, 1/3 gallon of Limestone F (2.75 L Limestone F/kg limestone) should be applied.

If the pH needs to be adjusted up and only calcium is needed, the crop can be watered with a hydrated lime solution, also known as builder's lime (calcium hydroxide). This material can be injurious to green tissue and in large quantities can damage roots. A rate of 1.5 lb/100 ft^2 (75 g/m^2) is recommended for fresh-flower crops. The hydrated lime can be applied dry to the substrate surface and then immediately syringed from the plant surfaces and watered into the substrate. Hydrated lime can also be applied to fresh-flower and pot crops by mixing 1 lb in 5 gal of water (24 g/L) and applying the dissolved portion to 20 ft^2 (10 L/m^2) of root substrate. Hydrated lime is more soluble than ground limestone but is not completely water soluble. It reacts much more quickly than ground limestone yet has a shorter residual effect in the root substrate. If one application does not solve the problem, a second one can be made after a few weeks.

The high pH level of hydrated lime can cause a conversion of ammonium nitrogen to ammonia gas. This gas is injurious to roots and foliage. Hydrated lime should not be used when ammonium-containing slow-release fertilizer such as MagAmp or Osmocote is in the root substrate or when a high proportion (over 50 percent) of ammonium is used in a liquid fertilizer program.

Fertilizers

Greenhouse fertilizers can range from an acidity level of 1,700 lb of potential acidity to a basicity of 420 lb of potential basicity per ton (850 kg potential acidity to 210 kg potential basicity/metric ton) (Table 9-5). The direction of pH shift can be regulated by selection of an acid or basic fertilizer and the rate of shift can be controlled by the level of potential acidity or basicity. Selection of the appropriate pH adjustment strength of a fertilizer should be based first on the alkalinity level of the irrigation water. When a rapid substrate pH rise is expected, due to high water alkalinity, fertilizers with high potential acidity levels should be selected. A good choice would be 20-2-20 that has equal concentrations of nitrogen and potassium and a potential acidity of 800. A 15-16-17 fertilizer with a potential acidity of 165 would lower substrate pH, but much more slowly. Growers who mix their own substrate, or purchase it from a supplier who is willing to alter the limestone content, should counteract any adverse water alkalinity and calcium-to-magnesium balance by first altering liming materials in the substrate. Then, they should turn to fertilizer selection to maintain the desired pH regime. When it is not practical to alter liming materials in the substrate, fertilizer selection becomes the first line of attack.

When selecting a fertilizer formula, it is important to check the calcium and magnesium content. Remember, the need to adjust substrate pH is often associated with the need to supply and/or adjust the ratio of calcium and magnesium.

Lowering Substrate pH

It may be necessary to lower the pH level of root substrates. This can be accomplished by the use of sulfur, iron sulfate, or aluminum sulfate. All of these sources react in the soil to ultimately form sulfuric acid. Rates of application are listed in Table 9-11. These recommendations are designed to lower the pH level to 5.0, a level required for producing a crop of blue-flowered hydrangeas. The table is useful for determining other changes. For example, to lower the pH level from 6.5 to 6.0, one would use 0.5 pound of sulfur per cubic yard (0.3 kg/m^3) of substrate, which is the difference between 2.0 and 1.5 lb as listed in column 2 (1.2 and 0.9 kg in column 3); alternately, one could also use 1.5 lb of iron sulfate per cubic yard (0.9 kg/m^3) of substrate, or the difference between 5.25 and 3.75 lb as listed in column 4 (3.1 and 2.2 kg in column 5).

Any one of the three materials can be mixed into the substrate dry as a preplant amendment. Iron sulfate and aluminum sulfate, being water soluble, may be dissolved in water and applied to the root substrate surface of an existing crop. Sulfur is not soluble. However, there is a liquid flowable sulfur product on the

Table 9-11 **Quantities of Sulfur, Iron Sulfate, or Aluminum Sulfate Necessary to Lower the pH Level of Greenhouse Root Substrate from Various Levels to 5.0**[1]

	Sulfur		*Iron Sulfate or Aluminum Sulfate*	
pH Change	*lb/yd^3*	*kg/m^3*	*lb/yd^3*	*kg/m^3*
8.0 to 5.0	3.5	2.1	8.75	5.2
7.5 to 5.0	3.25	1.9	7.75	4.6
7.0 to 5.0	2.5	1.5	6.5	3.9
6.5 to 5.0	2.0	1.2	5.25	3.1
6.0 to 5.0	1.5	0.9	3.75	2.2
5.5 to 5.0	0.75	0.5	2.0	1.2

[1]From Tayama (1966).

market (Sulfur F by Cleary Chemical Corporation, Somerset, NJ 08875) that can be injected into the greenhouse water lines for application to an established crop. The sulfates react within a few days, while sulfur must be oxidized by soil microbes, a process requiring several weeks or more if the substrate has been pasteurized.

Summary: pH, Calcium, and Magnesium Control

Control of substrate pH is the single most important step in a fertilization program. Intimately associated with pH is the level and balance of calcium and magnesium. Control of these factors begins with analysis of irrigation water. Water alkalinity is equivalent to that of limestone and may need to be counteracted with acidifying measures. The first measure that can be taken is a reduction in the amount of limestone incorporated into root substrates during formulation. The second measure is selection of a fertilizer formula with a suitable level of potential acidity. Alkalinity in water is balanced very often by calcium and magnesium. This brings up the possibility of an imbalance of these two nutrients. In the event of an imbalance, selection of the proper form of limestone and formula of fertilizer is important. Calcitic limestone supplies calcium only, while dolomitic limestone provides calcium and magnesium. Fertilizers can lack or contain calcium and magnesium, and can vary in the proportion of these. When adjustments in liming materials and fertilizer formulas do not keep substrate pH within specifications, flowable limestone or hydrated lime adjustments can be made during the crop. These are drastic, onetime measures. When pre-plant liming materials and fertilizers cannot hold pH down to the recommended upper critical level, it is time to consider injection of acid into the irrigation water. This will eliminate the excessive alkalinity but will not alter the calcium and magnesium content of the water. It is still important to check and, if necessary, alter the ratio of these two nutrients. Water alkalinity can be high enough to render acid injection unsuitable. In this case, the components of alkalinity and other salts need to be removed from water by a reverse osmosis apparatus. See the "Water Quality" section in Chapter 8 for more details.

Slow-Release Fertilizers

Many of the nutrient sources used in the early days of greenhouse culture were, in effect, slow-release fertilizers. They were mainly organic materials of plant and animal origin, which, upon degradation, slowly gave up their nutrient content to the soil. These were materials such as bone meal, seaweed, and fish. Today, synthetically produced slow-release fertilizers have slow, sustained release patterns ranging from three months to several years. For greenhouse culture, the three-to-four-month release period is most popular. There are five common categories of these fertilizers. Some, when incorporated into the soil prior to planting, will provide all the necessary nitrogen, phosphorus, and potassium for the entire crop period, thus eliminating the need for a post-plant fertilization program. Others provide micronutrients in an equally effective manner. The five common categories are as follows:

1. Plastic-encapsulated fertilizers.
2. Slowly soluble fertilizers.
3. Urea aldehydes.
4. Sulfur-coated fertilizers.
5. Chelated micronutrients.

Slow-release fertilizers are in one sense a form of automation, since they eliminate the need for a continual input of labor into fertilization. These fertilizers are more efficient than water-soluble fertilizers in that a greater percentage of applied nutrients is utilized by the plant. Conversely, fewer nutrients leach from the root zone into the water table. This factor is important today as pollution guidelines and regulations are being developed by government agencies. Although crops can be fertilized exclusively with slow-release fertilizers, many growers currently use these fertilizers in conjunction with a continual fertilization program merely as an insurance program against nutrient shortage. The reason for the limited use of slow-release fertilizers as the sole source of nitrogen and potassium is the fear of not being able to slow the rate of release, should it be necessary during the crop.

Plastic-Encapsulated Fertilizers

Notable examples of plastic-encapsulated slow-release fertilizers are the products under the trade names Osmocote and Sierra. They consist of plastic-coated spheres of dry, water-soluble fertilizers formulated from such carriers as potassium nitrate and ammonium sulfate. Particle diameters are about 1/8 inch (3 mm) or less.

These fertilizers are mixed into root substrates prior to planting. Water vapor in the soil atmosphere penetrates the capsule wall. Once inside, the water vapor condenses on the fertilizer surface because the fertilizer lowers the vapor pressure. This reduces the moisture content of the atmosphere inside the capsule to a level lower than that in the moist soil atmosphere outside. As a result, water vapor continues to diffuse into the capsule in an attempt to equalize the moisture content of the atmosphere on both sides of the plastic film. Soon, sufficient water has condensed inside to dissolve the fertilizer. As water continues to enter and pressure builds up inside, the walls of the capsule enlarge, forming fissures through which the fertilizer solution passes to the soil solution, where it can be taken up by plant roots. The longevity of this process is controlled by the composition and thickness of the plastic coating.

Osmocote is available for greenhouse fertilization in a variety of formulas that roughly encompass the ratios of 1-1-1, 2-1-1, and 3-1-2. The release periods range from 3 to 4 months, which is suitable for most greenhouse crops, to 14 to 16 months for long-term crops such as carnation and rose. Listed in Table 9-12 are the various Osmocote and Sierra controlled-release fertilizers.

Another product in the plastic-encapsulated category is Nutricote. It is also available under the trade name Florikan CRF. Again, it consists of a variety of solid, soluble fertilizers coated with a plastic polyolefin resin. The resin can vary in composition to yield products with release periods of 40, 70, 100, 140, 180, 270, or 360 days at a root-substrate temperature of 77°F (25°C). Many analyses are available, but the three most popular are 13-13-13, 14-14-14, and 18-6-8 (see Table 9-13 for rates). The 14-14-14 formula does not contain micronutrients, while the other two do. The Nutricote coating is different from the previous fertilizers in that it remains flexible after production. Water diffuses from the soil

Table 9-12 **Osmocote and Sierra Controlled-Release Fertilizers and Their Release Periods[1]**

Analysis	*Longevity (months)*[2]	*Product Name*[3]
14-14-14	3–4	Osmocote
19-6-12	3–4	Osmocote
13-13-13	8–9	Osmocote
18-6-12	8–9	Osmocote
17-7-12	2–14	Osmocote
18-6-12	—	Osmocote Micro-prill
15-11-13	3–4	Osmocote Plus
15-10-12	5–6	Osmocote Plus
16-8-12	8–9	Osmocote Plus
15-8-11	12–14	Osmocote Plus
15-9-11	12–14	Osmocote Plus
15-9-10	16–18	Osmocote Plus
15-10-12	3–4	Sierra General-Purpose
15-10-10	5–6	Sierra General-Purpose
17-6-12	3–4	Sierra Plus Minors
17-6-10	8–9	Sierra Plus Minors
16-6-10	2–14	Sierra Plus Minors
15-10-12	3–4	Sierra Plus Minors Greenhouse Mix
15-10-10	5–6	Sierra Plus Minors Greenhouse Mix
16-8-12	8–9	Sierra Tablets

[1]From Scotts-Sierra Horticultural Products Co., Marysville, OH 43041.

[2]At an average root substrate temperature of 70°F (21°C).

[3]When the product name includes the designation "Plus Minors," micronutrients are included; otherwise they are not.

Table 9-13 **Rates in lb/yd^3 (kg/m^3) for Incorporation of Three of the Most Popular Formulations of Nutricote[1] into Greenhouse Root Substrates[2]**

Release Type (days)[3]	*Sensitive Crops*		*Medium-Feeding Crops*		*Heavy-Feeding Crops*	
			13-13-13			
70	2.5	(1.5)	5	(3.0)	8.5	(5.1)
100	3.5	(2.1)	7.5	(4.5)	12	(7.1)
140	5	(3.0)	9	(5.4)	13	(7.8)
180	6	(3.6)	11	(6.6)	17	(10.2)
270	8	(4.8)	13	(7.8)	21	(12.6)
360	11	(6.6)	15	(9.0)	25	(15.0)
			14-14-14			
40	2	(1.2)	5	(3.0)	8	(4.7)
70	4	(2.4)	9	(5.4)	14	(8.3)
100	5	(3.0)	12	(7.1)	20	(11.9)
140	8	(4.7)	15	(9.0)	22	(13.0)
180	12	(7.1)	20	(11.9)	28	(16.6)
270	16	(9.5)	24	(14.2)	32	(19.0)
360	20	(11.9)	28	(16.6)	36	(21.3)
			18-6-8			
70	2	(1.2)	4.5	(2.7)	7.5	(4.5)
100	3	(1.8)	6.5	(3.9)	11	(6.6)
140	4.5	(2.7)	8	(4.8)	12	(7.2)
180	6	(3.6)	11	(6.6)	14	(8.4)
270	8	(4.8)	13	(7.8)	16	(12.0)
360	11	(6.6)	15	(9.0)	18	(13.8)

[1]Also sold under the trade name Florikan CRF containing Nutricote.

[2]The Florikan company recommends incorporating magnesium sulfate (Epsom salt) into the root substrate at a rate of 3 to 5 lb/yd^3 (1.8 to 3.0 kg/m^3) to supply sulfur when using any of these products.

[3]Based on a root substrate temperature of 77°F (25°C).

through a mazelike pattern of molecule-sized passages in the resin coating. Inside, fertilizer is dissolved and pressure builds up. The nutrient solution then flows back out through the passages to the root substrate.

An important precaution for Osmocote, Sierra, and Nutricote is that they should not be steam pasteurized. This could result in excessive nutrient release and an ensuing plant injury. The ammonium and nitrate forms of nitrogen in Osmocote, Sierra, and Nutricote are well balanced.

Slowly Soluble Fertilizers

Limestone is an example of a fertilizer with limited solubility. When it is applied to root substrate, a small percentage becomes available. As the initially available quantity is depleted through plant utilization or leaching, more is released to replace it.

A good example of a complete fertilizer in this category is MagAmp. It is a coprecipitate of magnesium ammonium phosphate and magnesium potassium phosphate. MagAmp has the grade 7-40-6 and is an effective source of nitrogen, phosphate, and potassium for 3 to 4 months in the medium grade and 8 to 10 months in the coarse grade. It is designed to be mixed into the root substrate. Recommended crops and rates of application appear in Table 9-14. The unusually high level of phosphate in MagAmp can result in reduced availability of iron, manganese, copper, and zinc in the root substrate. As a result, careful attention should be paid to the micronutrient fertilizer program. This fertilizer also contains 12 percent magnesium, which is high enough to antagonize the uptake of calcium in some plants. Special care should be exercised to maintain calcium at a moderately high level in the root substrate to avert a deficiency. This problem is particularly important in areas where the water supply has a low calcium content. Ammonium toxicity can be a problem in low-pH substrates when MagAmp is the sole source of nitrogen. This situation can be rectified by using MagAmp at one-third the normally recommended rate and supplementing with periodic liquid fertilizer. The supplemental liquid fertilizer should contain nitrogen in the nitrate form to compensate for nitrogen in MagAmp, which is all ammonium plus urea.

Urea Aldehydes

Most notable in the urea aldehyde category is urea formaldehyde. This slow-release nitrogen-containing fertilizer is sold under several trade names, including Borden's 38, Ureaform, and Uramite. It contains 36 percent nitrogen and becomes available slowly, with about 65 percent released the first year, 25 percent the second year, and 10 percent the third year. This fertilizer has gained considerable prominence as a source of nitrogen for home lawns and golf courses. Much of urea formaldehyde exists in long chemical chains that cannot be taken up by plant roots. Once it is in the soil, microorganisms feed upon these chains, break-

Table 9-14 **Recommended Uses and Rates of Application for MagAmp Slow-Release Fertilizer**[1]

Crop	*Rate*	*MagAmp Grade*
Seedling and cutting beds	5–7 lb/100 ft^2 (245–340 g/m^2)	Medium
Bedding plants	8 lb/yd^3 (4.8 g/L)	Medium
Greenhouse pot plants	15 lb/yd^3 (9 g/L)	Medium

[1]From Scotts-Sierra Horticultural Products Co., Marysville, OH 43041.

ing them down into smaller pieces, some of which are urea. Urea is a form of nitrogen readily utilized by the plant. This breakdown process occurs slowly over a long period of time.

A second urea aldehyde that is often used in the nursery trade is IBDU (isobutylidene diurea). This urea aldehyde depends on chemical hydrolysis rather than microbial attack for the release of nitrogen.

Urea aldehydes have not been used extensively in greenhouse culture, except in mixed formulations, for two reasons. First, they provide only the nutrient nitrogen, leaving the need for a potassium and phosphate source. Second, they provide nitrogen in the form of urea, which is ultimately converted to ammonium either in the substrate or in the plant. As discussed earlier, most greenhouse plants do not respond well to ammonium nitrogen exclusively. Urea formaldehyde is used by some azalea growers as a top dressing to guarantee against nitrogen deficiency because of the high requirement of this crop for nitrogen relative to potassium. It is used at the rate of 1 rounded teaspoon (5 g) per 6-inch (15-cm) pot at two-month intervals during periods of heavy growth. A continual fertilization program is used in addition to this application (see Table 9-3).

Sulfur-Coated Fertilizers

Prills (small spheres) of various fertilizers including urea, ammonium polyphosphate, triple superphosphate, potassium sulfate, and potassium chloride are coated individually with a combination of sulfur, a waxlike sealant, and possibly a conditioner such as diatomaceous earth. Various combinations of these sulfur-coated materials are blended to yield a multitude of grades such as 13-13-13, 21-6-12, and 7-34-0. Release of nutrients is dependent upon soil microorganisms, which convert the insoluble elemental sulfur coating to soluble sulfate. When this happens, water enters the capsule and dissolves the fertilizer contained in it. The release period is typically three to four months, but coating alterations can extend the release period to one year. The sulfur-coated products have gained wide acceptance outdoors for turf, nursery, and landscape uses. They are not used to any extent in greenhouses because all of the nitrogen is ammonium and/or urea. While nitrates could theoretically be coated, they have not been because of the danger of explosion when the molten sulfur is sprayed on the prills of nitrate.

Chelated Micronutrients

The word *chelate* is derived from a Greek word meaning "claw." It is appropriate because chelates are large, organic chemical structures that encircle and tightly hold the micronutrients iron, manganese, zinc, and copper. Plant roots can absorb the micronutrient-chelate combination. The micronutrient is then released inside the plant. Alternatively, plants can absorb micronutrients alone as they are slowly released into the root substrate solution by chelates. When the substrate pH is higher than that desired for a specific crop, it contains high levels of hydroxide and possibly carbonates that will precipitate iron, manganese, zinc, and copper. Precipitated nutrients are insoluble and unavailable to plants. Chelated micronutrients are protected from precipitation. However, micronutrients are slowly released from chelates, after which they can be precipitated. The value of chelates stems in great measure from their long release period.

Roses were traditionally grown in a moderately acid substrate of pH 5.5 to 6.0. With time, it was learned that the benefit of the low pH level was the higher solubility of iron, a feature necessary for this poor accumulator of iron. Actually, the rose grows better at a higher pH level, provided that sufficient iron is available. Today, iron is routinely applied in the chelated form at the rate of 1 lb/1,000 ft^2 (4.9 g/m^2) of bed at a frequency of about every three months. The alternative source of iron, iron sulfate, is much cheaper by weight. However, under adversely high pH conditions, it can be more expensive because it needs to be applied at a greater frequency than chelated iron. Iron sulfate rapidly releases iron into the soil solution where it precipitates.

The chelated forms of iron, manganese, zinc, and copper are the most common forms used in premium greenhouse fertilizers. Their high solubility compared to alternative forms makes them very desirable to fertilizer formulators. In correcting individual micronutrient deficiencies, there is usually an advantage to using chelated iron when the substrate pH level is adversely high for the given crop. If the pH level is normal, the cheaper iron sulfate (ferrous sulfate) form serves well. However, chelated manganese, zinc, and copper do not show as great an advantage. Unless the soil pH is extremely high, the extra expense of the chelated forms of manganese, zinc, and copper is not usually warranted. The sulfate form works well for reasons of economics and effectiveness.

Nutritional Monitoring

Nutritional problems will develop even in the best of fertilization programs. The careful grower makes use of three systems for monitoring nutrient status and is thus able to forecast problems as well as develop remedies. These systems are (1) visual diagnosis, (2) soil testing, and (3) foliar (leaf) analysis. Each test provides some information that is not provided by the others.

Visual Diagnosis

Visual diagnosis can be employed only after damage has occurred. Often, the damage is only partially reversible. Therefore, one should not rely on visual diagnosis as a routine measure of nutrient status. Each nutrient deficiency has several symptoms that are common to many crops. These symptoms are listed in Table 9-15 and are shown in Figures 9-6 through 9-21. It should be noted that some crops will not develop these symptoms and many crops, in addition to these symptoms, will develop others. Only the more typical symptoms are presented here.

A few definitions will be of help in reading Table 9-15. *Chlorosis* refers to a process whereby green chlorophyll is lost. The leaf tissue turns progressively lighter green and finally yellow. *Necrosis* refers to the death of cells and is manifested as various shades of brown. *Interveinal chlorosis* is chlorosis occurring between the veins (vascular tissue) of the leaf. The veins remain green in color. A *witch's broom* is a typical boron deficiency symptom because more boron is required for flower-bud formation than for vegetative-shoot development. When the plant reaches the stage of flower-bud formation, it aborts, giving rise to lateral vegetable shoots. These develop but, in turn, abort as flower buds are initiated. This process continues until a proliferation of developing shoots gives the appearance of a broom. *Strap leaves*, typical of calcium deficiency, are long, thin leaves having the appearance of a strap.

Table 9-15 Key to the Classical Symptoms of Various Nutrient Deficiencies

Deficiency Symptoms	*Deficient Nutrient*	*Reference*
a. The dominant symptom is chlorotic foliage.		
b. Entire leaf blades are chlorotic.		
c. Only the lower leaves are chlorotic followed by necrosis and leaf drop.	Nitrogen	Figure 9-6
cc. Leaves on all parts of the plant are affected and sometimes have a beige cast.	Sulfur	Figure 9-12
bb. Yellowing of leaves takes the form of interveinal chlorosis.		
c. Only recently mature or older leaves exhibit interveinal chlorosis.	Magnesium	Figure 9-11
cc. Only younger leaves exhibit interveinal chlorosis. This is the only symptom.	Iron	Figure 9-13
d. In addition to interveinal chlorosis on young leaves, gray or tan necrotic spots develop in chlorotic areas.	Manganese	Figure 9-15
dd. While younger leaves have interveinal chlorosis, the tips and lobes of leaves remain green, followed by veinal chlorosis and rapid, extensive necrosis of leaf blades.	Copper	Figure 9-18
ddd. Young leaves are very small, sometimes missing leaf blades altogether, and internodes are short, giving a rosette appearance.	Zinc	Figure 9-17
aa. Leaf chlorosis is not the dominant symptom.		
b. Symptoms appear at the base of the plant.		

Table 9-15 (continued)

Deficiency Symptoms	Deficient Nutrient	Reference
c. At first all leaves are dark green, and then growth is stunted. Purple pigment often develops in leaves, particularly older leaves.	Phosphorus	Figures 9-6, 9-8
cc. Margins of older leaves become chlorotic and then burn, or small chlorotic spots progressing to necrosis appear scattered on old leaf blades.	Potassium	Figure 9-9
bb. Symptoms appear at the top of the plant.		
c. Terminal buds die, giving rise to a witch's broom. Young leaves become very thick, leathery, and chlorotic. Rust-colored cracks and corking occur on young stems, petioles, and flower stalks. Young leaves are crinkled.	Boron	Figure 9-19
cc. Margins of young leaves fail to form, sometimes yielding strap leaves. The growing point ceases to develop, leaving a blunt end. Light green color or uneven chlorosis of young tissue develops. Root growth is poor in that roots are short and thickened.	Calcium	Figure 9-10

Figure 9-6 **Nitrogen and phosphorus deficiency: The marigold seedling on the left is normal. The center plant is deficient in nitrogen and has uniformly chlorotic leaves with the symptoms more intense at the base of the plant than farther up. The plant on the right is deficient in phosphorus. It is stunted, but unlike the nitrogen-deficient plant, it has a normal green color in this intermediate stage of deficiency.**

(a) (b) (c) (d)

Figure 9-7 Ammonium toxicity: (a) The tomato plant displays curling of the margins of older leaves, followed by irregular chlorosis and necrosis. (b) Leaves on the chrysanthemum plant have a leathery texture and are curled. Chlorosis of various patterns develops on older leaves, followed by necrotic spots and finally burning of the leaf tips. The terminal ends of many roots are burned. (c) Early symptoms in the gloxinia plant include downward rolling and chlorosis of the lower leaf margins. (d) Later symptoms in gloxinia are interveinal chlorosis of the older leaves followed by necrosis.

(a)

(b)

Figure 9-8 Phosphorus deficiency: (a) Normal petunia seedlings on the left are in sharp contrast with plants in an early stage of phosphorus deficiency on the right. The deficient plants are heavily stunted and deeper green in color. (b) A normal chrysanthemum plant on the left compared to a plant in late stages of deficiency on the right. Late symptoms include older leaves with uniform chlorosis, followed by necrosis and a lighter green color of the entire plant.

(a) (b)

Figure 9-9 Potassium deficiency: (a) Symptoms in begonia seedlings first appear as necrotic spots along the distal margin of older leaves. (b) A Rieger begonia with leaves dissected. The lowest leaf is completely burned, the next leaf up has a continually burned margin, the middle leaf has necrotic spots mainly along the margin, and the upper two leaves are normal.

(a) (b)

Figure 9-10 Calcium deficiency: This deficiency is generally expressed at the top of the plant as a rather irregular chlorosis of foliage and incomplete formation of tissue. (a) Leaves of the chrysanthemum plant are incompletely formed, giving the appearance of long, narrow "strap" leaves. (b) Often, the growing point (meristem) stops developing and takes on a blunt appearance, as seen in this rose plant. Note the incompletely formed leaves on the lower left side of the rose shoot. (c) Flower tissue may also be incompletely formed, as in the case of this petunia plant. Collapse of tissue in the petal lobes and corolla tube is evident. (d) Young leaves of poinsettia often become chlorotic in irregular patterns, misshapen, and necrotic along the margin.

(c)

(d)

Figure 9-10 *(continued)*

Figure 9-11 Magnesium deficiency: Symptoms of magnesium deficiency, like those of nitrogen and potassium deficiencies, begin at the base of the plant and progress upward. Interveinal chlorosis of foliage is the predominant symptom, as seen in this petunia plant.

Figure 9-12 Sulfur deficiency: The plant as a whole is affected. Foliage becomes uniformly lighter green in color. Chrysanthemum leaves here show symptoms increasing in intensity from left to right. Some plants develop a beige cast in addition to chlorosis. (*Photo courtesy of A. M. Kofranek, University of California, Davis, CA.*)

(a) (b)

Figure 9-13 Iron deficiency: Symptoms of iron deficiency are similar to those of magnesium deficiency in that interveinal chlorosis is the principal symptom, but they differ in that iron deficiency appears at the top of the plant first. (a) Interveinal chlorosis on hydrangea. (b) Interveinal chlorosis of the young leaves just below the bracts on poinsettia.

(a) (b)

Figure 9-14 Iron toxicity: Low root substrate pH fosters excessive uptake of iron and possibly manganese, copper, and zinc. Marigold and seed geranium are particularly susceptible. (a) Marigold exhibits bronze speckling, similar to spider mite damage, on the youngest fully expanded and lower leaves. (b) Iron toxicity appears on the older leaves of seed geranium as interveinal chlorosis, bronze speckling in the chlorotic areas, and necrosis of the leaf margins.

Figure 9-15 Manganese deficiency: Symptoms of manganese deficiency start out the same as iron deficiency with interveinal chlorosis of young foliage. In the later stages of manganese deficiency, however, tan to gray spots develop in the chlorotic areas, as seen in these chrysanthemum leaves. *(Photo courtesy of A. M. Kofranek, University of California, Davis, CA.)*

Figure 9-16 Manganese toxicity: Symptoms appear on lower leaves as either necrosis of the leaf tip or purplish spots scattered across the leaf blade, as in these Rhapis palm leaflets.

(a)

(b)

Figure 9-17 Zinc deficiency: Reduced leaf size is very typical of zinc deficiency; in fact, zinc deficiency is frequently called "little leaf disease" in field crops. Shortened internodes and irregular chlorosis of young foliage are also typical. (a) Zinc-deficient carnation shoots on either side of a normal shoot. Symptoms include small leaves and very short internodes. (b) Kalanchoe is more prone to zinc deficiency than most other greenhouse crops. Prolific branching from a broad, flattened stem is indicative of this deficiency in kalanchoe. This condition is termed *fasciation.* The stem on the left is normal, while that on the right is deficient in zinc.

Figure 9-18 Copper deficiency: (a) The early symptom is interveinal chlorosis of young foliage. It is different from that caused by iron deficiency because lobes and points at the leaf margin tend to be deeper green than the inner portion of the leaf blade, as seen in these chrysanthemum leaves. (b) When copper deficiency progresses in chrysanthemum, the first fully expanded leaves suddenly turn necrotic as the upper leaves partially regain a green color. With time, the chlorosis of the plant tip intensifies and necrosis occurs again. (c) Some crops exhibit veinal chlorosis as a late stage of copper deficiency, as seen in this chrysanthemum plant. (d) Copper deficiency in rose is seen as irregular interveinal chlorosis of young leaves. When leaves are forming, necrosis occurs on the leaf blade, resulting in the development of very small leaves. This pattern of small leaves may be cyclic, giving an hourglass effect.

Figure 9-19 Boron deficiency: (a) Leaves become thickened and leathery in some plant species, and irregular chlorosis develops as in the case of this petunia. (b) Other plant species develop crinkled leaves with irregular chlorosis of young leaves, as in this Rieger begonia plant. Rust-colored cracks are common on stems, petioles, flower stalks, and sometimes on the leaf blades of many plants. Cork develops over the cracks. A cross-section of tissues of many boron-deficient plants would show rust coloration and a breakdown of vascular tissue. The boron requirement of reproductive (floral) growth is greater than that for vegetative growth; thus, abortion of the meristem often occurs when the flower bud begins to form. Side shoots develop and in turn abort, only to give rise to more and more side shoots. (c) Eventually, a "witch's broom" forms, as in these carnation shoots. If flowers form, they are usually incomplete. Carnation flowers frequently have split calyx and a low petal count. (d) The gladiolus flower on the left shows incompletely formed petals as a result of boron deficiency.

Figure 9-20 Boron toxicity: Symptoms appear on older leaves as a reddish-brown necrosis at the margins, as in this Rieger begonia leaf.

Figure 9-21 Molybdenum deficiency: This deficiency is rare in greenhouse crops except for poinsettia, where it is common. As pictured here, median poinsettia leaves develop a chlorotic margin that turns yellow and later necrotic. These symptoms spread inward, eventually killing the entire leaf.

Soil Testing

Analyses of greenhouse root substrates will generally include a measurement of the pH and soluble-salt levels. Neither of these can be determined by visual diagnosis or foliar analysis procedures. In addition, soil tests give concentrations of ammoniacal nitrogen, nitrate nitrogen, phosphate, potassium, calcium, magnesium, manganese, and sometimes sulfate, iron, copper, and zinc. Only a portion of most nutrients in the soil is immediately available to a plant. Soil-testing procedures give an estimate of the proportion of each nutrient that is available. One could say that soil testing is futuristic in that it estimates what will be taken up by the plant. Foliar analysis differs in that it provides a measurement of nutrients in the plant tissue that were taken up in the past. These two greatly different views work well together for diagnosing the nutritional status of a crop. Soil testing is commonly practiced during crop growth but is equally valuable when it is done prior to planting. The pH level of the greenhouse substrate is best adjusted prior to planting because of the need to thoroughly mix into it the limestone or sulfur used to adjust pH.

Each soil-testing laboratory provides an interpretive chart of values for its tests. Even though two labs may use the same test, their interpretive values may differ. This could be caused by subtle differences in a procedure within the test. Most labs use the saturated-paste procedure for greenhouse substrates (Table 9-16). Some labs use procedures designed for field crops and do not have standards for greenhouse crops.

Growers who use tests for which there are no standards should record and relate over the years their crop responses to fertilizer applications and soil-test results. This procedure ultimately indicates to the grower the levels of each nutrient that should be maintained for best growth. Whether interpretive tables exist or not, one should take monthly soil samples and should keep a log book in

Table 9-16 **Nutrient Concentration (ppm) Guidelines for Greenhouse Substrate from Michigan State University for the Saturated-Media Extract (Saturated-Paste) Method[1] and from The Ohio State University for the Saturated-Paste Extract Method[2] for Crops in General[3]**

Interpretation	*Nitrate-N*	*Phosphate-P*	*Potassium*	*Calcium*	*Magnesium*
		Michigan State University			
Low	0–39	0–2	0–59	0–79	0–29
Acceptable	40–99	3–5	60–149	80–199	30–69
Optimum	100–199	6–10	150–249	200+	70+
High	200–299	11–18	250–349	—	—
Very High	300+	19+	350+	—	—
		The Ohio State University			
Extremely low	0–29	0–3.9	0–74	0–99	0–29
Very low	30–39	4.0–4.9	75–99	100–149	30–49
Low	40–59	5.0–5.9	100–149	150–199	50–69
Slightly low	60–99	6.0–7.9	150–174	200–249	70–79
Optimum	100–174	8.0–13.9	175–244	250–324	80–124
Slightly high	175–199	14.0–15.9	225–249	325–349	—
High	200–249	16.0–19.9	250–299	350–399	125–134
Very high	250–274	20.0–40.0	300–349	400–499	135–174
Excessively high	275–299	40.0+	350+	500+	175+

[1]From Warncke and Krauskopf (1983).

[2]From Peterson (1984).

[3]Values are listed in parts per million (ppm).

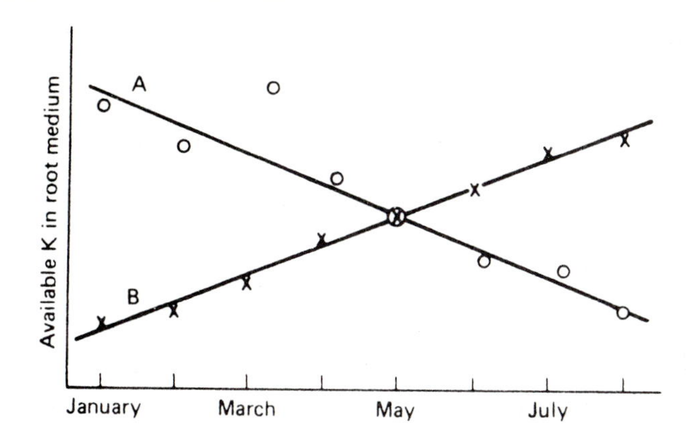

Figure 9-22 An illustration of the value of monthly soil sampling as opposed to a single sample date. Samples A and B drawn in May would indicate similar nutrient situations. This is erroneous, since the root substrate represented by sample A is decreasing in K level while that represented by sample B is increasing. These facts are borne out only by sequential sampling.

which are recorded crop responses; soil-testing and foliar analysis results; application dates of water, fertilizer, and pesticides; and any other factors affecting growth. Quite often, it is necessary to adapt values in the interpretive table to individual situations. This can be accomplished through the logging procedure.

After an individualized interpretive table is developed, monthly soil sampling should be continued. This practice permits identification of faulty samples, which might infrequently occur as a result of poor sampling procedure or an error in testing. Note in Figure 9-22 the March value for sample A, which is abnormally high. This value is an indication of a faulty sample. It also indicates the direction in which nutrient levels are changing in the root substrate. Note again in Figure 9-22 that the potassium levels in the root substrates represented by samples A and B in May are equal. If this level were desirable and May was the only month in which a sample was taken, no action would be taken. Yet, from the series of sample results, it can be seen that in one substrate the level of potassium application should be diminished at this time, while in the other it should be increased so that imbalances do not occur the following month. Only sequential sampling will bear out these facts.

Test Meters Most soil testing is performed by institutional and commercial laboratories. Growers can purchase equipment to test their own root substrates (Figure 9-23). Inexpensive pocket meters for testing pH and EC begin at about $45 each. These are accurate enough for soil testing. More accurate laboratory meters for these two tests begin around $250 and are used by larger firms. Specific-nutrient tests, including tests for nitrate, potassium, and sodium, can be conducted with Cardy meters, which are available at approximately $340 each. All of these tests can be run directly on a soil extract in a matter of minutes. No chemical alterations of the soil extract are necessary. All of these meters are available from Spectrum Technologies, Inc., 12010 S. Aero Dr., Plainfield, IL 60544. The pH and EC meters are also available from greenhouse-supply companies. Every greenhouse needs to have at least a pH and an EC meter. The pH level can give an estimate of the quantity of limestone to incorporate into sub-

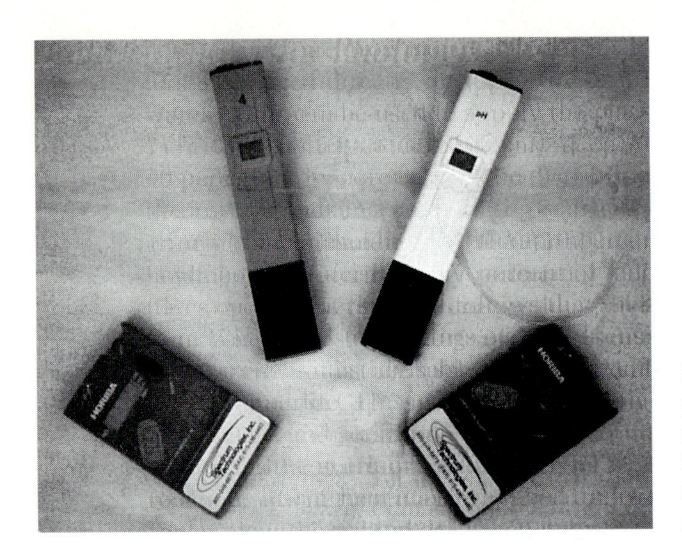

Figure 9-23 **Meters frequently purchased by greenhouse firms for testing their root substrate. From top left clockwise: EC meter, pH meter, and Cardy potassium and nitrate ion meters.**

strate during its formulation. Later during the crop, the pH level gives an indication of the acidity or basicity level of fertilizer to use. The EC level indicates whether an inadequate, adequate, or excessive level of fertilizer exists in the root substrate. Larger firms with specific-nutrient Cardy meters will be able to determine the balance of nitrate and potassium in the substrate without waiting for soil test results to come back from the testing lab. This latter information is used to make adjustments in the fertilizer ratio.

Sampling Procedure An important decision in soil testing is that of the number of samples to be drawn. No set area can be assigned to a sample. To determine the boundaries of the area included in one sample, one should consider the origin of the substrate and its fertilization history. The wide variety of components in greenhouse substrates react differently with plant nutrients. Clays tend to retain potassium, while pine bark– and peat moss–based soilless substrates are often associated with micronutrient deficiencies, particularly iron. Although two substrates might be handled under the same fertilization program, in time the available levels of nutrients will vary so much that separate soil tests will be needed. On the other hand, a single substrate used for a single crop might have been subjected to two fertilization programs applied to two different crops previously grown in this substrate. It will be necessary to take two soil samples because residual nutrients such as phosphorus might have been applied in greater quantity to one previous crop than to the other. Generally speaking, a soil sample area could be as small as a single bench of plants and as large as several acres.

Other more subtle factors to consider in soil testing will become recognizable as one becomes familiar with a greenhouse range. For example, a chrysanthemum grower experienced root injury, necrosis of leaf margins, and overall stunting of plants in specific sections of outdoor ground beds during rainy seasons.

The problem was due to poor drainage in the low spots of the beds. Under this condition of low soil-oxygen content, an excessive proportion of the large manganese reserve of the substrate was converted to an available form and, in turn, resulted in manganese toxicity. The higher areas of the beds were sufficiently drained and aerated to prevent an excessive conversion of unavailable to available manganese. To identify this problem, it was necessary to recognize and sample the problem section only.

State soil-testing laboratories in most states offer services for no fee or for a nominal cost. Many commercial laboratories also offer soil testing, foliar analysis, and consulting services. The customary volume of soil submitted is 1 pint (500 cc). It is important that this sample be collected properly so that it is truly representative of the total area of crop sampled. A soil-sampling tool such as the one shown in Figure 9-24 should be used. The top 1/2 inch (1.3 cm) of soil is scraped aside, and the soil-sampling tube is then pressed into the substrate until it makes contact with the bottom of the pot or bench. In this way, the entire root zone is sampled. The top 1/2 inch (1.3 cm) is avoided because abnormal levels of fertilizer salts build up there as a result of water evaporation and also because roots rarely grow where there is rapid drying. At least 10 substrate cores should be taken for one sample. Some should be taken from the edge of the bench and others from the center, since drying conditions, which affect salt accumulation, differ in these locations. The 10 cores should be collected from all sectors of the sample area.

The substrate sample should be mailed to the testing laboratory without delay. If the substrate is wet enough to cause a breakdown of the package during shipment, it should be partially dried in the sun or on a warm surface, such as a

Figure 9-24 **A sampling tool used for obtaining cores of substrate from pots or beds in greenhouses for soil-testing purposes. The side of the tube is cut away to permit removal of the substrate core.**

boiler, prior to shipping. It is very important that information sheets supplied by the testing laboratory be completed and sent with the sample. This information aids in identifying problems and developing recommendations.

Substrate Extraction Procedures Growers who have their own test equipment can use one of two substrate extraction methods. The 1:2 extract method calls for placing one volume of dry substrate, compacted to the same density that it was in the pot or bench, into a watertight container such as a cup and then adding two volumes of water. The mixture is thoroughly stirred and allowed to sit for one to two hours. Then the substrate solution is squeezed from the slurry through cheesecloth. This extract can be used for pH and EC determinations. Alternately, the saturated-paste extract procedure calls for placing any volume of substrate in a watertight container. The substrate can be at any moisture level. Water is slowly added with stirring until the surface of the substrate just begins to glisten with free water. After 15 minutes, more water is added with stirring if necessary to return the substrate surface to a glistening state. After a total of one to two hours' contact between the water and the substrate, the soil solution is squeezed from the slurry through cheesecloth. This extract can be used for pH, EC, and specific-nutrient tests.

In both extract procedures, it is best to use distilled water to prevent the addition of salts from low-quality water. Distilled water can be purchased at most grocery stores. The pH level will be slightly higher in the 1:2 extract due to dilution of the hydrogen ion that constitutes acidity. The saturated-paste pH is more indicative of the true situation of the plant root since there is less excess dilution of the substrate solution in this extract. In each extract, there is a subjective step that must be carefully executed. In the 1:2 procedure, the correct volume of substrate must be determined by compacting it to the same density that it was in the pot. In the saturated-paste extract, it is necessary to add the proper amount of water to just reach the glistening-paste state. For purposes of standardizing these steps, it is best to have a single individual run all of the extracts for a firm.

Soluble Salts A valuable test provided only through soil testing is the measure of the soluble-salt (EC) level. The method of measurement and effects of high levels on plants were discussed in the "Water Quality" section of Chapter 8. Interpretation charts are presented in Table 9-17 for three EC test procedures.

Soluble salts come from various sources. Irrigation water can contain salts. Soluble fertilizers are soluble salts. Initially insoluble, slow-release fertilizers dissolve with time, releasing nutrients into the root substrate that are in themselves soluble salts. Thus, some soluble salt must be present to ensure a proper level of fertilizer, but the level must not be too high. Other sources may not be desirable. Organic matter of high nitrogen content that undergoes rapid decomposition constitutes another source of soluble salts. Manure and highly decomposed peats may have sizable nitrogen contents that are rapidly released through degradation in the root substrate as ammoniacal nitrogen. Ammoniacal nitrogen can quickly build up to a toxic level because its positive electrical charge causes it to be held in the root substrate.

Table 9-17 Interpretation of Soluble-Salt Levels

Dilution[1]			*Saturated-Paste Extract,*	
1:2		*1:5*		
Soil	*Soilless*	*Soil*	*Soil & Soilless*[2]	*Interpretation*
0–25	0–?	0–10	0–0.75	Insufficient nutrition
26–50	?–100	11–25	0.75–2	Low fertility unless applied with every watering
100		50	—	Maximum for planting seedlings or rooted cuttings
51–125	100–175	26–60	2–4	Good for most crops
126–175	176–225	61–80	—	Good for established crops
176–200	225–350	81–100	4–8	Danger area
Over 200	Over 350	Over 100	Over 8	Usually injurious

[1]mho × 10^{-5}/cm. Some labs will report these values as mmho/cm (dS/m), in which case the decimal point in each of the first three columns is moved two places to the left. A value of 100 becomes 1 mmho/cm (1 dS/m).

[2]mmho/cm.

Seedlings are more sensitive to high soluble-salt levels than are established plants. Established plants vary in their resistance to high soluble-salt levels. African violet and azalea are particularly sensitive and should not be grown in substrates with a level exceeding 1.5 on the saturated paste test. Snapdragon is moderately sensitive and should be grown at levels below 2.5 on the saturated-paste test. Several houseplant crops are sensitive as well. Little research has been conducted on each of these many crops, and soluble-salt interpretations are missing for most.

Single upper critical EC levels probably do not exist for each crop. Plants can adapt to high substrate salt levels by synthesizing soluble organic solutes in the root cells. These substances raise the solute content of the root high enough to counteract the rise in salt in the substrate solution. Thus, water continues to be attracted from the substrate to the root. Time is required for the formation of these solutes. A chrysanthemum plant that has been fertilized weekly with a 600 ppm fertilizer solution and is growing at a desirable substrate salt level would probably be injured if it was suddenly fertilized with a 1,200-ppm fertilizer solution. However, this same plant would not be injured if the 600-ppm-nitrogen fertilizer solution was gradually raised in a series of fertilizer applications over a period of a several weeks to 1,200-ppm nitrogen. This would afford time for the root solute concentration to rise so that it still exceeded the substrate salt concentration derived from the 1,200-ppm nitrogen application.

Aside from plant adaptation, the exact EC level at which injury occurs depends to a large degree on watering practices. If the root substrate is not permitted to dry, then a high salt content may be tolerated. When the substrate dries, the salts become more concentrated than indicated by the test, and injuries may ensue. Ordinarily, it is not wise to maintain root substrates at a high moisture content, but between the time that a high salt level is identified and the cure is administered, it is expedient to do so.

Fortunately, soluble salts, as the name implies, are water soluble and can be leached from root substrates. The standard corrective recommendation calls for application of 1 gal of water per square foot (40 L/m^2) of root substrate for bench crops, or 2 gal/ft^3 (200 L/m^3) of root substrate for pot crops, a waiting period of a few hours, and then a second application of water at the rate of ½ gal/ft^2 (20 L/m^2) of root substrate surface. The waiting period gives the more slowly soluble salts time to dissolve.

Often, a root substrate with a soil base is adversely affected by the second application of water. The soil structure breaks down. This is particularly harmful in carnation and rose root substrates, where the crop is maintained for one, two, or five years without an opportunity to amend the root substrate. Researchers at the University of Connecticut have developed a more desirable procedure for leaching in these cases. Up to 5 gal of water are supplied to each square foot (200 L/m^2) of root substrate in one application, preferably with a trickle-type irrigation system. This procedure utilizes more water but eliminates the destructive second application of water, when the substrate is excessively wet and subject to breakdown of structure.

Foliar Analysis

Foliar analysis constitutes a third system for determining the nutrient status of crops. Like soil testing, it is valuable because it can be used to assess a problem before damage occurs. Foliar analysis is an analysis of representative leaves from a crop to determine the quantities of essential as well as potentially hazardous nonessential elements that the crop has taken up. The laboratory conducting the analysis compares the results to standards developed at many research institutions around the world and draws the necessary conclusions for the grower.

Foliar analysis works well because of the strong relationship between leaf composition and plant response, which is illustrated in Figure 9-25. Except in the zone of luxury consumption, where changes in leaf composition have no effect on growth, the nutrient content of the leaf can be used to predict the growth of the plant. It is fortunate that the zone of luxury consumption exists because it lessens the chance of injury from overapplication of fertilizer.

Foliar analysis offers more accurate testing for all micronutrients than soil testing. However, foliar analysis should be used in conjunction with soil testing to give two different measurements of nutrient status. Just as one would not enter a serious medical operation without two independent assessments of the situation, one should not base the nutritional future of crops on one system of assessment.

The strength of soil testing in combination with foliar analysis is seen in a situation that occurred in a carnation range some years ago in New York. The crop was growing slowly and showing symptoms of potassium deficiency. Contrary to

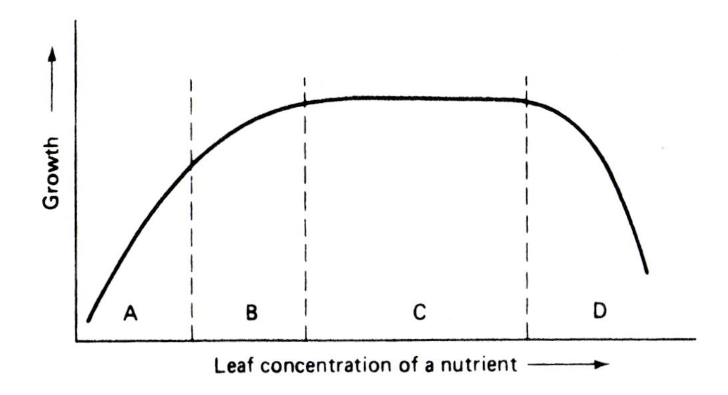

Figure 9-25 **A nutrient calibration curve. When all other factors are adequate, the leaf content of an essential nutrient strongly affects growth. Little growth occurs at low concentrations of the nutrient, but with small additions of the deficient nutrient, large increases in growth occur (zone A). When the rate of growth comes closer to the optimum level, increases in the leaf content of the deficient nutrient bring about continually diminishing growth responses (zone B), until a point is reached beyond which no growth response is caused by increases in leaf concentration of the nutrient (zone C). This is the zone of *luxury consumption.* Eventually, nutrient increases in the leaf reach a toxic level, resulting in decreases in growth and eventually in death (zone D).**

this observation, a soil test indicated that levels of nitrogen and potassium were high. Foliar analysis indicated that potassium was very deficient, while nitrogen was only moderately high. The combination of information led to the conclusion that potassium uptake was blocked by a high soil-nitrogen level. This relationship is an antagonism that frequently occurs. The conclusion was verified when, four weeks after reducing the rate of 20-20-20 fertilizer application to half of the previous level, the grower observed a twofold increase in the foliar level of potassium and the disappearance of potassium-deficiency symptoms.

Foliar analysis samples should be taken every four to six weeks and the results logged. The optimum level of nutrients in the foliage of each crop is different, necessitating the taking of a different sample for each crop. If one crop is growing in two dissimilar substrates or has been fertilized in two different manners, two samples must be taken for this crop. It is important that the correct leaves be sampled, since the nutrient level in each is different from the rest. The age of the crop is also important because the nutrient levels in each leaf change with time. The laboratory conducting the foliar analysis tests will provide a mailing envelope for the leaves and a set of instructions for collecting the proper leaves.

Rose plants are sampled by picking the two uppermost five-leaflet leaves on a stem whose flower calyx is cracking and whose color is just beginning to show. Thirty leaves with petioles attached should be collected.

Carnation plants that have not yet been pinched are sampled by collecting the fourth and fifth leaf pairs up from the base of the stem (area A in Figure 9-26).

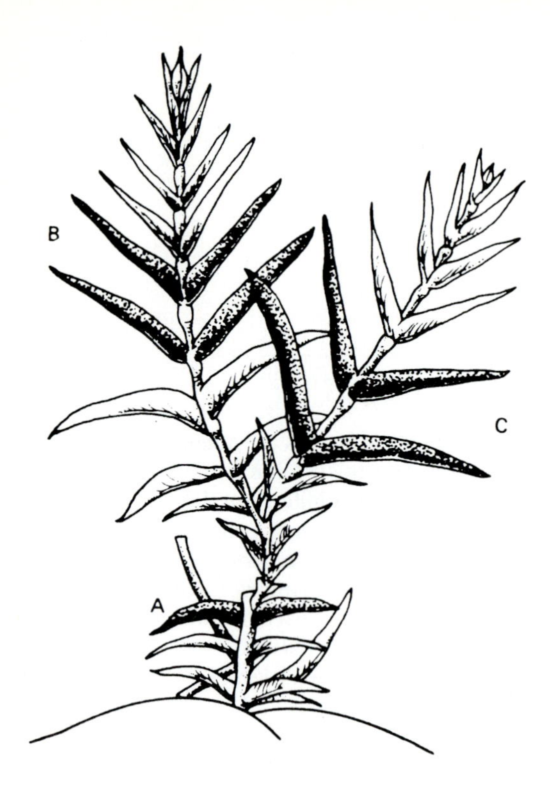

Figure 9-26 Leaf-sampling instructions for foliar analysis of florists' crops.

A–First sample area
B–Second sample area
C–Third sample area

Sampling continues in area A after pinching and until the resulting lateral shoots develop seven pairs of leaves. Then, the fifth or sixth leaf pairs are sampled on the new shoots (area B), counting the first pair of leaves to be separated for one-half the length of the leaves as the first leaf pair. Sampling continues in area B until a flower bud appears on this shoot. Then, sampling should be shifted to secondary lateral shoots, again using the fifth or sixth leaf pairs from the terminal end of the shoot (area C). When secondary lateral shoots develop flower buds, sampling is shifted to tertiary lateral shoots, and so on.

Chrysanthemums are typical of plants in general in that the youngest fully expanded leaves are sampled. These leaves are usually found on chrysanthemum plants about one-third of the distance down the stem from the top. Chrysanthemums are generally sampled five to six weeks after planting for a single-stem crop or five to six weeks after pinching for a pinched crop.

Foliar analysis kits contain an information form similar to the one shown in Figure 9-27. This form permits laboratory personnel to gather the information necessary to enable them to draw conclusions and make nutritional recommendations. It is best to rinse leaves for one minute in deionized or distilled water, available in grocery stores. The leaves should be blotted dry and placed in the paper envelope provided by the testing lab and mailed immediately. Do not place leaves in a plastic bag because they may rot during shipping. If distilled water is not available, leaves that are known to be free of fertilizer residues and soil can be shipped as is. Leaves contaminated with any nutrient-bearing material should

MAILING ADDRESS: PLANT ANALYSIS LABORATORY — DEPARTMENT OF HORTICULTURAL SCIENCE
N. C. STATE UNIVERSITY, RALEIGH, N. C. 27607

SAMPLE INFORMATION FOR FLORICULTURE CROPS | **LAB USE ONLY**

Name of Grower	County	Date Sample Received
Street Route	Date Sampled	Condition of Sample
City, State, Zip Code	Grower Sample No	Grower #
Telephone No	Name of Crop	County #
Identification of Field Sampled	Description of Sample Site (which field)	Sample #
		Date Analysis Received

SAMPLE INFORMATION

1. Name of crop ________ 2. Variety ________ 3. Date planted ________

4. Date pinched (if applicable) ________

5. This sample represents (a) an average condition ____ (b) a problem area ______

(c) a spotty or sporadic situation ____

6. Appearance or condition of plant and/or leaves ________________

7. Location of sample on plant ________________

8. Previous crop ________ 9. What is your fertilization program for the crop represented by this sample?

(List fertilizers used, rates, and frequency of application)

10. What ammendments did you use during preparation of the soil mix in which this crop is growing (List such materials as dolomitic limestone, superphosphate, fritted trace elements and the rates used).

North Carolina State University Agricultural Experiment Station
and North Carolina Agricultural Extension Service, Cooperating

FARMER OR GROWER COPY

Figure 9-27 A typical information form to be completed by the grower and submitted to the foliar analysis laboratory as an aid in the interpretation of results.

Table 9-18 **Minimum Critical Foliar Levels of Nutrients for Florists' Crops in General and for a Few Specific Crops[1]**

Nutrient	*General Crops*	*Rose*	*Carnation*	*Chrysanthemum*	*Poinsettia*	*Geranium*	*Rieger Begonia*
N (%)	—	3.0	3.0	4.5	3.5	2.4	4.7
P (%)	0.3	0.2	0.45	0.3	0.2	0.3	0.2
K (%)	—	1.8	3.0	3.5	1.0	0.6	0.95
Ca (%)	—	1.0	1.0	1.0	0.5	0.8	0.5
Mg (%)	0.3	0.25	0.3	0.3	0.2	0.14	0.25
Fe (ppm)	50–60	—	—	—	—	—	—
Mn (ppm)	30	—	—	—	—	—	—
Zn (ppm)	20	—	—	—	—	—	—
Cu (ppm)	5	—	—	7	—	—	—
B (ppm)	25	—	—	—	—	—	14

[1]Concentrations above these are sufficient, while those below are associated with deficiency. Macronutrient standards are specific to each crop. Few micronutrient standards have been developed for specific crops. Fortunately, micronutrient standards do not vary much among crops.

be washed with the cleanest water available. The charge for foliar analysis is considerably more than that for soil testing, ranging from $6 to $30 per sample. This charge is nominal, however, when one considers the total investment in a crop.

Considerable research has been conducted to determine the optimum levels of nutrients in foliage. Carefully developed standards exist for many florists' crops, and for other crops good estimates have been developed from years of observation. Table 9-18 lists the minimum critical levels for several nutrients for some of the major crops. Concentrations below these levels are associated with deficiency. Macronutrient standards vary sharply with each crop, whereas micronutrient standards remain rather constant for most crops. There are few crop variations for the general micronutrient standards listed in Table 9-18. Macronutrients are expressed as a percentage of the dry weight of the leaf tissue analyzed. This would not be convenient for micronutrients, since they constitute only a small fraction of 1 percent; instead, micronutrients are measured in parts per million (ppm). The conversion from percent to ppm is simple: 1 percent equals 10,000 ppm. Excellent sources of standard tissue concentrations of nutrients for greenhouse crops can be found in Mills and Jones (1996).

Testing laboratories report their results on a form more or less similar to that reproduced in Figure 9-28. Generally, the numerical level of each nutrient is reported along with an indication of which of five categories each of the nutrient levels fits:

1. Deficient—showing deficiency symptoms.
2. Low—hidden hunger.

PLANT ANALYSIS REPORT

Name of Grower: Paul Nelson	County: Wake	Date Sample Received: 3-19-85
Street, Route: 29 Main St.	Date Sampled: 3-17-85	Condition of Sample: Good
City, State, Zip: Raleigh, N. C. 27607	Grower Sample #: 2	Grower #: 27
Telephone No.: 737-3132	Name of Crop: Cut Chrysanthemum	County #: 13
Identification of Field Sampled: III	Description of Sample Site (Within The Field):	Lab Sample #: 1971 Date Analysis Received: 3-26-85

COUNTY SAMPLE NUMBER	FARMER OR GROWER SAMPLE NUMBER	N %	P %	K %	Na %	Ca %	Mg %	Mn ppm	Fe ppm	B ppm	Cu ppm	Mo ppm	Zn ppm	Al ppm
	2	4.50	0.85	2.10	0.02	1.2	0.38	125	150	23	9	1	44	

RANGE

	N	P	K	Na	Ca	Mg	Mn	Fe	B	Cu	Mo	Zn	Al
DEFICIENT			X										
LOW									X				
SUFFICIENT	X			X	X	X	X	X		X	X	X	
HIGH		X											
EXCESS													

COMMENTS AND RECOMMENDATIONS

1. Potassium is low. Apply potassium nitrate (13-0-44) for the next two weekly applications at the rate of 2 lbs. per 100 gal.

2. Phosphorus is too high. After correcting the potassium problem use a 1-0-1 ratio fertilizer rather than the 1-1-1 you have been using.

3. Boron is approaching the deficiency level. Apply one ounce of borax per 100 sq. ft. of bench space once.

Mailing Address: Plant Analysis Laboratory, Department of Horticultural Science, N. C. State University, Raleigh, N. C. 27607
North Carolina State University Agricultural Experiment Station and North Carolina Agricultural Extension Service, Cooperating.

FARMER OR GROWER COPY

Figure 9-28 **A typical foliar analysis report as it is received by the grower.**

3. Sufficient.
4. High—hidden toxicity.
5. Very high—showing toxicity symptoms.

Also included are recommended fertilization changes to correct any existent nutrient problems.

Corrective Procedures

Recommendations

Fertilization systems were recommended in the earlier part of this chapter. Under ideal conditions, these systems will work well, but conditions are not always ideal. The optimum rate of fertilizer application relates to the rate of plant growth, which, in turn, can be adversely affected by inclement weather, a poor root substrate that does not drain well, over- or underwatering, a dirty greenhouse covering, nutrient tie-up by constituents of the root substrate, antagonisms by other nutrients, and many other factors. Some nutrients are affected more than others; thus, it is important not only to adjust the rate of fertilization but also to change the ratio of nutrients in the fertilizer. Occasionally, a single nutrient will go far enough out of balance that it alone must be applied. Nitrogen and potassium ratio adjustments can be accomplished through alterations in the continual fertilizer formula (refer to Table 9-5). Corrective procedures for all nutrient deficiencies are listed in Table 9-19.

Phosphorus deficiency is uncommon but not altogether nonexistent. Its occurrence would indicate failure to incorporate sufficient phosphate into the root substrate prior to planting and could easily be corrected by switching to a complete, phosphate-containing fertilizer such as 20-10-20 in the continual fertilization program. Calcium and magnesium deficiencies do not often occur where the root substrate pH level has been properly adjusted with dolomitic limestone. Poinsettia was traditionally, and in many cases still is, grown in an acid substrate to minimize the development of root-rot organisms. Calcium and magnesium deficiencies often occur under these conditions. The new cultivars of poinsettia are prone to magnesium deficiency even when the root substrate pH level is adjusted to the recommended range. Easter lily is very susceptible to calcium deficiency. The use of calcium nitrate as a nitrogen source in the complete fertilizer can solve these calcium problems. An application of 2 lb of Epsom salt (magnesium sulfate) per 100 gal (2.4 g/L) of water is used to solve either a magnesium or a sulfur deficiency. Sulfur deficiencies have become more prevalent in recent years in soilless substrates that do not contain single superphosphate or gypsum.

It is generally safe to apply a micronutrient mixture when symptoms of a single micronutrient deficiency occur and there is evidence that no other micronutrients are present in high quantity. If this information is not known, the status of all micronutrients should be determined by a foliar analysis test. If all mi-

Table 9-19 Fertilizer Sources and Rates for Correction of Various Nutrient Deficiencies

Deficient Nutrient	*Fertilizer Source*	*Rate of Application*[1]	
		oz/100 gal	*g/L*
P	Switch to a complete fertilizer containing N-P-K for the continual program		
	or one application of diammonium phosphate or monopotassium phosphate	32	2.4
Ca	Switch part or all of the N source to calcium nitrate for a few weeks		
	or spray with 400 ppm Ca up to weekly from calcium nitrate	30	2.25
	or from calcium chloride (25% Ca)	20	1.5
Mg	Magnesium sulfate (Epsom salts)	32	2.4
S	Magnesium sulfate	32	2.4
	or switch N or K source to ammonium or potassium sulfate for a few weeks		
Fe	iron chelate (Sprint) or ferrous sulfate	4	0.3
	or foliar spray ferrous sulfate or Sprint iron chelate	4	0.3
Mn	Manganese sulfate	2	0.15
	or foliar spray manganese sulfate or manganese chelate	8	0.6
Zn	Zinc sulfate	2	0.15
	or zinc chelate	1	0.075
	or switch to the fungicide Zineb and spray at the recommended rate monthly		
Cu	Copper sulfate	2	0.150
	or copper chelate	1	0.075
	or foliar spray tri-base copper sulfate	4	0.3
B	Borax	0.5	0.038
	or Solubor	0.25	0.019
Mo	For soil-based substrate, drench once with sodium or ammonium molybdate	0.027[2]	0.002
	For soilless substrate, drench once with sodium or ammonium molybdate	2.67	0.200
	or foliar spray sodium or ammonium molybdate with a spreader-sticker	2	0.150

[1]These corrective procedures are to be applied once. Subsequent applications should be made only after soil and foliar analysis tests indicate the need. All fertilizers are to be applied to the root substrate unless foliar spray is specified.

[2]Dissolve 1 oz sodium or ammonium molybdate in 40 fl oz of water. Use 1 fl oz of this stock solution in each 100 gal of final-strength fertilizer solution.

cronutrients are present in moderate or low concentrations, then a micronutrient mix can be applied. Otherwise, only the deficient nutrient or nutrients should be applied. In the event that foliar analysis is not possible, small plots may be tested with the suspected deficient nutrient. Then, the deficient nutrient alone should be applied. Micronutrient excesses can be far more troublesome than deficiencies because micronutrients are difficult and sometimes impossible to remove from root substrate.

Iron deficiency is common in azalea, gloxinia, hydrangea, and rose. It also readily occurs in crops grown in soilless substrates if iron has not been incorporated into the substrate. The symptoms of manganese deficiency are similar to those of iron deficiency. Fortunately, manganese deficiency is fairly rare in the greenhouse. Zinc deficiency is rare in greenhouse crops except in the case of kalanchoe (Figure 9-17).

Copper deficiency occurs in specific soil types and thus follows geographical patterns. Soils of the southeastern United States are very prone to copper deficiency. The rose crop is an exception, in that the deficiency occurs almost universally. Certain cultivars, such as 'Golden Wave', 'White Butterfly', and 'Mary DeVor', are most sensitive. The cultivar 'Forever Yours' is moderately sensitive. Copper deficiency is most prevalent in old, established rose substrates that are high in humus.

Boron deficiency is a problem with carnation and snapdragon crops. The requirement for boron is similar in these and other crops, but the ability of these crops to take up boron is lower. Boron is readily taken up by chrysanthemum. When carnation or snapdragon crops follow a chrysanthemum crop, boron deficiency often occurs. The pink varieties of carnation are most prone.

Molybdenum deficiency often occurs in poinsettia but is practically nonexistent in other greenhouse crops. The symptoms for poinsettia are chlorosis along the margins of leaves of intermediate age. Chlorosis is rapidly followed by necrosis along the margins of these leaves. Affected leaves may be twisted or half moon–shaped. These symptoms spread to other leaves above and below on the plant.

Interactions

Before attempting to correct a nutrient deficiency, one should always be certain which nutrient is the cause of the problem. The carnation problem previously cited, in which potassium deficiency was induced by an excessive substrate-nitrogen level, is a good example. What first appeared to be a reasonable solution—to apply potassium fertilizer—would not have solved the problem; it might have led to an excessive soluble-salt level. The potassium deficiency was caused by an excessive level of nitrogen, and only a reduction of nitrogen in the soil would correct the deficiency. Such a relationship is known as an *antagonism.* Once the basic antagonisms are known, it is a simple matter to identify them in soil and foliar analysis reports.

The more common antagonisms are listed in Table 9-20. When a deficiency of one of the nutrients in the right column of Table 9-20 is identified, it should be determined whether an abnormally high level of the nutrient in the left column exists. If so, corrective action should involve reduction of the concentration of the nutrient in the left column. Note that some, but not all, of the nutrient antagonisms are reciprocal. For example, a high level of iron will reduce manganese uptake, and reciprocally a high level of manganese will reduce iron uptake.

Table 9-20 Common Antagonisms Occurring in Crops in General[1]

Nutrient in Excess	*Induced Deficiency*
N	K
NH_4, K, Ca, Mg, or Na[2]	NH_4, K, Ca, or Mg
Ca	Mg
Mg	Ca
Ca	B
PO_4	Fe, Mn, Zn, or Cu
Fe	Mn
Mn	Fe

[1]High root substrate levels of any nutrient in the left column can bring about a deficiency of the nutrients opposite it in the right column.

[2]An excess of any ion in the left column can induce a deficiency of any ion other than itself in the right column.

Summary

1. It is expedient to supply all essential nutrients, except possibly nitrogen and potassium, in sufficient quantity to last the full term of the crop at the time of preparing the root substrate. A pH adjustment with dolomitic limestone provides calcium and magnesium. An application of superphosphate supplies phosphorus, while gypsum is a good source of sulfur. The six fertilizer micronutrients can be incorporated into the substrate as a solid during mixing or immediately after planting as a single application of a liquid mixture.
2. The pH level of a root substrate is important because it regulates the availability of all essential nutrients to one degree or another.
3. Nitrogen and potassium are commonly applied as a liquid formulation with every watering or once per week. The rate and ratio of these two nutrients vary according to the crop. Crops in general respond best to a 1:1 N-to-K_2O ratio.
4. The form of nitrogen used affects the pH of the root substrate. Ammonium and urea tend to lower substrate pH. Nitrate tends to raise substrate pH.
5. The nitrogen and potassium formulation is prepared as a concentrate to conserve space and reduce the labor of mixing, and then is diluted and metered into the greenhouse water line by mechanical fertilizer proportioners. It is delivered to the bench or pots through the automatic watering system.
6. Nitrogen and potassium can be alternatively applied as a single application of a dry slow-release fertilizer that, depending on its formulation, can provide N-P-K for 3 to 14 months. Different formulations and types of slow-release fertilizers are available, eliminating the need for any regular fertilization during the crop schedule.

7. Identification of nutritional disorders is as important as the fertilization program itself. Visual diagnosis of disorders can be effective, but unfortunately it depends upon the presence of an injury that may not be completely reversible.
8. Soil testing is a valuable diagnostic tool in that it gives a measure of the root substrate pH and soluble-salt levels as well as a determination of the available levels yet to be taken up of many, but not all, nutrients. Some of the micronutrients are not included.
9. Foliar analysis is an excellent diagnostic tool to be used in conjunction with soil testing. Unlike soil testing, it provides a view of the quantities of nutrients already taken up by the plant. All essential nutrients are included in foliar analysis tests.
10. The problem of excessive soluble salts is prevalent in greenhouse culture because of heavy fertilization procedures. High soluble-salt concentrations result in reduced water availability to plants, which leads to desiccation and death of roots.
11. Soilless root substrates differ from soil-based root substrates in that a lower pH level is desirable, micronutrients are readily tied up, ammonium toxicity is more prevalent, and greater quantities of pre-plant phosphorus are necessary.

References

1. Aldrich, R. A., and J. W. Bartok, Jr. 1994. *Greenhouse engineering.* Pub. NRAES-33. Northeast Reg. Agr. Eng. Ser., Cornell Univ., 152 Riley-Robb Hall, Ithaca, NY 14853.
2. Bennett, W. F. 1993. *Nutrient deficiencies and toxicities in crop plants.* St. Paul, MN: APS Press.
3. Bould, C., E. J. Hewitt, and P. Needham. 1984. *Diagnosis of mineral disorders in plants.* Vol. 1., *Principles.* New York: Chemical Publishing.
4. Bunt, A. C. 1988. *Media and mixes for container-grown plants.* London: Unwin Hyman.
5. Chapman, H. D., ed. 1966. *Diagnostic criteria for plants and soils.* H. D. Chapman, 830 S. University Dr., Riverside, CA 92507.
6. Farnham, D. S., R. F. Hasek, and J. L. Paul. 1985. *Water quality: its effects on ornamental plants.* Univ. of California, Coop. Ext., Div. of Agr. Natural Resources, Leaflet 2995.
7. Koths, J. S., R. W. Judd, Jr., J. J. Maisano, G. F. Griffin, J. W. Bartok, Jr., and R. A. Ashley. 1980. *Nutrition of greenhouse crops.* Coop. Ext. Ser. of the Northeast States, NE 220.
8. Mills, H. A., and J. B. Jones, Jr. 1996. *Plant analysis handbook II.* Athens, GA: Micro-Macro Publishing.
9. Mortvedt, J. J., and C. Sine, eds. 1994. Fertilizer dictionary. In: *Farm chemicals handbook.* Willoughby, OH: Meister Publishing.
10. Peterson, J. C. 1982. *Effects of pH upon nutrient availability in a commercial soilless root medium utilized for floral crop production.* Ohio Agr. Res. and Devel. Center, Res. Cir. 268, pp. 16–19.

11. Peterson, J. C. 1984. Current evaluation ranges for the Ohio State floral crop growing medium analysis program. *Ohio Florists' Association Bul.* 654:7–8.
12. Plank, C. O. 1988. *Plant analysis handbook for Georgia.* College of Agr., Coop. Ext. Ser., Univ. of Georgia, Athens, GA.
13. Rader, L. F., Jr., L. M. White, and C. W. Whittaker. 1943. A measure of the effect of fertilizers on the concentration of the soil solution. *Soil Sci.* 55:201–208.
14. Reuter, D. J., and J. B. Robinson. 1986. *Plant analysis.* Sydney, Australia: Inkata Press.
15. Roorda van Eysinga, J. P. N. L., and K. W. Smilde. 1980. *Nutritional disorders in chrysanthemum.* Wageningen, The Netherlands: Center for Agr. Publishing and Documentation.
16. Roorda van Eysinga, J. P. N. L., and K. W. Smilde. 1981. *Nutritional disorders in glasshouse tomatoes, cucumbers, and lettuce.* Wageningen, The Netherlands: Center for Agr. Publishing and Documentation.
17. Scaife, A., and M. Turner. 1984. *Diagnosis of mineral disorders in plants.* Vol. 2, *Vegetables.* New York: Chemical Publishing.
18. Sprague, H. B. 1964. *Hunger signs in crops,* 3rd ed. New York: David McKay.
19. Tayama, H. K. 1966. Extension slants—production pointers. *Ohio Florists' Association Bul.* 442:9.
20. Truog, E. 1948. Lime in relation to availability of plant nutrients. *Soil Sci.* 65:1–7.
21. Warncke, D. D., and D. M. Krauskopf. 1983. *Greenhouse growth media: testing and nutrition guidelines.* Michigan State Univ. Agr. Ext. Bul. E-1736.
22. Westerman, R. L., ed. 1990. *Soil testing and plant analysis.* Madison, WI: Soil Sci. Soc. of Amer.
23. Winsor, G., and P. Adams. 1987. *Diagnosis of mineral disorders in plants.* Vol. 3, *Glasshouse crops.* London: Her Majesty's Stationery Office.

CHAPTER 10

Alternative Cropping Systems

Since the beginning of commercial greenhouse crop production, our intrigue with futuristic technology, desire to direct nature, and aspiration for achievement have led to the trial of many innovative cultural systems. Recently, two separate powerful forces have come together to enhance the quest for these systems. First, internationalization of greenhouse production has given birth to a level of competition that demands reduced costs of production and marketing. Solutions are being sought through automation and subsequent computerized control of these emerging high-technology cropping systems. Second, antipollution regulations demand that cropping systems be developed that reduce the amounts of nutrients and pesticides in greenhouse effluent. Most attention is being given to closed cultural systems in which water, with its contained nutrients and pesticides, is captured, treated, and reused. Open cultural systems, where excess water from the bottom of pots or beds is allowed to flow into the ground, are being modified to reduce the volume of effluent.

Solutions for both of the current needs of production efficiency and pollution abatement are found simultaneously in a number of new systems. Two systems commercialized in the early 1970s employ the principles of *nutriculture* (hydroponics) and are used for fresh-flower and vegetable production. These systems are the *nutrient film technique* (NFT) and *rock wool culture.* Remaining alternatives include the ebb-and-flood system for pot plants and bedding plants, trough culture for pot plants, the float system for tobacco seedlings, and *whole-firm recirculation* for all crop types. The former three systems were discussed in Chapter 8, "Watering." Ebb-and-flood, trough, float, and whole-firm recirculation are not forms of nutriculture, since conventional root substrate and pots or flats are used. However, the methods for delivering nutrient solution in these systems make use of the principles of nutriculture.

Nutriculture involves the culture of plants in an inert substrate such as water (hydroponics), gravel (gravelculture), sand (sandculture), rock wool, or air (aeroponics). An inert substrate is one that neither contributes nor alters the form of plant nutrients. Soil, peat moss, and bark are examples of non-inert substrates that are both biologically and chemically active. These substrates contribute nutrients that are held on their negative exchange sites and others that are released during substrate weathering and decomposition. In addition, these substrates are handled in a manner in which large populations of microorganisms develop in them, which can change the form of applied nutrients (for example, convert ammonium to nitrate). Inert substrates, such as rock wool or water in the NFT system, afford much greater control over plant nutrition than is possible in today's non-inert substrates. Before the previously mentioned cultural systems are described, it will be useful to explore their origins and the subsequent developments that led to the present situation.

Historical Background

Setting the Stage (1860–1928)

The origin of the current commercialization of nutriculture is better identified with the period around 1860 than with any other time during the 295-year history of nutriculture research. While the six essential macronutrients and iron had been identified by 1844, it was not until 1860 that von Sachs introduced in Germany a complete nutrient formula for growing plants hydroponically. In 1861, Knop described an improved formula that is used to this day. Through the late 1800s and the first three decades of the 20th century, considerable effort was directed at improving hydroponic cultural systems and nutrient formulas and at discovering most of the remaining micronutrients. The importance of solution aeration and periodic replacement of the nutrient solution was yet to be realized. Throughout this period, nutriculture remained a technique for research purposes.

Early Commercial Attempts (1928–1970)

The first published attempt to develop the commercial potential of hydroponics occurred in the United States in 1929. Gericke (1929) built an experimental nutrient-solution tank covered with wire netting, canvas, and ½ inch (1.3 cm) of sand. Plants were anchored in the sand. A 2-acre (1-ha) commercial planting followed.

Sandculture dates back to studies of Salm-Horstmar (1849), who introduced the idea of using sand as well as other inert media. Commercial impetus goes back to McCall (1916) in the United States, who saw an advantage in the nutritional control that hydroponics offers and the physical properties of support and aeration provided by sand. Robbins (1928) worked out a method for growing a number of crops in sand in the greenhouse environment. Laurie (1931) indicated the commercial potential for sandculture. In 1935, a procedure for growing carnations in sandculture was developed at the New Jersey Agricultural Experiment Station (Bickart and Connors 1935). Refinements in the physical system for sandculture and gravelculture followed (Eaton 1936; Withrow and Biebel 1936; Shive and Robbins 1937; Chapman and Liebig 1938). The American technology was duplicated and advanced in England by Templeman and Watson (1938). They did not find superior growth in nutriculture.

The next notable application of nutriculture came during World War II, when both Japan and the United States used sandculture and gravelculture to produce fresh vegetables for the war effort (Ticquet 1952). The first American installation was created in early 1945 on Ascension Island, an island nearly devoid of soil. Twenty-five 3-foot-by-400-foot (0.9-m-by-122-m) sandculture beds produced 94,000 lb (43,000 kg) of salad vegetables the first year. Additional installations were created at Atkinson Field in British Guiana and on Iwo Jima later that year. At the same time, Japan constructed 5 acres (2 ha) under glass and 50 acres (20 ha) outside at Chofu and 25 acres (10 ha) outside at Otsu. These installations later were used by American troops during the war in Korea.

Hoagland and Arnon (1950) developed the famous Hoagland's solution, which is commonly used to this day in research and commerce (Table 10-1). They found that tomato growth was equal in soil, sand, and water culture and stated that the use of hydroponics would be dictated by economic considerations.

The most common commercial nutriculture systems used until the early 1970s were sandculture and gravelculture (Kiplinger 1956; Weinard and Fosler

Table 10-1 **Chemical Composition of Hoagland's All-Nitrate Nutrient Solution[1]**

		Weight	
Chemical	*Formula*	*mg/L*	*oz/100 gal*
Potassium dihydrogen phosphate	KH_2PO_4	136.0	1.81
Potassium nitrate	KNO_3	505.0	6.73
Calcium nitrate	$Ca(NO_3)_2 \cdot 4H_2O$	1,180.0	15.73
Magnesium sulfate	$MgSO_4 \cdot 7H_2O$	492.0	6.55
Iron tartarate[2]	$FeC_4H_6O_6$	5.0	0.067
Manganous chloride	$MnCl_2 \cdot 4H_2O$	1.81	0.024
Zinc sulfate	$ZnSO_4 \cdot 7H_2O$	0.22	0.003
Copper sulfate	$CuSO_4 \cdot 5H_2O$	0.08	0.001
Boric acid	H_3BO_3	2.86	0.038
Molybdic acid	$H_2MoO_4 \cdot H_2O$	0.02	0.0003

[1]From Hoagland and Arnon (1950). The following concentrations, in parts per million, are achieved in this formulation: N—210, P—31, K—234, Ca—200, Mg—48, S—64, Fe—1.4, Mn—0.5, Zn—0.05, Cu—0.02, B—0.5, and Mo—0.01.

[2]Today, iron chelates are substituted for iron tartarate. Iron DTPA is commonly used at rates from 2 to 4 ppm iron (20 to 40 mg iron DTPA/L).

1962; Epstein and Krantz 1965; Hewitt 1966; Marvel 1966; Maynard and Barker 1970; Maas and Adamson 1971; Sowell 1972). Watertight beds or benches, often constructed from concrete, served to hold gravel. Nutrient solution, generally similar to Hoagland's solution, was flushed through the gravel one to four times per day and less often in sand, depending on season and size of crop. Between flushes, the nutrient solution was stored in tanks. Periodic analysis and adjustment of the solution for volume, pH level, and nutrient concentration permitted continual use for many weeks before replacement. Aeration presented no problem, since the aggregate pores were filled with nutrient solution for only brief periods. The primary problems that emerged were nutrient imbalances, particularly of micronutrients, and disease. Firms with access to rapid solution analyses and with a staff member well-versed in nutritional chemistry did not find nutrition to be a problem. Rapid spread of diseases was a constant threat for everyone. Many firms maintained diazoben in the nutrient solution to suppress development of pathogens.

A modest number of firms around the world substituted coarse (0.5-inch, 1.3-cm) Palabora vermiculite from Africa for gravel (Bentley 1955, 1959). Its light weight made it attractive for use in movable benches. The high water- and nutrient-holding capacity allowed operators to reduce nutrient-solution application to three or four times per week. This cultural system became known as *verniculaponics.*

Throughout this period, there was no major commercial adoption of a nutriculture system. Major deterring factors were the absence of plastics for inert, watertight construction; insufficient equipment for automated monitoring and control of nutrient solutions; and the lack of computers for overall control. However, considerable technology was developed in this period to fuel the large-scale adoption of nutriculture about to come.

Large-Scale Successes (1970–Present)

The first nutriculture system to gain wide acceptance in greenhouse culture and to prove itself economically efficient was the nutrient film technique (NFT). Its background is interesting. DeStigter (1961, 1969), at the Plant Physiological Research Center in The Netherlands, developed the prototype for this system. For research purposes, he developed a method of growing roots in a thin, retrievable layer to facilitate making autoradiographs of them. Cooper (1973), in England, through communications with DeStigter, saw the commercial potential in this procedure and developed the NFT system from it.

NFT is a form of hydroponics in which plants are grown in narrow, sloped channels (Figure 10-1). A thin film of recirculating nutrient solution flows through the roots in the channels. With NFT, unlike the classical hydroponic systems, aeration is not a problem because the nutrient solution is confined to a depth of 1/8 inch (3 mm). Commercial installations of NFT began in the early 1970s. In 1982, there were approximately 125 acres (50 ha) in England and considerably more area in The Netherlands. Although the area in the United States is unknown, several hundred small firms are producing vegetables by the NFT system in greenhouses.

The most extensive development of NFT occurred in The Netherlands partly because of pasteurization problems. Most greenhouses there heat with hot water, which is inappropriate for pasteurization; thus, methyl bromide is very popular. In the sandy soils of the concentrated greenhouse region known as the Westland, methyl bromide readily permeates the soil as well as the plastic walls of water pipes lying within these soils. Contaminated drinking water prompted a restriction in the dosage rate of methyl bromide and threatened its future use altogether. Without steam to fall back on, NFT became very attractive.

NFT production in northern Europe peaked and began to decline in the early 1980s. A strong reason for the change was disease buildup in the closed recirculating solution. Contamination from one plant soon reached all other plants. The high initial cost of an NFT system and its continual consumption of electrical energy might also have contributed to this shift (Dungey 1983).

Open cultural systems, in which nutrient solution makes a single pass by the roots and then on to waste, were sought next to get around the disease problem while still accommodating the need for automation. The system of rock wool culture had reached a sufficient stage of development by the early 1980s to take over as the predominant alternative cropping method. Rock wool consists of fibers formed from melted rock that, in final form, resemble fiberglass insulation. These fibers are formed into cubes for propagating plants and into slabs for growing plants on to maturity.

Rock wool culture was pioneered in Denmark during the late 1950s. By the early 1970s, horticultural rock wool was in production in Denmark. Today, nearly all greenhouse cucumber and tomato plants in Denmark are produced in rock wool. Greenhouse area in rock wool production has grown in The Netherlands from an estimated 450 acres (180 ha) in 1980 to 2,500 acres (1,000 ha) in 1983 and to 5,000 acres (2,000 ha) in 1988. In England, while there were only 15 acres (6 ha) in 1978, there were 63 acres (25 ha) in 1982 (Hanger 1982). The areas of rock wool production in Belgium and West Germany in 1988 were 1,500 acres (600 ha) and 125 acres (50 ha), respectively (Molitor 1990). Rock wool appeared in the American market in the early 1980s. The first crops grown in rock wool were the full range of greenhouse vegetables. More recently, carnation, chrysanthemum, gerbera, and rose have been grown in it.

Antipollution Legislation

The history of pesticide regulation is a short one. The first legislation in the United States included the Insecticide Act of 1910 and the Federal Insecticide, Fungicide, and Rodenticide Act (FIFRA) of 1947. Both were intended to protect the purchaser against fraud relative to the effectiveness of the pesticide purchased. This direction is understandable when one realizes that the importance of synthetic organic pesticides began around 1940 with worldwide use of the insecticide DDT for purposes such as controlling malaria-carrying mosquitoes.

However, by 1984, there were about 600 basic pesticide chemicals on the market in 45,000 to 50,000 formulations (Johnson 1984).

The legislative shift to protection of public health and the environment came in 1952 with an amendment to the Federal Food, Drug, and Cosmetic Act. The amendment established a procedure for setting tolerance levels for pesticides in food, feed, and fiber. Social sensitivities were piqued in their early stages by Rachel Carson's book *Silent Spring* (1962), which dramatically pointed out the dangers of pesticides to people and to the environment. This was augmented by the worldwide environmental movement of the 1960s. In 1964, FIFRA was amended to encompass safety considerations in the labeling of pesticides. The year 1970 was a milestone in that the first Earth Day was held and the Environmental Protection Agency (EPA) was established. FIFRA was again amended to include a mandate for the protection of public health and the environment as a guiding principle for the use of pesticides. The mandate allowed for a reasonable balance between economic, social, and environmental costs and the benefits of the use of any pesticide. Such a balance of risk and benefit was important, given the estimate of a 30 percent loss in crop productivity from pests if pesticides were not used (Wilkinson 1987). In today's world, with its high rate of death from malnutrition and its rapid rise in population, such a crop reduction would be devastating.

Current initiatives to clean up water drew strength in the late 1970s when the following pesticides were found in groundwater supplies: aldicarb on Long Island and in Wisconsin; atrazine in Iowa; and ethylene dibromide in California. The EPA subsequently formulated plans to assess groundwater contamination and in 1984, 1986, and 1988 released plans and updates. The 1988 update, "Proposed Pesticide Strategy," indicated the need to evaluate each pesticide independently and set a standard of acceptable risk for it. Other legislation, including the Water Quality Act of 1987, FIFRA, and the Safe Drinking Water Act, provides mandates and authority to federal, state, and local agencies to assess, monitor, and regulate contamination in water supplies. These responsibilities have not been placed on any one level of government. Agencies at federal, state, and municipal levels are involved in all three functions at the present. In 1992, the EPA developed the Federal Worker Protection Standard. This set of rules defines and governs safe procedures for handling pesticides. Any pesticide with a label that refers to the Federal Worker Protection Standard must be handled in accordance with the rules of this legislation. These rules are covered in Chapter 14, "Insect Control." While national standards will continue to emerge from federal agencies such as the EPA, state and municipal agencies are empowered to issue stricter standards for their jurisdictions.

Nutrient pollution can take several forms, including excessive levels of nitrate, phosphate, total salt, or heavy metals such as copper or zinc. A national limit has been set for nitrate in effluent waters at 10 ppm nitrate nitrogen, which equates to 44.3 ppm nitrate. Similar limits have been set in many Western European countries. While a strict national limit has not been set for phosphate, regional measures have been taken to reduce its use. Bans on phosphate detergents are one example of initiatives being taken. Phosphates in surface water foster algal growth, which, in turn, robs water of oxygen and leads ultimately to fish death. Limits on phosphate in greenhouse effluent will undoubtedly occur. Likewise, limits on total salt level and individual heavy-metal micronutrients are not universally set, but probably will be one day.

NFT Systems

Cultural Procedures

Cooper (1979) thoroughly outlined the NFT system in his book. The system begins with a channel, free of valleys and peaks, laid out on a 1 percent slope. The channel must be level across its width to ensure that the whole floor will be covered with nutrient solution. Such channels may be molded into a concrete floor or may be situated on raised platforms. For crops such as tomato and cucumber, a channel is typically about 9 in (23 cm) wide and 2 in (5 cm) high. It may be constructed from wood, plastic, metal, or concrete. Lettuce, chrysanthemum, snapdragon, and other fresh flowers are more often grown in a bed fashioned from several closely spaced parallel channels (Figure 10-1).

The channel needs to be watertight. Channels constructed from leaky materials are lined with a film plastic sufficiently large to cover the top of each channel, including the plant-propagation blocks. The film plastic should be 5 mil (0.13 mm) thick or thicker; otherwise, the plastic will adhere to the roots, which will cause it to ripple along the bottom of the channel. This, in turn, will force the solution to puddle in spots and to flow around roots in other places. Watertight channels, while not requiring a lining, do need a covering. This may be formed from a solid material or a film plastic.

Channel coverings serve to (1) prevent water loss through evaporation; (2) restrict light entry to prevent algal growth, which would remove nutrients and plug the system; and (3) help control root temperature. The outer surface of the covering should be white or silver to reduce heat absorption and to reflect light to the plants for better growth. Air inside a black channel would become hot enough to burn roots on warm, bright days. White plastic does not sufficiently restrict light; therefore, film plastic is sold with one (inner) surface black and the other surface white. In regions of temperature extremes, an insulated channel covering may be constructed by using two film plastic coverings with a dead-air space between them.

The nutrient solution is handled in a closed recirculating system (see Figure 10-1). A tank, usually built into the floor, collects solution by gravity flow from the ends of the channels. Solution is pumped from the tank to a header pipe that runs perpendicular to the upper ends of the channels. Small tubes running from the header pipe supply each channel. The flow rate should be sufficient to maintain a nutrient film thickness of not more than 1/8 inch (3 mm) over the entire bottom surface of the channel. Greater depth will exclude oxygen from the roots. A flow rate of about 0.5 gpm (2 L/min) per channel is required. In some systems, the solution is constantly recirculated. More commonly in the United States, the solution is circulated for 10 minutes out of every 15 minutes to increase aeration

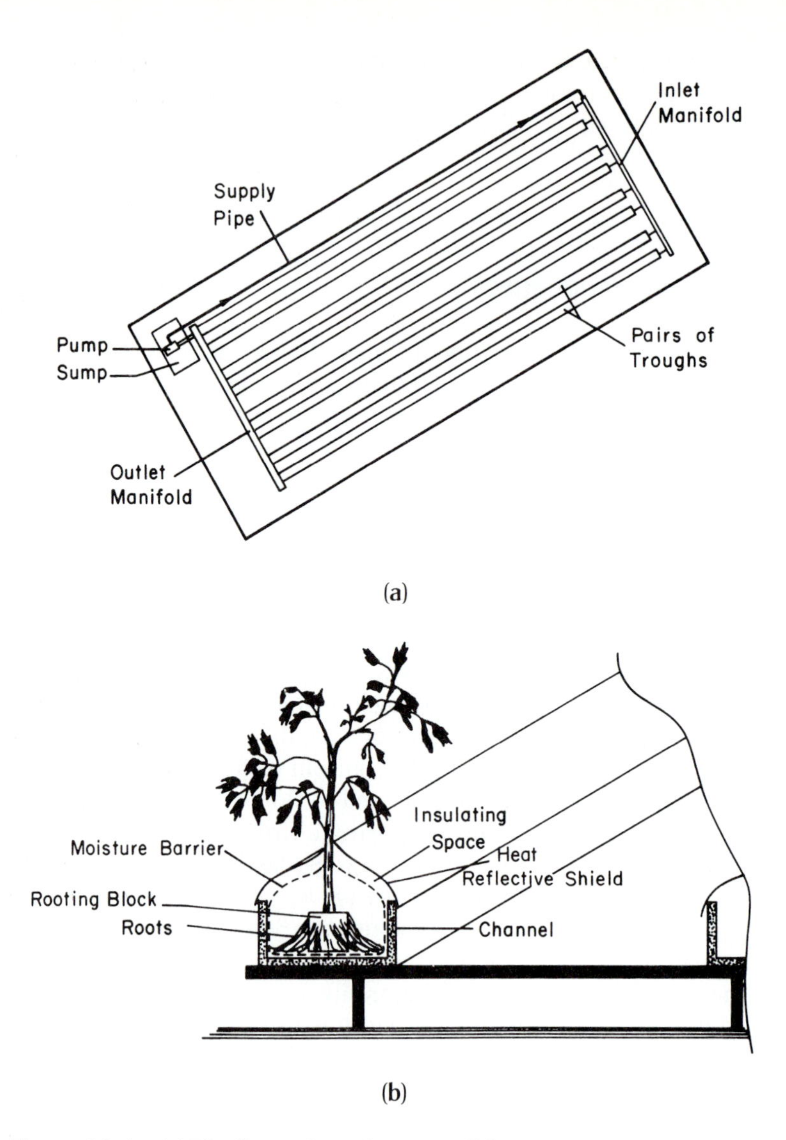

Figure 10-1 (a) The floor plan of one possible NFT system. Nutrient solution is pumped from the sump tank to an inlet manifold at the upper end of the NFT troughs. From there, the solution flows by gravity to an outlet manifold at the lower end of the troughs and finally back to the sump. (b) A cross-sectional view of a trough, showing plant placement and the arrangement of the film plastic moisture barrier and thermal barrier. The outer thermal barrier is optional.

Table 10-2 Theoretically Ideal Concentrations of Elements in Nutrient Solution for NFT Cropping[1]

Element	*Symbol*	*Concentration (ppm)*
Nitrogen	N	200
Phosphorus	P	60
Potassium	K	300
Calcium	Ca	170
Magnesium	Mg	50
Iron	Fe	12
Manganese	Mn	2
Boron	B	0.3
Copper	Cu	0.1
Molybdenum	Mo	0.2
Zinc	Zn	0.1

[1]From Cooper (1979).

of the roots. A considerable volume of water will be lost through transpiration, necessitating continual additions to the holding tank. This can be automatically handled by installing a float valve on a water inlet line in the tank.

Cooper (1979) suggests the nutrient concentrations listed in Table 10-2 as being ideal for NFT culture. He has grown over 50 species of ornamental, fruit, and vegetable plants in this solution for three continuous years without problems. Sources of nutrients and required weights are presented in Table 10-3. It is not necessary for a firm to formulate its own nutrient solution. Various com-

Table 10-3 Weights of Chemical Compounds Required to Give Theoretically Ideal NFT Cropping Concentrations[1]

		Weight	
Chemical	*Formula*	*g/1,000 L*	*oz/100 gal*
Potassium dihydrogen phosphate	KH_2PO_4	263	3.51
Potassium nitrate	KNO_3	583	7.77
Calcium nitrate	$Ca(NO_3)_2 \cdot 4H_2O$	1,003	13.37
Magnesium sulfate	$MgSO_4 \cdot 7H_2O$	513	6.84
EDTA iron	$[CH_2 \cdot N(CH_2 \cdot COO)_2]_2FeNa$	79	1.05
Manganous sulfate	$MnSO_4 \cdot H_2O$	6.1	0.081
Boric acid	H_3BO_3	1.7	0.023
Copper sulfate	$CuSO_4 \cdot 5H_2O$	0.39	0.005
Ammonium molybdate	$(NH_4)_6Mo_7O_{24} \cdot 4H_2O$	0.37	0.005
Zinc sulfate	$ZnSO_4 \cdot 7H_2O$	0.44	0.006

[1]From Cooper (1979).

panies sell NFT fertilizers. Generally, they come in two or three packages that must be added separately to the tank to prevent precipitation.

The solution is used in most European systems for many months before replacement. In several American systems, it has been replaced at two-week intervals. This latter practice is inconsistent with present needs to comply with antipollution rules. In either case, it is necessary to test the solution for pH and EC (soluble-salt) levels at least daily. The pH level should remain in the range of 5.8 to 6.5. When it decreases, potassium hydroxide is added; when it increases, sulfuric acid (battery acid) is added. Different fertilizer formulations will have different EC levels. For Cooper's solution, the level should start at 3 dS/m (3 mmho/cm). When it drops to 2 dS/m, all nutrients should be added in sufficient quantity to restore the level to 3 dS/m.

Computerized equipment is available that will automatically sample nutrient solution from the holding tank, analyze it for pH, EC, and individual nutrients, and make the appropriate additions of acid, base, or fertilizer. This equipment includes concentrate tanks of acid, base, and individual fertilizer salts such as potassium nitrate and magnesium sulfate. The test results are processed in the computer to determine which nutrient salts and how much of each will be injected into the NFT solution. These additions are then automatically made. Concentrate from any individual tank or combination of tanks can be added to the single-strength nutrient tank in order to hold all nutrients in balance. Such a system permits the use of a nutrient solution for a considerably longer time before it is discarded. The automated system also allows the pH and nutrient concentrations of the solution to be maintained more precisely than by merely analyzing solutions once per day.

Plants to be set in an NFT system are propagated in containers such as blocks of rock wool, foam cubes, or in netlike pots containing soilless substrate. It is important that the propagation unit not contribute peat moss or other loose substances that will plug the system. A propagation area is set up for establishing plants under conditions more ideal to this stage. It also permits growing at high plant densities to cut overhead costs. Young plants are often grown in channels in the propagation area; however, plants within the channels as well as the channels themselves are placed much closer together than they are in the finishing greenhouse. When established plants are then moved to channels in the finishing greenhouse (Figure 10-2), there can be a problem of nutrient solution meandering along the plastic and missing some root blocks. This problem can be solved by placing sticks beneath the plastic just below each plant to form a dam for puddling water around each block. In a week or two, when roots develop across the channel, the sticks can be removed.

Advantages

The growth in popularity of NFT can be attributed to several factors:

1. The NFT system eliminates the materials and labor costs for steam or methyl bromide pasteurization between crops, as well as the period of 10 to 14 days

Figure 10-2 Carnations growing in an NFT system. Note header pipes running across the upper end of the troughs for delivering nutrient solution.

required for methyl bromide application and aeration. If the channel in which plants are grown is formed from film plastic, the plastic is gathered up with the crop and discarded, and new plastic is laid out. Permanent channels not lined with film plastic are rinsed with a sterilant such as bleach between crops.

2. NFT has the potential for conserving water and nutrients. The nutrient solution is recirculated in a mostly closed system where little evaporation occurs and excess water and nutrients are reused.
3. Recirculation of solution provides an excellent method for reducing nutrient and pesticide effluent from greenhouses.
4. NFT has the very attractive advantage of the potential for automation. Formulation, testing, and adjustment of nutrient solutions can be handled at a central point, and even this can be done automatically. The solutions are mechanically delivered to the crop. Some of the heat may likewise be delivered in the nutrient solution. Heavy root substrates and the handling of them are eliminated.

In-Line Pasteurization

Closed-system NFT lost ground to open-system rock wool culture in the 1980s because economically acceptable means of pasteurization of the nutrient solution were not available. Six options for pasteurization are receiving current attention. Their adoption not only will clear the way for further development of NFT but also will permit operation of the rock wool system in a closed fashion rather than in the open system that is most common today. These pasteurization systems will also have application in all other types of recirculating crop systems.

Equipment is available for one method of pasteurization that heats nutrient solution while it is recirculating between the crop and the sump tank. The solution is heated into the range of 203 to 221°F (95 to 105°C) (depending upon the equipment purchased) in 30 seconds and held at this temperature range for 10 to 30 seconds, after which it is cooled at an equally rapid pace.

Other pasteurization systems are available that are installed in the recirculation system in a similar position. One system produces ozone to destroy microorganisms. An additional benefit from the ozone is that it replenishes oxygen in the solution. Another system is an ionization device in which microorganisms are killed by copper and silver ions. In yet another system, chlorine or bromine is injected as a microbicide. When chlorine is used, it is important to filter organic matter out of the solution, since chlorine will bind to it and thus become ineffective.

Ultraviolet (UV) light is used to destroy microorganisms in a fifth system. UV lamps are contained inside a tube. The recirculating nutrient solution flows through the tube, over the lamps. A 2.5 kW lamp can treat 2,640 gal/hr (10,000 L/hr) and has a life expectancy of 8,000 hours. Smaller units are available that treat 400 gal/hr (1.5 m^3/hr).

Filtration is a sixth method for excluding disease organisms from the solution. Membrane filters, fine enough to trap bacteria, are contained in cartridges. One cartridge can treat nearly 3,200 gal/day (12,000 L/day). A 5-acre (2-ha) greenhouse could be serviced by six cartridges. Filters must be cleaned periodically with chemicals or high-pressure flushing, depending on water quality. The life expectancy of filter cartridges is three to four years.

Rock Wool Culture

Manufacture

Rock wool is produced by burning a mixture of coke, basalt, limestone, and possibly slag from iron production. At a 2,900°F (1,600°C) temperature in the furnace, the rock minerals melt. This liquid is tapped from the base of the furnace. A stream flows onto a high-speed rotor. Droplets thrown from the rotor lengthen into fibers. The fibers are sprayed with a binding agent in an airstream, which cools and carries them to a conveyor, where they are deposited onto a belt. The pad of fibers is then compressed between rollers to a specified density. Finally, it is cut into desired dimensions.

Product Description

Insulation- and acoustical-grade rock wool is not suitable for plant growth. Horticultural-grade rock wool is formulated to a prescribed higher density to provide the air- and water-holding requirements of plants. Rock wool used in cubes for propagation and in slabs for finishing crops is, unlike industrial rock wool, treated with surfactants to improve water absorbance. It contains about 3 percent solid and 97 percent pore space.

Rock wool is not biodegradable, but it does slowly weather. Slabs are often used for two years of cropping before disposal. Initially, rock wool contains no significant quantity of soluble materials. However, fibers can contain calcium, magnesium, iron, manganese, copper, and zinc, and there is evidence that these can slowly be released for plant uptake (Rupp and Dudley 1988). Since the cation exchange capacity is negligible, applied nutrients are not adsorbed. Nutrient availability is dictated by the nutrient solution applied. The pH of rock wool is between 7.0 and 8.5 (often 8.0), but is not buffered. It is important that the nutrient solution applied have a pH level in the range of 5.5 to 6.0. The pH level of the rock wool will adjust to the nutrient-solution pH level after one application.

Horticultural rock wool is available in 0.7- to 4-inch (18 - to 100-mm) cubes with or without predrilled holes for propagation of seed or cuttings (Figure 10-3).

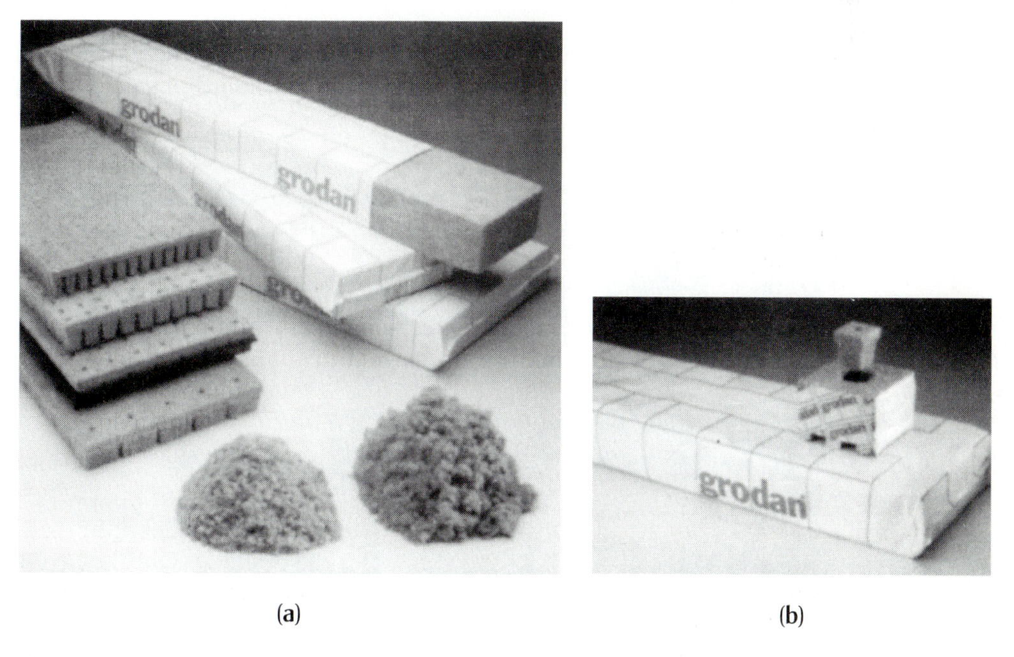

Figure 10-3 Various rock wool products used in the greenhouse. (a) Blocks of cells of various dimensions for propagation of seeds and cuttings, polyethylene-wrapped slabs of rock wool for supporting vegetable and fresh-flower crops, and loose rock wool to be used as a component in root substrate for pots. (b) An individual propagation cell in which a seed or cutting would be propagated, a block with a predrilled hole to accommodate the propagation cell, and a slab on which the block will ultimately be placed after plants have been grown for an interim period of time at close spacing in the block. *(Photos courtesy of Grodania A/S, Hovedgaden 483, 2640 Hedehusene, Denmark; distributed in North America by Agro Dynamics, 10 Alvin Ct., East Brunswick, NJ 08816.)*

The smaller cubes can be obtained unwrapped in blocks suitable for use in trays. Larger cubes are often wrapped on the vertical sides with polyethylene to prevent evaporation and spread of roots into adjacent cubes and are sold in single-row strips. The 3-inch and 4-inch (76-mm and 100-mm) cubes can be obtained with a depression in the top of suitable size to insert smaller cubes into them for transplanting purposes. For purposes of finishing crops, slabs of rock wool are available. These can be obtained unwrapped or wrapped in white polyethylene. Slab widths can be 6, 8, 12, or 18 in (15, 20, 30, or 46 cm), lengths can be 30 or 39 in (76 or 100 cm), and the height is usually 3 in (7.6 cm). Granulated rock wool is also available for potting substrate (as described in Chapter 6).

Cultural Procedures

As in the case of NFT, plants can be propagated at a high density in small cubes in a specially regulated environment. They may then be transplanted into larger blocks and spaced out into a moderately high-density nursery area to cut down overhead costs. The large cubes are finally set on top of slabs for final production (Figure 10-4).

Evenness of the floor is not as important for rock wool culture as it is for NFT. Where there is a slope along the length of the bed, solution in individually

Figure 10-4 Pictured here is a tomato crop growing in a rock wool system. The 3-inch (7.6 cm) rock wool propagation cubes are sitting on rock wool slabs 6 in (15 cm) wide by 3 in (7.6 cm) deep. The slabs are wrapped in white polyethylene film. Nutrients and water are provided to the propagation cubes through plastic microtubes fed from PVC pipe running along the sides of the slabs.

Figure 10-5 **Pictured here are roses growing in slabs of rock wool that are individually wrapped in polyethylene film and placed end to end. Nutrient solution is delivered from the black plastic pipes above the slabs through microtubes to individual plants.**

wrapped slabs cannot drain from the high end to the low end of the bed. Narrow slabs are placed end to end in a double row for cucumber, tomato, and rose crops to form a bed (Figure 10-5). A space is generally left between the two rows for vegetables. For fresh flowers, the wider slabs are used to line a bench. Ideally, for double-row culture, the floor should be level from end to end but sloped at a 2° angle to a drain between the two rows of the bed.

Cubes with individual plants are placed on top of the slabs at the desired final spacing for the crop involved. Roots penetrate into the slabs below the cubes in two to four days. During this period of adaptation, it is necessary to water the plants frequently. The additional height of the cube plus the slab causes the cube to drain excessively and remain too dry. In temperate climates, slabs are covered with polyethylene. The cover may be omitted in tropical climates to allow for evaporative cooling of the root zone. When wrapped slabs are used, three drain holes must be cut along one side—the low side if the floor is sloped to a drain between double rows. A 1.5-inch (4-cm) slit is made starting at the bottom of the side and extending upward at a 45° angle from the floor.

Nutrient solution is applied with each watering. Solutions used for NFT and other nutriculture systems are appropriate. Crops can require 3 to 10 applications per day. This frequency will vary with plant size and weather conditions. When only two or three plants are grown on a slab, as in the case of tomato or cucumber, nutrient solution is delivered to each cube through a microtube supplied from a plastic pipe running along the bed. This is not feasible when large numbers of cubes are grown on a slab. In this instance, three or four fertilizer emitters are placed directly on top of the slab. Traditional fertilizer injectors and water mains can be used to supply the plastic pipes.

Rock wool in a 6-inch (15-cm) -deep configuration can hold water in about 50 percent of its pore space. The distribution of water is unequal, with pores in the lower 1 inch (2.5 cm) holding nearly 100 percent water and those in the top 1 inch (2.5 cm) holding less than 10 percent. A 3-inch (7.6-cm) -deep slab will hold enough water to fill 77 percent of its pore space, leaving 23 percent of the pores open for aeration. It follows that the small cubes will have even less aeration. When seeds or cuttings of plants sensitive to oxygen stress are propagated in these cubes, it is advisable to place the cubes on a well-drained substance such as sand or perlite. This increases the effective depth of the cube and consequently its drainage and aeration.

Just as the nutrient solution in NFT systems is sometimes heated to keep the root zone warm, heat may be applied under rock wool slabs. A polystyrene board is placed on the ground. It has a notch at the middle of its top side running along its length. A plastic hot-water pipe is placed in this notch. Rock wool is placed directly on the pipe. By warming the root zone, cooler air temperatures can be maintained in the greenhouse, which results in fuel savings.

The salt content of rock wool can be lowered for the next crop by applying only water during the last days of the previous crop. Old crops can be removed by twisting the cubes off the slab. Rock wool may be used for one year for cucumber and two years for tomato and floral crops less sensitive to oxygen stress. If the rock wool is to be used for a second year, it is advisable to pasteurize it. Most of the water can be removed prior to pasteurization by cutting off the water supply during the last days of the final crop. The slabs are then stacked and covered with a pasteurization cover, and steam is applied for 30 minutes. Methyl bromide may be used as well. It can be readily washed out of the rock wool after fumigation. Repeated use of rock wool results in collapse of its structure and a buildup of organic matter from roots with a resultant loss of aeration. Cucumber is very sensitive to this problem.

Advantages

Rock wool offers several advantages:

1. *Elimination of pasteurization.* As in the case of NFT, pasteurization can be eliminated in rock wool culture, unless it is used for a second year.
2. *Production efficiency.* Rock wool is an excellent inert substrate for open-system nutriculture. This greatly reduces the chances for the spread of disease, since the nutrients are not recirculated. It is, however, possible to recirculate nutrient solution in a rock wool system. In this case, slabs are placed on a sloped, paved floor. Nutrient solution is collected at the low point and is handled in the same way as in NFT.

Figure 10-6 **An effluent water treatment system at Monrovia Nursery Co. in Azusa, Calif. Effluent from plant pots during watering or fertilization is channeled throughout the entire firm to a settling pond, where large particles settle to the bottom. A coagulant is added to aggregate fine particles and these are filtered. Chlorine is added to eliminate disease organisms. After testing, appropriate nutrients and/or water are added to the effluent to restore the desired fertilizer concentration and nutrient balance for subsequent application.**

3. *Reduced production space.* Rock wool is lightweight and self-contained, which allows movement of plants into different environments and densities in different stages for faster crop production and lower overhead cost. The light weight further permits growth of crops on movable benches. Both of these factors reduce the overhead costs of the crop by reducing average growing space.

Whole-Firm Recirculation

Some firms, such as those in California, are faced with a zero tolerance of nitrate in the effluent water from their property. This, in effect, means there can be no runoff. One way of achieving this is to recirculate all water from the firm (Figure 10-6). Whole-firm recirculation is being done commercially (Skimina 1986).

Plants in the greenhouse or the field are grown on a plastic-lined or paved surface. Water or nutrient solution is applied in the conventional manner to the top of the pot, flat, or bench if it is a fresh-flower crop. Leachate passing out of the bottom of the container is caught on the greenhouse floor and flows to a lined ditch. A network of ditches from each growing area carries water to a set of settling ponds. Much of the sediment in the leachate settles out in these ponds. From these ponds, water is pumped to an equalization pond to better establish an average level of fertilizer in the leachate coming from the previous applications of water or fertilizer to the crops. A flocculant, such as alum, is added to cause remaining, suspended solids to flocculate (gather together) so that they can settle out. Clear leachate is drawn from this pond above the sediment and is then injected with chlorine, which pasteurizes it to ensure that disease pathogens are not returned to the crop. Chlorine can also cause iron and manganese to precipitate. The leachate is filtered to remove any remaining solids that might otherwise have clogged small orifices in the watering system. The cleaned leachate is tested for pH level and individual nutrient concentrations. Acid or base is added to adjust the pH level, and water or individual nutrient concentrates are added as indicated by the tests. Since much of the water and some of the nutrients were used by the crop in the previous applications, new nutrient solution is made and added to restore the volume of the recycled leachate. This solution is held in a reservoir, from which it will be drawn when needed for the crop again. If only water is next required on the crop, clean water is used and the leachate from it is captured in the same system as just described.

Summary

1. Current antipollution legislation regarding nutrients and pesticides in greenhouse effluent water and the need to reduce production costs to meet intense international competition are resulting in the development of several alternative crop production systems for greenhouses.
2. NFT (nutrient film technique) is a closed production system in that leachate from root substrate is recycled and does not go to the environment. It is used for vegetable and fresh-flower production. Plants are grown with bare roots in covered, sloped channels. Nutrient solution, containing all essential fertilizer nutrients, is recirculated through the roots in a thin layer about 1/8 inch (3 mm) deep. The nutrient solution must be tested repeatedly and altered for pH level and nutrient concentrations.
3. Rock wool culture can be used for growing vegetables or fresh flowers in either a closed or an open production system. In the open production system, nutrient solution passes over the roots and then is released to the environment for disposal. Rock wool, which bears a resemblance to fiberglass, is made by melting various forms of rock and spinning it into fibers. The fibers are formed into cubes for propagating plants and into slabs for growing plants on to maturity. Nutrient solution containing all essential fertilizer nutrients is passed frequently through the rock wool from top to bottom.
4. Whole-firm recirculation is a closed production system for all greenhouse crops. Plants can be grown in conventional containers and root substrate. Con-

ventional fertilizer solutions are applied to the surface of the substrate. Effluent from the bottom of the pots or beds throughout the entire firm is captured and directed to a single holding pond. The effluent is pasteurized, tested, altered to correct the pH level and fertilizer concentrations, and then reused.

References

1. Bentley, M. 1955. *Growing plants without soil.* Johannesburg, South Africa: Hydro Chemical Industries Ltd.
2. ———. 1959. *Commercial hydroponics.* Johannesburg, South Africa: Bendon Books.
3. Bickart, H. M., and C. H. Connors. 1935. *The greenhouse culture of carnations in sand.* New Jersey Agr. Exp. Sta. Bul. 588.
4. Biggs, T. 1982. Rockwool in horticulture—European experiences. *Australian Hort.* (July):18–21.
5. Carson, R. 1962. *Silent spring.* Boston: Houghton Mifflin.
6. Chapman, H. D., and G. F. Liebig. 1938. Adaptation and use of automatically operated sand-culture equipment. *J. Agr. Res.* 56:73–80.
7. Cooper, A. J. 1973. Rapid crop turn-round is possible with experimental nutrient film technique. *Grower* 79:1048–1052.
8. ———. 1979. *The ABC of NFT.* London: Grower Books.
9. DeStigter, H. C. M. 1961. Translocation of C^{14} photosynthates in the graft muskmelon, Cucurbita ficifolia. *Acta Botanica Neerlandica* 10:466–473.
10. ———. 1969. A versatile irrigation-type water-culture for root-growth studies. *Zeitschrift für Pflanzenphysiologie* 60:289–295.
11. Dungey, N. O. 1983. Assessing the future of NFT and rockwool. *Hort. Now* 12:17–18.
12. Eaton, F. M. 1936. Automatically operated sand-culture equipment. *J. Agr. Res.* 53:433–444.
13. Epstein, E., and B. A. Krantz. 1965. *Growing plants in solution culture.* Univ. of California Agr. Ext. Ser. Bul. 196.
14. Gericke, W. F. 1929. Aquaculture: a means of crop production. *Amer. J. Bot.* 16:862.
15. Hanger, B. 1982. Rockwool in horticulture: a review. *Australian Hort.* (May):7–16.
16. Hewitt, E. J. 1966. *Sand and water culture methods used in the study for plant nutrition.* London: Eastern Press.
17. Hoagland, D. R., and D. I. Arnon. 1950. *The water-culture method for growing plants without soil,* rev. ed. Univ. of California Agr. Exp. Sta. Cir. 347.
18. Hurd, R. G., ed. 1980. Symposium on research on recirculating water culture. *Acta Hort.* No. 98.
19. Ito, T., ed. 1994. Hydroponics and transplant production. Acta Hort. No. 396.
20. Johnson, E. 1984. EPA and pesticides: an interview with Edwin Johnson. *Environmental Protection Agency J.,*10(5):4–8
21. Kiplinger, D. C. 1956. *Growing ornamental greenhouse crops in gravel culture.* Ohio Agr. Exp. Sta. Cir. 92.
22. Knop, W. 1860. Über die Ernährung der Pflanzen durch Wässerige Lösungen bei Ausschluss des Bodens. *Landw. Vers.-Stat.* 2:65.
23. Krause, W. 1983. Rockwool development. *Grower* 99(16):43–44.
24. Laurie, A. 1931. The use of washed sand as a substitute for soil in greenhouse culture. *Proc. Amer. Soc. Hort. Sci.* 28:427–431.

25. Maas, E. F., and R. M. Adamson. 1971. *Soilless culture and commercial greenhouse tomatoes.* Pub. 1460. Canada Dept. of Agr.
26. Marvel, M. E. 1966. *Hydroponic culture of vegetable crops.* Florida Agr. Ext. Ser. Cir. 192-B.
27. Maynard, D. N., and A. V. Barker. 1970. Nutriculture: a guide to the soilless culture of plants. Pub. 41. Massachusetts Coop. Ext. Ser. and USDA.
28. McCall, A. G. 1916. The physiological balance of nutrient solutions for plants in water culture. *Soil Sci.* 2:207–253.
29. Molitor, H. 1990. Irrigation, nutrition, and growth media: the European perspective with emphasis on subirrigation and recirculation of water and nutrients. *Acta Hort.* No. 272.
30. A new twist for hydroponics. 1976. *Amer. Vegetable Grower* (November):21–23.
31. Robbins, W. R. 1928. The possibilities of sand culture for research and commercial work in horticulture. *Proc. Amer. Soc. Hort. Sci.* 25:368–370.
32. Rober, R., ed. 1983. Nutrient film technique and substrates. *Acta Hort.* No. 133.
33. Rupp, L. A., and L. M. Dudley. 1988. Rockwool: how inert is it? *Greenhouse Grower* 6(11):17–18, 20.
34. Sachs, J. von. 1887. *Lectures on the physiology of plants.* Oxford, England: Clarendon Press.
35. Salm-Horstmar, F. 1849. Versuche über die Notwendigen Aschenbestandteile einer Pflanzen-Species. *J. Prakt. Chem.* 1.
36. Serra, G., and F. Tognoni, eds. International symposium on new cultivation systems in the greenhouse. *Acta Hort.* No. 361.
37. Shive, J. W., and W. R. Robbins. 1937. *Methods of growing plants in solution and sand cultures.* New Jersey Agr. Exp. Sta. Bul. 636.
38. Skimina, C. A. 1986. Recycling irrigation runoff on container ornamentals. *HortScience* 21(1):32–34.
39. Sowell, W. F. 1972. *Hydroponics: growing plants without soil.* Auburn Univ. Coop. Ext. Ser. Cir. P-1.
40. Templeman, W. G., and F. J. Watson. 1938. Growing plants without soil by nutrient solution methods. *J. Ministry of Agr.* 45:771–781.
41. Ticquet, C. E. 1952. *Successful gardening without soil.* London: Arthur Peterson.
42. Weinard, F. F., and G. M. Fosler. 1962. *Hydroponics as a hobby.* Univ. of Illinois Agr. Ext. Ser. Cir. 844.
43. Wilkinson, C. F. 1987. The science and politics of pesticides. In Marco, G. C., R. M. Hollingworth, and W. Durham, eds. *Silent spring revisited,* pp. 25–46. Washington, DC: Amer. Chem. Soc.
44. Withrow, R. B., and J. B. Biebel. 1936. A sub-irrigation method of supplying nutrient solutions to plants growing under commercial and experimental conditions. *J. Agr. Res.* 53:693–701.
45. Withrow, R. B., and A. P. Withrow. 1948. *Nutriculture.* Purdue Univ. Agr. Exp. Sta. Sp. Cir. 328.

 CHAPTER 11

Carbon Dioxide Fertilization

Role of Carbon

Carbon is an essential plant nutrient and is present in the plant in greater quantity than any other nutrient. About 40 percent of the dry matter of plants is composed of carbon. Plants obtain carbon from carbon dioxide gas (CO_2) in the air. For the most part, CO_2 gas diffuses through the stomatal openings in leaves when they are open. Once inside the leaf, carbon from CO_2 gas moves into the cells, where, in the presence of energy from the sun, it is used to make carbohydrates (sugars). The carbohydrates are translocated to various parts of the plant and transformed into other compounds needed for growth or maintenance of the plant. The process whereby CO_2 is utilized by the plant is known as *photosynthesis* and occurs in the green chloroplasts within cells. The process is summarized in the following equation:

$$CO_2 + \text{water} + \text{energy from sunlight} \rightarrow \text{carbohydrate} + \text{oxygen} \qquad (11\text{-}1)$$

Air, on the average, contains slightly more than 0.03 percent (300 ppm) CO_2. The average level at the present time is 345 ppm; the level of CO_2 in air outdoors can vary from 200 to 400 ppm. Levels of 400 ppm are common in industrial areas where fuels are combusted. The carbon in fuels is converted to CO_2 during the process of combustion. Due to combustion and deforestation, the level of CO_2 has been increasing since around 1880, when the average level was about 294 ppm. The present rate of increase is 1 to 2 ppm per year. The CO_2 level will also be higher in areas such as swamps and riverbeds, where large quantities of plant material are decomposing. Microorganisms feeding upon plant or animal remains respire CO_2 gas, much as we humans do when we utilize plant- and animal-derived foods. This CO_2 gas is evolved through a process called *respiration,* which is summarized as follows for carbohydrates:

$$\text{carbohydrate} + \text{oxygen} \rightarrow CO_2 + \text{energy} + \text{water} \qquad (11\text{-}2)$$

Respiration is the opposite of photosynthesis. It is a process that releases energy originally captured from sunlight in the process of photosynthesis. The energy released is used by the plant for various functions of growth, such as nutrient uptake.

A CO_2 level of 300 ppm is sufficient to support plant growth as we know it in the world today. Most plants, however, have the capacity to utilize greater concentrations of CO_2 and, in turn, attain more rapid growth. This genetic capability apparently stems back to primordial times, when plants adapted to CO_2 levels 10 to 100 times the level that currently exists.

Carbon Deficiency

In the winter, greenhouses may be closed during the day to conserve heat. This situation may occur for several consecutive days during periods of inclement weather in northern production areas. During daylight hours, CO_2 is removed from the air by plants through the process of photosynthesis. The level continually drops in a closed greenhouse, and the rate of photosynthesis decreases until a point is reached at which growth stops.

It has been reported that an active sunflower leaf can consume the CO_2 in a column of air 8 ft (2.4 m) above it in an hour. Not all crops utilize CO_2 at this rate. However, in a matter of a few hours, the CO_2 level in a closed greenhouse can drop to the compensation point where growth stops. The actual level at which this happens varies for different greenhouse crops, but, in general, it occurs at levels of 50 to 125 ppm CO_2. Carbon deficiency can occur for several days at a time, prolonging the culture time of the crop by the same number of days or reducing the quality of the crop. Deficiency of CO_2 can be even more pronounced inside the plant canopy. Circulation of air in the greenhouse can help to alleviate this problem.

Carbon Dioxide Injection

Effects on Plants

Researchers were surprised to find that plant responses continued to increase as CO_2 levels were raised above the 300-ppm level present in the atmosphere. Levels of 2,000 ppm continued to evoke growth responses in some crops. Apparently, as previously stated, this stems back to the earlier adaptation of plants to higher CO_2 levels in primordial times.

For most greenhouse crops tested over a wide range of geographical latitudes, a response has been reported for increased CO_2 levels up to the range of 1,000 to 1,500 ppm. At levels of 1,500 ppm and greater, the results vary from a positive to a negative response. This is not surprising, since the level of CO_2 required for maximum photosynthesis is related to other factors that can control photosynthesis. Lettuce growing in the winter in England or The Netherlands at 50°N latitude will have a lower potential photosynthetic rate because of lower available sunlight than a crop in Spain or in the southern United States. A higher level of CO_2 will be required to support the higher photosynthetic rate in the brighter region. Negative responses appear to be caused by CO_2 toxicity, which is manifested in lower yield as well as chlorosis (sometimes interveinal) and necrosis of lower leaves. These symptoms resemble those of magnesium and potassium deficiencies. Upper threshold levels of CO_2 are crop specific—for example, 2,200 ppm for tomato, 1,500 for cucumber, and 1,200 for gerbera and chrysanthemum. The commonly injected levels of CO_2 in greenhouses around the world today are 1,000 to 1,500 ppm. This level is not generally considered harmful to people, although much higher levels can have adverse effects. The maximum level tolerated in submarines is 5,000 ppm.

Weight increases in lettuce of 31 percent have been reported from the use of 1,600 ppm CO_2. In other studies, this has translated into a 20 percent earlier harvest. A 48 percent increase in tomato production has been reported as a result of injection of 1,000 ppm CO_2. Injection of 1,000 ppm CO_2 in England has resulted in a 23 percent increase in cucumber fruit weight.

Specific effects from CO_2 injection on rose crops include a decrease in the number of blind shoots (shoots that fail to develop), increased stem length and weight, a greater number of petals, and a shorter cropping time in the winter. A test in Massachusetts (42°N latitude) showed a 53 percent increase in weight of roses cut when 1,000 ppm CO_2 was injected.

Chrysanthemum yields increase in the form of thicker stems and greater height when CO_2 is injected (Figure 11-1). Excessive stem lengths reduce the value of pot mums and, in the case of cut mums, are left behind in the bench when the flowers are cut. The increased height, however, can be translated into a reduction in the length of time required to flower the crop. Because the flower date is controlled by manipulating the length of day, it is possible to program chrysanthemums to flower up to two weeks earlier without a reduction in height when CO_2 is injected. This represents a considerable savings in production time, considering a normal crop time of 12 to 16 weeks.

Carnation yields have been increased up to 38 percent by CO_2 injection. The weight of flowers and strength of stems have been increased, and the time required for shoots to reach flowering has been reduced by as much as two weeks. Equally beneficial effects have been obtained for carnation stock plants. Cuttings of greater quality and number have been produced, and the useful life of the stock plant has been increased as well.

CO_2 injection has caused a variety of beneficial effects on a number of other crops. Fall crops of snapdragons were of better quality, while spring crops were reported to have flowered 13 days early. Rooting of geranium cuttings was im-

Figure 11-1 **Dramatic increases in growth can be achieved by enriching the CO_2 level of the greenhouse atmosphere, as seen in these pot mums.** *(Photo courtesy of R. A. Larson, Department of Horticultural Science, North Carolina State University, Raleigh, NC 27695-7609.)*

proved, and the height as well as the number of branches on subsequent plants was increased. Blindness was decreased in Dutch iris. The number, quality, and size of blooms on orchid plants were increased, as were poinsettia bract diameters. Other crops reported to benefit from CO_2 injection include African violet, *Campanula isophylla, kalanchoe,* and poinsettia.

Crop responses vary according to the extent to which elevated levels of CO_2 can be maintained. CO_2 levels drop appreciably when ventilators are open more than 2 in (5 cm) and even more so when cooling fans are on. Our own studies, where CO_2 was injected only when the ventilators were open less than 2 in (5 cm), demonstrated that it was uneconomical for a number of crops in climatic zone 8 (Raleigh, North Carolina, at 35 °N latitude) and points farther south (Nelson and Larson, 1969). There were too few hours when elevated CO_2 levels could be maintained.

Recent studies in England and The Netherlands have investigated summer injection of CO_2. When ventilators are open more than 5 percent of their capacity or fans are on, the atmosphere is naturally enriched by incoming air to the ambient level of 345 ppm. When ventilators are open less than 5 percent, levels up to 1,000 ppm are set for the CO_2 injection equipment. Results indicate an economic advantage from injecting CO_2 during low-level ventilation. These procedures would work well for cooler locations above 40 °N latitude where greenhouses are frequently opened and closed during the summer months. In warmer climates where fans are used for cooling continuously throughout the summer, closed-loop heating and cooling systems (as described in Chapter 3) may in the future afford an opportunity to inject CO_2 for a greater portion of the summer.

Increased growth stimulated by CO_2 injection has necessitated other changes in the cultural programs of some crops. Growers often fertilize lightly in the winter because slow growth is expected. Enrichment of the atmosphere with CO_2 leads to heavier rates of growth and ultimately a nutrient shortage. If a heavier rate of fertilizer is applied, the growth rate continues to increase in response to applied CO_2. Fertilization should probably be increased by an amount equal to the increase in the rate of growth.

Light is often another limiting factor. When light intensity is low, the rate of photosynthesis is slowed down. Once sufficient CO_2 is added to achieve the maximum rate of photosynthesis at the low light intensity, further additions of CO_2 have no effect. If light intensity is increased by cleaning the glass on the greenhouse or by using supplemental lights during the daytime, higher levels of CO_2 will stimulate further increases in growth. CO_2 can be injected when the light intensity is above 500 fc (5,380 lux) but is not economically feasible at lower intensities.

Heat is another limiting factor. Temperatures established before the era of CO_2 injection are not always adequate, now that the limiting factor of CO_2 has been eliminated. Raising daytime temperatures for crops fertilized with CO_2 has been generally beneficial, while raising nighttime temperatures has not. An increase of as much as 10°F (6°C) has been recommended for roses. Geranium, snapdragon, and chrysanthemum respond well to a 5 to 10°F (3 to 6°C) increase. The increase for carnation should be 5°F (3°C) or less, since this is a cool-temperature crop having a maximum beneficial daytime temperature as found by Holley, Goldsberry, and Juengling (1964) to be 69°F (20°C). Temperatures for vegetables, including tomato, cucumber, lettuce, and pepper, can be raised 5 to 10°F (3 to 6°C).

Today's grower, in injecting CO_2, should be certain that the greenhouse covering is clean enough to ensure the maximum light intensity possible that can be tolerated by the crop. The grower should experiment with raising the daytime temperature 5 to 10°F (3 to 6°C) as well as increasing the fertilization rate, and, in general, should ensure that all cultural procedures are practiced in a manner that promotes optimal growth.

Method of Carbon Dioxide Injection

Since CO_2 injection is effective only during the daylight hours when photosynthesis occurs, it should be injected from sunrise until one hour before sunset. It should be injected only when the ventilation fans are off or, in the case of greenhouses cooled by ventilators, when the roof ventilators are open less than 2 in (5 cm). CO_2 cannot be injected during the warm seasons, because greenhouse cooling generally coincides with daylight hours. Depending on the latitude where

Figure 11-2 **A CO_2 generator used for enriching the greenhouse atmosphere with CO_2 for the purpose of increasing photosynthesis and growth.** ***(Photo courtesy of Johnson Gas Appliance Co., Cedar Rapids, IA 52405.)***

the greenhouse is located, the season for CO_2 injection will begin between late September and early November and extend to April or early May.

A few brands of CO_2 generators are popular today (Figure 11-2). One unit sells for about $475 with a gas pressure gauge and a 24-volt solenoid valve. This unit provides 1,500 ppm CO_2 in a typical greenhouse of 5,000 ft^2 (465 m^2) floor area. It burns LP or natural gas. This generator has a burner range up to 60,000 Btu/hr (15,120 kcal/hr or 17,580 W); thus, it can consume 60 ft^3 (1.7 m^3) of natural gas per hour. A control package that automatically turns up to three units on at sunrise and off at sunset is available for $125. Another popular brand can be obtained in various models that burn either natural gas, propane, or kerosene. The larger model with a natural gas burner can provide up to 1,200 ppm CO_2 in a typical greenhouse of 25,000 ft^2 (2,323 m^2) floor area and sells for about $2,000.

CO_2 generators are hung above head height along the center of the greenhouse. Within each is a precisely calibrated burner with an open flame. Under conditions of complete combustion, gas is converted to CO_2 and water. CO_2 produced in some units rises out of the burner into the greenhouse atmosphere, where convection currents move the gas about the greenhouse. In other products, the CO_2 is positively displaced from the burner by a fan.

Gas consumed in the CO_2 generator must be of a high purity level, since any sulfur contained in it is converted to sulfur dioxide gas. When sulfur dioxide comes in contact with moisture on plant surfaces, it is converted to sulfurous acid and eventually to sulfuric acid. This burns the plant (see Figure 3-2). The sulfur content of natural gas and propane should not exceed 0.02 percent by weight (Blom, Straver, and Ingratta 1984), while the sulfur content of kerosene should not exceed 0.06 percent (Hand 1971).

Incomplete combustion will cause the formation of ethylene and carbon monoxide gases, which are injurious to a plant (see Figure 3-1). Internodes on the plant become shortened, branching increases, and flowers become distorted and injured from ethylene. The upper limit of ethylene for plants is 0.05 ppm. Carbon monoxide is harmful to humans. The upper average limit of carbon monoxide for humans is 50 ppm (American Conference of Government Industrial Hygienists 1986). Therefore, only a burner designed for CO_2 production should be used inside a greenhouse, and it should be periodically calibrated. The burner should be kept clean and adjusted to a clear blue flame. The plumbing should be checked for gas leaks, since unburned fuel may be injurious to plants. Manufactured gases, and to a degree natural gas, can contain propylene and butylene, which are injurious to plants, causing symptoms similar to ethylene. The threshold for propylene, above which plant injury occurs, is 10 ppm (Hicklenton 1988).

It is equally important to provide sufficient oxygen to support complete combustion of the fuel. In a film plastic greenhouse or a glass greenhouse located where it is prone to ice formation on the surface, an air inlet must be provided. The rule for heating systems applies here, where 1 in^2 of opening is provided per 2,500 Btu of burner capacity per hour (1 cm^2 per 100 kcal/hr or per 733 W).

Considerable progress has been made since injection of CO_2 was commercialized in the early 1960s. The concept actually dates back to the earlier part of this century, but it was not until commercial methods of application were available that extensive efforts were put forth to develop a system. Early research by Professor Holley at Colorado State University in the late 1950s, as well as work in The Netherlands and England, led to commercial systems using liquid CO_2 or dry ice, which is solid (frozen) CO_2. (See Hicklenton 1988 for historical details.)

CO_2 gas under pressure becomes liquid. At a low temperature, it can be solidified into dry ice. In the early 1960s, liquid CO_2 tanks were installed at a greenhouse range and were serviced by CO_2 distributors. CO_2 gas formed above the liquid in these tanks and was carried by metal tubing to the greenhouses. A set of pressure-regulating valves reduced the pressure to a low level. Once in the greenhouse, the gas was distributed along the length of the greenhouse in a plastic tube 1/8 to 1/4 inch (3 to 6 mm) in diameter with needle holes every 12 in (30 cm) along the length.

Liquid and solid CO_2 proved a more expensive source of CO_2 in the 1960s than the combustion of fuels. Burners were developed. Some early equipment was large and had to be located outside the greenhouse. The exhaust, essentially pure CO_2, was brought into the greenhouse through a duct and distributed along the length of the greenhouse through the conventional winter convection-tube ventilation system (as described in Chapter 4).

With time, smaller generators were developed that were installed overhead in the greenhouse. These produced CO_2 in open-flame burners using kerosene, propane, or natural gas. Being simpler systems, they cost less to purchase. Depending upon the equipment purchased, they could handle greenhouse areas of 5,000 to 15,000 ft^2 (365 to 1,400 m^2). This generation of CO_2 generators is in common use today in the United States.

Partly because of recent shifts in fuel costs, larger greenhouse firms find the cost of CO_2 from liquid CO_2 comparable to that of CO_2 generated from the combustion of fuel. A large consumption volume is necessary in order to negotiate an economical, steady source of liquid CO_2 from the supplier. Liquid CO_2 has the advantage of purity. Unlike CO_2 generation in burners, the use of liquid CO_2 releases no heat, which can be a disadvantage during the winter and an advantage at the beginning and end of the CO_2 injection season.

Measurement and Control of Carbon Dioxide Levels

The simpler forms of CO_2 injection control use either a time clock or a light sensor to turn the CO_2 generator on in the morning and off in the evening. During the day, the CO_2 generator is automatically turned off when the ventilating fans come on. In the event of roof ventilation, mechanical switches are installed on the ventilators to allow the CO_2 generator to operate only when the vents are open less than 2 in (5 cm).

When controlled by simple on-off switches, CO_2 generators will not have the same net effect in all greenhouses. The CO_2 level will be lower in glass greenhouses, where air leaks exist, than in film plastic greenhouses. In order to adjust the fuel pressure on some generators to compensate for this variation, it is important to know what level is being maintained in the greenhouse atmosphere. Simple CO_2 testers can be purchased for \$235 to \$325. These consist of a small hand pump that is stroked a given number of times to pass air through a tube. The tube contains a CO_2-sensitive chemical that changes color as CO_2 is absorbed. The length of the tube that changes color is measured on a scale that directly indicates the level of CO_2 in the air passed through the tube. The tubes are disposable and sell for about \$45 for a box of 10. (Testers are available from Fisher Scientific Co., 711 Forbes Ave., Pittsburgh, PA 15219, and from Hydro Gardens, Inc., P.O. Box 9707, Colorado Springs, CO 80932).

The current, more sophisticated generation of CO_2-control systems is based on CO_2 sensors. These sensors continually monitor the CO_2 level in the greenhouse. The signal from the sensor is used to control the CO_2 generator so that a constant CO_2 level can be maintained. CO_2 sensors are priced in the range of $750 to $1,500. A sensor can be connected to several greenhouses by sampling tubes, through which air is drawn by a pump. Typically, air is sampled and tested for one minute from each greenhouse. Information from the single sensor is received by a computer, which in turn controls CO_2 generators in each greenhouse.

Current computer systems can program the daily on and off times to coordinate with the changing day length of the season. This ensures that the full advantage of CO_2 enrichment is realized and that no unneeded CO_2 is used. It is a simple matter for these systems to further coordinate injection of CO_2 with ventilation events. With a computer system it is possible to have the greenhouse CO_2 concentration vary in accordance with the intensity of light in the greenhouse. In this way, CO_2 is conserved during times of low light incidence, yet on bright days the full potential of CO_2 can be achieved. A computer can effectively integrate light intensity, CO_2 level, and temperature. This will ensure that daytime temperatures are raised in proportion to the level of CO_2 being injected. A computerized CO_2 system makes it possible to have a continually adjusted optimum balance of light intensity, temperature, and CO_2 level. This will result in conservation of CO_2 when no enrichment is needed or during periods when a lower-than-standard level of enrichment is required. On the other hand, it will ensure that yield, as a function of CO_2 enrichment, is always optimized. Future growers will have computer systems that bring even more growth inputs into the CO_2 equation. These systems might also base the decision of which inputs and the level of each to apply on the economic return expected. This technology will bring crops closer to their photosynthetic potential than is possible today. It will also result in increased profit since heat, CO_2, and other inputs will be applied only in accordance with projected profitability.

Economics of Carbon Dioxide Injection

In cool climates, it is common to inject CO_2 for an average of 5 hours per day over a six-month period (900 hours). At an output of 60,000 Btu/hr (15,120 kcal/hr or 6.3 MJ/hr), this amounts to 54 million Btu or 540 therms (100,000 Btu/therm) of natural gas per burner. Since one burner handles about 5,000 ft^2 (464.7 m^2) of greenhouse, the natural-gas consumption is 0.108 therm/ft^2 (29,300 kcal or 11.4 MJ/m^2) of greenhouse area per year. At $0.60 per therm, this carries a fuel cost of $.065/$ft^2$ ($0.70/$m^2$) of greenhouse area per year. Depreciation on the equipment adds very little to the cost of injecting CO_2. The higher yields and shorter production time of crops more than justify the costs.

Professor Koths at the University of Connecticut points out an additional benefit from CO_2 injection. Many crops are grown at a 5°F (3°C) warmer daytime temperature when CO_2 is injected. The greenhouse acts as a solar collector. Heat is stored in the greenhouse structure, soil, plants, and benches. At night, the extra heat resulting from the higher daytime temperature is released. This conserves heating fuel and thereby pays part of the cost of CO_2 injection.

Summary

1. Carbon is an essential plant nutrient and is supplied as CO_2 gas in the atmosphere. A concentration of about 345 ppm is present in the atmosphere.
2. CO_2 is used during daylight hours in the process of photosynthesis. When the greenhouse is closed on cold winter days, the CO_2 concentration in the air inside the greenhouse can be lowered in a few hours to a level where the rate of carbohydrate produced in photosynthesis equals the rate of carbohydrate breakdown through respiration. Net growth ceases at this point, delaying the crop or reducing quality.
3. CO_2 is often added to the greenhouse atmosphere during daylight hours of months when the greenhouse is not continuously ventilated. Common methods of addition today are either (1) the burning of kerosene, LP gas, or natural gas in special burners inside the greenhouse, or (2) the release of gas from liquid CO_2.
4. Reestablishment of a normal level of CO_2 (about 350 ppm) results in dramatic growth responses. Interestingly, further increases in CO_2 concentration up to 2,000 ppm or higher induce even greater growth responses. Concentrations of 1,000 to 1,500 ppm are the levels generally established in the greenhouse.

References

1. Aldrich, R. A., and J. W. Bartok, Jr. 1994. *Greenhouse engineering.* Pub. NRAES-33. Northeast Reg. Agr. Eng. Ser., Cornell Univ., 152 Riley-Robb Hall, Ithaca, NY 14853.
2. American Conference of Government Industrial Hygienists. 1986. Carbon dioxide: documentation of the threshold limit values and biological exposure indices. *Amer. Conf. Govt. Industrial Hygienists,* pp. 102–103. Cincinnati, OH.
3. Bauerle, W. L., and T. H. Short. 1984. Carbon dioxide depletion effects in energy efficient greenhouses. In Short, T. H., ed. Energy in protected cultivation III. *Acta Hort.* No. 148.
4. Blom, T., W. Straver, and F. J. Ingratta. 1984. *Using carbon dioxide in greenhouses.* Ontario Ministry of Agr. and Food, Factsheet 290-27.
5. Freeman, R. 1991. The importance of carbon dioxide. In Ball, V., ed. *The Ball Red Book,* 15th ed., pp. 229–236. West Chicago, IL: Geo. J. Ball.
6. Gaastra, P. 1966. Some physiological aspects of CO_2 application in glasshouse culture. In Hardh, J. E., ed. Symposium on vegetable growing under glass. *Acta Hort.* 4:111–116.
7. Hand, D. W. 1971. CO_2 and hydrocarbon fuels. *ADAS Qtr. Rev.* 1:18–23.
8. Hicklenton, P. R. 1988. *Grower handbook series.* Vol. 2, *CO_2 enrichment in the greenhouse.* Portland, OR: Timber Press.

9. Hicklenton, P. R., and P. A. Jolliffe. 1978. Effects of greenhouse CO_2 enrichment on the yield and photosynthetic physiology of tomato plants. *Canadian J. Plant Sci.* 58:801–817.
10. Holley, W. D. 1975. The CO_2 story. In Ball, V., ed. *The Ball Red Book,* 13th ed., pp. 156–159. West Chicago, IL: Geo. J. Ball.
11. Holley, W. D., K. L. Goldsberry, and C. Juengling. 1964. Effects of CO_2 concentration and temperature on carnations. *Colorado Flower Growers Association Bul.* 174:1–5.
12. Mastalerz, J. W. 1969. Environmental factors: light, temperature, carbon dioxide. In Mastalerz, J. W., and R. W. Langhans, eds. *Roses: a manual on the culture, management, diseases, insects, economics and breeding of greenhouse roses,* pp. 95–108. Pennsylvania Flower Growers' Association, New York State Flower Growers' Association, Inc., and Roses, Inc.
13. Nelson, P. V., and R. A. Larson. 1969. *The effects of increased CO_2 concentration on chrysanthemum and snapdragon.* North Carolina Agr. Exp. Sta. Tech. Bul. 194.
14. Shaw, R. J., and M. N. Rogers. 1964. Interaction between elevated carbon dioxide levels and greenhouse temperatures on the growth of roses, chrysanthemums, carnations, geraniums, snapdragons, and African violets. *Florists' Rev.* 135(3486):23–24, 88–89; (3487):21–22, 82; (3488):73–74, 95–96; (3499):21, 59–60; (3491):19, 37–39.
15. Wittwer, S. H. 1966. Carbon dioxide and its role in plant growth. *Proc. 17th International Hort. Cong.* 3:311–322.
16. Wittwer, S. H., and W. M. Robb. 1964. Carbon dioxide enrichment of greenhouse atmospheres for food crop production. *Economic Bot.* 18:34–56.

CHAPTER 12

Light and Temperature

Light Intensity for Photosynthesis

Photosynthesis

Visible light constitutes a source of energy for plants. Light energy, carbon dioxide (CO_2), and water all enter into the process of photosynthesis through which carbohydrates are formed:

$$CO_2 + \text{water} + \text{light energy} \rightarrow \text{carbohydrate} + \text{oxygen} \quad (12\text{-}1)$$

Considerable energy is required to reduce the carbon that is combined with oxygen in CO_2 gas to the state in which it exists in the carbohydrate. The light energy thus utilized is trapped in the carbohydrate. Later, the carbohydrate can be translocated (moved) from the green stem and leaf cells, where photosynthesis occurs, to all other parts of the plant. The carbohydrate can be converted into all other compounds needed in the plant. Amino acids may be formed and then combined into protein chains. Fats may be formed from carbohydrates. From all these compounds, other compounds arise, such as cellulose for cell walls, pectin to cement the walls together, hormones to regulate growth, and DNA to constitute chromosomes. The sun's energy is passed along in all of these compounds. These processes result in growth of the plant, which can be detected as an increase in dry matter.

Energy must be liberated at times to power other processes in the plant. The uptake of nutrients, formation of proteins, division of cells, maintenance of membranes, and several other processes require an input of energy. This energy is obtained when compounds formed as a direct or indirect result of photosynthesis are broken down in very much the reverse process of photosynthesis. This is the process of respiration:

$$\text{carbohydrate} + \text{oxygen} \rightarrow CO_2 + \text{water} + \text{energy} \quad (12\text{-}2)$$

Respiration occurs in all living organisms at all times. It is temperature dependent, increasing with increases in temperature. When animals eat plants, they obtain energy from the compounds they ingest. This energy was originally derived from light through photosynthesis. It can be released from these compounds by the animals through respiration. The same holds true for humans when we eat animal or plant tissue. Thus, we see that most living organisms are ultimately dependent upon light energy.

When all factors such as CO_2 level, temperature, and water are optimized for photosynthesis, an optimum light intensity can be determined. If the light intensity is diminished, photosynthesis (and growth) slows down. If higher-than-optimal light intensities are provided, growth again slows down because the chloroplasts are injured. Chloroplasts are the organelles within green cells in which photosynthesis occurs.

Greenhouse crops are subjected to light intensities as high as 12,000 fc (129.6 klux) on clear summer days to below 300 fc (3.2 klux) on cloudy winter days. For most crops, neither condition is ideal. Many crops become light-saturated (that is, photosynthesis does not increase at higher light intensities) at about 3,000 fc (32.3 klux). Of course, this is assuming that all leaves are exposed to an intensity of 3,000 fc (32.3 klux), which is rarely the case. Upper leaves cast shadows on lower leaves, thus reducing the light intensity at the lower leaves. As illustrated in Figure 12-1, an individual leaf at the top of the plant may saturate at

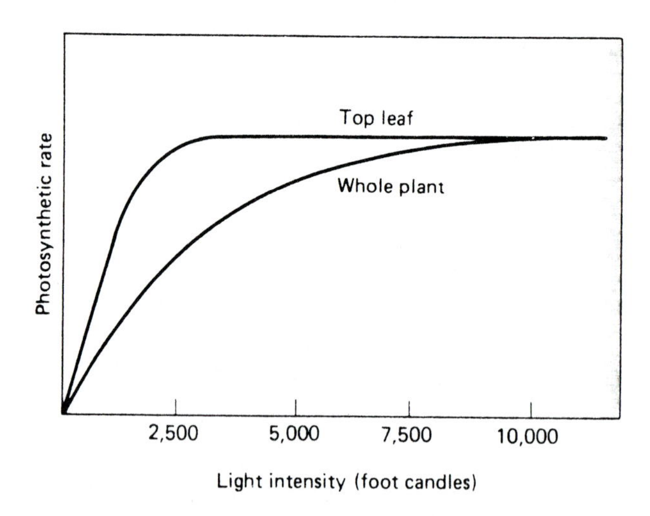

Figure 12-1 **The effect of light intensity on the rate of photosynthesis of a single leaf at the top of a plant and of the whole plant. While the single leaf reaches its maximum rate of photosynthesis at 3,000 fc (32.3 klux), an intensity of 10,000 fc (108 klux) might be required for the whole plant in order to raise the light intensity within the leaf canopy to 3,000 fc (32.3 klux).**

3,000 fc (32.3 klux), while the plant as a whole may not reach light saturation until 10,000 fc (108.0 klux).

Rose and carnation plants will grow well under full summer light intensities. Poinsettia foliage is deeper green if the greenhouse is shaded to the extent of about 40 percent from midspring to midfall. This is typical of most crops. In addition to shading crops to prevent chloroplast suppression, crops such as chrysanthemum and geranium are shaded to prevent petal burn; high light intensity is believed to raise the temperature of the petal tissue to an injurious level. Other crops require even more shading. Foliage plants are burned at light intensities over 2,000 to 3,000 fc (21.5 to 32.3 klux), and African violet loses chlorophyll at intensities of 1,500 fc (16.1 klux) and higher. The optimum light intensity for African violets is near 1,000 fc (10.8 klux). As will be seen later in this chapter, African violet, gloxinia, many foliage plants, and annual seedlings can be grown quite satisfactorily in growth rooms at a light intensity of 600 fc (6.5 klux). Thus, it is apparent that light-intensity requirements of photosynthesis vary considerably from crop to crop.

Light Quality

Not all light is useful in photosynthesis. Light is classified according to its wavelength in nanometers (nm). This classification is referred to as *quality*. Ultraviolet (UV) light has short wavelengths, less than 400 nm (Figure 12-2). For the most part, UV light cannot be seen by the human eye. In large quantities, it is harmful to plants. Glass screens out most UV light and all light below a wavelength of 325 nm. Visible, or white, light occurs between the wavelengths of 400 to 700 nm. At the shortest wavelength, visible light appears violet. Blue, green, yellow, orange, and red light occur around wavelengths of 460, 510, 570, 610, and 650 nm, respectively. "Far-red" light (700 to 750 nm) occurs at the limit of our visual perception and has an influence on plants other than through photosynthesis. Infrared energy occurs at longer wavelengths and is not involved in plant processes.

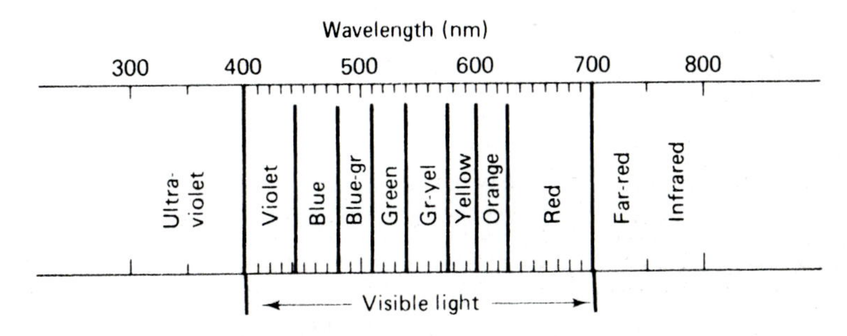

Figure 12-2 Types of radiant energy having wavelengths of 300 to 800 nm. Visible light is in the range of 400 to 700 nm.

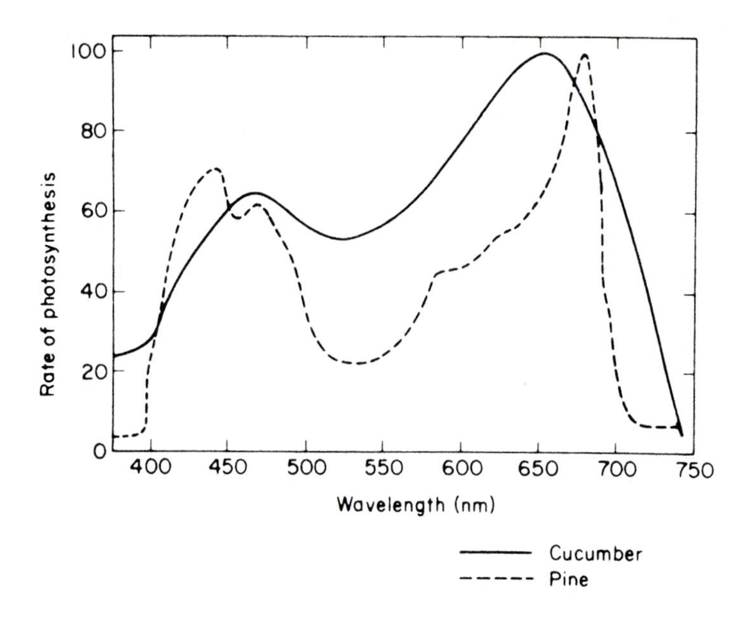

Figure 12-3 **Rates of photosynthetic activity occurring under different qualities of light between the ultraviolet wavelength of 350 nm and the far-red wavelength of 750 nm.** *(From Electricity Council [1972].)*

It is primarily the visible spectrum of light that is used in photosynthesis (Figure 12-3). There are peaks in the blue and red bands where photosynthetic activity is higher. When blue light alone is supplied to plants, growth is shortened, hard, and dark in color. When plants are grown in red light, growth is soft and internodes are long, resulting in tall plants. Figure 12-3 clearly shows that all visible light qualities (wavelengths) are readily utilized in photosynthesis.

Maximizing Light Intensity

It is important to ensure the highest light intensity possible during the dark portion of the year, from midfall through early spring, for all crops except the low-light group already mentioned. In this way, growth is maximized.

Greenhouse Design Maximization of light begins in the planning stage of the greenhouse range. The simpler the frame and the farther apart the sash bars, the greater the light intensity inside. A very significant stride forward was made when all-metal greenhouses were popularized in the 1950s. Because of the strength of the metal members of these greenhouses, fewer sash bars were required to support the heavy weight of the glass. When glass widths increased from 16 to 24 in (40 to 61 cm), the number of sash bars was reduced by one-third. Today, glass widths of 3 ft and even 6 ft (0.9 m and 1.8 m) reduce the number of sash bars to about one-quarter of the original number.

Frame simplicity is particularly important in film plastic greenhouses. The wooden-frame plastic greenhouses (which have almost entirely passed out of the picture now) required very massive frames that greatly reduced interior light intensity. It is very important to keep the wood of such structures painted white so that it reflects light into the greenhouse rather than absorbing it. The same is true of wooden sash bars on glass greenhouses. They should be painted every other year on the outside and about every five years (or as needed) on the inside. The pipe-frame Quonset and all-metal gutter-connected plastic greenhouses, with their minimal frames, are very good in terms of maximizing light intensity.

The importance of greenhouse design can be seen in light-transmission figures presented by Professor Holley of Colorado State University. The frame blocks 10 percent of the sunlight, the sash bars another 5 percent, and the glass another 7 percent. Actually, the 78 percent light-transmission figure for this overall greenhouse is unusually high. Figures near 65 percent would not be uncommon. Other factors that further decrease the transmission level are (1) overhead equipment such as automatic shading, heating and cooling systems, plumbing, and plant supports, and (2) the geographical orientation of the greenhouse, which was discussed in Chapter 2.

The covering material is another consideration. Standard float glass transmits 88 percent of light impinging on it, while low-iron glass transmits up to 92 percent. The moderately higher price of low-iron glass is justified when high-light-requiring crops, such as rose, are grown. The double-layer polyethylene covering is very popular because of the 40 percent savings in heating energy compared to single coverings of glass or polyethylene. However, similar savings can be achieved in a single-layer glass greenhouse when a thermal screen is used within it at night. The single glass layer transmits 88 percent of light, while the double-layer polyethylene cover transmits only 76 percent of light.

Clean Glass Many greenhouses are shaded during the summer to reduce light intensity. A residue of shade may still remain in winter. In addition, dust accumulates on the glass. These deposits reduce light intensity; 20 percent reductions commonly occur. Dirty glass should be washed as the dark season approaches (usually in October or November in the Northern Hemisphere). Commercial glass-cleaning products are available through greenhouse-supply companies. Some can be used on glass, rigid plastic, and film plastic. One make-your-own formula calls for dissolving 11 lb of oxalic acid in 33 gal of water. The greenhouse is sprayed with this solution when the greenhouse is damp. A good time to spray is in the morning after a heavy dew or after a light shower. If the weather is dry, the greenhouse should be hosed down first. The solution should remain on the glass for three days, after which it can be rinsed off or the rain can be allowed to remove it.

FRP greenhouses need cleaning as well. A household detergent can be applied with a sponge or rag at the end of a pole. Commercial materials are also available. Sometimes, the inside of the glass becomes dirty as well. The formulation given here can be sprayed inside the glass greenhouse and hosed off, providing it does not contact the plants. Benches containing plants should be covered with plastic film during cleaning.

Plant Spacing Plants tend to proliferate within a bench until the available light energy is fully utilized. In other words, an equal amount of dry matter will be produced in a bench whether plants such as chrysanthemum are spaced on 5- or 7-inch (13- or 18-cm) centers. In the former case, smaller stems and blooms are produced. The size and quality of product desired will dictate the proper plant spacing.

Generally, a greater amount of space per plant is provided in the winter than in the summer because of less available light. Catalogs provided by suppliers of plant material will indicate the proper spacing for various seasons of the year. It is best to follow their recommendations.

Some growers of fresh flowers have found it best to leave an open space along the center of the bench from end to end, as pictured in Figure 12-4. This permits

Figure 12-4 A winter planting arrangement for chrysanthemum that allows for a space along the center of the bench to increase light intensity at that point with the resultant effect of improved quality.

light to enter the center of the bench where it would normally be darkest. There is a resultant increase in overall quality. The same number of plants are used in a bench in this system. They are simply spaced closer together to compensate for the open space in the center.

Reducing Light Intensity

The need for reducing light intensity from mid spring to early fall has already been pointed out. This may be accomplished in two ways: (1) by spraying a shading compound on the greenhouse, or (2) by installing a screen fabric over the greenhouse or inside the greenhouse above head height.

When the entire greenhouse range needs shading, some growers use the spray method because it is less expensive. Commercial shading compounds can be purchased from florist-supply companies or can be made on the premises by mixing white latex paint with water. One part paint in 10 parts water provides a very heavy shade, while 1 part paint in 15 to 20 parts water provides a standard shade. The shading compound can be sprayed on from the ground by means of a pesticide sprayer. In some large operations, it is sprayed on from the air by a helicopter. Most of the shading compound will wear off by early fall. If it does not, it needs to be washed off.

When shade is desired only to protect flowers, sheets of screening are sometimes used only where they are needed. Chrysanthemum and geranium may be grown at full light intensity in northern areas, but the flowers must be protected from sunburn. Cheesecloth was commonly used years ago and is still used when it affords a price advantage. Longer-lasting synthetic fabrics are more popular today, including such materials as polypropylene, polyester, saran, and aluminum-coated polyester. The former three can be purchased in different densities of weave providing many shade values from 20 to 90 percent, although 50 percent is commonly used. Aluminized polyester sheets are constructed from thin strips of clear polyester and aluminum-coated polyester sewn together. The ratio of coated to clear strips determines the degree of shading. Clear polyester plastic is good at keeping radiant heat out during a bright summer day as well as keeping heat in during a cold winter night. The aluminum coating reflects this form of heat and thus substantially improves this barrier.

The problem with spraying shading on the outside of the greenhouse or installing a fixed sunscreen in the greenhouse is that the barrier is still in place when the light intensity is low on cloudy days and early in the morning or late in the afternoon on all days. Thus, periods of inadequate light intensity occur, causing reduced growth and delayed crops.

Modern greenhouses have automated equipment to draw sunscreens across the greenhouse in response to photocells. In this way, screening is applied only during the hours when it is needed. If the sunscreens are judiciously selected, they can also serve as thermal blankets for retaining heat on winter nights. The same apparatus used for drawing the screens in the daytime during the bright months can be used for drawing them during the nights in the winter. Manual operation of sunscreens would require constant vigilance seven days a week, which is out of the question. Photocells can perform this task, just as time clocks can perform the task of drawing thermal blankets on winter nights. Computers make the job even easier. They can be preprogrammed long in advance and for a large number of different crop zones, each with its own separate events and dates.

Supplemental Lighting

During the dark seasons of the year, light intensity is below optimum for most crops in most greenhouse production areas of the world. While the previously discussed methods for maximizing light intensity help, they do not completely solve the problem. The growth of affected crops is erratically slow because of day-to-day variations in weather conditions, and final quality is reduced. There is an increase in blindness (failure of shoots to develop) on crops such as rose and orchid. The size (grade) of fresh flowers is smaller, stems are thinner, and plants can be adversely tall. This situation can be rectified by using supplemental lighting in the greenhouse to increase the rate of photosynthesis.

Lamp Types Many types of lamps have been used in the greenhouse. Basically, they fall into three groups: (1) incandescent, (2) fluorescent, and (3) high-intensity-discharge (high-pressure mercury, metal halide, low-pressure sodium, and high-pressure sodium). Light emissions typical of each type can be seen in Figure 12-5.

Incandescent (tungsten-filament) lamps (Figure 12-5a) are generally not used for supplemental lighting because of excessive heat, poor light quality, and low efficiency. In order to avoid excess heat, the light intensity must be kept too low. For most plants, the high proportion of red and far-red light emitted causes tall, soft growth and other changes in plant form. These lamps are also very inefficient, converting only 7 percent of the electrical energy consumed into light energy. Much of the energy is converted into heat, which at times is of value and at other times is a detriment. While incandescent lamps are of little value for supplemental lighting, they are the lamp of choice for photoperiodic lighting (which will be discussed later in this chapter). For this application, a low-intensity light (10 fc, 108 lux) is applied for a short duration during the middle of the night in the winter.

Fluorescent lamps are the most common lamps used in growing rooms and over small germination areas in the greenhouse. (This application will be discussed later in the "Growth Rooms" section.) They are rarely used for finishing crops. In separate trials, Bickford and Dunn (1972) and J. W. Mastalerz (1969) increased the yield of roses using Gro-Lux (plant growth A) fluorescent lamps. Best results were obtained by placing lamps without reflectors between the plants. These and other commercial trials have shown positive results on the crops, but the economics are questionable because of the large number of lamps

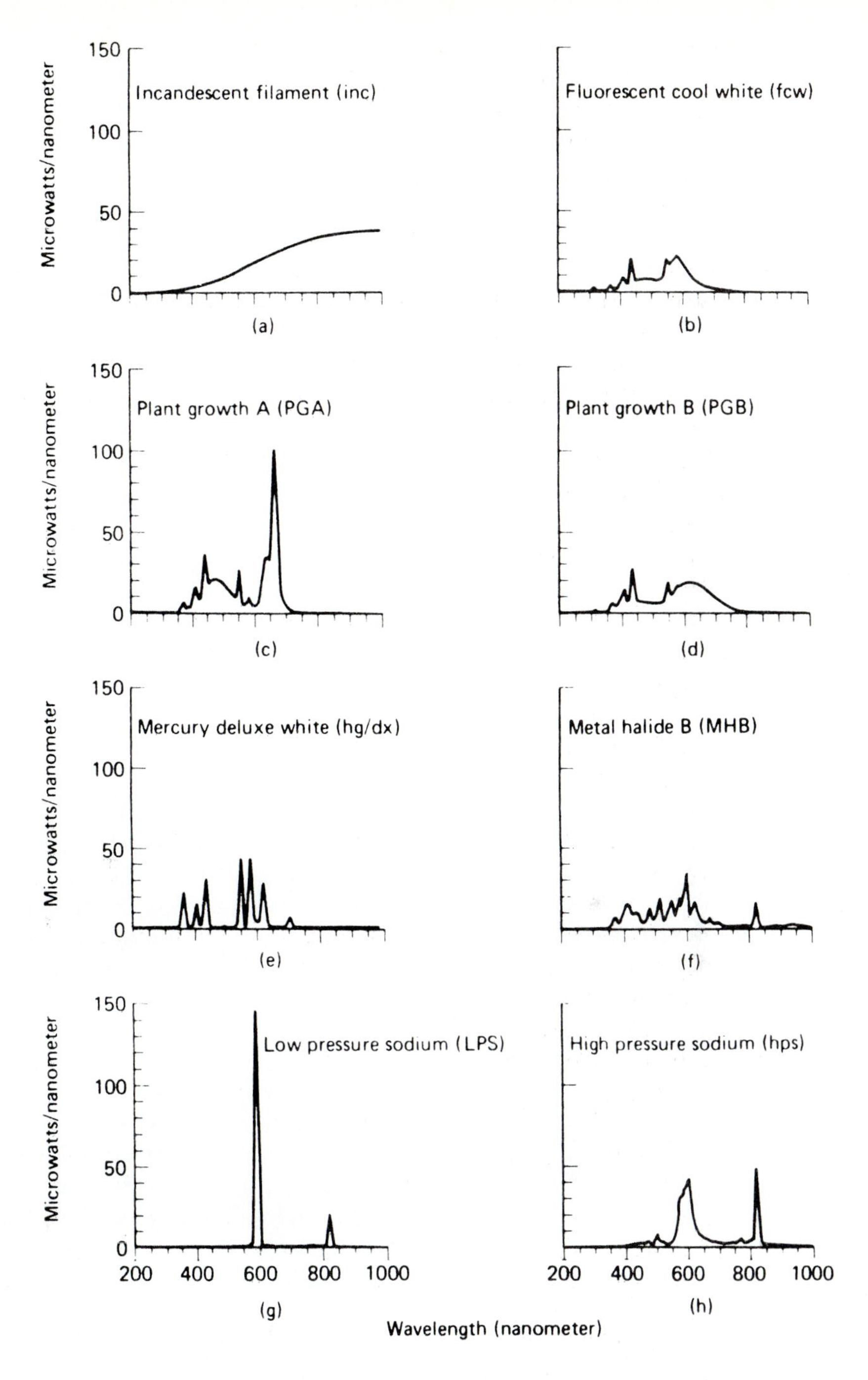

Figure 12-5 **The spectrum of light emissions measured as radiant power per lumen from eight types of lamps considered for use in greenhouses.** *(From Campbell, Thimijan, and Cathey [1975].)*

required. The low power (wattage) of the lamps increases the number needed and, consequently, the total cost of fixtures and wiring required to do the job; also, the larger number of fixtures increases the area of shadows cast on the crop.

Among the more efficient of the fluorescent lamps are the cool white and warm white tube types. These lamps convert 20 percent of electrical energy consumed by them and the ballast to visible-light energy and have similar spectral-light emissions. Cool white lamps (Figure 12-5b) are perhaps the most commonly used fluorescent lamps for plant growth. Light emitted tends to predominate in the blue region. A number of other fluorescent lamps with special phosphors are used for emitting a spectrum of wavelengths more in line with the requirements of photosynthesis. These are categorized into two groups: plant growth A (Figure 12-5c) includes the earlier lamps with enhanced radiation in the red range, while plant growth B (Figure 12-5d) includes the later generation of lamps with extended spectral emission beyond 700 nm.

High-intensity-discharge (HID) lamps are the preferred lamps for finishing crops in greenhouses today (Figure 12-6). High-pressure mercury-type HID lamps were once more common in Europe than in the United States. These lamps have been mostly replaced by metal-halide lamps and, more recently by high-pressure sodium-type HID lamps. Light emissions from high-pressure mercury lamps are somewhat similar to those from fluorescent tubes. Model MBFR/U (formerly popular in Europe) has fluorescent powder on the inner surface of the

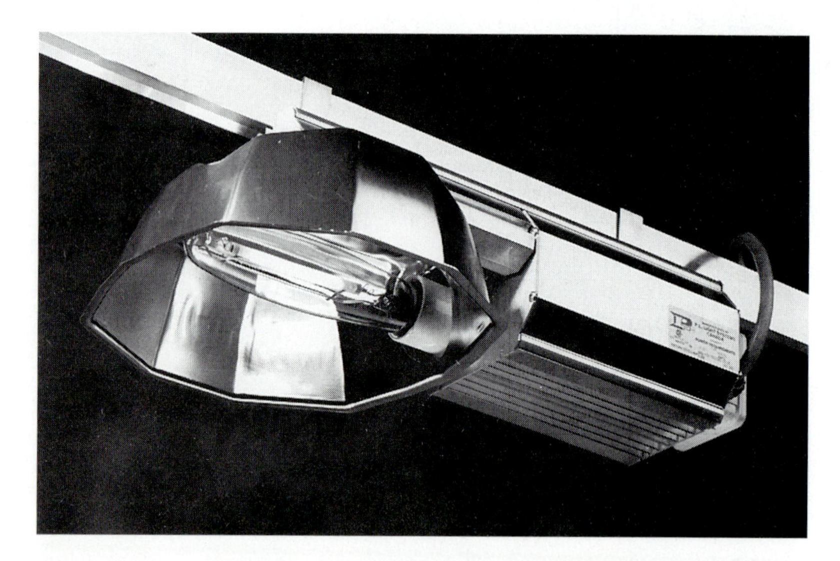

Figure 12-6 A high pressure sodium light assembly including lamp, reflector, and ballast designed exclusively for greenhouse installation. Unlike the light assemblies sold for other commercial applications the reflector pictured above maximizes uniformity of light intensity across the plant zone. (*Photo courtesy of P.L. Light Systems Canada Inc., P.O. Box 206, Grimsby, Ontario, Canada L3M 4G3.*)

glass bulb, which converts much UV light to visible wavelengths, particularly red. This feature makes the lamp more desirable for plant growth and increases its efficiency to 13 percent of the electrical energy input into the lamp and ballast. Similar American lamps are the Mercury Clear and Mercury Deluxe White models (Figure 12-5e), which are often seen along roadways. These lamps are available in sizes up to 1,000 W. They have been used for up to 10,000 hours, at which point they still had 70 percent of their original output.

High-pressure metal halide–type HID lamps (Figure 12-5f) were at one time also more commonly used in Europe than in the United States. These lamps are available in sizes up to 2,000 W and can convert 20 percent of total electrical input energy into light in the 400-to-700-nm band. These lamps cost more than high-pressure mercury lamps, have a shorter life, and lose their output level faster, but they more efficiently utilize electricity.

Low-pressure sodium (LPS) -type HID lamps (Figure 12-5g), available in 35-, 55-, 135-, and 180-W sizes, were popular for a while. As refinements came along in the metal-halide and high-pressure sodium lamps, LPS lamps lost popularity in the greenhouse. LPS lamps are the most efficient, with 27 percent of the electrical input into the lamp and ballast being converted to visible radiation. These lamps have a life expectancy of 18,000 hours.

LPS lamps emit most of their light in a narrow band around 589 nm. Because little light is emitted in the 700-to-850-nm range, there are adverse effects on some crops when these lamps are the only source of light. In tests of African violet, lettuce, and petunia, grown under LPS lamps only, the plants developed pale green foliage. Lettuce plants were smaller in comparison to plants under both LPS and incandescent lamps. If 10 percent of the total light is supplied from incandescent lights or from natural daylight, the problem is averted. Lettuce plants grown under LPS lamps in northern Europe developed strap leaves. This problem does not occur in the United States, where winter light intensities are higher and days are longer. The problem is believed to be caused by low levels of blue light. These problems are not encountered with high-pressure sodium (HPS) lamps because of the broader spectrum of light emitted. Reasons given in the greenhouse industry for not using LPS lamps include their bulky reflectors, which cast shadows, and the higher cost of purchasing and installing the large number of LPS lamps required relative to HPS lamps.

HPS-type HID lamps (Figure 12-5h) gained more popularity in the United States than the high-pressure mercury lamps because they are cheaper to purchase and operate. The majority of lamps being installed in greenhouses today worldwide are HPS lamps. The light-emission spectrum predominates more in the higher wavelengths, with a peak at 589 nm (yellow). The light-emission spectrum extends beyond the visible range (400 to 700 nm) into the 700-to-850-nm range. Radiation in this latter range is required for stem elongation, increased fresh weight, and early flowering of most plants. HPS lamps are very efficient, converting 25 percent of the electrical input into the lamp and ballast into visible radiation. Suitable models of HPS lamps for greenhouse use are available in 400-, 430-, 600-, and 1,000-W sizes. The life expectancy of HPS lamps can be as much as 24,000 hours.

Commercial Application While supplemental lighting is common in the northern latitudes above 40°N in North America and 50°N in Europe, it is also gaining popularity in more southern latitudes. Although little has been done academically to evaluate the economics of supplemental lighting, years of commercial success indicate that it is profitable, particularly with decreasing winter light. It costs about $2.65/ft^2 ($28.50/m^2) of greenhouse floor to purchase and install an HPS system. The price of a 430-W lamp and fixture, including ballast, is around $185. The price of wire and installation is about $100 per lamp. One lamp will provide a light intensity of 400 fc (4.3 klux) over a plant area of 108 ft^2 (10 m^2). A second cost to consider is the consumption of electricity. Lamps of 400-, 430-, and 600-W sizes consume 465, 490, and 675 W of energy per hour, respectively.

Supplemental lighting is used for most crops but is particularly popular for chrysanthemum and geranium stock plants, Elatior begonia, rose, and plug seedlings. Light intensities of 300 to 600 fc (3.2 to 6.5 klux) at plant height are generally used for seedlings and ornamental plants, with 400 fc (4.3 klux) being the most common level. Intensities of 600 to 1,000 fc (6.5 to 10.8 klux) are used for vegetable crops.

The various manufacturers of lamps determine for growers the height and spacing of lamps according to the desired light intensity and the configuration of the greenhouse. The floor area that can be serviced by a lamp is determined by dividing the number of lumens of light emitted from the lamp toward the crop by the desired number of foot-candles of supplemental light. The number of lumens of light received by the crop depends on the brand of the lamp. Reflector design is a principal factor in determining this level. For the most popular greenhouse brand of 430-W lamp used to supply 400 fc of light, the 38,000-lumen output of this lamp is divided by 400 fc to yield a floor area of 108 ft^2 (10 m^2). In the recent past the 400-W lamps, and currently the 430-W lamps, are used almost exclusively because of the better uniformity in light intensity across plants that can be achieved with this size of lamp within the confines of the greenhouse. Because of the increased height of greenhouse gutters, the 600-W lamps have recently begun to gain acceptance. The 1,000-W lamps are rarely used because few greenhouses are sufficiently tall to accommodate them effectively.

Equally important to the lamp is the fixture that holds it and reflects the light to the plant area. Special horticultural fixtures are used. They are designed to spread the light in a square pattern as broadly and as uniformly as possible and with a minimum size of fixture, to reduce shading. The 430-W greenhouse lamp mentioned previously has a total output of 53,000 lumens, with 43,100 lumens directed toward the crop. Lamps designed for other industrial purposes direct a much lower percentage of their light toward the crop. Specialty designs include a reflector for lamps hung over an aisle to minimize the light in the aisle and direct it toward the two adjacent beds instead. Likewise, reflectors are available for use along the sides and ends of greenhouses to direct light into the greenhouse that would otherwise strike the wall.

Crops receiving supplemental lighting are generally given a light period of 16 to 18 hours. This includes the time when the lamps are on as well as any part of the day when it is bright enough to turn the lamps off. Generally, the lamps are turned off when the natural light intensity exceeds a set point, often twice the intensity provided by the supplemental lamps. The yield benefit of extending the 18 hours of light to 24 hours is small for many crops. The postproduction life of some crops is reduced when they are grown under 24 hours of light. In the case of cut roses, many growers see a disadvantage to applying 24 hours of light because during the night, blooms develop beyond the desired stage for cutting in the morning when the work crew arrives. Lighting roses for 18 hours speeds up flower development to the point where successive harvests can be made for the four dates of maximum profit: Christmas, Valentine's Day, Easter (if it falls on the right date), and Mother's Day. In addition, a 50 percent or better increase in yield can be obtained.

Plant response to supplemental light is greatest in the young-plant stage, beginning with the first true leaves, and diminishes with time. This is fortuitous because plants can be grown at a higher density when they are young, and thus it is possible to light a relatively small area of the greenhouse. A good example would be bedding plants, which are often lighted when they are in plug trays, where there can be as many as 648 seedlings per tray. Much more light would be required later, when these seedlings are transplanted into flats similar in size to the plug trays but holding only one-tenth or fewer plants.

Tomato seedlings in flats or soil blocks are often lighted starting at the time of germination for a period of two to three weeks at an intensity of 465 fc (5.0 klux). In some cases, light is applied for 12 hours per day, enabling the lighting arrangement to be drawn on tracks to a second batch of seedlings each day. The switching of lights occurs at midnight and noon. In this way, each batch of seedlings can be exposed to a 16-hour day length, since an additional 4 hours of daylight will be obtained when they are not under the lights. Day lengths of more than 16 hours are avoided, since they retard growth and flowering. "Five-week-old" plants can be produced in less than half the time with this method.

Cucumber seedlings for greenhouse fruit production are started under supplemental light intensities of 280 to 465 fc (3.0 to 5.0 klux) from November to February in the Northern Hemisphere. Lettuce seedlings produced for growing in the greenhouse can require from two weeks in the summer to eight weeks in the winter to produce under natural light in England. If they are lighted at 700 fc (7.5 klux), they can be produced in 11 days in the winter.

Some pot chrysanthemum growers set up two zones for producing this crop. In the first zone, plants are spaced tightly and are provided a warmer temperature, elevated CO_2, and supplemental lighting. The plant is most responsive to all of these environmental factors in the early stages. Thus, the higher costs of providing this environment can be confined to a short period when the crop occupies minimal space.

Growth Rooms

Growth rooms are used for producing seedlings (Figure 12-7). They may be constructed in the headhouse, in a barn, or in the greenhouse. Many materials can

Figure 12-7 **A growth room for starting seedlings. Plants are grown on tiered shelves to conserve space. Light is supplied entirely by fluorescent lamps above each shelf.** *(Photo courtesy of George J. Ball, Inc., West Chicago, IL 60185.)*

be used for construction, including waterproof plywood. The chamber should be well insulated with a material such as polyurethane board and have a moisture-tight barrier on the inside.

Shelves may be built inside in tiers, up to 2 ft (61 cm) apart, although they can be as close as 9 in (23 cm). Space between shelves should be sufficient to permit a minimum space of 6 in (15 cm) between the lamps and the top of the plants when low-energy, 40-to-75-W lamps are used. If 8-foot (244 cm) long, high-energy 215-W lamps are used, the shelves should be 2 ft (61 cm) apart. Attached to the bottom of the shelf above are the lamps. Fluorescent lamps are best for growth rooms because of the uniform light intensity they emit over a wide area and the low heat level they emit. High-intensity lamps would be difficult to use in such close proximity to the plants, although they are used in some chambers. Cool white fluorescent lamps are commonly used.

A light intensity of 500 to 1,400 fc (5.4 to 15.0 klux) is used, depending on the crop. Light intensities of 500, 750, and 1,400 fc (5.4, 8.1, and 15.0 klux) can be provided by 8-foot (244-cm) -long, (244-cm), 125-W lamps installed 2 ft (61 cm) above the shelf at spacings of 8.9, 6.0, and 3.2 in (22.6, 15.2, and 8.1 cm), respectively. A sheet of aluminum foil or a coat of aluminum paint should be provided above the tubes to maximize the light intensity below them, even when reflectorized tubes are used.

Light is generally applied for 16 hours per day; however, some crops, such as many of the bedding plants and lettuce, will respond to 24 hours of illumination. During the illumination period, the temperature is held between 70 and 80°F (21 and 27°C), depending on the type of seed being germinated.

The lamps will provide most of the heat needed. Only during exceptionally cold periods is it necessary to provide supplemental heat. A thermostatically controlled heater can be installed for this purpose. The greatest requirement regarding temperature is that of maintaining a uniform temperature. Hot and cold spots will form if the air within the room is not circulated. The two walls running the length of the room should be constructed of perforated material such as pegboard. The pegboard should be set 6 to 8 inches (15 to 20 cm) in from the outer wall of the room to provide a chamber for air movement behind the pegboard (Figure 12-8). A false ceiling provides a chamber overhead that is connected to the wall chambers. A fan is placed overhead that will cause air to be drawn in one wall and expelled through the other, thus setting up an airflow pattern across the growth room. The heater can be installed in the space over the ceiling in front of the fan. Should the room become too hot, a thermostatically activated motorized ventilator is used in the air duct behind the wall or over the ceiling. For very fine details of growth room designs, see Mastalerz (1985).

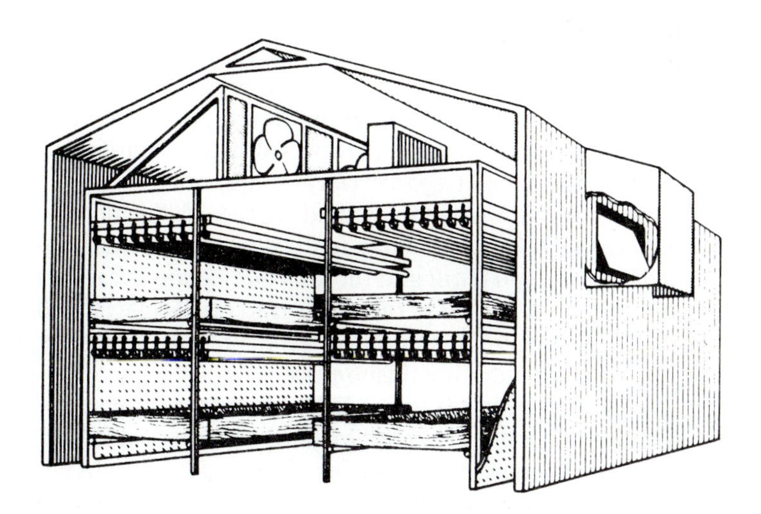

Figure 12-8 **Arrangement of benches, fluorescent tubes, air-circulation system, heater, and ventilator in a six-bench growing room.** *(From Electricity Council [1972].)*

Light Duration for Photoperiodism

What Is Photoperiodism?

We have just taken a look at one dimension of light: its intensity. Now we focus on a second dimension: its duration. Living organisms are innately aware of rhythmic forces in their environment. Fiddler crabs at Cape Cod, Massachusetts, like any other of their species in the world, will feed when the tide is low. It is then that food is trapped in small pools in which they can maneuver. When the tide comes in, it is time to sleep. If these crabs are placed in a tank of seawater in the darkness on Cape Cod, they will continue to feed when the tide goes out and sleep when it comes in on the beaches outside the tank. Upon being moved to Chicago, they eventually change their feeding time to the time when the tide would be low if there were an ocean in Chicago. Clearly, the crab is responding to the gravitational forces of the moon, which regulate the tides.

Plants and animals respond to many such rhythmic forces. The rate of metabolism of an earthworm has predictable peaks and valleys in accordance with the lunar month and the 24-hour solar day. Japanese industries have gone so far as to plot efficient and inefficient days in the lives of some of their key employees. What emerges from these data is a recognizable rhythm, such that one can be assured that on certain days an individual will perform at peak potential while on other days it might be better that he or she is not at work.

There is a mechanism in plants that tracks time. It is highly precise and can discern a five-minute difference within a 24-hour cycle. It is called *photoperiodism* because it is locked into the 24-hour solar day and is based upon the light-dark cycle. Photoperiodism is the response of a plant to the day-night cycle. Response can mean many things, including rosette growth versus bolting in lettuce, bulb formation versus leaf and stem formation in onion, tuber formation in dahlia, flowering of chrysanthemum, downward flagging of leaves of bean, a change in the shape of newly forming leaves, red pigmentation in bracts (leaves) of poinsettia, the formation of plantlets along the margins of leaves of *Kalanchoe daigremontiana* (Hamet and Perrier), and so on.

Plants are customarily classified in regard to photoperiodism as long-day plants, short-day plants, and day-neutral plants. In this book, the long-day plant is called a *short-night plant* and the short-day plant is called a *long-night plant* because the mechanism that permits plants to track time actually measures the dark period. This change of terminology avoids the need for reciprocal thinking and corrects two long-standing misnomers.

Long-night plants are ones that undergo a response such as flowering only when the night length becomes longer than a critical length. Poinsettias require about 12 hours of darkness to flower. This length of night occurs in the latter part of September in the Northern Hemisphere. Prior to September 15, since the nights are too short to afford a 12-hour dark period, plants grow vegetatively. In the latter part of September and on later dates, the nights are long enough to permit at least 12 hours of darkness; thus, the poinsettia buds change from vegetative buds that form leaves and stems to reproductive buds that form flower parts. Of course, several weeks must pass before these flower parts become large enough to be seen. Azalea, chrysanthemum, kalanchoe, and Lorraine and Rieger begonia are all long-night plants in terms of the flowering response. Tuber formation in dahlia and tuberous begonia is a long-night response.

Short-night plants undergo a response when the nights are shorter than a critical length. Aster forms a rosette type of growth when nights are longer than a critical length, and develops tall stems and initiates flower buds under shorter nights. Short-night conditions prevent tuber formation in dahlia and tuberous begonia and thereby encourage flowering. Calceolaria and cineraria initiate flower buds at low temperatures. After this point, short nights hasten flowering. Short nights increase the height of Easter lily. Plantlets form along the margins of some bryophyllum leaves under short-night conditions.

Day-neutral plants, such as rose, do not respond to the relative length of the light and dark periods. Other forces determine when a response will occur in these plants. Some require a certain level of maturity before they flower, while others must accumulate a specific quantity of solar energy. Some varieties of chrysanthemum as well as Mathiola incana (stock), calceolaria, and cineraria initiate flower buds when a sufficient length of time at a cool temperature has passed. Calceolaria and cineraria require four to six weeks at 50°F (10°C) for flower initiation. In terms of the length of the night and these particular responses, these are all day-neutral plants.

Some plants will flower at any night length but do so faster at a particular night length. Carnation will flower at any night length, but flowers fastest under short-night conditions. Rieger begonia will flower at any night length, but flowers fastest under long-night conditions. These are called *facultative short-night* and *facultative long-night* plants, respectively.

The critical night length is not any set figure. It is different for each plant species and can be different for cultivars within a species. Take as an example the single species of chrysanthemum classified as Dendranthema X grandiflorum (Ramat.) Kitamura. The hardy garden varieties can have a critical night length of 8 hours, while many greenhouse-forcing varieties have a critical night length of 9.5 hours. Since a night length of 9 hours is below the critical length for the greenhouse varieties, they remain vegetative. Since 9 hours is more than the critical length for the garden varieties, they initiate flower buds and proceed to flower. Kalanchoe has a critical night length of about 11.5 hours; poinsettia, about 12 hours. Flowering occurs in each case at night lengths greater than these critical lengths.

The critical night length is also dependent upon temperature. The night temperature is more important than the day temperature. Table 12-1 shows the effect of night temperature on the critical night length of the chrysanthemum cultivar 'Encore'. As the night temperature goes up, the critical night length gets shorter. Thus, it is clear that a garden chrysanthemum will initiate flower buds sooner during a hot summer than during a cold summer.

Table 12-1 **Critical Night Length for the Chrysanthemum Cultivar 'Encore' at Each of Three Night Temperatures**

Night Temperature		
°F	*°C*	*Critical Night Length (hours)*
50	10	10.25
60	16	9.5
80	27	8.75

The Mechanism of Photoperiodism

The pigment in photoperiodic plants known as *phytochrome* serves as the light receptor. When the plant is in daylight or artificial light, phytochrome exists in a form known as *Pfr,* which is sensitive to light in the far-red region with a peak response at 735 nm. If the plant is exposed to far-red light, Pfr phytochrome will quickly change to *Pr* phytochrome, which is sensitive to red light with a peak response at 660 nm. This same response will occur when the plant is placed in darkness, but it occurs very slowly under this condition. The Pr form developed in darkness or under far-red light rapidly returns to the Pfr form when the plant is exposed to daylight again. Levels of Pfr might look like those proposed in Figure 12-9 during summer and winter daily cycles.

The important point is that the Pfr form is rapidly produced in the light, and the Pr form is slowly produced in the dark. The Pfr form is the active form that controls the photoperiodic response. It inhibits flowering in long-night plants and promotes flowering in short-night plants. During the summer when nights are short, the level of Pfr in a long-night plant such as chrysanthemum does not become low enough to permit flowering. During a long winter night, the level of Pfr does become low enough to permit flowering.

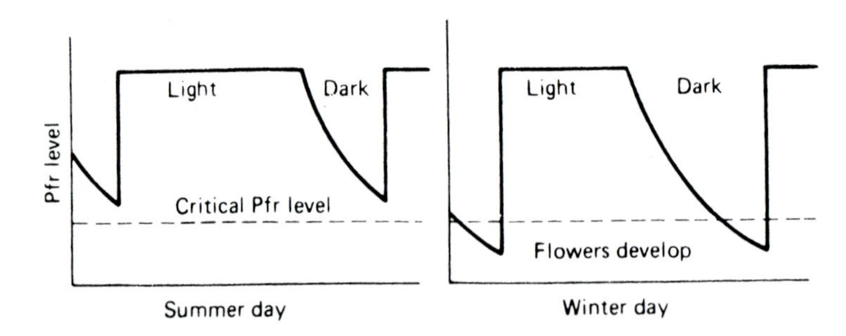

Figure 12-9 **The effects of a summer and winter day-night cycle on the Pfr phytochrome level in a plant. Pfr quickly builds up in the light period and slowly diminishes in the dark period. A long-night plant will flower only when the Pfr level falls below the critical level, as shown, for the long winter night.** ***(Adapted from Electricity Council [1973].)***

Methods of Photoperiodic Control

Day-length control for the greenhouse chrysanthemum is typical of long-night crops as a whole. Chrysanthemum is grown for an initial period under short-night conditions to develop a plant of suitable size that will support large blooms and tall stems; then the plants are grown under long-night conditions to induce flower-bud initiation and subsequent development.

Short-Night Treatment Depending on the season of the year and the variety of chrysanthemum, the short-night treatment can last from two to eight weeks. If this stage occurs during the summer, there is no need to do anything but grow the plants under the natural short nights. However, if the crop is planted during the winter when nights are long, it will then be necessary to shorten the dark period by turning lights on during the night. Lights may be turned on in the late afternoon to extend the day into the evening, or they may be turned on during the middle of the night to break the dark period. Fewer hours of lighting are required if the dark period is interrupted in the middle of the night, and this procedure is commonly used. The light break in the middle of the night restores the Pfr phytochrome level, and, since neither of the two dark periods before or after the light break is very long, the Pfr level does not diminish sufficiently to permit flowering.

Since the dark period in the Northern Hemisphere becomes longer as December 21 approaches, the number of hours of supplemental light required increases. The number of hours of light to apply for any given month at 40°N latitude is presented in Table 12-2. A word of caution is needed here: The night length and, consequently, the amount of light to apply depend upon the latitude at which one is located on the earth. The shortest night of the year occurs around June 21 for the Northern Hemisphere. On this day, the dark period is 12 hours long near the equator, while no darkness occurs at the North Pole. Thus, the farther north one is located, the shorter the night is. The longest night of the year occurs around December 21. There are 24 hours of darkness at the North Pole on this day and 12 hours of darkness near the equator. In this case, the farther north one goes, the longer the night is. Northern latitudes have shorter summer nights and longer winter nights than points farther south. The light period near the equator is always 12 hours long. All places in the world reach a midway point

Table 12-2 **Duration of Light to Apply during Night for Different Months to Ensure Short-Night Conditions at a Latitude of 40°N**

Month	*Hours of Light*
June–July	0
May–August	2
March–April and September–October	3
November–February	4

around March 21 and September 21, at the vernal and autumnal equinox, when the light period is 12 hours everywhere. These relationships can be seen in Figure 12-10.

From this discussion, it should be apparent that the period of the year in which light must be applied and the duration needed on any given night will depend upon the latitude where one is located. Some companies providing chrysanthemum cuttings make available excellent catalogs presenting cultural techniques as well as lighting and shading schedules for this crop. The schedules are given according to the zone in which one lives.

Incandescent lamps work best for extending the day or reducing the night length because a large percentage of the light emitted is in the red zone, which is required by Pr phytochrome. The required light intensity is very low, and most plants respond to 1 to 2 fc (11 to 22 lux). A minimum intensity of 10 fc (108 lux) should be provided, however, to avoid any failure. The most important parts of the plant to illuminate are the recently mature leaves.

To provide the required intensity of light for a 4-foot (1.2-m) -wide bed, one string of 60-W bulbs 4 ft (1.2 m) apart should be installed not more than 5 ft (1.5 m) above the soil along the middle of the bed. Two beds can be lighted by installing a row of 100-W bulbs 6 ft (1.8 m) apart and not more than 6 ft (1.8 m) above the soil between the beds. Larger incandescent floodlight bulbs can be installed along the center ridge of the greenhouse to light the entire width. A minimum of 1 W is required for each square foot of ground lighted (10.8 W/m^2). Occasionally, the larger lights are installed in clusters to reduce the cost of wiring.

The cost of lighting a crop can be reduced by *flash-lighting* (cyclic lighting). As little as one second of light at 10 fc (108 lux) every five seconds will keep phytochrome in the Pfr form and cause some chrysanthemum cultivars to remain vegetative. This frequency requires heavy-duty switches; thus, longer light periods are generally used.

If the standard program calls for four hours of light in the middle of the night,

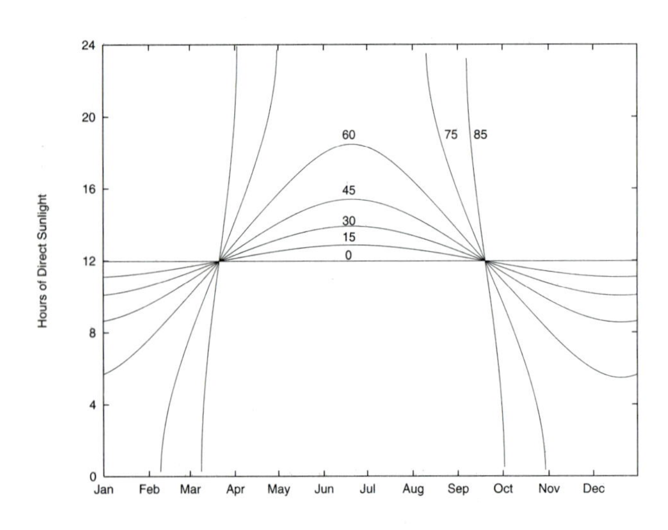

Figure 12-10 **The daily length of direct sunlight received at northern latitudes of 0°, 15°, 45°, 60°, 75°, and 85° throughout the year. At the North Pole (90°N latitude) vertical lines would exist at the vernal and autumnal equinoxes (approximately March 21 and September 21) indicating a shift from 0 to 24 hours of direct sunlight or vice versa on these days.** ***(Graph courtesy of Michael Owen, North Carolina State University, Astrophysics.)***

from perhaps 10 P.M. to 2 A.M., one can divide this duration into 30-minute periods and apply light for 20 percent of each period. In this way, one would apply light for 6 minutes out of every 30 minutes between 10 P.M. and 2 A.M. Shorter periods will work as well, as long as light is applied 20 percent of the time. It is essential that a minimum light intensity of 10 fc (108 lux) be applied in this system.

A greenhouse range using cyclic lighting could be divided into five zones, and all could receive their light requirement during one four-hour period. This would reduce the consumption of electricity by 70 to 75 percent and would permit the use of lighter main wiring, which would reduce the initial wiring cost. Time clocks are available that can control such a system.

Long-Night Treatment After a period of short nights has been provided to establish the plant, a period of long nights must be provided to bring about flower-bud initiation and development. During the winter, since the nights are naturally long enough, nothing is done. When this stage of growth occurs in the summer, however, it is necessary to pull an opaque screen over the plants in late afternoon and off again in the morning. The screen should be applied from 7 P.M. to 7 A.M. Some growers apply it at 5 P.M. before their workforce leaves, which can have harmful effects during the summer if heat builds up underneath. Flowering will be delayed, and, at higher temperatures, flower buds will abort. If the screen must be pulled at 5 P.M., the sides should be left up for air circulation and someone should return at 7 P.M. to lower them. It is necessary to continue pulling the screen until color shows in the buds. Beyond that time, it need not be applied. The screen should be applied every day of the week. For each day that is skipped every week, the crop will flower a day or so later.

A good grade of sateen (cotton) cloth works well. More recently, polyester cloth has been used for light screening. It carries the advantage of resisting rot. Cloth should be dense enough to reduce the light intensity beneath to 2 fc (22 lux) when the intensity outside is 5,000 fc (53.8 klux). Cloth should be sufficiently porous to permit water to penetrate if it is used in a leaky greenhouse or in the field. Black polyethylene also may be used as long as water is not a problem. Tears in the cover should be immediately repaired to prevent light leaks. Wherever light leaks in, plants will develop with incomplete or hollow flower buds called *crown buds.* Commercial opaque screens are made of a number of materials. Some have an aluminized outer side to reflect heat and keep plants cooler under it. In the summer, heat can build up under the screen, since it has to be pulled before sunset and removed after sunrise. Such heat buildup can lead to delayed flowering and even bud abortion.

The expenditure in labor to manually pull black cloth over frames, as pictured in Figure 12-11, is considerable. Larger growers use power-operated shading (Figure 12-12). Some make this apparatus themselves. An electric motor turns a pipe shaft along one side of the greenhouse. Cables attached to the shaft run across the greenhouse to a shaft on the other side and back again. Cloth is attached to the cables, enabling it to be drawn across the greenhouse and back again.

A number of commercial systems are available for automatically pulling photoperiodic shade fabric as well as thermal blankets (see the "Heat Conservation"

Figure 12-11 Manual pulling of black cloth in the early evening during the summer to establish long-night conditions for photoperiodic control of flowering. The light bulbs are used during the middle of winter nights to give a short-night effect. *(Photo courtesy of J. W. Love, Department of Horticultural Science, North Carolina State University, Raleigh, NC 27695-7609.)*

Figure 12-12 A commercially available power-operated system for shading plants to create long-night conditions for photoperiodic control of flowering during the summer and for retaining heat during winter nights at White's Nursery in Cheasapeake, Va.

section in Chapter 3) and sunscreens (see the "Reducing Light Intensity" section earlier in this chapter). The same fabric used for heat retention during winter nights may be used for photoperiodic shading in the summer. Fabric with an aluminized or reflective outer surface is desirable.

The cost might seem high, but the advantages that go with it easily sell these systems. At the flip of a switch, one person can cover an acre (0.4 ha) of greenhouse in a few minutes, eliminating perhaps as much as two to three hours of manual labor. The system can be further controlled by an automatic time clock, which eliminates the need for any person. This savings occurs twice a day. Because a single sheet of fabric is suspended overhead, it is possible to operate the fan-and-pad cooling system, pulling the cool air beneath the cover. This prevents excess buildup of heat, which can occur even at 7 P.M. in the summer, and, in turn, prevents reduction in plant quality and delays in flowering.

Temperature

We have already taken a long look at heat in terms of providing it in the greenhouse, removing it from the greenhouse, and controlling it at the desired level. Now we look briefly at its influence on crops. Temperature is a measure of the level of heat present. All crops have a temperature range in which they can grow. Below this range, processes necessary for life stop; ice forms within the tissue, tying up water necessary for life processes; and cells are possibly punctured by ice crystals. At the upper extreme, enzymes become inactive, and again processes essential to life stop.

All biochemical reactions in the plant are controlled by enzymes. Enzymes are heat sensitive. The rate of reactions controlled by them will often double or triple each time the temperature is increased by 18°F (10°C), until an optimum temperature is reached. Further increases in temperature begin to suppress the reaction until it stops.

Numerous biochemical reactions are involved in the process of photosynthesis. These all have the net effect of building carbohydrate and storing energy. Photosynthesis occurs during the daylight hours because of its dependence on light. Another extensive set of biochemical reactions is involved in the overall process of respiration. The net effect here is a breakdown of carbohydrate and a release of energy. Respiration occurs in all living cells at all times.

When photosynthesis exceeds respiration, net growth occurs. When they equal each other, net growth stops. If respiration exceeds photosynthesis, the plant declines in vigor and will eventually die. To ensure that photosynthesis exceeds respiration, plants are grown in cool temperatures at night to keep the respiration rate down and in warm temperatures by day to enhance photosynthesis.

As a general rule, greenhouse crops are grown at a day temperature 5 to 10°F (3 to 6°C) higher than the night temperature on cloudy days and 15°F (8°C) higher on clear days. With CO_2 enrichment, the day temperatures may be an additional 5°F (3°C) higher. The night temperature of greenhouse crops is generally in the range of 45 to 70°F (7 to 21°C). Primula, *Mathiola incana*, and calceolaria grow best at 45°F (7°C); carnation and cineraria, at 50°F (10°C); rose, at 60°F (16°C); chrysanthemum and poinsettia, at 62 to 64°F (17 to 18°C); and African violet at 70 to 72°F (21 to 22°C).

Temperature Interrelationships

A rule by F. F. Blackman, in essence, states that the rate of any process that is governed by two or more factors will be limited by the factor in least supply. Photosynthesis is a good case in point. It is dependent upon heat, light, CO_2, and other factors. On cloudy days, it is futile to raise the temperature more than 5 to 10°F (3 to 6°C) above the night temperature because the low light intensity will limit the rate of photosynthesis, and any additional heat applied will be without beneficial effect. On bright days, light does not limit photosynthesis; thus, if the temperature is not raised, heat may become the limiting factor for photosynthesis. Even on dark days, the rate of photosynthesis will increase with CO_2 enrichment of the greenhouse atmosphere.

Light intensity is higher in the summer than in the winter, and photosynthetic rates can be expected to be higher in the summer. This is very fortunate, since it calls for higher daytime temperatures in the summer than in the winter to prevent heat from becoming the limiting factor. Cooling fans can be set at a higher temperature in the summer—as high as 80 to 85°F (27 to 29°C)—which saves considerable electrical energy.

Blackman's law is well illustrated in the curves of Figure 12-13, which were developed by Gaastra (1962). In the lowest curve, the rate of photosynthesis

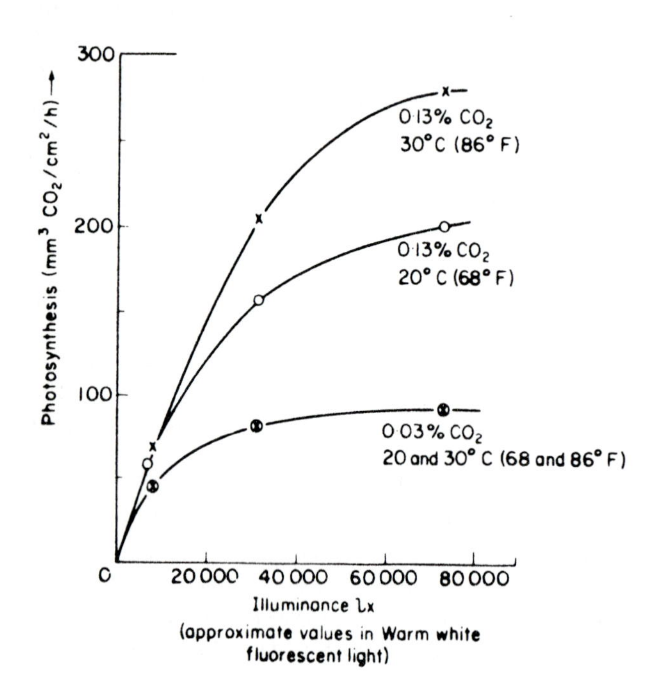

Figure 12-13 Effects of CO_2 concentration, light intensity, and leaf temperature on photosynthesis in cucumber. *(From Gaastra [1962].)*

began to plateau at about 3,800 fc (40,000 lux), regardless of whether the temperature was at 68°F or 86°F. The 300-ppm (0.03 percent) level of CO_2 became a limiting factor at that point. When the temperature was held at 68°F and the CO_2 level was increased to 1,300 ppm (0.13 percent), the rate of photosynthesis increased. Then the 68°F temperature became the limiting factor, because the increase in temperature to 86°F at the same 1,300-ppm CO_2 level brought about another increase in photosynthesis.

This interaction of CO_2, light intensity, and temperature is reminiscent of observations in Chapter 11. It was stated there that increases in the CO_2 level in the greenhouse brought about beneficial effects from raising the daytime temperature above that normally maintained for many crops. When CO_2 is eliminated as a limiting factor for photosynthesis, a daytime temperature increase of 5°F (3°C) can often be profitable.

One must be careful when determining how high to raise the temperature, because it affects processes in addition to photosynthesis. Generally, higher temperature results in faster growth, but with it a reduction in quality can occur. Longer stems, thinner stems, and smaller flowers may occur. Quality and quantity must always be weighed in making such a decision. In the previous discussion about raising the temperature 5°F (3°C) along with an increase in the CO_2 level, no adverse loss in quality is to be expected.

DIF–The Day-to-Night Temperature Relationship

Effects of daytime temperatures versus nighttime temperatures on growth and flowering of floral crops have been reported for some time (Cathey 1954; Parups 1978; Cockshull, Hand, and Langton 1981; Parups and Butler 1982). More recently, scientists at Michigan State University, including Drs. Heins, Karlsson, Erwin, and Berghage, have uncovered a practical relationship between plant height and the day-to-night temperature differential. They gave the acronym *DIF* to this temperature differential. DIF refers to the differential obtained when one subtracts the night temperature from the day temperature:

$$\text{DIF} = \text{day temperature} - \text{night temperature} \qquad (12\text{-}3)$$

The DIF values are +10, 0, and −10 for day and night temperature combinations of 70°F and 60°F, 65°F and 65°F, and 60°F and 70°F, respectively. The information that follows is drawn from the work of this Michigan State University team.

Height Control by DIF Plant height can be controlled by DIF. A shift from a positive DIF toward a zero DIF results in a large reduction in height (Figure 12-14). While height continues to decline as the DIF value is shifted from zero to negative values, it is of smaller magnitude. Two underlying relationships are involved. Plant height can be decreased by decreasing the day temperature and

Figure 12-14 **The effect of decreasing DIF values (from left to right) on decreasing final Easter lily height. For all five plants, the day temperature was 26°C (79°F), but the night temperatures were 14, 18, 22, 26, and 30°C (57, 64, 72, 79, and 86°F). From left to right, the DIF values were 12, 8, 4, 0, –4°C (29, 22, 16, 7, and 0°F).** ***(From Heins and Erwin [1990].)***

also by increasing the night temperature. Conversely, plant height is increased by increases in day temperature as well as by decreases in night temperature. The effect is upon the length of stem internodes rather than the number of leaves. Controlling plant height by altering DIF is being used commercially on a wide range of crops. Large responses have been achieved in Asiatic lily, *Celosia,* chrysanthemum, dianthus, Easter lily, fuchsia, geranium, gerbera, hypoestes, impatiens, Oriental lily, petunia, poinsettia, portulaca, rose, salvia, snap bean, snapdragon, sweet corn, tomato, and watermelon. Small or no response has been obtained in aster, French marigold, hyacinth, narcissus, platycodon, squash, and tulip. DIF provides an effective means for controlling height because a shift in growth can be seen within one to two days of a shift in DIF.

Controlling plant height through environmental modification rather than by chemical height retardants is particularly attractive in this era when all synthetic chemicals are under scrutiny. A second plus is the cost savings realized by reducing and, in many cases, dropping the use of chemical height retardants.

Flowering Time The rate of growth and maturation of plants is generally temperature dependent. Fortunately, it is the 24-hour temperature average and not just the day or night temperature alone that usually controls the rate of development. Because of this, tall, intermediate, or short plants may be produced for the same flowering date by reducing the DIF values in a way that the daily average temperature remains constant. The + 10, 0, and –10 DIF values referred to earlier all have the same average daily temperature of 65 °F if the day and night lengths are taken to be the same length. The day-night temperature combinations were 70°F/60°F, 65°F/65°F, and 60°F/70°F. Each combination has an average of 65°F; thus, these three DIF values offer a choice of three heights of plants for the same market date.

Two factors limit the range of temperatures that can be selected for developing a value of DIF. Each crop plant has a unique temperature-to-growth relationship, as shown in Figure 12-15. Temperatures may be selected within the linear range of the curve (the straight dashed line) and up to the optimum temperature (50 to 80°F in this curve). Higher and lower temperatures should not be selected, because unacceptably low growth rates occur at these temperatures. Both the timing and the quality of the crop could be jeopardized. The temperature range for "warm" crops, including hibiscus and poinsettia, is 50 to 80°F (10 to 27°C). The range for crops tolerant of cool temperatures, such as chrysanthemum, Easter lily, and petunia, is 40 to 80°F (4 to 27°C).

The second limit to the range of temperatures available for selection is the temperature requirement for flower initiation and development. At a night temperature of 73°F (23°C), poinsettia flower development is inhibited. While day temperatures may be higher, night temperatures may not exceed this level. Night temperatures above 72 to 75°F (22 to 24°C) can cause heat delay in flowering of chrysanthemums. It may be necessary to raise night temperatures to 75°F (24°C) to slow height development in Easter lily, but one pays a price in increased flower abortion at temperatures above 70°F (21°C). Such temperature sensitivities are greatest at the time of flower initiation and the early stages of flower-bud devel-

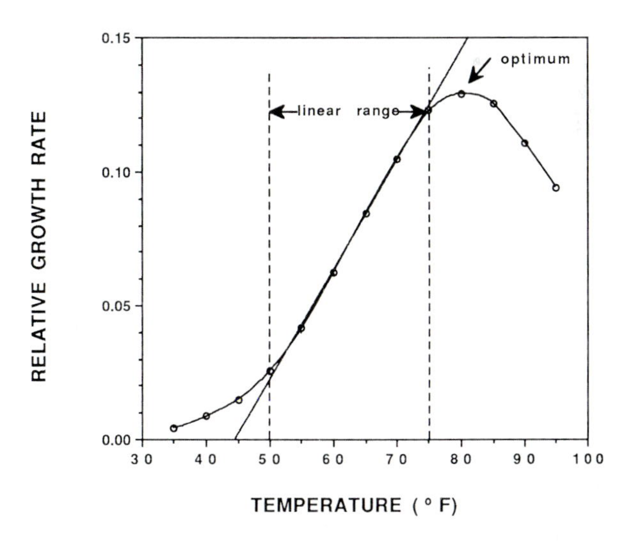

Figure 12-15 **Hypothetical effect of temperature on relative growth rate. A straight line has been superimposed on the curve to show the range of linear response, which lies between 50 and 75°F (10 and 24°C). Most processes in the plant have a similar relationship with temperature.**

opment. Once flower buds are visible, the plant is much less sensitive to heat delay. Higher night temperatures can be attempted at these times to achieve large negative DIF values when plants are excessively tall.

Side Effects The lower the DIF value, particularly in the negative range, the greater the chance of chlorosis of leaves. Chlorosis occurs on young, immature leaves. If the low DIF treatment is applied correctly, normal green color will return as these affected leaves mature. Chlorosis becomes a permanent problem when the plant is treated at too early an age. Plug seedlings treated during the first week have only immature leaves; thus, the whole plant turns completely chlorotic, and growth of the plant is severely reduced. Such stunting is not later corrected. Depending on species, a DIF value lower than –2 to –3 should not be applied to plug seedlings during the first one to three weeks.

A second side effect of very low DIF treatments is downward curling of leaves. This problem is particularly pronounced on Easter lily. If the curled leaves are not mature, they will uncurl when returned to a normal positive DIF value.

Implementation of DIF in Warm Seasons There is little difficulty in lowering the day temperature in northern locations during the winter. Later in the spring and even in the winter in warm climates, it may appear impossible to do so. Low DIF effects can still be produced by lowering the day temperature for the first two hours of the day immediately after sunrise. This is based on the fact that the greatest rate of internode elongation occurs during the night, with a maximum peak at sunrise. Greater height suppression of Easter lily was achieved by applying a negative DIF only during the two hours after sunrise than by applying negative DIF for seven hours beginning two hours after sunrise and ending at sunset.

It is important that the temperature be at the lower setting as soon as the shift from dark to light occurs. For a hot-water heating system, this means that the thermostat should be set lower perhaps 45 minutes before sunrise to allow sufficient time for the system to cool. The thermostat could be set lower 15 minutes before sunrise for faster-reacting heating systems, such as unit heaters.

Graphical Tracking The day-to-day implementation of DIF to control height is best handled through *graphical tracking* (Heins and Erwin 1989; Carlson and Heins 1990). A good example of such a graph for Easter lily can be seen in Figure 12-16. In this case, lilies were grown in a 6-inch (15-cm) standard pot having a height of 6 in (15 cm). The final plant-plus-pot height desired was in the range of 22 to 24 in (56 to 61 cm). Based on previous experience, the plant was expected to double in height after the point of visible bud. Thus, for the date of emergence the height was plotted as 6 in (15 cm) (the height of the pot), and at market stage it was plotted as 22 to 24 in (56 to 61 cm). Typically, the Easter lily is at half of its final height at the first point of visible flower bud. For the date of visible flower bud, minimum and maximum heights were plotted as 14 and 15 in (36 and 38 cm) (half the final plant height plus the 6-inch, 15-cm pot height). Growth between each of these pairs of dates proceeds in a straight line; thus, points were connected by straight lines to form the minimum and maximum

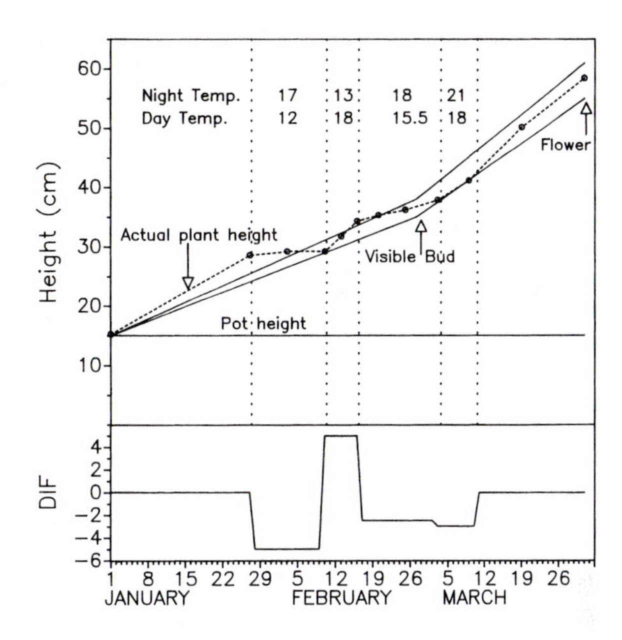

Figure 12-16 **Graphical tracking plot for Easter lily. Solid curves depict the range of desired heights for various dates (the tracking window). The dashed line shows the actual measured heights of a commercial crop in which DIF was used to control height. Day and night temperatures applied to the commercial crop are shown in degrees Celsius. Temperature conversions for 12, 13, 15.5, 17, 18, and 21°C are 54, 55, 60, 63, 64, and 70°F, respectively. The actual DIF values applied are given in degrees Celsius in the curve at the bottom of the figure.** ***(From Heins and Erwin [1989].)***

height curves. Twice a week, the grower measured the height of the plants and plotted the results on the graph. On January 27, plants were too tall and a –10°F (–5°C) DIF treatment was applied (53°F day/63°F night,12/17°C). By February 10, plant height gain had been suppressed to where it was at the minimum acceptable level. DIF was increased to +9°F (5°C) and resulted in an increase in the rate of height rise. On three subsequent dates, further adjustments in DIF were made. In the end, the crop reached a desired height without the use of chemical growth regulators.

Unlike Easter lily, growth of chrysanthemum and poinsettia proceeds in a sigmoidal curve fashion, as shown by Figure 12-17 (Carlson and Heins 1990). In this curve, the percentage of final plant height (not including the pot height) is plotted on the vertical axis against the relative time to flower on the horizontal axis. Relative time to flower is used because different cultivars of mums and poinsettias can require a different number of weeks to flower. If the crop requires 10 weeks to flower, its relative time to flower after 5 weeks of growth will be 0.5 (50

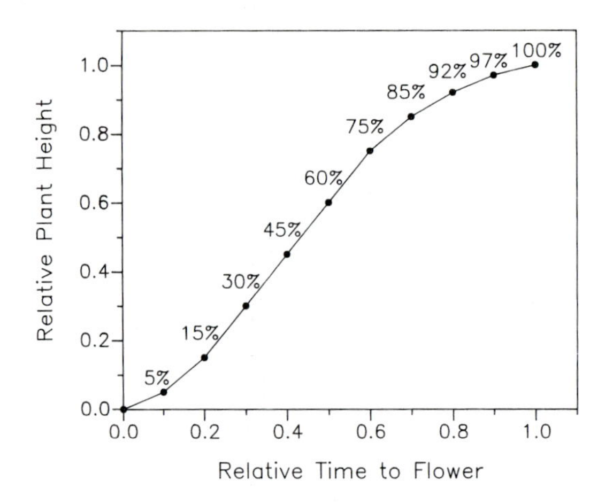

Figure 12-17 **Growth curve for chrysanthemum and poinsettia beginning at the date of pinch and ending at market date.** *(From Carlson and Heins [1990].)*

percent). The relative time to flower begins at the date of pinch. To develop a theoretical tracking curve for chrysanthemum or poinsettia, the final plant height desired is selected. Based on the percentages in Figure 12-17, the height the crop should be at each week is determined, and these heights are plotted on the vertical axis of a graph against weeks of growth on the horizontal axis. Then the actual height of the crop is measured twice a week and plotted on the same graph as the theoretically computed height curve. When actual crop height differs from the theoretical height, the DIF value is altered to bring the crop back into line.

Computer Control Each year the need for an environmental computer to monitor and control the greenhouse environment becomes more critical to making a profit. Implementation of height control through DIF is a good example. DIF plans can be handled through thermostats and manual settings. However, increases in the number of crop zones requiring different DIF values and complications brought on by warm-weather restriction of low day temperatures to the first two hours of the day makes it nearly impossible to handle the task manually. Computerized control is the most efficient way to go. Such a system can calculate the average daily temperature needed for controlling the date of crop maturity. In the future, growth curves will probably be available in computer software for all crops. Such programs will take over temperature control of the greenhouse. The crop manager will simply enter into the program the dates of planting, pinching, and harvesting; the final desired height; and biweekly height measurements. The computer will do the rest, and with less labor input as well as greater conservation of heating and cooling energy.

Summary

1. Light plays two general roles in the growth and development of plants. Light is a source of energy for the process of photosynthesis, in which carbon is fixed into carbohydrates and ultimately all organic compounds of the plant. A relatively high intensity of light in the energy spectrum of 400 to 700 nm is required. Light also regulates the developmental forms of some plants—for example, vegetative versus reproductive growth. It is the duration of light and not so much its intensity that is important in this process.
2. Light is often a limiting factor to photosynthesis and growth during the winter in northern latitudes. In order to maximize interior light intensity, single greenhouses can be oriented with the ridge east to west and ridge-and-furrow greenhouses with the ridge north to south. The glass or FRP covering can be washed. Plants should be spaced farther apart in the winter to increase the amount of light per plant.
3. Supplemental lighting during the daylight hours to enhance photosynthesis is highly effective. The economics, however, bear scrutiny. There are situations of high-density plantings such as rooting and seedling beds and the production of young plants where it is most profitable. High-pressure sodium lamps are most commonly used today. Continued research is needed to reduce the electrical costs through lamps of increased efficiency and to minimize the size of the fixture for reasons of shading.
4. A number of seedling producers construct growth rooms. Plants are grown on tiered shelves with a bank of fluorescent lights above each shelf. This is the sole source of light for photosynthesis. A growth room can be better insulated than a greenhouse. Often, the heat from the lamps is sufficient to meet the heat requirement of the room.
5. Light plays its second role in photoperiodism, which is the response of an organism to the day-night cycle. The relative length of the light and dark periods governs a number of responses, including flowering, leaf shape, stem elongation, bulb formation, and pigmentation. In terms of flowering, long-night plants are those that initiate and develop flower buds when the night is longer than a critical length. Conversely, short-night plants are those that initiate and develop flower buds when the night is shorter than a critical length. The critical night length varies among plant species and even among cultivars within a species. Not all plants are photoperiodic. Those that do not respond to the day-night cycle are day-neutral plants.
6. Long nights are established in the greenhouse during the summer by covering the plants with a plastic or cloth fabric in the early evening (about 7 P.M.) and removing it in the morning (7 or 8 A.M.). The cover should be capable of reducing the light intensity beneath to 2 fc (22 lux) when the outside intensity is 5,000 fc (53.8 klux). Automatic equipment is available for this operation. Short-night conditions can be established in the winter by providing 10 fc (108 lux) of illumination for a period of one to four hours during the middle of the night. Incandescent lights serve this purpose best.

7. Phytochrome is the receptor pigment in young tissue that responds to the light-dark cycle. The Pfr form of phytochrome rapidly builds up during the light period and is slowly converted to the Pr form during the dark period. The Pfr form is the active form that inhibits flowering in long-night plants and promotes flowering in short-night plants. A long night is required to lower the level of Pfr phytochrome to the point where flowering can occur in a long-night plant.
8. Heat is a form of energy and a factor essential to growth. Deleterious effects result from levels that are too high or too low. Heat is just one factor governing growth. The rate of growth is limited by the factor in shortest supply. It is not always economically feasible to optimize all factors affecting growth in a greenhouse; thus, the best temperature for a crop will depend upon the following factors:
    a. Light is often limiting in the winter. On low-light-intensity (cloudy) days, a day temperature 5 to 10°F (3 to 6°C) above the night temperature is maintained; on brighter winter days, a day temperature 15°F (8°C) higher day temperature is beneficial to growth. Although even higher temperatures would not be beneficial to growth in the winter, they are beneficial in the summer when light intensity is higher and not limiting to growth.
    b. The CO_2 level inside greenhouses often limits growth. When it is raised, growth increases to a point at which previously adequate temperatures become the limiting factor. A 5°F (3°C) rise in day temperature is often beneficial when the greenhouse atmosphere is enriched with CO_2.
9. Plant height can be controlled by adjusting the day-to-night temperature ratio. The term *DIF* refers to the temperature difference obtained by subtracting the night temperature from the day temperature. The rate of stem internode elongation is increased by increases in day temperatures and by decreases in night temperatures. Therefore, when DIF is highly positive (day temperatures are much higher than night temperatures), plants become tall. Large reductions in plant height are achieved by reducing DIF from positive to zero values; further, but more modest, growth reductions are obtained by continuing to reduce DIF to negative values. The concept of DIF works best when plants are young and their rate of growth is rapid.

References

1. Bickford, E. D., and S. Dunn. 1972. *Lighting for plant growth.* Kent, OH: The Kent State Univ. Press.
2. Campbell, L. E., R. W. Thimijan, and H. M. Cathey. 1975. Spectral radiant power of lamps used in horticulture. *Trans. Amer. Soc. Agr. Engineers* 18(5):952–956.
3. Carlson, W. H., and R. Heins. 1990. Get the plant height you want with graphical tracking. *Grower Talks* 53(9):62–63, 65, 67–68.
4. Cathey, H. M. 1954. Chrysanthemum temperature study. C. The effect of night, day, and mean temperature upon the flowering of *Chrysanthemum morifolium. Proc. Amer. Soc. Hort. Sci.* 64:499–502.

5. Cathey, H. M. and L. E. Campbell. 1979. Relative efficiency of high- and low-pressure sodium and incandescent filament lamps used to supplement natural winter light in greenhouses. *J. Amer. Soc. Hort. Sci.* 104:812–825.
6. Cockshull, K. E., D. W. Hand, and F. A. Langton. 1981. The effects of day and night temperature on flower initiation and development in chrysanthemum. *Acta Hort.* 25:101–110.
7. Downs, R. J. 1975. *Controlled environments for plant research.* New York: Columbia Univ. Press.
8. Electricity Council. 1972. *Growelectric handbook no. 1: growing rooms.* Electricity Council, 30 Millbank, London SW1P4RD.
9. Electricity Council. 1973. *Growelectric handbook no. 2: lighting in greenhouses.* Electricity Council, 30 Millbank, London SW1P4RD.
10. Gaastra, P. 1962. Photosynthesis of leaves and field crops. *Netherlands J. Agr. Sci.* 10(5):311–324.
11. Garner, W. W., and H. A. Allard. 1920. Effect of the relative length of day and night and other factors of the environment on growth and reproduction in plants. *J. Agr. Res.* 18:553–607.
12. GTE Products Corp. 1980. *Horticultural lighting.* Eng. Bul. 0-351. Danvers, MA: GTE Products Corp., Sylvania Lighting Center.
13. Heins, R. D. 1990. Choosing the best temperature for growth and flowering. *Greenhouse Grower* 8(4):57–64.
14. Heins, R., and J. Erwin. 1989. Tracking Easter lily height with graphs: Easter lily response to temperature during forcing. Part 2. *Grower Talks* 53:64, 66, 68.
15. Heins, R., and J. Erwin. 1990. Understanding and applying DIF. *Greenhouse Grower* 8(2):73–78.
16. Mastalerz, J. W. 1969. Environmental factors: light, temperature, carbon dioxide. In Mastalerz, J. W., and R. W. Langhans, eds. Roses: a manual on the culture, management, diseases, insects, economics and breeding of greenhouse roses, pp. 95–108. Pennsylvania Flower Growers' Association, New York State Flower Growers' Association, Inc., and Roses, Inc. (Available from R. W. Langhans, Dept. of Flor. and Orn. Hort., Cornell Univ., Ithaca, NY 14853.)
17. Mastalerz, J. W. 1985. Growth rooms. In Mastalerz, J. W., and E. J. Holcomb, eds. *Bedding plants.* Vol. 3, pp. 141–150. Pennsylvania Flower Growers' Association. (Available from E. J. Holcomb, Dept. of Hort., The Pennsylvania State Univ., University Park, PA.)
18. Parups, E. V. 1978. Chrysanthemum growth at cool night temperatures. *J. Amer. Soc. Hort. Sci.* 103:839–842.
19. Parups, E. V., and G. Butler. 1982. Comparative growth of chrysanthemum at different night temperatures. *J. Amer. Soc. Hort. Sci.* 107:600–604.
20. Phillips Gloeilampenfabrieken. 1982. *Artificial lighting in horticulture.* Eindhoven, The Netherlands: Phillips Gloeilampenfabrieken, Lighting Design and Eng. Center, Lighting Div.
21. Sage, L. C. 1993. Pigment of the imagination. New York: Academic Press.
22. Stolze, J. A. B., J. Meulenbelt, and J. Poot, eds. 1993. *Application of growlight in greenhouses.* PL Light Systems Canada Inc., P.O. Box 206, Grimsby, Ontario L3M 4G3, Canada.
23. Templing, B. C., and M. A. Verbruggen, eds. 1977. *Lighting technology in horticulture,* 2nd ed. Eindhoven, The Netherlands: Phillips Gloeilampenfabrieken, Lighting Design and Eng. Center, Lighting Div.

 CHAPTER 13

Chemical Growth Regulation

Floriculture is unlike other areas of agriculture in that an entire plant, or at least a major portion of the plant, is appraised according to its aesthetic value. While minor insect damage, leaf blemishes, or unusually tall height may not affect the yield or value of a bean crop, it does reduce the value of an ornamental plant. Several chemicals are used by greenhouse growers to control growth in one or another of its many forms to give the desired aesthetic effect. For example, final plant height can be made shorter than the natural height, terminal buds can be pinched (destroyed), the cold requirement of crops such as azalea can be chemically substituted, and rooting can be promoted. Hopefully, lateral shoots will soon be chemically disbudded (removed).

Classification

Chemicals used to control growth are either naturally occurring plant hormones or synthetically produced compounds. *Hormones* are compounds produced in the plant at one site and then transported to a different part of the plant, where they affect growth. There are five categories of plant hormones: (1) auxins, (2) gibberellins, (3) cytokinins, (4) ethylene, and (5) inhibitors.

Auxins promote growth primarily through cell enlargement. The major auxin produced in plants is indole-3-acetic acid (IAA). Synthetic auxins include indole-3-butyric acid (IBA), indolepropionic acid (IPA), and naphthalene acetic acid (NAA). Auxins play an important commercial role in plant propagation.

Gibberellic acid (GA) likewise promotes growth through cell enlargement. Various gibberellins have been isolated from species of the fungus *Gibberella.* This fungus attacks rice plants and causes them to grow tall and threadlike. Gibberellins promote growth, but unlike that produced by auxins, the growth is uniform throughout the plant tissue. The commercial roles of gibberellins are varied.

Cytokinins are associated with rapidly growing tissue. They might be thought of as the juvenility or antisenescence hormone. Cytokinins enhance growth mainly through cell division. These hormones are used in tissue culture preparations to stimulate callus cell growth. Otherwise, they do not play a commercial role in greenhouse crop production.

Ethylene is naturally produced in fruits, seeds, flowers, stems, leaves, and roots, and controls a multitude of processes. Ethylene lends itself to numerous commercial applications. In some cases, ethylene gas or a synthetic ethylene-producing chemical, ethephon, is applied to enhance ethylene levels for purposes such as height retardation, prevention of stem topple, flower promotion, color formation in fruit, and fruit ripening. In other situations, ethylene levels are reduced through the application of silver thiosulfate (STS) to prolong flowering and prevent petal or floret abscission.

The best-known hormonal inhibitor is *abscisic acid* (ABA). ABA promotes abscission of leaves and petals as well as a number of other processes. It is not a major hormone in the vegetative stages of growth but comes into play in the later stages of maturity and senescence. ABA is not commercially important in greenhouse crop production.

A number of synthetic compounds, often known by their trade names, also exist for control of greenhouse plant growth. These include the height-retarding chemicals A-Rest, B-Nine, Bonzi, Cycocel, and Sumagic; the chemical pinching agents Atrimmec and Off-Shoot-O; and the ethylene producer Florel.

Growth-Regulating Compounds

Only the commercially important hormones and synthetic plant-growth-regulating compounds will be covered in this section. Headings follow compound names rather than roles because of the overlapping roles of these materials. Recommendations tabulated by Bailey at North Carolina State University for the use of these and other growth regulators appear in Table 13-1 (Bailey 1997).

Auxins

Auxins are involved in *tropistic* growth movements. Such movements include the downward growth of roots, the upward growth of shoots, and the growth of shoots and leaves toward the light. It is believed that shoots grow toward the light source because auxin is inactivated by light. This occurs more on the bright side of the stem; thus, there is greater promotion of growth on the darker side.

Auxins also inhibit lateral shoot development. When the top of the main shoot of a plant is removed, the source of auxin is lost from that shoot, and lateral shoots are free to develop. This is why pinching (the removal of shoot tips) is practiced on some floral crops; multiple lateral shoots are promoted. A plant is

Table 13-1 Growth Regulators for Floricultural Crops[1]

Crop	Purpose	Chemical	Rate	Precautions and Remarks
Ageratum	To control plant height	A-Rest	7 to 26 ppm spray (3.4 to 12.6 fl oz/gal)	Plug culture and flat culture differ in recommended rates. The rates shown in this table include both plug (lower rates) and flat culture (higher rates) recommendations. Apply ALL foliar sprays of plant growth regulators using 0.5 gallon per 100 square feet of bench area. Growers should refer to Horticulture Information Leaflet #528, Height Control of Greenhouse Crops, for application techniques and timing for growth regulators on floricultural crops. Contact floricultural specialists at NC State University for further application information.
		B-Nine	2,500 to 5,000 ppm spray (0.39 to 0.79 oz/gal)	
		Bonzi	5 to 45 ppm spray (0.16 to 1.44 fl oz/gal)	
		Cycocel	400 to 3,000 ppm spray (0.43 to 3.25 fl oz/gal)	
		Sumagic	20 to 30 ppm spray (5.12 to 7.68 fl oz/gal)	
Alternanthera (Joseph's-Coat)	To control plant height	A-Rest	25 to 132 ppm spray (12.1 to 64 fl oz/gal)	
			0.25 to 0.50 mg a.i. drench for a 6 inch pot (1 to 2 fl oz/gal of drench solution; apply 4 fl oz/6 inch pot)	Drench volumes and mg a.i. vary with pot size. Contact floricultural specialists at NC State University.
Amaryllis	To control plant height	Bonzi	23.66 mg a.i. drench for a 6 inch pot (6.4 fl oz/gal of drench solution; apply 4 fl oz/6 inch pot)	Drench volumes and mg a.i. vary with pot size. Contact floricultural specialists at NC State University.
Aster	To control plant height	B-Nine	5,000 ppm spray (0.79 oz/gal)	See Ageratum.
Azalea	To control plant height	A-Rest	26 ppm spray (12.6 oz/gal)	Contact floricultural specialists at NC State University.
	To promote flower initiation	B-Nine	1,500 to 2,500 ppm spray (0.24 to 0.39 oz/gal)	Apply solution when new growth from final pinch is 1 to 2 inches long.
		Cycocel	1,000 to 4,000 ppm spray (1.08 to 4.34 fl oz/gal)	Optimum rates are generally between 1,000 and 2,000 ppm. Two to six multiple sprays may be needed. Apply first application when new growth is approximately 2 inches long.
	To promote lateral shoot growth on vegetative plants	Off-Shoot-O	Use a 3 to 5% solution (8.6 to 14 fl oz/gal) solution in greenhouses; use 5 to 7% (14 to 20 fl oz/gal) outdoors. Apply as a foliar spray.	Efficacy is related to relative humidity and temperature. Spray a few plants to check activity prior treating the entire crop; effects should be visible in about 1 hour. Be certain chemical covers shoot tip. Ineffective if microscopic flower buds are present.
	To increase lateral branching	Atrimmec	3,125 to 6,250 ppm spray (2 to 4 fl oz/gal)	Contact floricultural specialists at NC State University.
		Florel	2,471 to 4,943 ppm spray (8 to 16 fl oz/gal)	
	To control plant height, reduce bypass shoot elongation, and promote flower bud initiation	Bonzi	100 to 200 ppm spray (3.2 to 6.4 fl oz/gal)	To control plant height and promote flower bud initiation, apply after final shaping, when new growth is 1.5 to 2 inches long. To reduce bypass shoot development, apply after bud set, when bypass shoots are barely visible.
			0.59 to 1.77 mg a.i. drench for a 6 inch pot (0.16 to 0.48 fl oz/gal of drench solution; apply 4 fl oz/6 inch pot)	Drench volumes and mg a.i. vary with pot size. Contact floricultural specialists at NC State University.
	To control plant height	Sumagic	10 to 15 ppm spray (2.56 to 3.84 fl oz/gal)	Apply at 1.5 quarts per 100 square feet of bench area. Contact floricultural specialists at NC State University for further application information.

[1]From Bailey (1997).
[2]a.i. = active ingredient

Crop	Purpose	Chemical	Rate	Precautions and Remarks
Azalea, continued	For partial or full substitution of cold	GibGro	250 to 1,000 ppm spray (1 to 4 fl oz/gal)	GibGro 4LS has 24(c) registration for distribution and greenhouse use only within North Carolina. Spray timing, concentration, and number of applications varies with cultivar as well as intended degree of cold substitution. Consult the label for exact recommendations.
	To prevent flower bud initiation during vegetative growth	GibGro	100 to 750 ppm spray (0.4 to 3 fl oz/gal)	GibGro 4LS has 24(c) registration for distribution and greenhouse use only within North Carolina. Apply two to three sprays at 2 to 3 weeks intervals after each pinch.
Bedding Plants (Not specifically listed in this table)	To control plant height	A-Rest	6 to 66 ppm spray (2.9 to 32 fl oz/gal)	See Ageratum.
			0.06 to 0.12 mg a.i. drench for a 4 inch pot (0.5 to 1 fl oz/gal of drench solution; apply 2 fl oz/4 inch pot)	Drench volumes and mg a.i. vary with pot size. Contact floricultural specialists at NC State University.
		B-Nine + Cycocel	800 to 5,000 ppm B-Nine (0.13 to 0.79 oz/gal) + 1,000 to 1,500 ppm Cycocel (1.08 to 1.63 fl oz/gal) applied as a tank mix spray	It is recommended to use the highest rate of Cycocel that does not cause excessive leaf yellowing, and then adjust the B-Nine rate up and down within the labeled range to attain desired level of height control.
		Bonzi	30 ppm spray (0.96 fl oz/gal)	Users should conduct trials on a small number of plants, adjusting the rates as needed for desired final plant height and duration of height control.
			0.118 mg a.i. drench for a 6 inch pot (0.032 fl oz/gal of drench solution; apply 4 fl oz/6 inch pot)	Drench applications are recommended only for bedding plants in 6 inch or larger containers.
		Cycocel	800 to 3,000 ppm spray (0.87 to 3.25 fl oz/gal)	Users should conduct trials on a small number of plants, adjusting the rates as needed for desired final plant height and duration of height control.
Bedding Plant Plugs (Not specifically listed in this table)	To control plant height	A-Rest	3 to 35 ppm spray (1.5 to 17 fl oz/gal)	See Ageratum.
			Drench plug flats with a 0.5 to 1 ppm solution (0.25 to 1 fl oz/ gal)	For uniform application, use a subirrigation delivery system. Plug trays should not be excessively dry prior to the subirrigation treatment.
		B-Nine + Cycocel	800 to 5,000 ppm B-Nine (0.13 to 0.79 oz/gal) + 1,000 to 1,500 ppm Cycocel (1.08 to 1.63 fl oz/gal) applied as a tank mix spray	It is recommended to use the highest rate of Cycocel that does not cause excessive leaf yellowing, and then adjust the B-Nine rate up and down within the labeled range to attain desired level of height control.
		Bonzi	5 ppm spray (0.16 fl oz/gal)	Users should conduct trials on a small number of plants, adjusting the rate as needed for desired final plant height and duration of height control. Plants should develop 1 to 2 true leaves prior to first application.
		Cycocel	400 to 1,500 ppm spray (0.43 to 1.63 fl oz/gal)	Users should conduct trials on a small number of plants, adjusting the rates as needed for desired final plant height and duration of height control.
Begonia	To control plant height	A-Rest	3 to 15 ppm spray (1.5 to 7.3 fl oz/gal)	See Ageratum.
		B-Nine	5,000 ppm spray (0.79 oz/gal)	See Ageratum.
Begonia, Elatior	To increase lateral branching	Atrimmec	781 to 1,562 ppm spray (0.5 to 1.0 fl oz/gal)	
Bleeding Heart	To control plant height	A-Rest	65 to 132 ppm spray (31.5 to 64 fl oz/gal)	

Table 13-1 (continued)

Crop	Purpose	Chemical	Rate	Precautions and Remarks
Bleeding Heart, continued	To control plant height	A-Rest	0.25 to 0.50 mg a.i. drench for a 6 inch pot (1 to 2 fl oz/gal of drench solution; apply 4 fl oz/6 inch pot)	Drench volumes and mg a.i. vary with pot size. Contact floricultural specialists at NC State University.
Bougainvillea	To increase lateral branching	Atrimmec	1,562 ppm spray (1 fl oz/gal)	
Bromeliad	To promote flower initiation	Florel	2,471 ppm spray (8 fl oz/gal)	Contact floricultural specialists at NC State University.
Bulb Crops (Not specifically listed in this table)	To control plant height	A-Rest	25 to 50 ppm spray (12.1 to 24.2 fl oz/gal)	
			0.50 mg a.i. drench for a 6 inch pot (2 fl oz/gal of drench solution; apply 4 fl oz/6 inch pot)	Drench volumes and mg a.i. vary with pot size. Contact floricultural specialists at NC State University.
		Bonzi	100 ppm spray (3.2 fl oz/gal)	Users should conduct trials on a small number of plants, adjusting the rate as needed for desired final plant height and length of height control.
			1.183 mg a.i. drench for a 6 inch pot (0.32 fl oz/gal of drench solution; apply 4 fl oz per 6 inch pot)	Drench volumes and mg a.i. vary with pot size. Contact floricultural specialists at NC State University.
			20 ppm bulb soak (0.64 fl oz/gal)	Soak for 15 minutes. Users should conduct trials on a small number of bulbs, adjusting the rate and soaking period (up to 1 hour) as needed for desired final plant height.
Caladium	To control plant height	Bonzi	100 to 200 ppm spray (3.2 to 6.4 fl oz/gal)	First spray applications should be made when plants are 2 to 4 inches tall.
			1.183 to 2.366 mg a.i. drench for a 6 inch pot (0.32 to 0.64 fl oz/gal of drench solution; apply 4 fl oz/6 inch pot)	First drench applications should be made when plants are 1 to 2 inches tall. Drench volumes and mg a.i. vary with pot size.
Calla Lily	To control plant height	Bonzi	1.183 to 3.549 mg a.i. drench for a 6 inch pot (0.32 to 0.96 fl oz/gal of drench solution; apply 4 fl oz/6 inch pot)	See Caladium.
			20 ppm rhizome/tuber soak (0.64 fl oz/gal)	Soak the rhizomes/tubers for 15 minutes prior to planting.
Celosia	To control plant height	A-Rest	7 to 26 ppm spray (3.4 to 12.6 fl oz/gal)	See Ageratum.
		B-Nine	2,500 ppm spray (0.39 oz/gal)	
		Bonzi	4 to 50 ppm spray (0.13 to 1.60 fl oz/gal)	
		Cycocel	400 to 3,000 ppm spray (0.43 to 3.25 fl oz/gal)	
		Sumagic	10 to 20 ppm spray (2.56 to 5.12 fl oz/gal)	
China Aster	To control plant height	A-Rest	7 to 26 ppm spray (3.4 to 12.6 fl oz/gal)	
Chrysanthemum, Cut	To reduce "neck" stretching	B-Nine	2,500 ppm spray (0.39 oz/gal)	Spray upper foliage 5 weeks after start of short-day treatment.
Chrysanthemum, Potted	To control plant height	A-Rest	25 to 50 ppm spray (12.1 to 24.2 fl oz/gal)	Contact floricultural specialists at NC State University.
			0.25 to 0.50 mg a.i. drench for a 6 inch pot (1 to 2 fl oz/gal of drench solution; apply 4 fl oz/6 inch pot)	Drench volumes and mg a.i. vary with pot size. Contact floricultural specialists at NC State University.
		B-Nine	1,000 ppm preplant foliar dip (0.16 oz/gal)	Contact floricultural specialists at NC State University.

Table 13-1 *(continued)*

Crop	Purpose	Chemical	Rate	Precautions and Remarks
Chrysanthemum, Potted, continued	To control plant height	B-Nine	1,250 to 5,000 ppm spray (0.20 to 0.79 oz/gal)	Spray when new growth from pinch is 1 to 2 inches long. Some varieties may require another application 3 weeks later.
		Bonzi	50 to 200 ppm spray (1.6 to 6.4 fl oz/gal)	Contact floricultural specialists at NC State University.
			0.118 to 0.473 mg a.i. drench for a 6 inch pot (0.032 to 0.128 fl oz/gal of drench solution; apply 4 fl oz/6 inch pot)	Drench volumes and mg a.i. vary with pot size. Contact floricultural specialists at NC State University.
		Sumagic	2.5 to 10 ppm spray (0.64 to 2.56 fl oz/gal)	Contact floricultural specialists at NC State University.
Chrysanthemum, Garden	To increase lateral branching	Florel	500 ppm spray (1.619 fl oz/gal)	Florel applications will provide some growth retardant effects. A delay in flowering will also occur with the use of Florel. Read the label for restrictions on timing of applications.
Clematis	To control plant height	A-Rest	25 to 132 ppm spray (12.1 to 64 fl oz/gal)	
			0.25 to 0.50 mg a.i. drench for a 6 inch pot (1 to 2 fl oz/gal of drench solution; apply 4 fl oz/6 inch pot)	Drench volumes and mg a.i. vary with pot size. Contact floricultural specialists at NC State University.
Cleome	To control plant height	A-Rest	7 to 26 ppm spray (3.4 to 12.6 fl oz/gal)	See Ageratum.
		Cycocel	400 to 3,000 ppm spray (0.43 to 3.25 fl oz/gal)	
Clerodendrum	To increase lateral branching	Atrimmec	1,042 to 2,083 ppm spray (0.67 to 1.33 fl oz/gal)	
Coleus	To control plant height	Bonzi	5 to 45 ppm spray (0.16 to 1.44 fl oz/gal)	See Ageratum.
		Cycocel	400 to 3,000 ppm spray (0.43 to 3.25 fl oz/gal)	
		Sumagic	10 to 20 ppm spray (2.56 to 5.12 fl oz/gal)	
Columbine	To control plant height	A-Rest	65 to 132 ppm spray (31.5 to 64 fl oz/gal)	
			0.25 to 0.50 mg a.i. drench for a 6 inch pot (1 to 2 fl oz/gal of drench solution; apply 4 fl oz/6 inch pot)	Drench volumes and mg a.i. vary with pot size. Contact floricultural specialists at NC State University.
Cornflower	To control plant height	A-Rest	7 to 26 ppm spray (3.4 to 12.6 fl oz/gal)	See Ageratum.
Cosmos	To control plant height	B-Nine	5,000 ppm spray (0.79 oz/gal)	See Ageratum.
Crossandra	To control plant height	B-Nine	2,500 ppm spray (0.39 oz/gal)	
Daffodil	To control plant height	Bonzi	2.366 to 4.732 mg a.i. drench for a 6 inch pot (0.64 to 1.28 fl oz/gal of drench solution; apply 4 fl oz/6 inch pot)	See Caladium.
		Florel	1,000 to 2,000 ppm spray (3.24 to 6.47 fl oz/gal)	Contact floricultural specialists at NC State University.
Dahlia	To control plant height	A-Rest	7 to 26 ppm spray (3.4 to 12.6 fl oz/gal)	See Ageratum.
			0.25 to 0.50 mg a.i. drench for a 6 inch pot (1 to 2 fl oz/gal of drench solution; apply 4 fl oz/6 inch pot)	Drench volumes and mg a.i. vary with pot size. Contact floricultural specialists at NC State University.
		B-Nine	5,000 ppm spray (0.79 oz/gal)	See Ageratum.

Crop	Purpose	Chemical	Rate	Precautions and Remarks
Dahlia, continued	To control plant height	Bonzi	5 to 45 ppm spray (0.16 to 1.44 fl oz/gal)	See Ageratum.
		Cycocel	400 to 3,000 ppm spray (0.43 to 3.25 fl oz/gal)	
		Sumagic	10 to 20 ppm spray (2.56 to 5.12 fl oz/gal)	
Delphinum	To control plant height	A-Rest	35 to 132 ppm spray (17 to 64 fl oz/gal)	See Ageratum.
			0.25 to 0.50 mg a.i. drench for a 6 inch pot (1 to 2 fl oz/gal of drench solution; apply 4 fl oz/6 inch pot)	Drench volumes and mg a.i. vary with pot size. Contact floricultural specialists at NC State University.
Dianthus	To control plant height	A-Rest	7 to 26 ppm spray (3.4 to 12.6 fl oz/gal)	See Ageratum.
		Bonzi	5 to 60 ppm spray (0.16 to 1.92 fl oz/gal)	
		Cycocel	400 to 3,000 ppm spray (0.43 to 3.25 fl oz/gal)	
Dracaena	To control plant height	A-Rest	25 to 132 ppm spray (12.1 to 64 fl oz/gal)	
			0.25 to 0.50 mg a.i. drench for a 6 inch pot (1 to 2 fl oz/gal of drench solution; apply 4 fl oz/6 inch pot)	Drench volumes and mg a.i. vary with pot size. Contact floricultural specialists at NC State University.
Dusty Miller	To control plant height	B-Nine	5,000 ppm spray (0.79 oz/gal)	See Ageratum.
Easter Lily	To control plant height	A-Rest	50 ppm spray (24.2 fl oz/gal)	Contact floricultural specialists at NC State University.
			0.25 to 0.5 mg a.i. drench for a 6 inch pot (1 to 2 fl oz/gal of drench solution; apply 4 fl oz/6 inch pot)	Drench volumes and mg a.i. vary with pot size. Contact floricultural specialists at NC State University.
		Sumagic	10 to 25 ppm spray (2.56 to 6.4 fl oz/gal)	Contact floricultural specialists at NC State University.
			0.03 to 0.06 mg a.i. drench for a 6 inch pot (0.065 to 0.13 fl oz/gal of drench solution; apply 4 fl oz/6 inch pot)	Drench volumes and mg a.i. vary with pot size. Contact floricultural specialists at NC State University.
Exacum	To control plant height	B-Nine	2,500 ppm spray (0.39 oz/gal)	
Fatshedera	To control plant height	A-Rest	65 to 132 ppm spray (31.5 to 64 fl oz/gal)	
			0.25 to 0.50 mg a.i. drench for a 6 inch pot (1 to 2 fl oz/gal of drench solution; apply 4 fl oz/6 inch pot)	Drench volumes and mg a.i. vary with pot size. Contact floricultural specialists at NC State University.
Flowering/Foliage Plants, Herbaceous Species (Not specifically listed in this table)	To control plant height	A-Rest	20 to 50 ppm spray (9.7 to 24.2 fl oz/gal)	Recommended starting rate for an A-Rest spray on a new herbaceous flowering or foliage species is 33 ppm (16 fl oz/gal).
			0.125 to 0.25 mg a.i. drench for a 6 inch pot (0.5 to 1 fl oz/gal of drench solution; apply 4 fl oz/6 inch pot)	Drench volumes and mg a.i. vary with pot size. Contact floricultural specialists at NC State University.
		Bonzi	30 ppm spray (0.96 fl oz/gal)	Users should conduct trials on a small number of plants, adjusting the rate as needed for desired final plant height and length of height control.
			0.118 mg a.i. drench for a 6 inch pot (0.032 fl oz/gal of drench solution; apply 4 fl oz/6 inch pot)	Drench volumes and mg a.i. vary with pot size. Contact floricultural specialists at NC State University.

Crop	Purpose	Chemical	Rate	Precautions and Remarks
Flowering/Foliage Plants, Herbaceous Species (Not specifically listed in this table), continued	To control plant height	Cycocel	800 to 4,000 ppm spray (0.87 to 4.34 fl oz/gal)	Optimum rate depends on species, desired amount of height control, and environmental conditions. The suggested initial rate for small-scale trials is 1,250 ppm. Example herbaceous species known to respond to cycocel are Achimenes, Aster, Astilbe, Begonia (hiemalis), Begonia (tuberous), Calceolaria, Carnation, Chrysanthemum, Columbine, Easter lily, *Gynura aurantiaca*, Ivy, Kalanchoe, *Lilium* spp., Morning glory, Pachystachys, *Pilea* spp., Pentas, *Salvia* spp., Schefflera, *Sedum* spp., and Sunflower.
			2,000 to 4,000 ppm drench	Drench volumes vary with pot size. See label for recommended volumes. Herbaceous species known to respond to cycocel are listed above.
Flowering/Foliage Plants, Woody Species (Not specifically listed in this table)	To control plant height	A-Rest	50 ppm spray (24.2 fl oz/gal)	
			0.25 mg a.i. drench for a 6 inch pot (1 fl oz/gal of drench solution; apply 4 fl oz/6 inch pot)	Drench volumes and mg a.i. vary with pot size. Contact floricultural specialists at NC State University.
		Bonzi	50 ppm spray (1.6 fl oz/gal)	Users should conduct trials on a small number of plants, adjusting the rate as needed for desired final plant height and length of height control.
			0.237 mg a.i. drench for a 6 inch pot (0.064 fl oz/gal of drench solution; apply 4 fl oz/6 inch pot)	Drench volumes and mg a.i. vary with pot size. Contact floricultural specialists at NC State University.
		Cycocel	800 to 4,000 ppm spray (0.87 to 4.34 fl oz/gal)	Optimum rate depends on species, desired amount of height control, and environmental conditions. The suggested initial rate for small-scale trials is 1,250 ppm. Example woody species known to respond to cycocel are *Baleria cristata*, Bougainvillea, Camellia, Gardenia, Fuchsia, Hollies, Hydrangea, Lantana, *Pseuderanthemum lactifolia*, Rhododendron, and Roses (potted).
			2,000 to 4,000 ppm drench	Drench volumes vary with pot size. See label for recommended volumes. Woody species known to respond to cycocel are listed above.
Freesia	To control plant height	Bonzi	100 to 300 ppm corm soak (3.2 to 9.6 fl oz/gal)	Soak corms in the solution for 1 hour before planting.
Fuchsia	To increase lateral branching	Atrimmec	781 to 2,343 ppm spray (0.5 to 1.5 fl oz/gal)	
		Florel	500 ppm spray (1.619 fl oz/gal)	Florel applications will provide some growth retardant effects. A delay in flowering will also occur with the use of Florel. Read the label for restrictions on timing of applications.
Gardenia	To control plant height	A-Rest	50 ppm spray (24.2 fl oz/gal)	
			0.25 mg a.i. drench for a 6 inch pot (1 fl oz/gal of drench solution; apply 4 fl oz/6 inch pot)	Drench volumes and mg a.i. vary with pot size. Contact floricultural specialists at NC State University.
		B-Nine	5,000 ppm spray (0.79 oz/gal)	
	To increase lateral branching	Atrimmec	2,343 to 4,687 ppm spray (1.5 to 3.0 fl oz/gal)	
Geranium	To control plant height	A-Rest	26 to 66 ppm spray (12.6 to 32 fl oz/gal)	See Ageratum.
		Bonzi	10 to 30 ppm spray (0.32 to 0.96 fl oz/gal)	Apply to zonal geraniums when new growth is 1.5 to 2 inches long. Apply to seed geraniums approximately 2 to 4 weeks after transplanting.

Crop	Purpose	Chemical	Rate	Precautions and Remarks
Geranium, continued	To control plant height	Cycocel	800 to 1,500 ppm spray (0.87 to 1.63 fl oz/gal)	First application should be made 2 to 4 weeks after planting plugs or rooted cuttings (after stems have started elongating). Multiple applications may be needed.
		Sumagic	3 to 6 ppm spray (0.77 to 1.54 fl oz/gal) for cutting geraniums and 2 to 4 ppm spray (0.51 to 1.02 fl oz/gal) for seed geraniums	See Ageratum.
	To promote earlier flowering in seed geraniums	Cycocel	1,500 ppm spray (1.63 fl oz/gal)	Make two applications at 35 and 42 days after seeding. Treated plants should flower earlier, be more compact, and more well-branched than untreated plants.
	To increase lateral branching	Atrimmec	1,562 ppm spray (1 fl oz/gal)	Labeled for ivy geraniums only.
		Florel	500 to 1,000 ppm spray (1.619 to 3.24 fl oz/gal)	Labeled for zonal and ivy geraniums. Use the lower concentration for ivy geraniums. Florel will also provide some growth retardant effect. A delay in flowering will also occur with the use of Florel. Read the label for restrictions on timing of applications.
Gerbera Daisy	To control plant height	A-Rest	25 to 132 ppm spray (12.1 to 64 fl oz/gal)	
			0.25 to 0.50 mg a.i. drench for a 6 inch pot (1 to 2 fl oz/gal of drench solution; apply 4 fl oz/6 inch pot)	Drench volumes and mg a.i. vary with pot size. Contact floricultural specialists at NC State University.
Gomphrena	To control plant height	Cycocel	400 to 3,000 ppm spray (0.43 to 3.25 fl oz/gal)	See Ageratum.
Grape Ivy	To increase lateral branching	Atrimmec	781 to 1,562 ppm spray (0.5 to 1 fl oz/gal)	
Hibiscus	To control plant height	Bonzi	30 to 150 ppm spray (0.96 to 4.8 fl oz/gal)	Application should be made when laterals are 1 to 4 inches long. Single applications control lateral growth for 3 to 6 months.
		Cycocel	200 to 600 ppm spray (0.22 to 0.65 fl oz/gal)	Multiple applications starting prior to first pinch are recommended.
Holly	To control plant height	A-Rest	50 ppm spray (24.2 fl oz/gal)	
			0.25 mg a.i. drench for a 6 inch pot (1 fl oz/gal of drench solution; apply 4 fl oz/6 inch pot)	Drench volumes and mg a.i. vary with pot size. Contact floricultural specialists at NC State University.
Hyacinth	To reduce stem topple	Florel	1,000 ppm spray (3.24 fl oz/gal)	Contact floricultural specialists at NC State University.
Hybrid Lily	To control plant height	Bonzi	250 to 500 ppm spray (8.0 to 16.0 fl oz/gal)	See Caladium.
			1.183 to 2.366 mg a.i. drench for a 6 inch pot (0.32 to 0.64 fl oz/gal of drench solution; apply 4 fl oz/6 inch pot)	
			20 to 30 ppm bulb soak (0.64 to 0.96 fl oz/gal)	Soak bulbs in the solution for 15 minutes prior to planting.
Hydrangea	To control plant height	A-Rest	50 ppm spray (24.2 fl oz/gal)	
			0.25 mg a.i. drench for a 6 inch pot (1 fl oz/gal of drench solution; apply 4 fl oz/6 inch pot)	Drench volumes and mg a.i. vary with pot size. Contact floricultural specialists at NC State University.
		B-Nine	2,500 to 7,500 ppm spray (0.39 to 1.18 oz/gal)	Contact floricultural specialists at NC State University.
Hypoestes	To control plant height	Cycocel	400 to 3,000 ppm spray (0.43 to 3.25 fl oz/gal)	See Ageratum.

Table 13-1 *(continued)*

Crop	Purpose	Chemical	Rate	Precautions and Remarks
Impatiens	To control plant height	A-Rest	10 to 44 ppm spray (4.8 to 21.3 fl oz/gal)	See Ageratum.
		Bonzi	5 to 45 ppm spray (1.44 to 2.02 fl oz/gal)	
		Sumagic	5 to 10 ppm spray (1.28 to 2.56 fl oz/gal)	
Jerusalem Cherry	To control plant height	Cycocel	400 to 3,000 ppm spray (0.43 to 3.25 fl oz/gal)	See Ageratum.
Kalanchoe	To increase lateral branching	Atrimmec	1,042 to 2,343 ppm spray (0.67 to 1.5 fl oz/gal)	
Lantana	To increase lateral branching	Atrimmec	781 to 1,562 ppm spray (0.5 to 1 fl oz/gal)	
		Florel	500 ppm spray (1.619 fl oz/gal)	Florel applications will provide some growth retardant effects. A delay in flowering will also occur with the use of Florel. Read the label for restrictions on timing of applications.
Liatris	To control plant height	A-Rest	25 to 132 ppm spray (12.1 to 64 fl oz/gal)	
			0.25 to 0.50 mg a.i. drench for a 6 inch pot (1 to 2 fl oz/gal of drench solution; apply 4 fl oz/6 inch pot)	Drench volumes and mg a.i. vary with pot size. Contact floricultural specialists at NC State University.
Lipstick Vine	To increase lateral branching	Atrimmec	521 to 1,042 ppm spray (0.33 to 0.67 fl oz/gal)	
Marigold	To control plant height	A-Rest	13 to 33 ppm spray (6.3 to 16 fl oz/gal)	See Ageratum.
		B-Nine	2,500 to 5,000 ppm spray (0.39 to 0.79 oz/gal)	
		Bonzi	10 to 60 ppm spray (0.32 to 1.92 fl oz/gal)	
		Cycocel	400 to 3,000 ppm spray (0.43 to 3.25 fl oz/gal)	
		Sumagic	10 to 20 ppm spray (2.56 to 5.12 fl oz/gal)	
Monstera	To control plant height	A-Rest	25 to 132 ppm spray (12.1 to 64 fl oz/gal)	
			0.25 to 0.50 mg a.i. drench for a 6 inch pot (1 to 2 fl oz/gal of drench solution; apply 4 fl oz/6 inch pot)	Drench volumes and mg a.i. vary with pot size. Contact floricultural specialists at NC State University.
Montbretia	To control plant height	Bonzi	20 to 30 ppm corm soak (0.64 to 0.96 fl oz/gal)	Soak corms in the solution for 15 minutes prior to planting.
Nastursium	To control plant height	Cycocel	400 to 3,000 ppm spray (0.43 to 3.25 fl oz/gal)	
Nepthytis, Green & Green Gold	To control plant height	A-Rest	25 to 132 ppm spray (12.1 to 64 fl oz/gal)	
			0.25 to 0.50 mg a.i. drench for a 6 inch pot (1 to 2 fl oz/gal of drench solution; apply 4 fl oz/6 inch pot)	Drench volumes and mg a.i. vary with pot size. Contact floricultural specialists at NC State University.
Pansy	To control plant height	A-Rest	3 to 15 ppm spray (1.5 to 7.3 fl oz/gal)	See Ageratum.
		Bonzi	1 to 15 ppm spray (0.03 to 0.48 fl oz/gal)	
		Sumagic	1 to 6 ppm spray (0.26 to 1.54 fl oz/gal)	See Ageratum.

Table 13-1 (continued)

Crop	Purpose	Chemical	Rate	Precautions and Remarks
Petunia	To control plant height	A-Rest	10 to 26 ppm spray (4.8 to 12.6 fl oz/gal)	See Ageratum.
		B-Nine	2,500 to 5,000 ppm spray (0.39 to 0.79 oz/gal)	
		Bonzi	5 to 60 ppm spray (0.16 to 1.92 fl oz/gal)	
		Sumagic	25 to 50 ppm spray (6.4 to 12.79 fl oz/gal)	
Philodendron	To control plant height	A-Rest	25 to 132 ppm spray (12.1 to 64 fl oz/gal)	
			0.25 to 0.50 mg a.i. drench for a 6 inch pot (1 to 2 fl oz/gal of drench solution; apply 4 fl oz/6 inch pot)	Drench volumes and mg a.i. vary with pot size. Contact floricultural specialists at NC State University.
Phlox	To control plant height	B-Nine	5,000 ppm spray (0.79 oz/gal)	
Pilea	To control plant height	A-Rest	25 to 132 ppm spray (12.1 to 64 fl oz/gal)	
			0.25 to 0.50 mg a.i. drench for a 6 inch pot (1 to 2 fl oz/gal of drench solution; apply 4 fl oz/6 inch pot)	Drench volumes and mg a.i. vary with pot size. Contact floricultural specialists at NC State University.
Poinsettia	To control plant height	A-Rest	0.06 to 0.25 mg a.i. drench for a 6 inch pot (0.25 to 1 fl oz/gal of drench solution; apply 4 fl oz/6 inch pot)	Contact floricultural specialists at NC State University.
		B-Nine	2,000 to 3,000 ppm spray (0.31 to 0.47 oz/gal)	Not effective in our studies.
		B-Nine + Cycocel	800 to 2,500 ppm B-Nine (0.13 to 0.39 oz/gal) + 1,000 to 1,500 ppm Cycocel (1.08 to 1.63 fl oz/gal) spray	Use the higher rates of this tank mix spray on stock plants and for finishing crops in very warm regions. Outside of very warm areas, growers should use the lower rates. Too late of an application can delay flowering and reduce bract size.
		Bonzi	10 to 30 ppm spray (0.32 to 0.96 fl oz/gal)	Contact floricultural specialists at NC State University.
			0.237 to 0.473 mg a.i. drench for a 6 inch pot (0.064 to 0.128 fl oz/gal of drench solution; apply 4 fl oz/6 inch pot)	Drench volume and mg a.i. vary with pot size. Consult the label for recommended volumes.
		Cycocel	800 to 1,500 ppm spray (0.87 to 1.63 fl oz/gal)	For natural season crops in N.C., do not apply cycocel after Nov. 1. Late applications can reduce bract size and delay flowering.
			3,000 to 4,000 ppm drench (3.25 to 4.34 fl oz/gal of drench solution)	Drench volume varies with pot size. Consult the label for recommended volumes.
		Sumagic	2.5 to 10 ppm spray (0.64 to 2.56 fl oz/gal)	Contact floricultural specialists at NC State University.
Portulaca	To control plant height	A-Rest	7 to 26 ppm spray (3.4 to 12.6 fl oz/gal)	See Ageratum.
Pothos	To control plant height	A-Rest	25 to 132 ppm spray (12.1 to 64 fl oz/gal)	
			0.25 to 0.50 mg a.i. drench for a 6 inch pot (1 to 2 fl oz/gal of drench solution; apply 4 fl oz/6 inch pot)	Drench volumes and mg a.i. vary with pot size. Contact floricultural specialists at NC State University.
Purple Passion (*Gynura aurantiaca*)	To control plant height	A-Rest	26 to 132 ppm spray (12.6 to 64 fl oz/gal)	

Crop	Purpose	Chemical	Rate	Precautions and Remarks
Purple Passion (*Gynura aurantiaca*) continued	To control plant height	A-Rest	0.25 to 0.50 mg a.i. drench for a 6 inch pot (1 to 2 fl oz/gal of drench solution; apply 4 fl oz/6 inch pot)	Drench volumes and mg a.i. vary with pot size. Contact floricultural specialists at NC State University.
Salvia	To control plant height	A-Rest	10 to 26 ppm spray (4.8 to 12.6 fl oz/gal)	See Ageratum.
		B-Nine	5,000 ppm spray (0.79 oz/gal)	
		Bonzi	5 to 60 ppm spray (0.16 to 1.92 fl oz/gal)	
		Cycocel	400 to 3,000 ppm spray (0.43 to 3.25 fl oz/gal)	
		Sumagic	5 to 10 ppm spray (1.28 to 2.56 fl oz/gal)	
Schefflera	To control plant height	A-Rest	25 to 132 ppm spray (12.1 to 64 fl oz/gal)	
			0.25 to 0.50 mg a.i. drench for a 6 inch pot (1 to 2 fl oz/gal of drench solution; apply 4 fl oz/6 inch pot)	Drench volumes and mg a.i. vary with pot size. Contact floricultural specialists at NC State University.
	To increase lateral branching	Atrimmec	3,125 ppm spray (2 fl oz/gal)	Labeled for *Schefflera arboricola* only.
Shrimp Plant	To increase lateral branching	Atrimmec	781 to 1,562 ppm spray (0.5 to 1 fl oz/gal)	
Snapdragon	To control plant height	A-Rest	10 to 26 ppm spray (4.8 to 12.6 fl oz/gal)	See Ageratum.
		Bonzi	5 to 90 ppm spray (0.16 to 2.88 fl oz/gal)	
		Sumagic	25 to 50 ppm spray (6.4 to 12.79 fl oz/gal)	
Spathiphyllum	To induce flower initiation	GibGro	250 ppm spray (1 fl oz/gal)	GibGro 4LS has 24(c) registration for distribution and greenhouse use only within North Carolina. One application should be made during the non-seasonal blooming period, typically June through January.
Sunflower	To control plant height	Cycocel	400 to 3,000 ppm spray (0.43 to 3.25 fl oz/gal)	
Tulip	To control plant height	A-Rest	0.125 to 0.5 mg a.i. drench for a 6 inch pot (0.5 to 2 fl oz/gal of drench solution; apply 4 fl oz/6 inch pot)	Drench volumes and mg a.i. vary with pot size. Contact floricultural specialists at NC State University.
		Bonzi	0.591 to 4.732 mg a.i. drench for a 6 inch pot (0.16 to 1.28 fl oz/gal of drench solution; apply 4 fl oz/6 inch pot)	Drench volumes and mg a.i. vary with pot size.
			2 to 5 ppm bulb soak (0.064 to 0.16 fl oz/gal)	Soak bulbs for 1 hour prior to planting.
Verbena	To control plant height	B-Nine	5,000 ppm spray (0.79 oz/gal)	See Ageratum.
		Cycocel	400 to 3,000 ppm spray (0.43 to 3.25 fl oz/gal)	
	To increase lateral branching	Atrimmec	521 to 1,042 ppm spray (0.33 to 0.67 fl oz/gal)	
		Florel	500 ppm spray (1.619 fl oz/gal)	Florel applications will provide some growth retardant effects. A delay in flowering will also occur with the use of Florel. Read the label for restrictions on timing of applications.
Vinca (*Catharanthus*)	To control plant height	A-Rest	5 to 18 ppm spray (2.4 to 8.7 fl oz/gal)	See Ageratum.
		B-Nine	2,500 ppm spray (0.39 oz/gal)	

Crop	Purpose	Chemical	Rate	Precautions and Remarks
Vinca (*Catharanthus*), continued	To control plant height	Cycocel	400 to 3,000 ppm spray (0.43 to 3.25 fl oz/gal)	
		Sumagic	1 to 3 ppm spray (0.26 to 0.77 fl oz/gal)	
Vinca Vine (*Vinca spp.*)	To increase lateral branching	Florel	500 ppm spray (1.619 fl oz/gal)	Florel applications will provide some growth retardant effects. A delay in flowering will also occur with the use of Florel. Read the label for restrictions on timing of applications.
Viola	To control plant height	Sumagic	1 to 5 ppm spray (0.26 to 1.28 fl oz/gal)	See Ageratum.
Wandering Jew	To control plant height	A-Rest	26 to 132 ppm spray (12.6 to 64 fl oz/gal)	
Woody Landscape Plants (Not specifically listed in this table)	To control plant height	A-Rest	50 ppm spray (24.2 fl oz/gal)	
			0.25 mg a.i. drench for a 6 inch pot (1 fl oz/gal of drench solution; apply 4 fl oz/6 inch pot)	Drench volumes and mg a.i. vary with pot size. Contact floricultural specialists at NC State University.
		Bonzi	0.473 mg a.i. drench for a 6 inch pot (0.128 fl oz/gal of drench solution; apply 4 fl oz/6 inch pot)	See Bedding Plants
			100 ppm spray (3.2 fl oz/gal)	
Zinnia	To control plant height	A-Rest	7 to 26 ppm spray (3.4 to 12.6 fl oz/gal)	See Ageratum.
		Cycocel	400 to 3,000 ppm spray (0.43 to 3.25 fl oz/gal)	

said to display *apical dominance* when only one shoot predominates. When apical dominance is lost, several lateral shoots usually develop simultaneously.

Auxins are effectively used for promoting root formation on cuttings. Materials commercially used include IBA, IPA, and NAA. IBA and NAA are often found in combination. Many types of cuttings benefit from the use of rooting substances. Root formation occurs faster, and in the end the root system is usually more extensive. The benefit is least on plant species that normally root quickly, and there are a few species where no benefit is seen.

Rooting compounds are very concentrated and so are always diluted. Talc powder is a customary diluent. Active-ingredient concentrations of 0.1 to 1.0 percent are used—the lower concentrations are used for easy-to-root soft cuttings and the higher concentrations are used for slower-to-root woody cuttings. The base of the cutting is dipped into the powder and then tapped to remove all but a thin film of powder. To reduce the possibility of disease transfer, a duster is often used.

Rooting compounds can be diluted by another method for use on woody cuttings where penetration is difficult. A concentrated stock solution of the rooting compound is made by dissolving it in alcohol. The stock solution is further diluted with water to a final concentration in the range of 500 to 5,000 ppm (0.05 to 0.5 percent). The cut end of cuttings is dipped in this solution for a short time and then "stuck" into propagation substrate in a propagation bed. The concentration of the solution and the length of dipping time (five seconds to a few minutes) are determined by the ease of rooting and the penetrability into the woody stem of the cutting.

Rooting compounds are a very common and valuable aid to the propagators of greenhouse crops, since so many crops are propagated by cuttings. African violet, azalea, begonia, carnation, chrysanthemum, geranium, hydrangea, kalanchoe, poinsettia, and many green plants are examples of plants that benefit from rooting compounds.

Gibberellins

GA inhibits root formation on leaves and stems; thus, it is not found in root-promoting products. It is used by gardeners for enlarging the size of camellia blooms. GA sprayed on geranium flowers at the time of first color appearance (at a concentration of 5 ppm) stimulates a 25 to 50 percent increase in flower size. The number of petals remains constant, but each petal is larger. When greater concentrations are applied, however, increased responses carry an adverse effect. Stems and flower stalks elongate and become thinner. Stems may become adversely weak; flowers that are normally flat may become undesirably spikelike.

Flowering of cyclamen can be accelerated by four to five weeks with a single spray of 50 ppm GA 60 to 75 days prior to the anticipated flower date (Widmer, Stephens, and Angell 1974). Higher concentrations result in adversely tall and weak flower stems. Lyons and Widmer (1983) suggest applying 0.25 ounce (8 mL) of 15 ppm GA_3 solution to the crown of the plant below the leaves 150 days after seed is sown.

Researchers have used gibberellins to replace the cold treatment of azalea. In the cold treatment, when the plant has reached sufficient size, it is pinched for the last time. New shoots are allowed to develop for about six weeks, and then flower-bud initiation is induced by about six weeks of long-night treatment. Once flower buds are established, a period of six weeks at a temperature of 45°F (7°C) or lower is required for development of flower buds. After this treatment, the plants are moved to the greenhouse and are forced into bloom in four to six weeks.

The cold treatment is expensive, requiring costly moving of plants and also cooler facilities (Figure 13-1). Considerable efforts have been made to reduce or eliminate the cold treatment (Boodley and Mastalerz 1959). Five weekly sprays of GA_{4+7} or GA_3 at a concentration of 1,000 ppm have proven effective (Figure 13-2) (Larson and Sydnor 1971; Nell and Larson 1974). The five consecutive weekly sprays begin when flower buds are well developed after the short-day treatment. Plants treated in this manner usually flower earlier and have larger blossoms than plants given the cold treatment. Most cultivars respond well; however, there can be some variation. For instance, flower pedicels (flower stems) may become too long, causing flowers to droop.

There have also been studies on partial replacement of the cold treatment. In one such study, after three weeks of cold treatment, plants were moved to the greenhouse for forcing, and three weekly sprays of GA_3 at 250 ppm were made.

Figure 13-1 A cooler is required for the growth of certain crops that need a cold period for flower-bud development. Such crops include azalea, Easter lily, flowering bulbs, and hydrangea.

Half of the cold treatment was eliminated, thereby permitting twice the volume of plants to be moved through the cooler facilities.

Hydrangea is also subjected to a period of cold storage. On occasion, the plants are removed prematurely, and slow development, small flowers, and short stems ensue. Research studies show promise of eliminating this situation by a spray of GA at a concentration of 5 to 50 ppm.

A 250-ppm GA spray applied to fuchsia four times at weekly intervals temporarily prevents flowering and stimulates rapid growth (Heins, Widmer, and Wilkins 1979). This could lend itself well to production of tree-type fuchsia. Tree-type geraniums can likewise be produced (Carlson 1982) by applying GA_3 as a spray to plants two weeks after potting. A total of five weekly applications of 250 ppm must be applied. A tolerable delay in flowering occurs. Excessive GA application results in distorted growth and poor plant quality.

Figure 13-2 Azalea 'Dogwood' plants during greenhouse forcing. The plant on the left received the standard cold treatment for flower-bud development. It is fully budded and will bloom in a few weeks. The plant on the right, rather than being placed in a cooler, was left in the greenhouse and sprayed five times at weekly intervals with gibberellic acid at a concentration of 1,000 ppm. In addition to replacing the cold treatment, this chemical hastened flowering and resulted in larger flowers. *(Photo courtesy of R. A. Larson, Department of Horticultural Science, North Carolina State University, Raleigh, NC 27695-7609.)*

Florel

Ethephon is the common name for the commercial product Florel (produced by Rhone-Poulenc Ag. Co., Research Triangle Park, NC 27709.) It is a 3.9% liquid concentrate of the chemical [(2-chloroethyl) phophonic acid]. Ethephon undergoes a chemical conversion that releases ethylene to the plant. Depending on the plant type and stage of growth, one or more of the following desirable responses can be induced: flowering, increased branching, height retardation, and leaf drop.

The commercial appeal of bromeliads is enhanced by the presence of a flower stalk. Flowering can be induced in two months' time by pouring 1/3 ounce (10 mL) of a diluted solution of ethephon (1.54 oz ethephon/gal water; 12 mL/L) into the vase of plants at least 18 to 24 months old (Heins, Widmer, and Wilkins 1979). Flowering of Dutch iris bulbs is likewise affected by ethephon. A number of bulbs, particularly smaller bulbs, fail to bloom when they are greenhouse forced. A spray of 156 ppm (4 mL ethephon/L water) applied to green plants in the Dutch bulb-production fields just prior to bulb harvest led to earlier flowering, less bud abortion, and fewer leaves during greenhouse forcing (Kamerbeek, Durieux, and Schipper 1980). British work showed that the reduction in leaf number permitted increased plant density from 14 to 30 bulbs/ft^2 (150 to 320 bulbs/m^2) (Krause 1984). Many Dutch iris bulbs are now treated by the producer in the storage area after harvest with 500 ppm ethylene gas for 24 hours to promote earlier and more extensive flowering on higher-quality plants.

As might be expected, ethylene plays a role in fruit maturation. Ethylene gas, or the ethylene-producing compound ethephon, is used to ripen apples, bananas, coffee, grapefruit, oranges, peppers, tobacco, and other fruit. Ethephon spray is applied to processing tomatoes in the field to hasten maturity. Fresh-use field tomatoes are treated after harvest in the mature-green stage with 200 ppm ethylene gas. This enhances color formation and hastens ripening by about two days (Lutz and Hardenburg 1968).

Leaf abscission, like flower and fruit formation, is part of the maturation process. It is likewise enhanced by ethephon. This has a commercial advantage in hydrangea production, where it is desirable to remove the leaves for the six-week cold treatment that occurs after flower-bud initiation and just prior to greenhouse forcing. An application of 1,000 to 5,000 ppm ethephon two weeks prior to the start of cold treatment has been shown to result in the defoliation of the cultivars 'Merville' and 'Rose Supreme' (Tjia and Buxton 1976).

Light intensity at the crown of rose bushes diminishes as the canopy grows larger over the years. This discourages the development of new canes from the base of the plant. Renewal of the plant is dependent in great part on the large, floriferous shoots that come from these basal breaks. The best hope for encouraging such breaks has come from scoring lower canes with a saw blade dipped in 7,500 ppm ethephon (Zeislin et al. 1972). Ethephon sprays have, in fact, been used commercially in Israel to stimulate basal branching of 'Baccara' rose.

Surprisingly, ethephon is used as a height retardant as well. Drenches or sprays serve well to control the height of narcissus (Briggs 1975; DeHertogh 1989; Moe 1980). Ethephon sprays prior to floret color result in shorter hyacinths and prevent stem topple, which is a problem with some cultivars (De-Hertogh 1989).

Cycocel

Pot plants must be grown to a height compatible with the environment in which they will be used. Many plants grow too tall if not checked. In past years, water and nutrients were withheld to reduce height, resulting in negative side effects in the appearance of the foliage and size of the bloom. Poinsettia stems were sometimes folded (Figure 13-3) to reduce height, which was an effective but time-consuming process. Cycocel is used today.

Height retardants, in general, result in shorter stem internodes but do not affect the number of leaves formed. Stems are thicker and leaves are deeper green because chlorophyll is more dense in the smaller cells. As a result, plants have a very pleasing appearance.

Cycocel [(2-chloroethyl) trimethylammonium chloride] is available in a liquid formulation containing 11.8% active ingredient. The common name of the chemical is *chlormequat.* (Cycocel is produced by American Cyanamid Company, P.O. Box 400, Princeton, NJ 08540.)

Figure 13-3 Years ago, poinsettia stems were often folded to reduce their height, as shown here. This time-consuming process has been replaced by the use of the chemical Cycocel. Chemically treated plants have the same number of leaves on shorter, thicker stems with deeper green foliage. *(Photo courtesy of J. W. Love, Department of Horticultural Science, North Carolina State University, Raleigh, NC 27695-7609.)*

Cycocel is applied as a spray to azaleas when about 1 inch (2.5 cm) of new growth occurs after the plants have been pinched for the last time. This checks growth and prompts early flower-bud initiation. Quite often, a larger number of flower buds develop. The retardant helps further by reducing the formation of vegetative shoots at the time of flower-bud development. These undesirable side shoots give the plant an unbalanced appearance.

Cycocel is recommended as a drench or a spray for poinsettia height retardation. Sprays can result in blotchy yellowing of foliage about 24 hours after application. This is a temporary situation and is not noticed at flowering time. Cycocel is also labeled for height control of geranium and hibiscus.

B-Nine SP

B-Nine SP (N-dimethylaminosuccinamic acid) is known under the common name *daminozide.* (B-Nine SP is produced by Uniroyal Chemical Co., World Headquarters, Middlebury, CT 06749.) It is an effective height retardant labeled for use on azalea, pot chrysanthemum, cut chrysanthemum, crossandra, exacum, foliage plants (except in California), gardenia, hydrangea, poinsettia, and bedding plants including ageratum, aster, begonia, *Celosia,* cosmos, dahlia, dusty miller, marigold, petunia, phlox, salvia, verbena, and vinca. B-Nine SP is sold as a soluble powder containing 85% active ingredient plus a wetting agent and is applied as a foliar spray to the upper leaf surfaces.

Azalea is treated with B-Nine SP for the same reason that Cycocel is used—to promote early and more extensive flower-bud set and to retard vegetative shoot development. Some cultivars of standard chrysanthemum develop a long pedicel, which is unattractive. A compact flower with a short pedicel can be produced by spraying the upper third of the foliage to the point of runoff two days after disbudding with a 0.25% concentration of B-Nine SP. Pot mums are sprayed when new shoots are 1.5 inches (4 cm) long, about two weeks after the pinch. No delay in flowering occurs. B-Nine SP is particularly useful in producing compact bedding plants but is not effective on coleus, French-type marigold, pansy, or snapdragon.

A-Rest

The chemical A-Rest [α-cylopropyl-α-(p-methoxyphenyl)-5-pyrimidinemethanol] (produced by SePRO, 11550 N. Meridian St., Suite 200, Carmel, IN 46032) takes the common name of *ancymidol.* A-Rest effectively controls the height of and is labeled for azalea, pot chrysanthemum (Figure 13-4), gardenia, geranium, gerbera, lilies (including Easter lily), poinsettia, tulip, several green (foliage) plants, and numerous annual and perennial plants. It is purchased as a solution containing 250 mg of active ingredients per quart (264 ppm). A-Rest can be applied as a spray or as a drench, depending on the crop.

A-Rest loses activity at low pH levels. Consequently, the effectiveness of A-Rest drenches in pine-bark substrates has been found to be poor (Larson,

Figure 13-4 **Chrysanthemum 'Nob Hill' treated with the chemical height retardant A-Rest at increasing concentrations from left to right. The plant on the extreme left received no chemical treatment. Those to the right have the same number of leaves but shorter internodes, thicker stems, and deeper green foliage.** *(Photo courtesy of V. P. Bonaminio, Department of Horticultural Science, North Carolina State University, Raleigh, NC 27695-7609.)*

Love, and Bonaminio 1974; Tschabold et al. 1975). A solution to the problem was found by Simmonds and Cumming (1977) by dipping hybrid lily bulbs in A-Rest solution. This procedure worked well for Easter lily as a dip prior to cold-storage treatment (Lewis and Lewis 1982). Larson (1985) refined the procedure, calling for a dip in 24 ppm A-Rest for 30 minutes after cold-storage treatment.

Another interesting function of A-Rest has been seen in research, where it was used to induce flowering of *Clerodendron* (Koranski, Struckmeyer, and Beck 1978). Retardation of vegetative growth prompts flowering in this plant. Cycocel has a similar effect in *Clerodendron* (Hildrum 1973).

Bonzi

Bonzi is a more recent height retardant. This product contains 0.4 percent of the active ingredient [(±)-(R*,R*)-β((4-chlorophenyl)methyl)-α-(1,1,-dimethylethyl)-1 H-1,2,4-triazole-1-ethanol]. This ingredient is a member of the triazine group of compounds; its common name is *paclobutrazol.* (Bonzi is distributed by Uniroyal Chemical Co., Ltd., World Headquarters, Middlebury, CT 06749.)

Bonzi can be absorbed by the roots from a soil drench or through the shoots from a spray. In either event, it is translocated to the upper portion of each shoot, where it reduces internode elongation. If a spray is used, it is very important that the spray coat all the stems of the plant; otherwise, some shoots will grow longer than others. As in the case of A-Rest, this material is less effective in pine bark–based root substrates. Higher rates must be used if it is applied as a drench to these root substrates. Higher rates are also needed when it is applied to very vigorous-growing varieties compared to those that are naturally shorter, and when it is applied during high-temperature periods.

Bonzi is labeled for spray application on geranium, hibiscus, poinsettia, and many bedding plants; for spray or drench application on pot chrysanthemum; for bulb dip application on pot freesia; and for spray, drench, and dip application on many bulb crops.

Sumagic

Sumagic is the most recent height retardant to be introduced. The common name for this chemical is *uniconazole.* Uniconazole is produced by Sumitomo Chemical Co., Ltd.; in the United States it is distributed as Sumagic for greenhouse use by Valent USA Corp., P.O. Box 8025, Walnut Creek, CA 94596. It has chemical properties closer to Bonzi than to the other height retardants and is also a member of the triazine chemical group. The active ingredient is (E)-(+)-(S)-1-(4-chlorophenyl)-4,4-dimethyl-2-(1,2,4-triazol-1-yl)-pent-1-ene-3-ol. Sumagic is labeled for use on several bedding-plant species, azalea, pot chrysanthemum, Easter lily, geranium, hibiscus, and poinsettia. Like Bonzi, it is effective at very low rates.

Off-Shoot-O

Off-Shoot-O is primarily composed of methyl octanoate and methyl decanoate in combination with an emulsifying agent. It is produced by Cochran Corp., P.O. Box 14603, Memphis, TN 38114. Off-Shoot-O is termed a chemical pinching agent because it causes death of the terminal bud on shoots, which, in turn, results in the development of side shoots (Figure 13-5). Often, more side shoots are produced from a chemical pinch than from a manual pinch. Off-Shoot-O is applied in a very fine spray to wet the shoot tips. The remainder of the plant need not be treated, and spraying is stopped before the point of runoff. Runoff increases the possibility of injury to lateral buds and leaves.

Azaleas are effectively pinched with this chemical. Considerable labor is saved, since azaleas must be pinched many times in order to produce a large plant with numerous shoots. Concentrations of 2 to 5 oz of product per quart (63 to 155 mL/L) are used, depending upon the cultivar. A concentration of 3.2 oz/qt (100 mL/L) is common. Several species of woody ornamentals can also be chemically pinched, including *Cotoneaster, Juniperus, Ligustrum, Rhamnus,* and *Taxus.*

Figure 13-5 **A greenhouse-forcing azalea plant sprayed with the chemical pinching agent Off-Shoot-O. Note the dead terminal buds and the resulting side shoots. This process is repeated several times over a period of 12 to 18 months to develop a highly branched plant of suitable size for forcing into bloom.** *(Photo courtesy of J. W. Love, Department of Horticultural Science, North Carolina State University, Raleigh, NC 27695-7609.)*

Atrimmec

Atrimmec [sodium salt of 2,3,:4,6-bis-O-(1-methylethylidene)-a-L-xylo-2-hexulofuranosonic acid] is known by the common name *dikegulac.* (Atrimmec is produced by PBI/Gordon Corp., P.O. Box 014090, Kansas City, MO 64101-0090.) The predecessor to this product was Atrinal. Atrimmec temporarily stops shoot elongation, thereby promoting lateral branching. It is thus a pinching agent for greenhouse crops, including azalea, Elatior begonia, bougainvillea, clerodendron, fuchsia, gardenia, grape ivy, ivy geranium, kalanchoe, lantana, lipstick vine, *Pachystachys lutea* (shrimp plant), *Schefflera arboricola,* and verbena. Branching can also be enhanced in 41 species of landscape ornamentals. Atrimmec is used to prevent flowering and ultimately fruiting in glossy privet, Japanese holly, multiflora rose, and ornamental olive.

Disbudding Agents

The growth suppressants that are under study today and that show promise of commercial application are the chemical disbudding agents (Figure 13-6). Stan-

Figure 13-6 **Various chemicals are under study to find a safe disbudding agent. Buds in each leaf axil of the chrysanthemum stem on the right were removed by hand. Buds were removed from the stem on the left by spraying it with an experimental disbudding agent.** ***(Photo courtesy of R. A. Larson, Department of Horticultural Science, North Carolina State University, Raleigh, NC 27695-7609.)***

dard chrysanthemum and most pot chrysanthemum cultivars require disbudding. This process is very time consuming in that all buds except the terminal flower bud are removed from each main stem. Some chemicals are now being tested that, when sprayed on the plant, will inhibit lateral bud development and leave the terminal bud unharmed. They are not yet ready for commercial use since there is still too close a margin of safety in timing and too much variation within cultivars. If they are applied too early, the terminal bud is injured; if they are applied too late, the effectiveness is reduced.

Cost of Materials

It is difficult to make cost comparisons for height retardants and pinching agents. Not all are effective on each crop. Some are applied as a drench, while others are applied as a spray. The number of pots that can be sprayed with a gallon depends upon the density of pots in the bench and the spray equipment; a high-pressure, fine-droplet spray covers more area. The concentration of growth regulator required varies according to the crop, its stage of growth, and the weather condi-

Table 13-2 **Costs of Various Height Retardants and Pinching Agents for Treating a Single Plant in a 6-Inch (15-cm) Pot**

Regulant	*Cost of Product*	*Method of Application*	*Rate of Product Dilution*	*Rate of Application*	*Cost per 6-Inch (15-cm) Pot*
A-Rest	$38.00/qt (liquid)	Drench	0.68 oz/gal	6 oz/pot (0.25 mg/pot)	5.4
		Spray	16 oz/gal	0.33 oz/pot (0.33 mg/pot)	7.1
Atrimmec	$204.00/gal (liquid)	Spray	3 oz/gal	0.67 oz/pot	3.9
B-Nine SP	$50.00/lb (powder)	Spray	0.4 oz/gal (0.25%)	0.67 oz/pot	0.91
Bonzi	$68.00/qt (liquid)	Drench	0.05 oz/gal	4 oz/pot (0.188 mg/pot	0.41
		Spray	1 oz/gal	0.33 oz/pot	0.69
Cycocel	$133.00/gal (liquid)	Drench	1 qt/10 gal (0.3%)	6 oz/pot	27.0
		Spray	1 qt/10 gal (0.3%)	0.67 oz/pot	2.9
Off-Shoot-O	$53.00/gal (liquid)	Spray	3.2 oz/qt	0.67 oz/pot	3.4
Sumagic	$73.00/qt (liquid)	Drench	0.1 oz/gal	4 oz/pot	0.69
		Spray	2.5 oz/gal	0.67 oz/pot	3.0

tions. Calculations presented in Table 13-2 are intended to give a rough idea of the cost of using these materials. You will have to correct these values for the concentrations and amounts applied in your operation. The resulting figures indicate that chemical height control and pinching is inexpensive, considering the improved quality of the crop that is achieved.

Summary

1. The five categories of plant hormones are auxins, gibberellins, cytokinins, ethylene, and inhibitors such as abscisic acid. Auxins, gibberellins, and ethylene are used in greenhouses as growth regulators. The other greenhouse plant-growth regulators (height retardants and pinching agents) are of synthetic origin.
2. The speed and extent of root formation of cuttings can be enhanced by the auxin IAA (indole-3-acetic acid). The related compounds IBA (indole-3-

butyric acid), IPA (indolepropionic acid), and NAA (naphthalene acetic acid) are used alone and in combination in commercial rooting products.

3. Gibberellic acid (GA) can play a variety of roles in the greenhouse. It may be used to substitute totally for cold treatment of azalea or partially for hydrangea. Cyclamen flowering can be hastened, while the flower size of geranium can be enlarged. GA can also be used to enhance growth and retard flowering of fuchsia and geranium for the purpose of developing tree-type forms.
4. The roles of ethylene are likewise varied. Flowering is induced in bromeliads by the application of ethylene or Florel, an ethylene-releasing compound. Treatment of Dutch iris bulbs with ethylene gas induces earlier and more extensive flowering. Both ethylene gas and Florel are used to hasten ripening of many fruits. Hydrangea leaf abscission can be accomplished by spraying with Florel. Florel has been used to stimulate basal shoot formation in rose. Finally, height of narcissus and hyacinth can be retarded with Florel.
5. Many greenhouse pot plants grow taller than desired. Plants with short internodes can be produced by treatment with chemical height retardants. Products include the following: A-Rest, B-Nine SP, Bonzi, Cycocel, and Florel.
6. It is necessary to stimulate branching in some greenhouse crops. This has been done in the past by manually pruning shoots. Off-Shoot-O sprays accomplish this in azalea by killing the growing points. Atrimmec suppresses growth of the growing points, thereby fostering branching of azalea, Elatior begonia, bougainvillea, clerodendron, fuchsia, gardenia, grape ivy, ivy geranium, kalanchoe, lantana, *Schefflera arboricola,* shrimp plant, and verbena. Florel stimulates branching in azalea, garden chrysanthemum, and geranium.
7. There are several other uses for which commercial growth regulants have been proven effective in research studies. However, they have not been labeled through the EPA for these uses, and, as a result, it is illegal to use them for these purposes. It is the intent of this book to recommend only those chemicals and uses that have been cleared by the EPA.

References

The manufacturers of growth regulators have technical literature available covering crop responses, methods of application, modes of action, and other background information. References with an asterisk before them lend themselves well to a general overview of the subject of chemical growth regulation.

1. Bailey, D. A. 1997. Growth regulators for floricultural crops. *North Carolina Commercial Flower Growers' Bul.* 42(5):1–12.
2. Boodley, J. W., and J. W. Mastalerz. 1959. The use of gibberellic acid to force azaleas without a cold temperature treatment. *Proc. Amer. Soc. Hort. Sci.* 74:681–685.
3. Briggs, J. R. 1975. The effects on growth and flowering of the chemical growth regulator ethephon on narcissus and ancymidol on tulip. *Acta Hort.* 47:287–296.

4. Carlson, W. H. 1982. Tree geraniums. In Mastalerz, J. W., and E. J. Holcomb, eds. *Geraniums.* Vol. 3, pp. 158–160. University Park, PA: Pennsylvania Flower Growers' Association.
*5. Cathey, H. M. 1975. Comparative plant growth-retarding activities of ancymidol with ACPC, phosphon, chlormequat, and SADH on ornamental plant species. *HortScience* 10:204–216.
6. DeHertogh, A. A. 1989. *Holland bulb forcers guide,* 4th ed. Hillegom, The Netherlands: The International Flower-Bulb Center. (Available from The Netherlands Flower-Bulb Info. Center, 250 W. 57th St., Suite 629, New York, NY 10019.)
7. Dicks, J. W. 1976. Chemical restriction of stem growth in ornamentals, cereals and tobacco. *Outlook on Agr.* 9(2):69–75.
8. Heins, R. D., R. E. Widmer, and H. F. Wilkins. 1979. *Growth regulators effective on floricultural crops.* Mimeo. Dept. of Hort. and Land Architecture, Univ. of Minnesota, St. Paul, MN.
9. Hildrum, H. 1973. The effect of daylength, source of light, and growth regulators on growth and flowering of *Clerodendron thomsonae* Ball. *Scientia Hort.* 1:1–11.
10. Kamerbeek, G. A., A. J. B. Durieux, and J. A. Schipper. 1980. An analysis of the influence of Ethrel® on flowering of iris 'Ideal': an associated morphogenic physiological approach. *Acta Hort.* 109:235–241.
11. Koranski, D. S., B. E. Struckmeyer, and G. E. Beck. 1978. The role of ancymidol in *Clerodendron* flower initiation and development. *J. Amer. Soc. Hort. Sci.* 103:813–815.
12. Krause, W. 1984. Experiments in early forcing show dramatic results. *Grower* 101(3):27–31.
*13. Larson, R. A. 1985. Growth regulators in floriculture. *Hort. Rev.* 7:399–481.
14. Larson, R. A., J. W. Love, and V. P. Bonaminio. 1974. Relationship of potting mediums and growth regulators in height control. *Florists' Rev.* 155(4017):21, 59, 62.
15. Larson, R. A., and T. D. Sydnor. 1971. Azalea flower bud development and dormancy as influenced by temperature and gibberellic acid. *J. Amer. Soc. Hort. Sci.* 96:786–788.
16. Leopold, A. C., and P. E. Kriedemann. 1975. *Plant growth and development,* 2nd ed. New York: McGraw-Hill.
17. Lewis, A. J., and J. S. Lewis. 1982. Height control of *Lilium longiflorum* Thunb. 'Ace' using ancymidol bulb dips. *HortScience* 17:336–337.
*18. Luckwill, L. C. 1981. *Growth regulators in crop production.* Studies in Biology 129. London: Edward Arnold.
19. Lutz, J. M., and R. E. Hardenburg. 1968. *The commercial storage of fruits, vegetables, and florist and nursery stocks.* USDA Agr. Handbook 66.
20. Lyons, R. E., and R. E. Widmer. 1983. Effects of GA_3 and NAA on leaf lamina unfolding and flowering of *Cyclamen persicum. J. Amer. Soc. Hort. Sci.* 108:759–763.
21. Mitchell, J. W., and G. A. Livingston. 1968. Methods of studying plant hormones and growth-regulating substances. USDA Agr. Res. Ser. Agr. Handbook 336.
22. Moe, R. 1980. The use of ethephon for control of plant height in daffodils and tulips. *Acta Hort.* 109:197–204.
23. Nell, T. A., and R. A. Larson. 1974. The influence of foliar applications of GA_3, GA_{4+7}, and PBA on breaking flower bud dormancy on azalea cultivars 'Redwing' and 'Dogwood'. *J. Hort. Sci.* 49:323–328.
*24. Nickell, L. G. 1982. *Plant growth regulating chemicals.* Berlin, Germany: Springer-Verlag.
*25. Nickell, L. G., ed. 1984. *Plant growth regulating chemicals.* Vols. 1 and 2. Boca Raton, FL: CRC Press.

*26. Sachs, R. M., and W. P. Hackett. 1977. Chemical control of flowering. *Acta Hort.* 68:29–49.
*27. Seeley, J. G. 1979. Interpretation of growth regulator research with floriculture crops. *Acta Hort.* 91:83–92.
28. Shanks, J. B. 1970. Chemical growth regulation for floricultural crops. *Florists' Rev.* 147:34–35, 50–58.
*29. Shanks, J. B. 1982. Growth regulating chemicals. In Mastalerz, J. W., and E. J. Holcomb, eds. *Geraniums.* Vol. 3, pp. 106–113. University Park, PA: Pennsylvania Flower Growers' Association.
30. Simmonds, J. A., and B. G. Cumming. 1977. Bulb-dip application of growth regulating chemicals for inhibiting stem elongation of 'Enchantment' and 'Harmony' lilies. *Scientia Hort.* 6:71–81.
31. Tayama, H. K., R. A. Larson, P. A. Hammer, and T. J. Roll, eds. 1992. *Tips on the use of chemical growth regulators on floriculture crops.* Ohio Florists' Association, 2130 Stella Ct., Suite 200, Columbus, OH 43215.
32. Thomas, T. H. 1976. Growth regulation in vegetable crops. *Outlook on Agr.* 9(2):62–68.
33. Tjia, B., and J. Buxton. 1976. Influence of ethephon spray on defoliation and subsequent growth on *Hydrangea macrophylla* Thunb. *HortScience* 11:487–488.
34. Tschabold, E. E., W. C. Meredith, L. R. Guse, and E. V. Krumkalns. 1975. Ancymidol performance as altered by potting media composition. *J. Amer. Soc. Hort. Sci.* 100:142–144.
*35. Weaver, R. J. 1972. *Plant growth substances in agriculture.* San Francisco: W. H. Freeman.
36. Widmer, R. E., L. C. Stephens, and M. V. Angell. 1974. Gibberellin accelerates flowering of *Cyclamen persicum* Mill. *HortScience* 9:476–477.
*37. Wilkins, H. F., W. E. Healy, and R. D. Heins. 1979. Past, present, future plant growth regulators. *Acta Hort.* 91:23–32.
38. Zeislin, H., A. N. Halevy, V. Mor, A. Brachrach, and I. Sapir. 1972. Promotion of renewal canes in roses by ethephon. *Hort. Sci.* 7:75–76.

 CHAPTER 14

Insect Control

Insects, mites, and other pests constitute an ever-present threat to the quality of greenhouse crops. Aside from the damage done to the crop, the presence of insects on the final product is disconcerting to the consumer. Most greenhouse crops are sold for aesthetic value. Yet regulations limiting the presence of pesticides in greenhouse effluent, the banning of specific pesticides, and the declining rate of new pesticide introductions for the greenhouse are challenging the grower's ability to control pests. A careful plan of prevention and control of insects must be developed for any greenhouse range.

Integrated Pest Management (IPM)

IPM is what its name implies—a pest control system in which preventive, surveillance, and corrective measures are integrated into a holistic program. The principal goals are to attain an acceptable level of pest control while reducing the use of pesticides and minimizing the impact of the overall program on the environment. The intent of IPM is not to eliminate the use of pesticides, but rather to restrict their use to strategic situations. IPM addresses all categories of pests, including insects, mites, animal pests, pathogenic diseases, and weeds. Steps in the process, in the order in which they would likely be enacted, are as follows:

1. Weed control in and around the greenhouse.
2. Sanitation, including cleaning and pasteurization prior to crop establishment.
3. Inspection and cleanup of newly acquired plants.
4. Screening of greenhouse entrances.

5. Routine surveillance for identification and quantification of pest appearances in the greenhouses as well as accurate recording of the same.
6. Adjustment of environmental conditions to render them suppressive to the pests at hand but not to the crop.
7. Pest eradication methods consisting of either biological control or pesticide application. (Pesticide methods of application, recommendations, and safety will be discussed later in this chapter.)

Weed Control

Insects abound in our world. During the warm seasons, they live in the soil and on vegetation outside the greenhouse. They crawl or fly into the greenhouse, or they may be carried in on clothing. Smaller insects such as thrips are carried in the airstream entering the greenhouse through open ventilators or the pads of the cooling system. Weeds harbor insects and diseases. They give the greenhouse establishment an unsightly appearance, which discourages customers and weakens the morale of employees. You can be sure that weeds reduce profitability. Weeds should be cleared away from the area surrounding the greenhouse and should also be eliminated in the greenhouse because they provide a hiding place and a source of food for insects. It does no good to spray crop plants in the bench if weeds under the bench are not sprayed; yet, to spray these weeds is a waste of time and money. Some growers keep the area bare around their greenhouses, while others maintain mowed grass. Thrips develop in grass flowers; hence, mowing is important. Mowing greatly reduces the types and numbers of insects invading the greenhouse.

Be very careful in selecting an herbicide (weed killer) for use adjacent to a greenhouse. Any herbicide that volatizes (converts to a gas) is potentially dangerous. Of the four herbicides that have federal (EPA) registration for use inside greenhouses, the first three in the following list have received grower acceptance while the fourth has just been registered.

1. Diquat (Reward) can be used in aisles and under benches as a contact killer for all weeds.
2. Glyphosate (Roundup) is a contact killer labeled for eliminating the broad spectrum of weeds in greenhouses, providing there is no crop inside.
3. Pelargonic acid (Scythe) can be used for contact weed control under benches.
4. Glufosinate-ammonium (Finale) is a broad-spectrum postemergent herbicide used on greenhouse floors, under benches and around foundations.

Sanitation

The best hope for pest control is prevention. There is no better time to take preventive measures than before the crop is planted. Measures can be taken then that are not possible later. Prior presence of disease may indicate the need to pasteurize root substrate. Pot-plant benches, watering systems, plant-support systems, tools, and used plant containers should be sterilized. (All of these measures were covered in Chapter 7 and will be covered again in Chapter 15.) Unmarketable plants, portions of plants remaining after fresh flowers are removed, and leaves on the floor should all be cleaned out. They are the perfect hosts for insects and disease and become vehicles for carryover to the next crop.

Algae should be killed in benches and on the floor. This is easily and safely accomplished with the bromine in products such as Dibrom and Agribrom. Algae, in addition to being a safety hazard by virtue of its slippery nature, harbors fungus gnats. Substrate should be hosed off pot-plant benches and walks so that it does not provide a habitat for insects or disease. Improperly functioning floor drains should be cleared so that wet conditions do not occur, which, in addition to providing an environment for insects and disease, also provide high-humidity conditions conducive to foliar disease development.

Plant Entry

Insects can be brought into the greenhouse on plants. Purchased seedlings and cuttings should be inspected carefully for insects and disease. If disease or insects are present, these plants should be isolated and treated with pesticides or rejected. Established plants brought to the greenhouse by customers are a common source of insect and disease problems. You will often be approached by friends and customers who want you to rejuvenate a plant for them, which often leads to more trouble than the plant is worth. Do not allow the entry of such plants into your production area.

Insect Screens

Insect screens offer a means for reducing the pest control program for greenhouses. In principle, screens placed over intake ventilators prevent the entry of insects and related pests. In actuality, not all insects are excluded. However, a high enough proportion are prevented entry to greatly reduce the pest control program. This form of pest control is a very valuable tool for counteracting insect resistance to pesticides and meeting the challenge of a growing sensitivity of personnel and the public to pesticide exposure. It is also an important tool for combating virus diseases, such as tomato spotted wilt and impatiens necrotic spot, carried and spread by thrips.

Decreasing hole size in screens bears a loose relationship to the degree of exclusion of insects. Some screens exclude insects that are smaller than the screen hole size. Two screen types having the same average hole size may not exclude insects to the same degree. One explanation may be the uniformity of hole sizes. Alternatively, one screen may be formed from a more pliable plastic that stretches as the insect squeezes through it. Likewise, the fibers in one screen may slip more easily to permit enlargement of the hole. The shape of the insect can also determine the degree of exclusion of a screen. Thrips are smaller than whiteflies, yet thrips are excluded to a higher extent by some screens than whiteflies. This probably relates to the smooth wings on whiteflies, which slip through the screen holes more easily than the fringed wings on thrips.

These discrepancies point out the need for screen efficacy comparisons. The results of a test of the effectiveness of 24 screen types for excluding thrips and whiteflies conducted by Bell and Baker (1997) is presented in Figure 14-1. Screens varied extensively in their effectiveness. Variations occurred in the relationship between hole size and effectiveness. As expected, low-air-resistance (large-hole) screens were least effective for excluding pests. Contrary to expectation, the most effective screens were found in both the moderate- and high-air-resistance screen groups. Screen selection was further complicated by the findings that not all of the best screens for excluding one type of pest were best for excluding the other pest.

When exhaust fans are operating, a negative pressure or vacuum develops in the greenhouse because of the difficult entry of replacement air through the cooling pads. When a screen is placed over the pad, the resistance to incoming air increases further. This increases the pressure drop in the greenhouse compared to outside pressure. As the pressure drop increases, the quantity of air removed from the greenhouse decreases. This reduces cooling system effectiveness and results in an increase in greenhouse temperature. To counteract some of the loss in cooling capacity, screen areas larger than the ventilator (pad) area may need

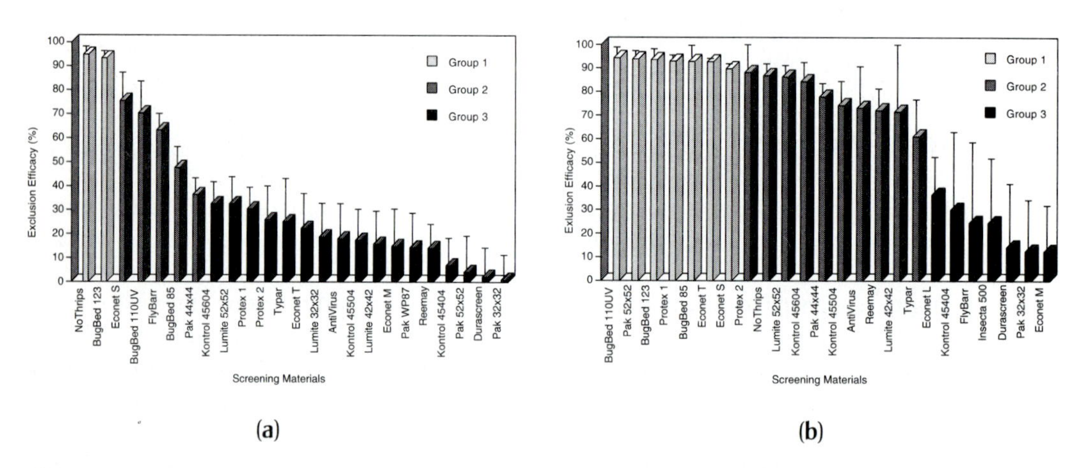

Figure 14-1 **A comparison of commercial greenhouse screens for their effectiveness in excluding (a) thrips and (b) whiteflies. Screens were tested on 0.5 × 0.5 × 1 m boxes. A screen was attached to the front and an exhaust fan to the rear of each box. The air approach velocity for each screen was 300 ft/min (92m/min). The exclusion efficiency values are the percent decrease in insects passing through the test screen when compared to a common fiberglass window screen. Group 1 screens did not differ significantly from the best screen in the comparison, Group 2 screens excluded more than the fiberglass control screen but less than the best screen, and Group 3 screens excluded a similar quantity of insects to the control screen. (*Graphs from Bell and Baker, 1997.*)**

to be used. The ratio of screen area to ventilator area can be as high as 5:1 for some screens.

Baker and Shearin (1995) recommend that the pressure drop in screened greenhouses should not exceed 0.1 in (2.5 mm) of water pressure. The pressure drop in unscreened, cooled greenhouses in North Carolina was reported to rarely exceed 0.03 in (0.76 mm) (Baker, Bethke, and Shearin 1995). If a screen adds a pressure drop equal to or smaller than the 0.1 in minus the pressure drop in the unscreened greenhouse, it is permissible to place the screen directly over the ventilator. In this case, it has an area equal to the ventilator area. On the other hand, if the resistance of the screen brings the total pressure drop in the greenhouse to more than 0.1 in, it is necessary to use more screen area than ventilator area. A procedure has been set forth for calculating the required ratio of screen-to-ventilator area (Baker and Shearin 1995). A program for use on IBM-compatible computers is also available for $10 (checks payable to the "North Carolina Agriculture Foundation") from James Baker, Entomology Extension, Box 7613, NCSU, Raleigh, NC 27695-7613.

Various arrangements for screen placement over ventilators are shown in Figure 14-2. Screens quickly become dirty. This reduces airflow and the capacity of the cooling system. Access should be provided to the screens so that they can be washed periodically from the inside. While screens can keep insects out of the greenhouse, they can likewise keep insects in. Workers should be trained to keep doors closed. Air leaks around doors, chimneys, and the like should be sealed.

There is another approach to screening specific insect-related pests without the use of screens. Slugs and snails can be kept out of benches by wrapping a band of thin copper foil tightly around each bench leg. These pests can gain access only by crawling into the bench or by being transported in on dirty pots, flats, or similar objects. Ants are a problem in a biological control system. They protect insects with high-sugar-content excretions (known as *honeydew*), such as aphids and white flies. Ants will kill predatory insects released for the purpose of controlling these pests. A sticky paste is sold that can be painted in a band on each bench leg. This will trap ants at that point.

Pest Surveillance

It is inevitable that insects and disease will enter the greenhouse. If they can be detected early, control measures can be taken before any significant damage is done and before any drastic pesticide applications are required. For one reason or another, insects will often establish themselves in a particular location in the greenhouse. It may be a warmer temperature zone or an air current that carries them there, or perhaps it is an area that is difficult to spray and thus not well protected. The grower should identify such spots and check them regularly. Insects often exhibit plant preferences. Some varieties of a crop are more attractive to them than others. These varieties should be watched carefully. Most insects seek the undersides of leaves. An inspection of plants from above may not reveal their presence. Plants should be picked up and turned on their sides to reveal the

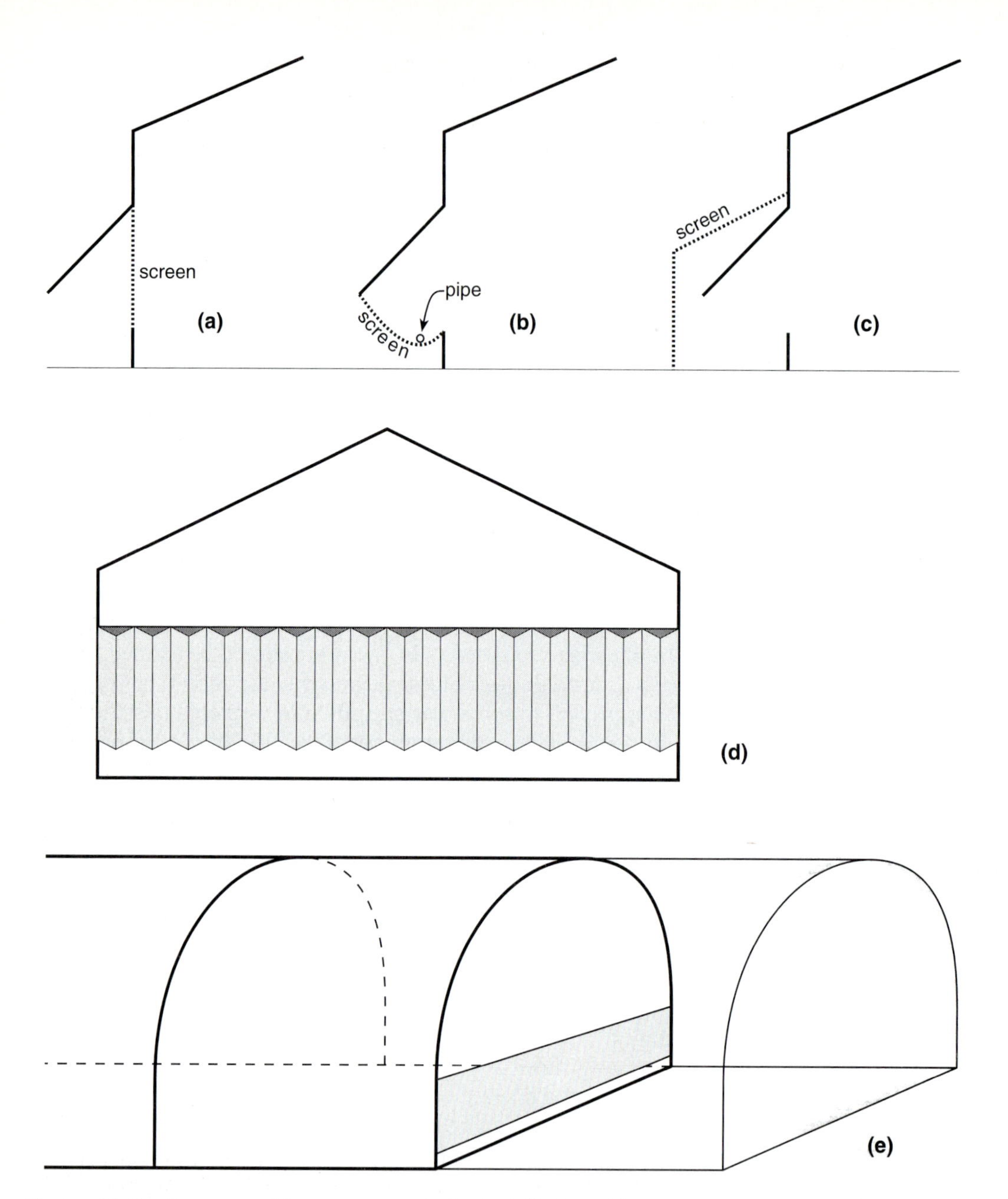

Figure 14-2 Some possible arrangements for installation of insect screens over ventilators in greenhouses. In the first situation (a and b), the screen has low air resistance and needs to cover only the same area as the ventilator; in the remaining situations the screen area needs to exceed the ventilator area and is accomplished by installing (c) a lean-to screen house over the ventilator (d) a pleated screen enclosure over the ventilator, and (e) adding one extra screen-clad section to the length of the greenhouse.

underside of the pot and the leaves. Other pests, such as slugs, hide beneath pots or pieces of bark or leaves on the surface of the root substrate during the day and come out to feed at night. You must be aware of these habits and look in the proper place for the pest or for the signs of the pest, such as slime trails.

The entire greenhouse should be inspected at least twice per week. It is recommended that three plants be examined on each bench. If insects are found, these plants are flagged for future observation to determine the speed of development of the insects. Unless a plant has been flagged, different plants should be inspected each time because insects are very erratic in their points of initial appearance. Yellow or blue sticky cards should be installed on sticks such that they can be raised as the crop grows to keep them just above the height of the plants. The blue card is particularly effective for western flower thrips. For small greenhouses, one card per 1,000 ft^2 (93 m^2) is sufficient. For large greenhouse blocks, one card might suffice for 10,000 ft^2 (930 m^2). Plants and cards should be inspected at the same time. It will help to have a 10X lens for identifying the insects. The type and number of insects should be determined and recorded. Cards need to be replaced weekly so that newer insect entries are not confused with older entries. It is important to inspect both plants and cards because flying stages of insects including aphids, fungus gnats, leaf miners, thrips, and whiteflies can be found on sticky cards but might leave plants when they are picked up. Conversely, slugs, spider mites, and immature stages of some insects would not be found on cards.

Records of the types of insects present, their numbers, their stages of development, and their rates of development are necessary for several reasons:

1. *Biological control:* As will be discussed later, this information is absolutely necessary if biological control is to be used. It is important to reduce the pest density to a specific level with safe pesticides such as insecticidal soap before releasing predators. Also, many biological organisms are specific to an individual type of pest.
2. *Pesticide selection:* Many pesticides are selective for certain insect species.
3. *Pesticide formulation:* Thrips located deep in flowers are often beyond the penetration of sprays. It would be necessary to use an aerosol or fog to penetrate the flowers.
4. *Application zone:* Surveillance makes it possible to distinguish the areas that need treatment from those that do not. Far less pesticide is used in this way.
5. *Pesticide timing:* Whiteflies are most vulnerable to sprays in the crawler through the second enstar stages. By the time they are in the fourth enstar stage, they have an outer exoskeleton that protects them from pesticide sprays. The systemic insecticide Oxamyl takes five days to get into the plant and then is active for about two weeks. It kills only insects that are feeding on the plant. This includes whiteflies in the crawler or second and third enstar stages but not in the egg, pupa, or adult stages.

Environmental Adjustments

Many pathogenic diseases require periods of high humidity and/or free water on foliage. When surveillance identifies the initial occupancy of such pests, the greenhouse environment should be adjusted to avoid such conditions. (The techniques are discussed in Chapter 15.) Insects likewise have specific requirements for rapid spread. The two-spotted spider mite has an optimum temperature requirement of 86°F (30°C) and develops best at a low air humidity. Frequent syringing of plants and an effective cooling system help to diminish a spider mite infestation.

Pest Eradication–Biological Control

Two schools of thought exist for pest eradication. One follows the more traditional policy of synthetic pesticide application, while the other minimizes the use of synthetic pesticides in favor of the release of biological organisms. Pesticide application issues will be discussed later in this chapter. Driving forces behind the shift to biological control include the following problems that are associated with primary reliance on pesticides for pest eradication: the large amount of pesticides entering the environment, the increased risk of litigation from accidental handling of pesticides, the development of resistance in pests that are continually exposed to the same pesticides, and the high cost of pesticides required.

Biological control is a system of reducing pest populations through the action of living organisms that are encouraged and released by humans. The living organisms used to reduce pest populations fall into three categories. *Predators* are insects and related animals that attack more than one host individual. Examples are ladybugs and predatory mites. *Parasites* attack a single host individual and complete their development in or on this individual. An example is the nematode *Steinernema feltiae,* which enters a leaf miner larva and develops within at the cost of life to the leaf miner. Predators and parasites are collectively termed *beneficial insects. Pathogens* are microorganisms that bring about disease in their host. *Bacillus thuringiensis* Berliner, which infects worm larvae, is a good example.

For biological control to work, it is necessary that a strict protocol be followed. The pest population must be at a relatively low level for effective control. If a high population is present, it is necessary to decrease it with a safe pesticide. The number of beneficial insects applied must be in a specified proportion to the number of pest individuals present. A surveillance program is necessary to determine the density of pests to meet both of these requirements. Pesticides toxic to the beneficial insects or pathogens must not be used during treatment, and residues of these cannot be present from previous applications. Some reasonably safe pesticides that can be used under specific conditions are listed in Table 14-1. Finally, environmental conditions may need to be altered in the greenhouse to favor the beneficial organisms and at the same time suppress the pests.

Most beneficial organisms are difficult to maintain and need to be ordered as required. Upon arrival, they are directly released in the crop. Again, this treat-

Table 14-1 Pesticides Commonly Considered Safe for Use in a Biological Control Program[1]

Common Name	*Trade Name*
Bacillus thuringiensis	Dipel, Victory
chlorothalonil	Daconil 2787, Termil
copper	copper (various forms)
dazomet	Basamid
DCNA	Botran
diflubenzuron	Dimilin
dodemorph	Milban
kinoprene	Enstar
oxamyl	Vydate
Streptomycin	Agrimycin
vinclozolin	Ornalin

[1]The above pesticides are considered safe for many biological control agents, however many of these pesticides have not been tested on all agents. Be certain to check with your supplier of predators and parasites to be certain that the pesticides you select are safe for the agents you use.

ment dictates the need for constant surveillance in order to have the few days' advance notice required for shipping time. Several insectaries produce beneficial organisms. (Sources in North America are listed in the bulletin *Suppliers of Beneficial Organisms in North America,* available from California Environmental Protection Agency, Department of Pesticide Regulation, Environmental Monitoring and Pest Management Branch, 1020 N Street, Room 161, Sacramento, CA 95814-5604.)

The beneficial organisms described next are some of those that are commercially available. Considerable private and governmental research is underway. New introductions are occurring and will undoubtedly increase in the near future. Such biological control offers excellent potential for coping with diminishing availability of pesticides and with governmental pollution regulations.

Aphid Predator The midge *Aphidoletes aphidimyza* feeds on several aphid species, including the green peach aphid. Larvae of this predator inject a toxin into the knee of aphids that paralyzes them. The predator then sucks out the body fluids through the thorax of its prey. One larva can kill 4 to 65 aphids per day. *Aphidoletes* are released at the rate of 1 pupa per 10 aphids or 1 pupa/yd^2 (1.2 pupa/m^2). The predator should be reintroduced three or four times every week or two. Optimum conditions include a temperature between 73°F and 77°F (23°C and 25°C) and a relative humidity of 80 to 90 percent. This midge will remain active down to a temperature of 59°F (15°C). However, it will hibernate in the winter when light intensity is low unless light is applied during the night.

Leaf Miner Parasites *Diglyphus isaea* is a black wasp about 1/12 inch (2 mm) long with short antennae. This parasite enters the leaf miner's tunnel, kills the pest, and lays an egg beside it. The developing predator feeds on the dead leaf miner. The parasite leaves the tunnel, kills other small leaf miner larvae for food, and eventually lays an egg in a tunnel to start the cycle over again.

Another black wasp, *Dacnusa sibirica,* slightly larger (1/8 inch, 3 mm) with long antennae, lays an egg in a leaf miner larva. The larva continues to live while the parasite develops within. After the larva pupates, only the parasite emerges from the leaf miner pupa.

A nematode parasite is also used for killing leaf miners. *Steinernema feltiae* enters the leaf miner larva through body openings such as the mouth. Inside, the nematode penetrates the gut wall and releases a bacterium, *Xenorhadbis nematophilus,* that kills the host. The nematode is sprayed on plants at night because it is killed by UV light during the day. The temperature should be about 70°F (21°C), and the drying time must be at least several hours to give the nematode time to swim to the host.

Mealybug Predator and Parasite Larvae of the Australian lady beetle, *Cryptolaemus montrouzieri,* kill mealybugs and will also kill aphids and scale insects if food is in short supply. The adult beetle lays eggs among the mealybugs. When the beetle larva emerges, it sucks the body contents out of mealybug eggs and young nymphs. This predator requires a large supply of food; thus, the pest population does not have to be reduced to a low level before its introduction. Optimum conditions for egg laying and larval development are a temperature of 72 to 77°F (22 to 25°C) and a relative humidity of 70 to 80 percent. As in the case of aphid predators, ants have to be controlled since they will protect these pests from predators. A boric acid bait can be used for controlling ants. The predator is released at the rate of 2 predators/yd^2 (2.4 predators/m^2).

There is also a parasite of mealybugs. The wasp *Leptomastix dactylopii* attacks only citrus mealybugs by laying an egg in either the third-stage nymph or the adult mealybug. The wasp develops inside the mealybug. This wasp develops best in sunny, warm, humid environments, but can tolerate dry conditions.

Mite Predators Several species of mites have been employed to kill damaging spider mites, primarily two-spotted spider mites. The predator mites draw out the body fluid of host mites. The predatory mites are released when the pest population is light, at the rate of 2 predators per mite-damaged leaf, or 2 predators per plant if plants are small, or 20 predators/yd^2 (24 predators/m^2) of infested plant area. Reintroductions are made at two-to-four-week intervals until control is achieved. Sometimes, two or more predatory mite species are introduced simultaneously to increase their surviving power and effectiveness.

Each predatory mite species has a different set of optimum conditions. *Phytoseiulus persimilis* develops best at 69 to 81°F (20 to 27°C) and a relative humidity of 60 to 90 percent. This predator avoids bright light and temperatures over 86°F. By comparison, the two-spotted spider mite pest develops best at 86°F (30°C). Thus, the predator is at a disadvantage in a hot greenhouse. Fortunately, a strain of this predator, collected in Israel, has been found to be active in greenhouse temperatures up to 105°F (41°C). *Phytoseiulus longipes* handles temperatures up to 100°F (38°C) and somewhat lower humidity levels. *Amblyseius caliornicus* develops best at more intermediate temperatures, up to 90°F (32°C), but at relative humidity levels above 60 percent. Predatory mites are shipped in drinking straws, capsules, or shaker bottles.

Thrips Predator The mite *Amblyseius cucumeris* attacks mites, including spider mites, as well as western flower thrips. The predatory mite feeds on young thrips larvae. Each predator can destroy one mite per day during its 30-day life. Favorable conditions for the predator include moderate temperatures with an optimum of 86°F (30°C) and a high humidity. Fortunately, in the case of spider mite pests, this is the opposite of the conditions favorable to the pest. When thrips are first detected, predator mites are released at the rate of 30 per plant. It is also possible to release this predator before an infestation occurs because it will feed on pollen. The predator is sold in cereal bran.

Whitefly Parasite The chalcidoid wasp *Encarsia formosa* is sold for parasitizing greenhouse whitefly pupae. The female adult lays eggs in 50 to 100 pupae. Most adult wasps are female. The egg hatches inside the pupa, and the parasite remains there until it has finished its development to the adult stage. The adult emerges from the dead, black pupa and immediately seeks other pupae to begin the life cycle over again.

Encarsia populations develop best when the temperature is at 73 to 81°F (23 to 27°C), the relative humidity is 50 to 70 percent, and the light intensity is 650 fc (7.0 klux) or higher. At 81°F, *Encarsia* produces twice as many eggs as the whitefly; below 70°F, the whitefly produces 10 times as many eggs. At relative humidity levels below 70 percent, the wasp does not develop as well as the whitefly. Under lower light intensities, *Encarsia* is not fully reproductive. This parasite is often available on paper tags containing parasitized whitefly pupae. The tags are hung on plants to be treated.

General Insect Pathogen and Parasite The pathogenic bacterium *Bacillus thuringiensis* Berliner is available commercially under several trade names. This bacterium is noninfectious to plants and humans but does kill the caterpillars (larvae) of many species of moths. The caterpillar ingests the bacterial spores. Inside the insect, the spore germinates and the bacteria enter the blood system. Shortly afterward, the insect stops feeding; after several days, it dies.

The parasitic nematode *Neoplectana carpocapsae* is one of several beneficial greenhouse nematodes and attacks the larval stage of over 250 insects including leaf miners, fungus gnats, thrips, and shoreflies. It can survive without a host for up to 90 days under ideal conditions. This nematode enters a body cavity of the host larva, usually the mouth, travels to the stomach, and there produces an associated bacterium, *Xenorhabdus nematophilus.* The bacterium breaks down the internal organs of the host. Required conditions for this nematode include very high humidity, no sunlight, and temperatures between 61 and 89°F (16 and 32°C). For this reason, soil application works well.

General Comments Many other biological organisms are commercially available for use in greenhouses. Companies supplying biologicals are a good source for a list of pests and the biological organisms effective against each.

By now, the importance of pest surveillance for biological pest control should be very apparent. The type of pest must be known to select the appropriate biological organism. The density of the pest must be determined to know whether a pesticide must be used before the biological release and to know what quantity of biological organisms to release.

The importance of environmental control should also be apparent by now. Unless a set of environmental conditions is established that favors the biological organism, and hopefully harms the pest, effective control may not be achieved. At the same time, the environmental conditions must be favorable to the crop. The time spent learning these environmental parameters can pay double dividends. More important than assuring efficacy of the biological organisms, these conditions can be used to discourage pest infestations and optimize crop productivity.

Biological control will not work in every instance. There will be times in a biological control program when synthetic pesticides will be required.

Insects and Other Pests in the Greenhouse

Numerous insects and related pests attack plants in the greenhouse. The life cycles of these pests must be understood in order to effectively control them. The majority of infestations are due to 10 types of pests. Before any control measures can be planned, the pest must be identified. While a few pesticides kill pests in general, most are made to kill one specific pest or a few particular pests. Spider mites, for instance, are controlled by a specific group of pesticides (miticides).

Feeding habits are very important. A pest such as the cyclamen mite, which feeds in the small scales of buds, is likely to escape injury unless a surfactant is added to the spray to ensure penetration into the bud. Whiteflies, which feed on the lower sides of leaves, point out the importance of spraying from beneath as well as from above. Knowledge of feeding habits also serves to tell the manager where to check for early detection of a problem.

The method of feeding by a pest will permit detection of those that are too small to be seen with the unaided eye or those that hide beneath clods of soil or under pots during the day. Slime trails are indicative of slugs and snails; distorted growth of new leaves can indicate cyclamen mites; light brown tracks in a leaf signify the presence of leaf miners; and patches of sandy-colored pinpoint spots on leaves are the result of a pest with piercing-sucking mouth parts, such as aphids and spider mites.

Many pesticides do not kill eggs; thus, the life cycle of a pest must be known. An insect such as the aphid, which can reproduce at 7 to 10 days of age, requires a more frequent spray program than one such as the mealybug, which takes six to eight weeks to mature. Some of the more important greenhouse pests are described in this section. Pertinent features about their identities and habits are presented to aid in an understanding of their control.

Aphids

Aphids attack a wide variety of greenhouse crops. The several types of aphids may be differentiated by color. The most prevalent greenhouse species is the green peach aphid (*Myzus persicae* Sulzer) (Figure 14-3). The wingless forms of the green peach aphid are yellowish-green in summer and pink to red in the fall and spring. The winged forms are brown. Aphids are 1/8 inch (3 mm) or less in length. They feed by inserting a tubelike piercing mouth part into the leaf and sucking out the sap. Feeding usually occurs in buds and undersides of leaves. Feeding on young bud leaves results in distorted leaves as they continue to grow. Older leaves may display patches of chlorotic pinpoint spots where cell contents have been drawn out.

Aphids excrete honeydew, which is rich in sugar. A black sooty mold often grows on the honeydew, marring the appearance of the plant. Ants collect the

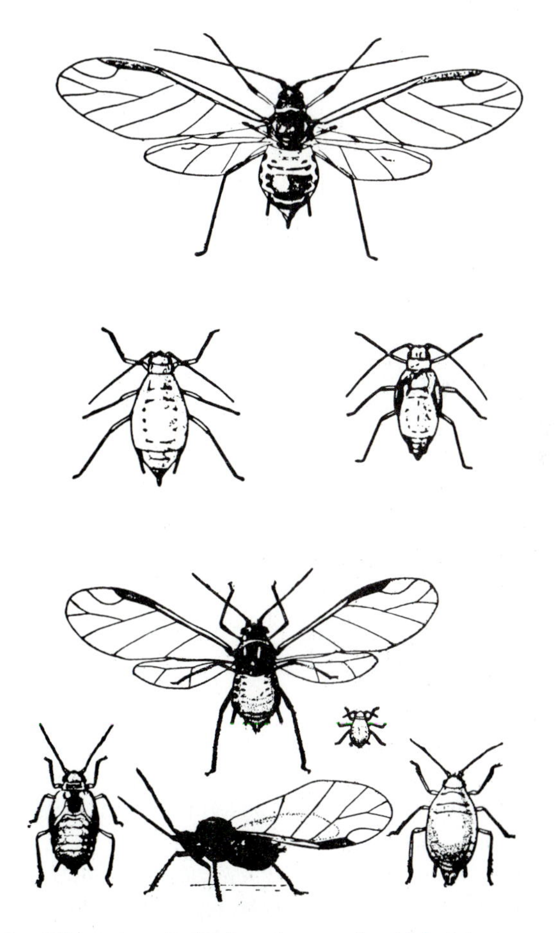

Figure 14-3 **Winged and wingless forms of aphids: (a) green peach aphids and (b) melon aphids. (*USDA sketch.*)**

honeydew and sometimes go as far as to farm aphids by moving them to suitable host plants and protecting them from predators.

Aphids usually give birth to living female nymphs. The nymphs give birth to successive generations of female nymphs in as short a time as 7 to 10 days. Each aphid reproduces for a period of 20 to 30 days. One aphid can give birth to 60 to 100 nymphs. The entire process takes place without mating. When the food supply becomes short or the colony becomes overcrowded, winged females appear and migrate. Male and female forms appear outdoors with advancing winter in the northern but not in the southern United States. They mate and lay eggs, which constitute an overwintering stage.

Fungus Gnats

Fungus gnats (*Bradysia* sp. and *Sciara* sp.) are thin, gray-colored flies with long, thin legs and antennae (Figure 14-4). They reside mainly on soil and will fly short distances when disturbed. The adult is about 1/8 inch (3 mm) long and has one pair of clear wings. There is a distinct Y-shaped vein at the tip of each wing. The body ranges in color from a black head to brownish-yellow legs and abdomen.

Shore flies are often thought to be fungus gnats. These flies also reside at the surface of the soil and show up on sticky traps. Shore flies are much larger and have darker eyes, legs, and wings. They also have pale spots on their wings, short antennae, and moderately long legs.

Damage is caused by the larvae, which are legless white worms with a black head and which grow up to 3/16 inch (5 mm) in length. These larvae normally feed on soil fungi and decaying organic matter. When population densities increase, fleshy storage organs such as bulbs and roots may be attacked. Delicate seedlings may be killed. A wide variety of crops are injured by these larvae. Infected plants may turn yellow, lack vigor, or wilt.

The female lays clusters of 20 to 30 eggs on moist soil surfaces. Soil rich in organic matter is preferred. As many as 300 eggs are laid during the 10-day life span of the adult. The egg hatches in about six days into a larva. The larva feeds for 12 to 14 days and then changes into a pupa in the soil. After five to six days, an adult fly emerges from the pupal stage. The life cycle from egg to adult requires about four weeks.

Leaf Miners

Leaf miners, in their larval (worm) stage, tunnel within leaves making unsightly tunnels (Figure 14-5). A heavy infestation renders the plant useless for sale. Leaf miner resistance to pesticides has led to devastating population buildups, particularly on chrysanthemum.

The adult female is a stocky fly about 1/12 inch (2 mm) long. It punctures the leaf surface with a tubelike appendage on the abdomen known as an *ovipositor* and inserts eggs through it. This activity leaves small white spots on the leaf. A larger number of punctures are made than the number of eggs deposited. Males

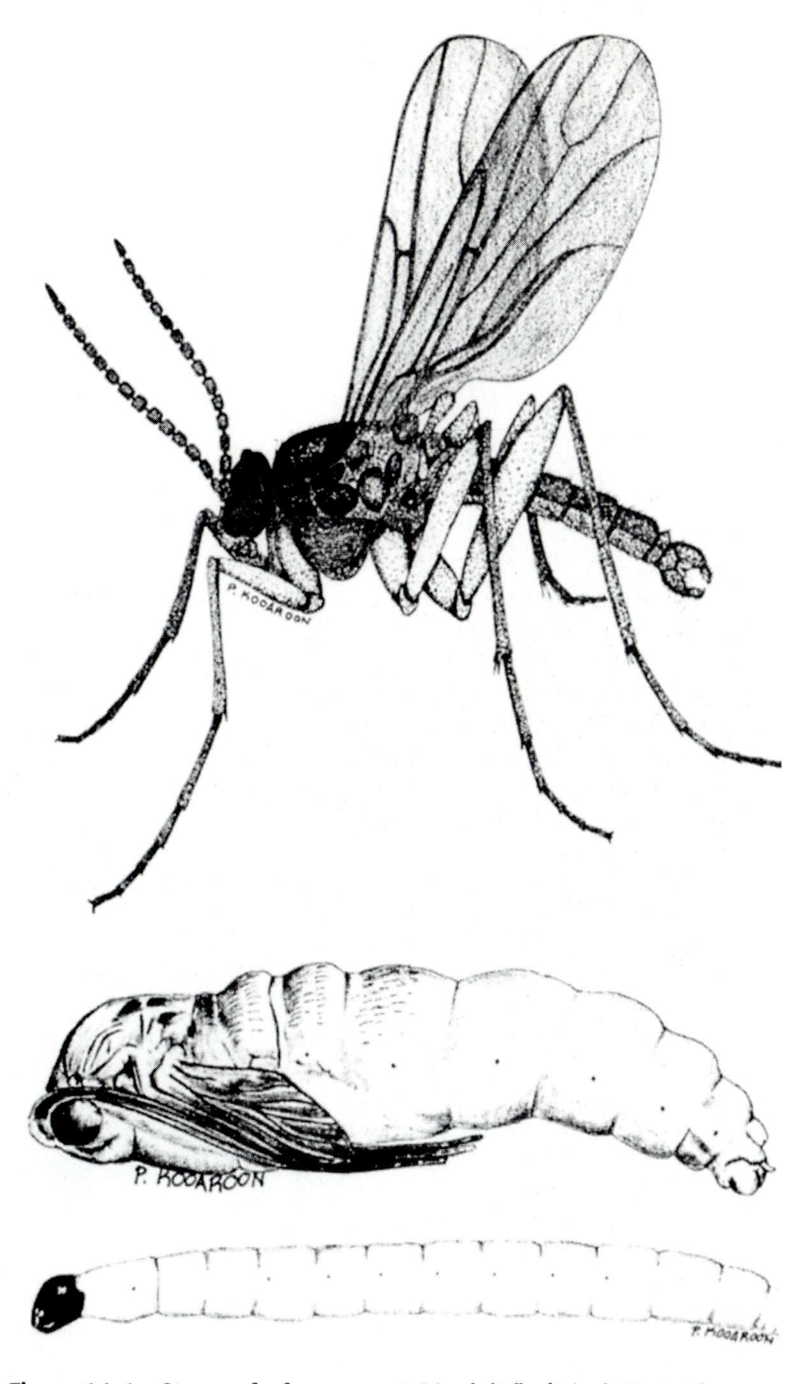

Figure 14-4 Stages of a fungus gnat: (a) adult fly, $^{1}/_{8}$ inch (3 mm) long; (b) pupa, $^{1}/_{8}$ inch (3 mm) long; and (c) larva, up to $^{3}/_{16}$ inch (5 mm) long. (*Sketches courtesy of J. R. Baker and P. Kooaroon, North Carolina Agricultural Extension Service, Raleigh, NC 27695-7613.*)

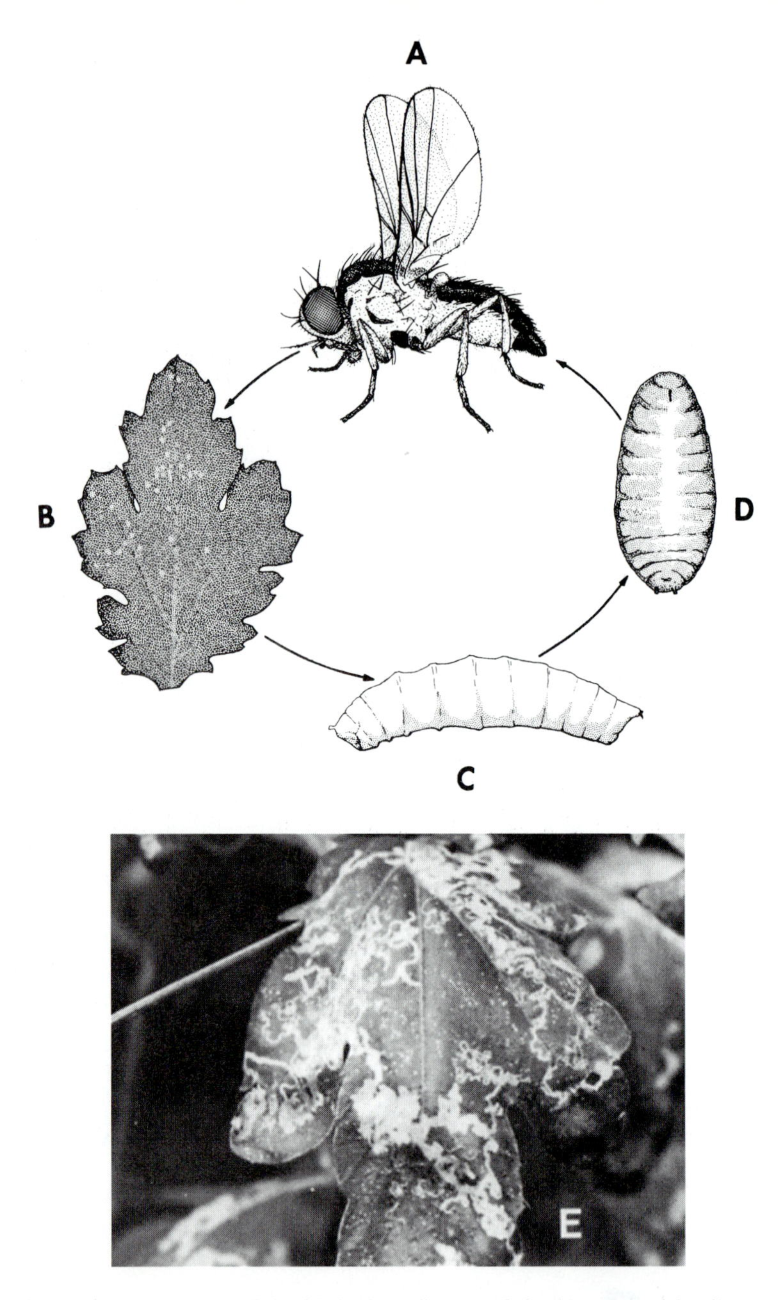

Figure 14-5 Stages of the leaf miner fly: (a) adult; (b) egg and feeding punctures; (c) larva; (d) pupa; (e) damage to chrysanthemum in a commercial greenhouse. (*Sketches courtesy of J. R. Baker, North Carolina Agricultural Extension Service, Raleigh, NC 27695-7613.*)

and females feed on the sap that oozes from the punctures. Each female lays about 100 eggs in its two-to-three-week life span. The egg hatches in five to six days into a soft white maggot, which reaches 1/8 inch (3 mm) long when mature. The maggot can tunnel for up to two weeks, at which time it drops out of the leaf into the bench or soil and turns into a pupa. After about two weeks, an adult fly emerges from the pupa, flies to a new leaf, and the life cycle begins again. About five weeks is required for the life cycle from egg to adult.

There are numerous species of leaf miners, but two are most prevalent in the greenhouse. A serpentine leaf miner adult (*Liriomyzia trifolii* Burgess) has a blackish body with yellow markings, a yellow head, and brown eyes. It makes serpentine mines. The chrysanthemum leaf miner adult (*Phytomyza atricornis* Meigen) is larger and black in color. Blotchy as well as serpentine mines are formed by these maggots.

Mealybugs

Mealybugs (*Pseudococcus*) are oval-shaped insects that appear white because of a waxlike powder that covers their bodies (Figure 14-6). The waxy deposit on their bodies includes filaments extending out around the periphery as well as some longer filaments up to 1/2 inch (13 mm) long extending from the back to give the appearance of a tail in the long-tailed mealybug. The actual insect is 1/5 to 1/3 inch (5 to 8 mm) long.

Mealybugs feed by means of a piercing-sucking mouth part. During feeding, citrus mealybugs inject a toxic substance into the plant. The plant becomes chlorotic and malformed. Like aphids, these insects also excrete honeydew, which provides a substrate for a black sooty mold to grow upon and further disfigure the plant. Ants sometimes farm these insects as they do aphids.

Long-tailed mealybugs give birth to living nymphs. Citrus mealybugs lay eggs that are deposited in a cottony sac. Several hundred yellowish or orange eggs are contained in a sac. The eggs can hatch in 5 to 10 days into nymphs. The nymphs move about and feed for six to eight weeks, at which time they finally become adults. The life cycle from egg to adult takes 7 to 10 weeks under favorable conditions.

Their waxy protective layer makes it difficult to control mealybugs. Surfactants help wettable-powder formulations of pesticides stick better. Aerosols play a useful role here as well. The nymphs, having a thinner protective coating, are easier to kill than the adults.

Mites

Mites are not insects. They belong to the class *Arachnida,* which includes spiders and scorpions. In the adult form, they have four pairs of legs. There are many species of mites, including several that attack crops.

Cyclamen mites (*Steneotarsonemus pallidus* Banks) are very small—about 1/100 inch (0.25 mm) long when fully grown. They cannot be seen with the un-

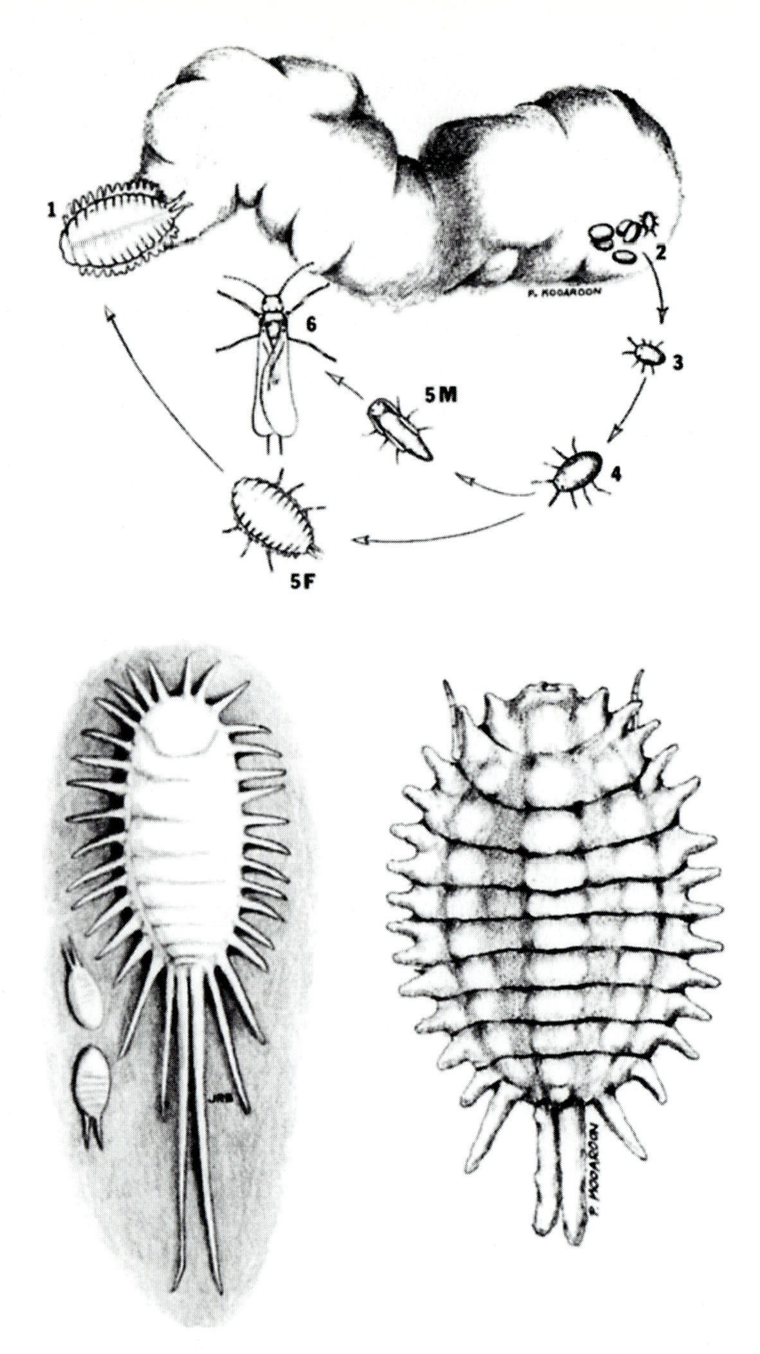

Figure 14-6 Various stages of mealybugs: (a) citrus mealybugs in all stages; (b) adult long-tailed mealybug; and (c) Mexican mealybug. (*Sketches courtesy of J. R. Baker and P. Kooaroon, North Carolina Agricultural Extension Service, Raleigh, NC 27695-7613.*)

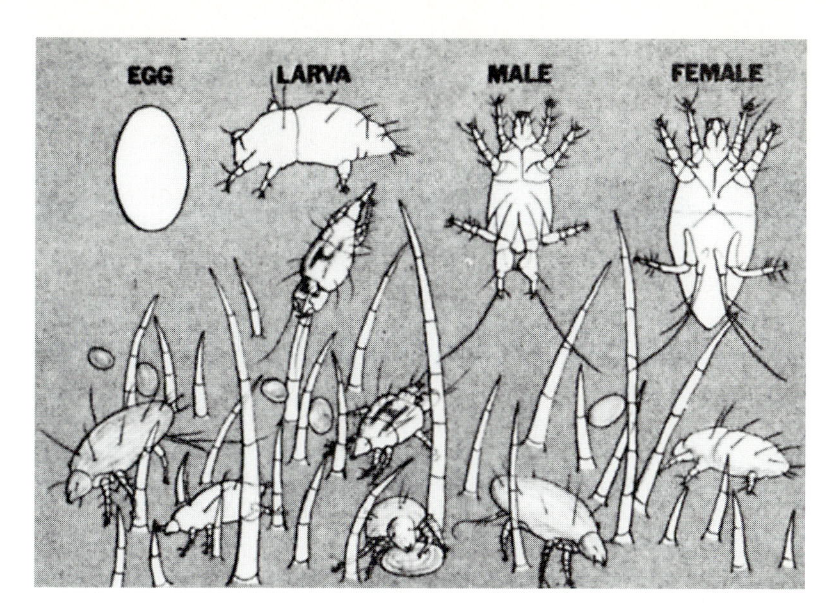

Figure 14-7 **All stages of the cyclamen mite shown to scale with the trichomes (leaf hairs) on a gloxinia leaf.** (*Sketch courtesy of J. R. Baker, North Carolina Agricultural Extension Service, Raleigh, NC 27695-7613.*)

aided eye. These mites (Figure 14-7) are semitransparent with a brownish tinge. Their development is favored by high humidity (80 percent or more) and low temperature (60°F, 16°C). The life cycle from egg to adult can occur in two weeks, but the adult female lives on for three to four weeks and lays up to 100 eggs.

Cyclamen mites affect a broad range of plants, many of which are green plants. The mite lives in and feeds on the bud and small adjacent leaves. It feeds by way of a piercing-sucking mouth part. Symptoms of infestation are curling of leaflets from the outside inward and distortion of young leaves such that small depressions are formed. Leaves may have a corky surface, often mistaken for boron deficiency. Flowers may also be distorted or fail to open altogether.

Two-spotted mites, or red spiders (*Tetranychus urticae* Koch), are perhaps the most troublesome of all greenhouse pests (Figure 14-8). These mites may be greenish, yellowish, or red in color and have two dark spots on their bodies. They are about 1/50 inch (0.5 mm) in length.

Two-spotted mites cause chlorotic stippling of leaves, as though a very fine tan-to-yellow sand has been sprinkled on them, when populations increase. However, on mums and other plants with thick leaves, chlorotic stippling may not be noticed. The spiders spin a silk strand that forms a web over leaves and flowers. Leaves and flowers soon begin to desiccate and consequently turn brown. These mites are most prevalent on the undersides of leaves and in flowers. They are difficult to control in flowers.

The time span from egg to adult is 10 days at 80°F (27°C), or 20 days at 70°F (21°C), or several months at lower temperatures. Low relative humidity favors

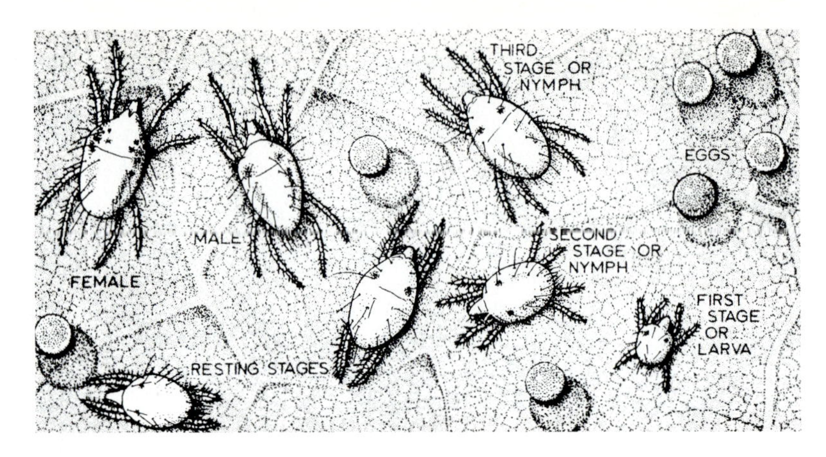

Figure 14-8 **All stages of the two-spotted spider mite.** (*Sketch courtesy of J. R. Baker, North Carolina Agricultural Extension Service, Raleigh, NC 27695-7613.*)

development of this mite. Eggs hatch in four to five days into six-legged nymphs, which feed for a short time. Next comes an inactive resting stage, a nympho-chrysalis, that lasts about 1½ days. This sequence is repeated for a total of three resting stages. The eight-legged adult emerges from the last resting stage.

During the egg stage and the resting stages, most miticides are ineffective, whether applied as an aerosol, smoke, or nonresidual spray. Since all stages are usually present simultaneously, several miticide applications are necessary. At high temperatures, it may be necessary to make applications as often as two days apart.

Scale Insects

There are several genera of scale insects, all of which belong to the superfamily Coccoidea (Figure 14-9). They vary in overall shell size up to ¼ inch (6 mm). Scale insects are similar to mealybugs and are included in the same superfamily. The unarmored types of scale insects (coccids) have a rubbery outer coating that cannot be detached. Some secrete wax. They may be flat, oval, or globular. Other unarmored scale insects secrete honeydew, which encourages development of a black sooty mold.

The many other types of scale insects (diaspids) are armored. The scale is not attached to the body and is composed of a wax secretion and cast-off skins of the immature stages of the insect. The scale covering can take on several configurations from round to oyster shell–shaped, can be several colors, and can be smooth or rough. Armored scale insects do not secrete honeydew.

The female armored scale insect does not move about as does the mealybug. It is saclike, wingless, usually legless, and feeds through a piercing-sucking mouth part that injects toxic saliva. Some scale insect species give birth to living young,

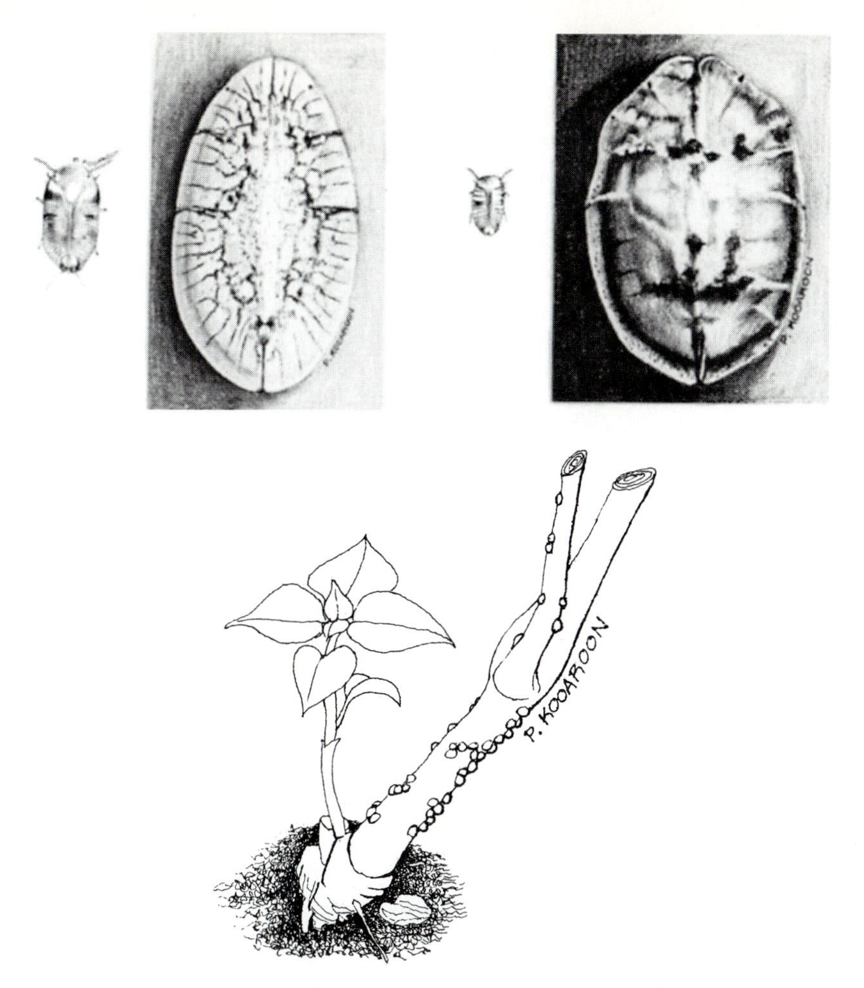

Figure 14-9 **Scale insect pests of ornamental plants: (a) brown soft scale (right) with its crawler; (b) hemispherical scale (right) with its crawler; and (c) bloodleaf plant infested with hemispherical scale.** (*Sketches courtesy of J. R. Baker and P. Kooaroon, North Carolina Agricultural Extension Service, Raleigh, NC 27695-7613.*)

while others lay eggs. The male has legs and one pair of wings, but no mouth parts (it does not feed). In some species, males are rare or absent, and virgin females give birth to young or lay eggs.

The first nymph (crawler) has legs and can travel about for two days in search of a suitable feeding area. It then inserts its mouth parts into the leaf and begins forming a shell. It remains here through several molts, losing its legs on the first molt. Eggs or young are produced under the body of the female and under the armor of the scales. Three to seven generations can develop in a year.

Slugs and Snails

Slugs and snails (Figure 14-10) are not insects. They are mollusks (a group of animals including snails, sea snails, clams, oysters, and octopuses). The slug lacks a shell, whereas the snail has a hard shell similar to those of some of its counterparts in the sea. Slugs range in size from ½ to 4 inches (1.3 to 10 cm) in length.

Slugs and snails have chewing mouth parts that allow them to eat seedlings and leaves. They feed by night and hide beneath pots, benches, or litter on the bench surface by day. Dark, moist hiding places are preferred. Slugs exude a slippery liquid as they move along. When it dries, shiny trails can be seen, which easily identifies their presence.

Slugs and snails may crawl into the greenhouse from vegetation or debris immediately outside the greenhouse. Thus, it is important to keep the greenhouse surroundings clear or mowed. They are also easily transported in on pots, flats, soil, or plants.

Slugs and snails lay clusters of 20 to 100 eggs in moist crevices along the soil or containers. Eggs can hatch in 10 days or less at temperatures above 50°F (10°C). Maturity occurs in three months to a year. Slugs and snails are usually killed with one of the two available baits, metaldehyde or methiocarb.

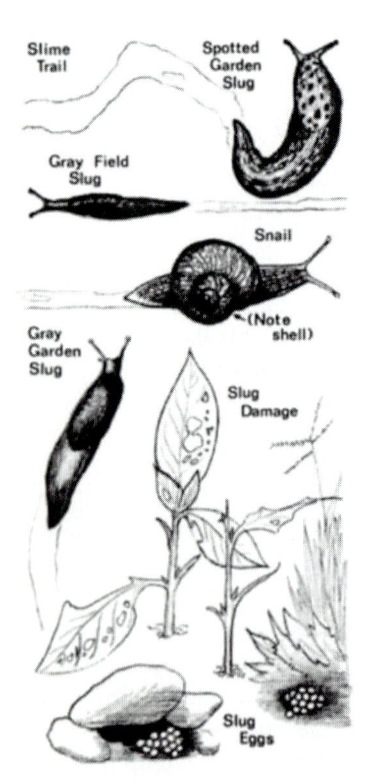

Figure 14-10 Various slugs and a snail. (*Sketches courtesy of J. R. Baker, North Carolina Agricultural Extension Service, Raleigh, NC 27695-7613.*)

Thrips

Thrips (Figure 14-11) are small—1/25 inch (1 mm) long—and have two pairs of fringed wings. They build up in large numbers outdoors and swarm into greenhouses during the warm months. Many are carried in air currents. Thrips feed on a broad range of crops. They are commonly found in buds, on flower petals, in axils of leaves, and between the scales of some bulbs. The adults are visible to the unaided eye but are usually hidden in buds or flowers. They can be detected by tapping buds or flowers over a sheet of white paper. Thrips will fall onto the paper and can be seen. The adults can be yellow, tan, brown, or black, depending on the species encountered.

Until recent years, the western flower thrips (*Frankliniella occidentalis*) was confined to areas west of the Rocky Mountains. Today, it is widespread across the United States and in Europe. It is a partner in what is probably the most serious pest problem of greenhouses today. Western flower thrips transmit impatiens necrotic spot virus and tomato spotted wilt virus (TSWV). These viruses infect a large number of greenhouse crops, destroy them, and have no cure. The only control is through elimination of the western flower thrips in the greenhouse. This has been exceedingly difficult to do because of the wide range of alternate hosts both inside and outside the greenhouse. A main driving force for the development of insect screens for greenhouses is the need to control these viruses.

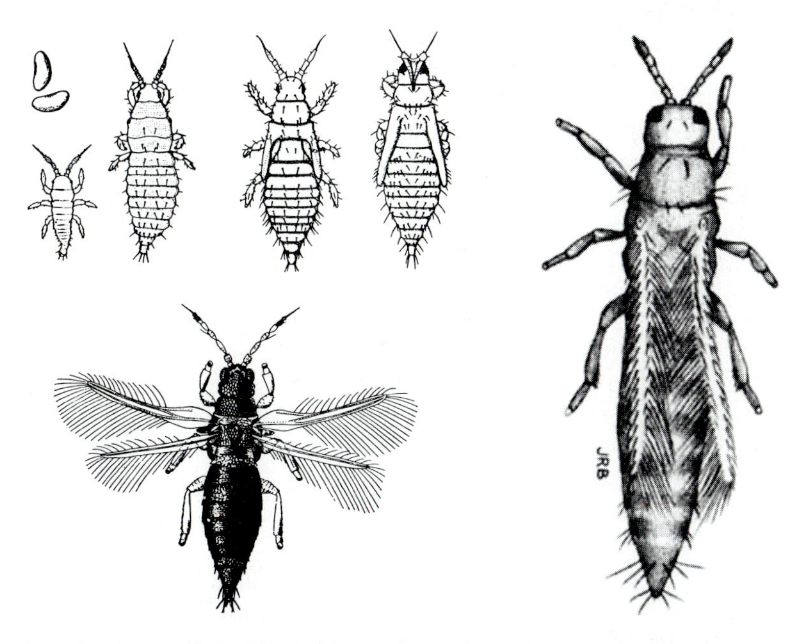

Figure 14-11 **Thrips (both species about 1/16 inch maximum): (a) stages of the greenhouse thrips and (b) an adult flower thrips.** (*Sketches courtesy of J. R. Baker, North Carolina Agricultural Extension Service, Raleigh, NC 27695-7613.*)

The adult female thrips cuts holes in leaves with a sawlike ovipositor on its abdomen and inserts eggs in them. Eggs hatch into nymphs in two to seven days and begin feeding. They have a rasping mouth part that is used to scrape the tender leaf or petal surface. They then suck the exuding plant sap, which causes a white (sometimes silver) discoloration. The injury occurs in streaks, rather than in a stippling pattern as is the case with mites or aphids. Later, the whitish areas turn tan or brown as the cells dry. The entire life cycle can occur in two weeks. Longer times are required at lower temperatures. The feeding nymphs and adults excrete brown droplets that turn black and can then be detected on petals and leaves.

Whiteflies

Two types of whiteflies constitute serious pests in the greenhouse: the greenhouse whitefly (*Trialeurodes vaporariorum* Westwood) and the silverleaf whitefly (*Bemisia argentifolii* Bellows and Perring). The silverleaf whitefly appeared in recent years as a major pest. It is highly prone to developing pesticide resistance, has a wide range of host plants, and is particularly adept at transmitting viruses.

Whiteflies are small insects—about 1/16 inch (1.5 mm) long—and have four wings (Figure 14-12). They are covered with a white waxy powder and resemble miniature moths. Whiteflies fly short distances when plant foliage is disturbed. They are found mainly on the undersides of young leaves. Crops on which whiteflies are particularly troublesome are ageratum, chrysanthemum, fuchsia, lantana, petunia, poinsettia, salvia, and tomato.

With the aid of a hand lens, it is possible to distinguish between the two types of whiteflies. The greenhouse whitefly adult is about 1/16 inch (1.5 mm) long, compared to the silverleaf adult, which is about 1/28 inch (0.9 mm) long. The greenhouse whitefly holds its wings flat over its abdomen, somewhat parallel to the leaf surface. The silverleaf whitefly holds its wings at a 45° angle with the leaf surface so that they form an A-frame "roof" over the abdomen.

Although whiteflies are tropical insects, they are always present in one greenhouse or another during the winter. This is a testimony to their ability to tolerate pesticides. During the summer, they fly into the greenhouse from numerous host plants outside; in the winter, they spread from one greenhouse range to another on plant material or on the clothing of workers. Yellow clothing is especially attractive to whiteflies.

Whiteflies feed through a piercing-sucking mouth part. They sometimes cause yellow stippling of leaves. They excrete honeydew, which supports growth of a black sooty mold.

The life cycles of the greenhouse and silverleaf whiteflies are fairly similar. The adult whitefly lays from just a few eggs to 20 eggs at a time, often in a circle. Up to 250 eggs are laid by each female. Each is attached to the leaf in an upright fashion by a thin stalk. The eggs of the greenhouse whitefly are pale green to purple, while the silverleaf whitefly eggs are whitish to light beige with the apex slightly darker. Newly hatched crawlers emerge from eggs in 5 to 10 days and

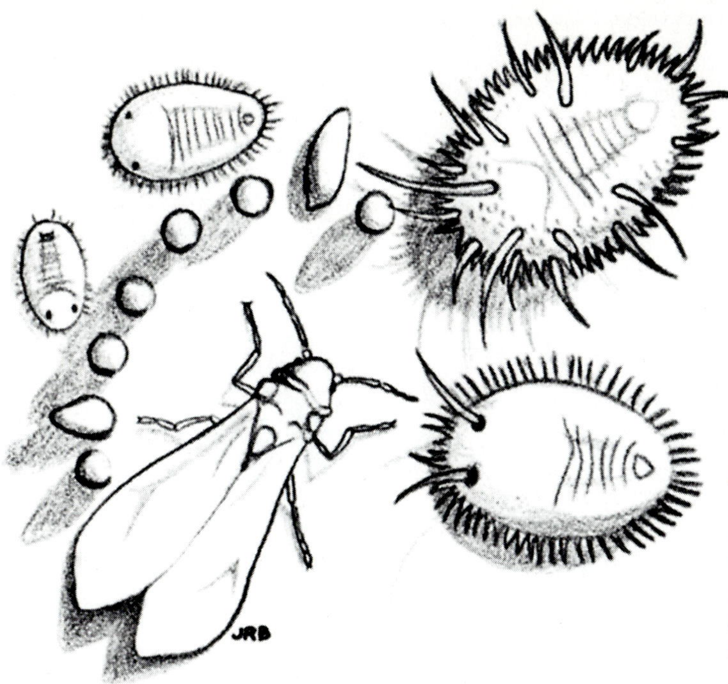

Figure 14-12 **The greenhouse whitefly: eggs, nymphs, pupa, and adult, all found on lower leaf surface.** (*Sketch courtesy of J. R. Baker, North Carolina Agricultural Extension Service, Raleigh, NC 27695-7613.*)

seek a feeding place. They insert their mouths into the leaf tissue and remain stationary for three weeks or so while they undergo three molts. During these stages, the crawler is a flat, scalelike insect and is transparent to greenish-yellow in color. At the end of this period, it transforms into a yellowish-green nonfeeding pupa. A week later, the winged adult emerges. Females begin laying eggs two to seven days later. Depending upon temperature, the whole life cycle can take four to five weeks.

Eggs, pupae, and to a degree the scalelike larvae in the stage prior to the pupal stage are not susceptible to pesticides. The adults are readily killed. Aerosols and smokes do a good job of knocking down adults, but a day later new adults may emerge. At warm temperatures, pesticides may need to be applied as often as three times per week. Sprays, because of their residual activity, are much more effective, but a continual program with spray applications possibly as often as every five days must be maintained through the period of one life cycle. If one application is missed such that adults are afforded an opportunity to lay eggs, the whole program must be started again.

Worms

Worms, the immature stages of several types of moths, attack crops in a variety of ways (Figure 14-13). Worms are a problem, particularly during the warm

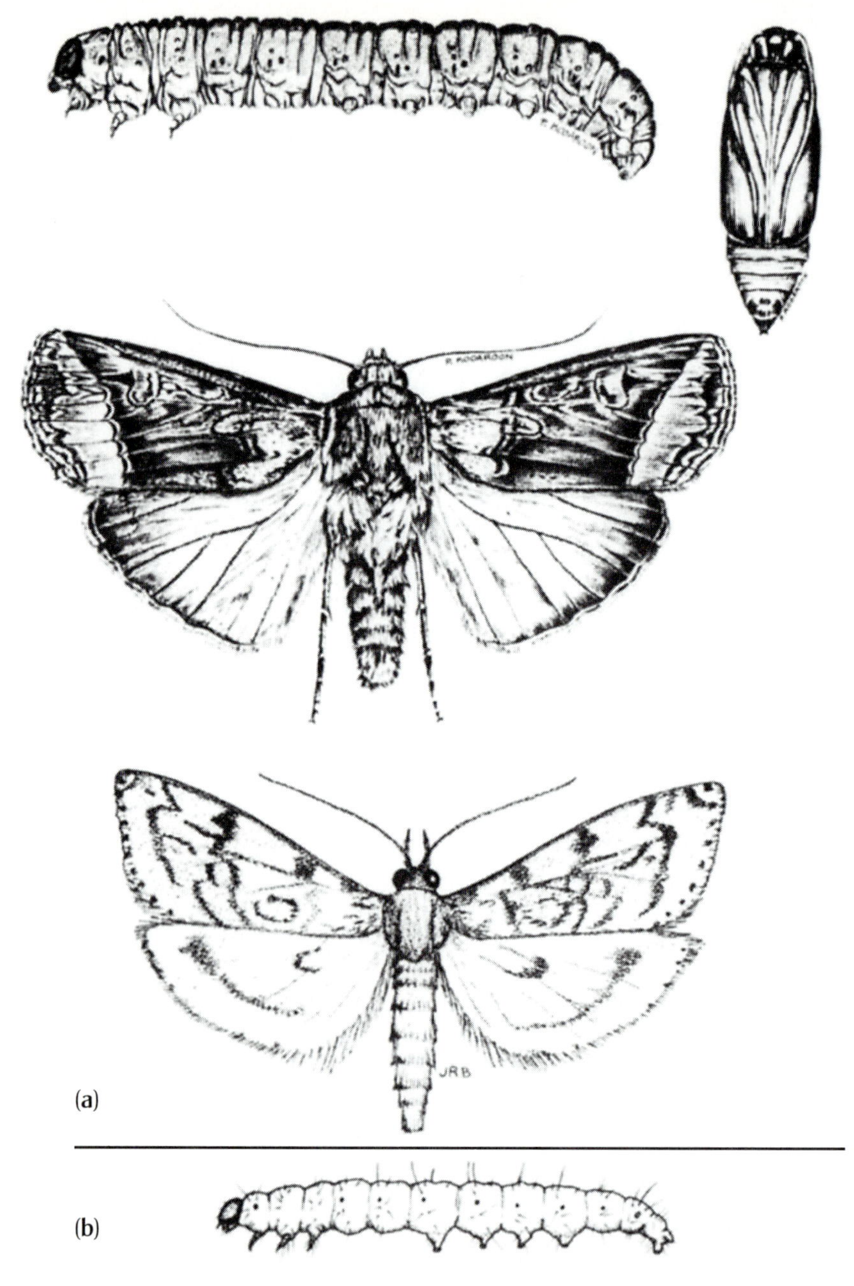

Figure 14-13 Various caterpillar pests in the greenhouse: (a) larva (top), pupa, and adult of the black cutworm; (b) larva and adult of the greenhouse leaf tier; (c) beet armyworm; and (d) corn earworm adult, eggs, larva, and pupa. (*Sketches courtesy of J. R. Baker and P. Kooaroon, North Carolina Agricultural Extension Service, Raleigh, NC 27695-7613.*)

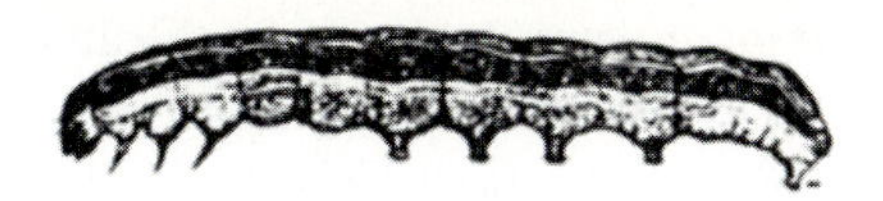

(c)

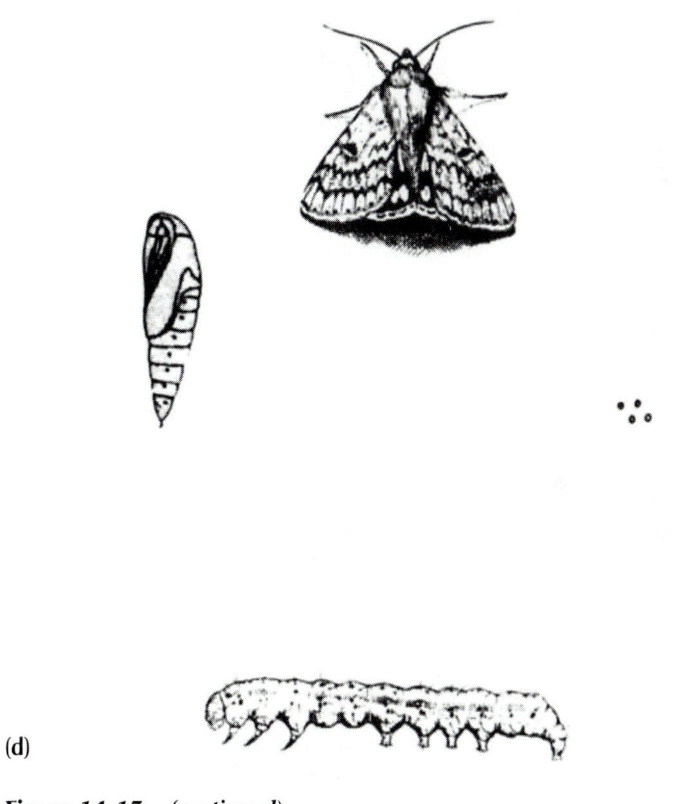

(d)

Figure 14-13 *(continued)*

months when moths abound outdoors. They randomly fly into the greenhouse or are attracted by lights. Once inside, moths lay eggs from which worms hatch.

Beet armyworms eat leaves of carnation, chrysanthemum, cyclamen, geranium, snapdragon, and other crops. Corn earworms eat the succulent plant parts (preferably the buds) of chrysanthemum, gladiolus, and rose. European corn borers tunnel into the stems of plants, particularly chrysanthemum. Cutworms feed on aerial plant parts. Leaf tiers and leaf rollers tie young leaves together or into a nest and feed on them from within.

Many other kinds of worms are occasionally encountered, but the control for all worms is similar. Stomach poisons with long residual lives are most effective, such as Mavrik and Lannate. Biological control is available through sprays of *Bacillus thuringiensis* Berliner (Bactospeine, Biotrol BTB, Dipel, Larvo BT, Thuricide, Vectobac, or Vegetable Insect Attack). This is a bacterium that attacks worms. Plants must be sprayed as new growth occurs to keep all surfaces protected. Once worms are inside the stem, bud, or nest of webbed leaves, control of these pests is almost impossible.

Methods of Pesticide Application

Eight methods for bringing pesticides into contact with insects or plants are widely used. For a given pest problem, generally two or more methods of control are available. Existing equipment, weather conditions, the type and stage of development of a crop or pest, susceptibility of a particular crop to injury, and economics will dictate which method to use.

Sprays and dusts leave a residue on the plant that continues to kill after the application. Aerosols, fogs, and smokes propel fine droplets of the pesticide into the air so that any existing insects, whether they are on the crop, beneath the bench, or elsewhere, are killed. But an effective residue is not left by these methods of applications. Repeated application is necessary. Otherwise, these systems must be used in conjunction with spraying or dusting. While aerosols, fogs, and smokes are simple to apply, a limited number of insecticides are available in such forms. Systemic granular insecticides, applied to the root substrate, are not the easiest to apply, but they offer a long residual period. A rather broad spectrum of insects can be killed with a material such as imidacloprid (Marathon) or oxamyl (Vydate) for a period of several weeks. On some crops, systemic pesticides eliminate the need for a tedious spray program.

High-Volume (HV) Spray

High-volume spraying has been the most common method of pesticide application in the greenhouse. This equipment is usually the least expensive of the various types of pesticide applicators. More pesticides are available for this form of application than any other form. Spray droplets are large, with most larger than 100 microns in diameter and many larger than 400 microns.

Several types of pesticide formulations are available for HV sprayers. Soluble powders are soluble enough to dissolve in water. The entire sealed packet can be dropped into the tank. The packaging dissolves along with the contents. This eliminates the possibility of the applicator coming into bodily contact with the pesticide. Wettable-powder (WP) pesticides consist of solid particles, usually clays, that are dispersed in water. WP formulations should be first mixed in a bucket and then poured into the sprayer tank. This is done to avoid solid deposits plugging the nozzle or an error in the concentration. The spray tank should have an agitator in it to prevent the particles from settling out. Emulsifiable concentrates (EC) are pesticides in an oily preparation with an emulsifying agent that renders the oil miscible in water. Without the emulsifying agent, the oil (which dissolves the pesticide) would not mix with water. Soluble-powder and WP formulations tend to be less toxic to plants than EC formulations of the same pesticide.

Leaf and stem surfaces have a waxy cuticle covering them. This surface has a tendency to repel water to varying degrees, depending on the plant species. When spray droplets bead up on the leaf surface, a considerable area is left unprotected and small insects are able to continue feeding. The problem can be remedied by the addition of a surfactant to the spray. Surfactants are also known as spreaders, wetters, or spreader-stickers. The surfactant lowers the surface tension, which increases the ability of the spray to spread out over the plant surface. A number of surfactants are on the market for this purpose, and among them are materials such as Du Pont Spreader-Sticker, Triton B-1956, Bio-Film, and Ortho R-77. The recommended rate should be used because too much surfactant can cause plant distortion. Mild liquid household detergents can also be used at the rate of 1 pint per 100 gal, or 1/2 teaspoon per gallon (1.25 mL/L). Incomplete wetting is mainly a problem of wettable powders, since EC formulations already contain an emulsifying agent that is itself a surfactant.

Sprayers can range from 1- to 4-gallon (4- to 15-L) hand-pumped units to electric or gasoline motor-powered sprayers of 10- to 200-gallon (40- to 750-L) capacity (Figure 14-14a). The larger sprayers are on wheels and are brought into the end or main aisles of the greenhouse. A hose attached to the sprayer is dragged along each aisle to permit spraying of the entire bench. A nozzle at the end of the hose is used to break the spray into fine droplets and direct the spray in a full pattern for maximum coverage of plants sprayed. A pressure range of 200 to 600 psi (1,380 to 4,140 kPa) is very common to ensure small droplet size. A pressure gauge should be installed at the sprayer or near the nozzle to ensure the correct pressure for even coverage.

To ensure that the undersides of leaves are sprayed, the nozzle should be at a 45 to 90° angle to the axis of the nozzle handle, or a sprayer with several nozzles at different angles, such as the Cornell nozzle (Figure 14-14b), should be used. The nozzle should be handled in a sweeping action to ensure that all of the plant is covered. Particular care should be taken to reach plants in the interior of a bench; otherwise, a point of reinfestation is left.

Some larger greenhouse ranges make use of a permanently installed plumbing system that brings spray from the mixing tank to any given section in the range. A single spray hose is attached at the point where spraying is desired. Consideration must be given to disposing of the pesticide in the line after each use.

Many sprays are corrosive to the spray equipment. They can be dangerous to workers if left in the hose or sprayer. With time, solid deposits can occur in the tank, lines, or nozzles. Some pesticides, upon standing for some time in the

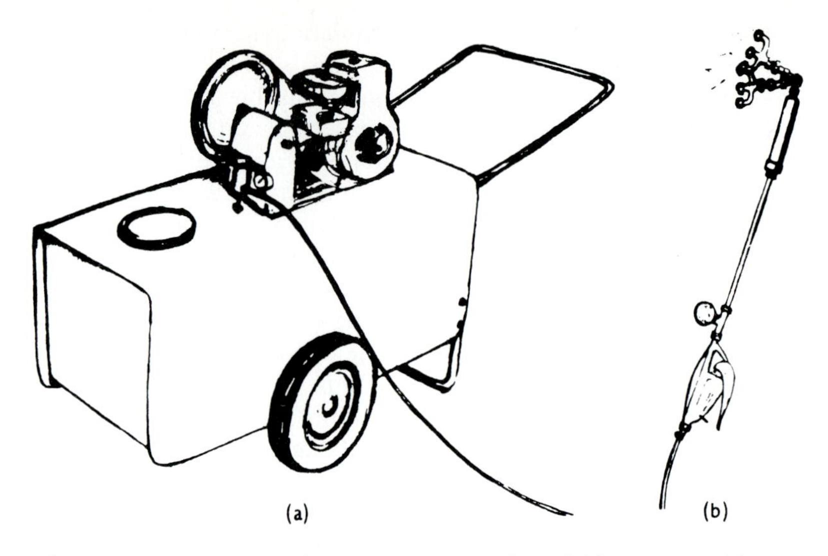

Figure 14-14 (a) A gasoline motor–powered pesticide sprayer with a 30-gal (110-L) tank capacity typical of many used for greenhouse pest control. Such sprayers commonly operate at 200 psi (1,380 kPa). (b) A Cornell nozzle used for spraying greenhouse crops. Six nozzles spray at a variety of angles to ensure complete coverage of upper and lower leaf surfaces. A gauge indicates spray pressure, and a lever gives the operator on-off control of the spray. (*Sketches courtesy of J. R. Baker, North Carolina Agricultural Extension Service, Raleigh, NC 27695-7613.*)

lines, break down to a form injurious to plants. For these reasons, the sprayer should be emptied after use and rinsed out. Clear water then should be pumped through the entire system to clean it.

Dust Application of Pesticide

A few pesticides can be obtained in a dust form. The active pesticide is mixed with talc, clay, diatomaceous earth, or similar filler. Dusters are used to apply dusts. These range from hand-cranked units for small greenhouse uses to large motorized dusters. Dusts, being lightweight, remain in the air for a period of time. Respirators or gas masks, depending on the toxicity of the pesticide, should be used to avoid inhaling them. Dusting is not a common method of pest control in greenhouses because of the visible residue left on the plants.

Low-Volume (LV) Sprayers

Droplet size is very important in terms of the volume of pesticide formulation needed and the pattern of leaf coverage. Large droplets of 100 microns and larger, typical of HV hydraulic sprayers, are propelled in a fairly straight line. These impact directly on the outer leaves and mainly on the upper surfaces. The smaller droplets generated in LV applicators reach inner foliage more effectively, as they are carried around the outer leaves by turbulence and eddy effects in the airstream.

Droplet size dictates the volume of spray liquid required for effective coverage of foliage. If the droplet size is cut in half from 400 microns to 200 microns, only one-eighth of the volume is required for effective foliar coverage. Halving droplet size again to 100 microns reduces the required volume again to one-eighth, for a total reduction to one-sixty-fourth of the original volume. Uniformity of droplet sizes is important so that they distribute uniformly over the foliage. Small droplets will not cover all of a given leaf. This is permissible be-

(a)

(b)

Figure 14-15 (a) A low volume pesticide applicator that generates a pressure up to 3,000 psi at the spray nozzle producing spray droplets of 30 to 60 micron diameter. With a tank capacity of 12 gallons (45 L), an individual can treat 45,000 ft^2 (4,180 m^2) in 45 minutes. (b) A low volume applicator that can be situated in a single position within a greenhouse and activated by timeclock or computer. This applicator produces droplet sizes of 0.5 to 10 microns diameter through a special air flow nozzle. A fan within this unit can distribute the spray throughout 30,000 ft^2 (2,787 m^2) of greenhouse. When used in conjunction with horizontal air fan or convection tube systems, one applicator can treat up to 70,000 ft^2 (6,503 m^2) of greenhouse. (*Photos courtesy of Dramm Corp., P.O. Box 1960, Manitowoc, WI 54221-1960.*)

cause there is a zone of protection extending out from each droplet. One large droplet in a non-uniform spray will contain the same volume of liquid as a high number of small droplets, but the large droplet greatly reduces the leaf area contacted. The volume of pesticide required, thus, goes up. Ideally, LV applicators should deliver small (less than 100 microns), uniform droplets.

Advantages of LV pesticide application include the possible use of less pesticide, a reduction in time required for application, and the possibility of highly uniform coverage. Disadvantages are found in the possibilities of applying too little or too much pesticide in local areas. It is difficult to see the pesticide emitted from some of this equipment. The handler must predetermine the volume of pesticide and length of time for application and then adhere strictly to this schedule.

Several types of LV sprayers are available (Weekman 1983). These use pesticides at 10 to 25 times the concentration encountered in HV hydraulic sprayers. Quantities of diluted pesticide formulation applied to an acre of greenhouse range from 0.5 to 15 gal (5 to 140 L/ha).

One type of LV sprayer is the spinning-disk applicator. These applicators offer an advantage in that a high percentage of droplets can be produced in the 50-to-100-micron range. This lowers the required volume to about 1 gallon per acre (10 L/ha). A concentrated pesticide is delivered to a spinning disk either by gravity flow or under low pressure. The liquid moves by centrifugal force to the outer perimeter of the disk, where small droplets shear off into the air. In some equipment, the centrifugal force continues to move the droplets out to the plant. In this case, a high proportion will land on the upper leaf surfaces unless the applicator is careful to direct the mist toward all surfaces. Other equipment uses a fan to generate an airstream that picks up the droplets leaving the spinning disk and directs them at the plant. Both leaf surfaces are easily contacted. Droplet size can be controlled by the pesticide flow rate to the disk or by the rotation speed of the disk. Large droplets are avoided in order to permit use of lower volumes of water with the pesticide, while ultrasmall droplets are avoided to prevent pesticide drift.

Other LV sprayers use the same hydraulic principle as the HV sprayers. These sprayers use a pressure of 1,000 to 3,000 psi (7,000 to 21,000 kPa) to break the spray droplets down to an average size of 30 microns (Figure 14-15a).

The air-assisted electrostatic sprayer produces small droplets and develops a negative electrical charge in them. Because the droplets are of similar charge, they repel one another, thereby distributing themselves uniformly in the spray pattern. The electrical charge of the spray induces a more positive charge in the plant, which attracts the spray particles to the surface of foliage. In this way, spray droplets are drawn to inner foliage and undersides of leaves. In this sprayer the air and pesticide formulation enter the nozzle in separate streams. As the pesticide formulation leaves the nozzle, it is sheared into 30-to-60-micron droplets by the airstream. The airstream then propels the droplets to the plant foliage.

Some LV sprayers are designed to be carried by the pesticide handler throughout the crop. Others are set in one location where they operate unassisted by anyone. These latter applicators are turned on and off, usually during the night, by a time clock or computer (Figure 14-15b). This is one more application that justifies use of a computer system for control of the greenhouse environment. At the end of the day the LV application equipment is positioned in a central location in the greenhouse, loaded with pesticide, and plugged into an electrical circuit. The computer is programmed to turn the equipment on for a prescribed length of time during the night. Then convection tubes or HAF fans are used to circulate the pesticide throughout the greenhouse for a period of time. At the end of this exposure period, the computer activates exhaust fans to purge the greenhouse prior to the arrival of the morning work crew. An advantage of the computer over a time clock in this application is the ability of the computer to integrate the pesticide application steps into complex systems of humidity and temperature control.

Aerosol Application

Aerosols constitute another method of LV pesticide application. A few insecticides (no fungicides) can be obtained in cylinders under pressure (Figure 14-16, Table 14-2). Propellants used in these cylinders include hydrocarbons such as isobutane and isopropane, fluorocarbon (Freon), and compressed carbon dioxide gas. The former propellants are under pressure in liquid form. When released into the lower pressure of the atmosphere, they expand into a gas and move at a

Figure 14-16 Application of an aerosol insecticide, resmethrin, to plants in a greenhouse. (*Photo courtesy of Whitmire Research Laboratories, Inc., St. Louis, MO.*)

Table 14-2 Pesticides Available in Aerosol, Fog, and Smoke Formulations

Pesticide	Formulation		
	Aerosol	*Fog*	*Smoke*
DDVP (Vapona)		X	X
Dithio (sulfotepp)		X	X
Endosulfan (Thiodan)		X	X
Lindane		X	X
Nicotine		X	X
Acephate (Orthene)	X	X	
Parathion			X
Resmethrin	X	X	
Tedion			X
Tedion-dithio			X

high velocity, carrying small droplets (15 to 20 microns) of insecticide with them. The carbon dioxide gas likewise expands when released.

The small liquid pesticide droplets are small enough to drift on the air currents, quickly dispersing throughout the entire greenhouse atmosphere. With time, they settle out, mostly onto the upper surfaces of plants. Since little residue accumulates on lower leaves and under surfaces, this form of pesticide application is used for immediate killing of existing insects and is very often used in combination with a spray or dust application that leaves a residue.

In some cases, the quantity of insecticide to be applied is related to the volume of the greenhouse. For other products, including resmethrin (SBP 1382) and acephate (Orthene), it is applied at a rate of 0.5 to 1 lb of product per 3,000 ft^2 of greenhouse floor area (0.8 to 1.6 g/m^2). It is usually sufficient to make one pass through the center aisle of narrow greenhouses—those up to 35 ft (11 m) wide—and two or more passes through wider greenhouses.

There are some very critical temperature requirements for the use of aerosol bombs. The temperature should be preferably 70 to 80°F (21 to 27°C). Below 60°F (16°C), improper distribution and reduced pest kill occur. At temperatures above 85°F (29°C), injury can occur to the plants. One temperature should be selected and maintained as closely as possible, since the dosage rate is dependent upon it. This temperature range is best held during the evening in the summer or during the late afternoon in the winter. Pyrethroid materials such as resmethrin break down in sunlight. It would be best to apply these in the evening.

It is also important that the aerosol be applied on a calm day so that it is not drawn out of the greenhouse or caused to distribute unevenly. The greenhouse must remain closed for at least two hours. Furthermore, moisture on the foliage during application can lead to injury. The plants should be well watered and the foliage dry at the time of application.

Fog Application

A limited number of insecticides and fungicides are available in oil-based carrier preparations for use in fogging equipment (see Table 14-2). Most are prepared at 10% insecticidal strength. In addition, carriers are available for mixing with most EC and WP insecticides and fungicides to permit their use in fog equipment. This LV method of application is very similar to that of aerosols. The same precautions must be followed. The pesticide is heated by a device to form small droplets (10 to 60 microns), which are then propelled into the greenhouse atmosphere. Unlike the aerosol, which cannot be seen, a white fog forms in this case and makes it easier to know whether all areas have been treated equally.

Various foggers may have gasoline motors or propane burners and may be carried on the back or pulled through the greenhouse on wheels. The pesticide formulation is injected into either a hot pipe, the exhaust, or a hot airstream to cause vaporization. For narrow greenhouses, one pass through the center aisle is sufficient; in wide greenhouses, the fogger should be moved through two or more aisles. Protective clothing and masks should be worn in accordance with the toxicity of the pesticide, since it is very difficult to avoid contact in this method of application.

The fog should not be directed toward the plants because heavy deposits as well as hot exhaust scales can be injurious to the plant. The oils used are injurious to some plant species. Flowers tend to be more susceptible than foliage. The fog should be aimed into the aisles.

Air leaks in the greenhouse and windy conditions outside make it difficult to do a uniform fogging job. Carbon buildup in the machine may also impair the effectiveness. The fogger should be cleaned periodically with wood alcohol or with a cleaner provided by the manufacturer.

Smoke Application

Smokes are another form of LV pesticide application. They are the simplest forms of pesticides to use. No special equipment is needed. Small containers of a combustible formulation containing the pesticide are placed along the center aisle of the greenhouse and ignited, usually with a sparkler. The pesticide is carried in the smoke throughout the greenhouse. Pesticides that can be used in smoke are limited to those few that can withstand the intense heat (see Table 14-2).

Smokes are generally less phytotoxic to plant foliage and flowers than aerosols and fogs, but the dose rate is as important. The volume of the greenhouse must be calculated and divided by the volume one can will treat to determine how many cans to use.

Further precautions are the same for smokes as they are for aerosols and fogs:

1. Do not use at temperatures above 85°F (29°C) or below 60°F (16°C).
2. Close the ventilators on the greenhouse and turn off fans.
3. Avoid using these methods on a windy day.
4. Be sure that the plants are well watered and the foliage is dry.
5. Once the pesticide is applied, follow the instructions on the label concerning warning signs on the greenhouse entrances.

Volatilization

According to current pesticide law, it is permissible for a grower to apply pesticides in a manner not stated on the label as long as it is not forbidden on the label. However, advisory personnel may not recommend to a grower a method of application not specifically on the label. Volatilization is one such method currently gaining acceptance among growers.

In this method, household frying pans are situated throughout the greenhouse above plant height. Pesticides such as the fungicide Bayleton, the insecticide resmethrin, and the miticide Pentac are added to the pan in the quantity labeled for spraying the crop in the area affected by the pan. The pan is wired to a time clock, which turns the system on during the night for a period of six to eight hours. A few hours before employees arrive in the morning, the system shuts off. Final dissipation of the pesticide is accomplished by turning on the exhaust fans. This system has been highly effective in controlling powdery mildew on roses. It has advantages of labor simplicity and very low residual levels for employee contact.

Root Substrate Application

Soil-inhabiting insects can be killed by drenching root substrate with insecticides such as resmethrin and insecticidal soap. Apart from these, the systemic insecticides imidacloprid (Marathon) and oxamyl (Vydate) warrant special mention. They are available in a granular formulation that is applied to the root substrate surface. When moistened, the insecticide is released into the root substrate, killing insects in it. From there, the insecticide is taken up by plants, rendering them toxic to insects feeding on them. These pesticides are effective against a broad range of insects.

Application to individual pots can be made with a small spoon or to large numbers of pots through applicator equipment such as the Perfect-A-Feed and the EZ Feeder Measure Master, which have been modified for metering out small quantities. These insecticides must be applied uniformly over the surface of the pot to ensure even uptake in all parts of the plants. Broadcast application can be made to pots from rim to rim or to fresh-flower beds. Holes are punched in the caps of plastic or metal cans with an ice pick. The insecticide is placed in the can and is shaken from it over the plants. The plants should be dry so that the granules settle on the root substrate.

The appropriate protective clothing and equipment specified on the label should be worn when applying these materials. Care should be taken to avoid spilling granules on nongrowing areas. Equipment that grinds granules should be avoided since safety rests in the entrapment of these insecticides in the granules. Once applied, they should be watered in well to begin the release process. For a period of one week, the root substrate should not be handled.

Systemic granular insecticides should not be used in root substrate prior to planting because this will bring workers into needless contact with them. Granules should not be applied prior to steam pasteurization, since this can cause a hazardous release.

Pesticide Recommendations

Pesticide recommendations can be obtained from a number of greenhouse grower associations and universities. All recommend the pest and crop for which they are registered with the EPA. The use of a pesticide for another crop constitutes a violation of the law, even though it may be safe for the crop. The pesticide may be applied against pests not named on the label unless the label specifically states that the pesticide must be used only for the pests named on the label.

Some materials are designated "Restricted Use Pesticide" because of their toxicity to humans or their danger to the environment. These pesticides may be found in each toxicity category. In all states, they can only be sold to and applied by a certified applicator. This is a person who has passed a course on pesticide handling and safety and who has been certified by the state. (Check with your cooperative extension service for further information.)

Pest Control Tables

Table 14-3 was developed by Baker and Bailey for greenhouse flower crops; Table 14-4 was developed by Sorensen, Walgenbach, and Kennedy for greenhouse vegetable crops. The former table was reproduced from the North Carolina Commercial Flower Growers Bulletin (Baker and Bailey, 1996) and the latter from the 1997 North Carolina Agricultural Chemicals Manual (NCSU College of Agriculture and Life Sciences, 1997).

Pest Resistance

A number of greenhouse pests have developed levels of resistance to pesticides that seriously challenge their control. These pests include aphids, leaf miners, spider mites, thrips, and whiteflies. Any given population of an insect is a rather heterogeneous mixture. Often, there will be a few individuals in the population that are resistant to whichever pesticide is used. The others will be killed, while these resistant strains will survive and multiply. Fortunately, these strains are less prolific than the susceptible strains, but occasionally they can build up to a serious level.

Table 14-3 Pesticides Labeled for Greenhouse Ornamental Insect and Related Pest Control[1]

This table lists pesticides currently labeled for use in commercial floriculture greenhouses for use on ornamental plants. Unless noted in the precautions and remarks column, the pesticide listed is labeled for any greenhouse flower or foliage plant. Read individual labels for any restrictions on use prior to application; labels do change. The abbreviation "NA" stands for not applicable; used when a column is not applicable for a product. Do not confuse volume rates (given in fluid ounces [fl oz]; 8 fl oz = 1 cup) with weight rates (given in avoirdupois ounces [oz]; 16 oz = 1 pound [lb]). Do not equate 1 fl oz with 1 oz; many dry products such as wettable powders should be weighed out for precise concentrations. Other volume and weight abbreviations used include: Tablespoon (Tbsp), teaspoon (tsp), pint (pt), gallon (gal), and pound (lb). Aerosol product and vaporized product application rates are based on either greenhouse area (ft^2) or greenhouse volume (ft^3).

†Pesticide classifications: BO—botanical; CA—carbamate; CH—chlorinated hydrocarbon; DI—Diphenyl; IGR—insect growth regulator; MI—microbial; OP—organophosphate; PY—pyrethroid

Pesticide formulations: A—aerosol (includes compressed total release products and smoke generator products); AS—aqueous suspension; D—dust; DF—dry flowable; EC—emulsifiable concentrate; F—flowable; G—granular; L—water-soluble liquid; ME—microencapsulated; P—pelleted; SP—soluble powder; WG—water-dispersible granules; WP—wettable powder; WSP—water-soluble packets.

Insect or Related Pest	Pesticide, Classification, and Formulation†	Amount of Formulation per Gallon of Spray	Amount per 100 Gallons of Water	Minimum Interval Between Application	Precautions and Remarks
Aphids	acephate -- OP 75 SP (Orthene Turf, Tree and Ornamental) 3 A (PT 300 TR Orthene & PT 1300 DS Orthene)	 1 to 2 tsp NA -- Aerosols	 1/3 to 2/3 lb NA -- Aerosols	 See product label See product label	Orthene 75 soluble powder is also available for use on tobacco. When spraying ornamental plants, be sure to use Orthene Tree, Turf and Ornamentals product. Apply total release aerosols using a rate of 1 lb per 1,500 to 3,000 ft^2. Treat as late in the day as possible. Apply directed spray aerosol using a release rate of 5 to 10 seconds per 100 ft^2.
	azadirachtin -- BO 0.3 F (Margosan-O) 3 EC (Azatin EC) 4.5 EC (Neemazad)	 2 1/2 to 5 tsp 1/2 tsp 1/8 to 1/4 tsp	 40 to 80 fl oz 8 fl oz 2 1/4 to 4 1/2 fl oz	 4 hours 4 hours 4 hours	
	Beauveria bassiana -- MI 7.16 EC (Naturalis-O)	 2 tsp to 2 Tbsp	 30 to 100 fl oz	 4 hours	Three to 5 sprays needed for adequate control.
	bendiocarb -- CA 76 WP (Dycarb, Ficam, Turcam)	 1 Tbsp	 12 to 20 oz	 See product label	Avoid excessive runoff. Labeled for chrysanthemums only.
	bifenthrin -- PY 10 WP (Talstar) 7.9 F (Talstar) 0.5 A (PT 1800 Attain)	 1 to 5 tsp 1/2 to 2 1/2 tsp NA -- Aerosol	 6.4 to 32 oz 2 to 10 fl oz NA -- Aerosol	 See product label See product label See product label	Apply aerosol using a rate of 1 lb per 1,500 to 3,000 ft^2. Treat as late in the day as possible. Building should be vented before reentry.
	chlorpyrifos -- OP 20 ME (PT 1325 DuraGuard)	 1/2 to 1 Tbsp	 1 1/2 to 3 pt	 12 hours	

[1]From Baker and Bailey (1996).

Insect or Related Pest	Pesticide, Classification, and Formulation†	Amount of Formulation per Gallon of Spray	Amount per 100 Gallons of Water	Minimum Interval Between Application	Precautions and Remarks
Aphids, continued	cyfluthrin -- PY 20 WP (Decathlon)	1/4 tsp	1.9 oz	See product label	
	diazinon -- OP 23 EC (Knox-Out)	1 to 2 Tbsp	3 to 6 pt	See product label	Diazinon should not be applied directly to open blooms; it is not labeled for use on poinsettias.
	dichlorvos -- OP 5 A (Vapona)	NA -- Smoke Generator	NA -- Smoke Generator	See product label	Dichlorvos may damage chrysanthemum varieties Shasta, Pink Champagne, and Nightingale.
	endosulfan -- CH 24.2 EC (Thiodan) 50 WP (Thiodan) 5 A (Thiodan)	2 tsp 1 Tbsp NA -- Smoke Generator	2 pt 1 lb NA -- Smoke Generator	24 hours 24 hours See product label	Apply as needed. Repeat applications may be necessary. Thiodan is phytotoxic to some varieties of geraniums and chrysanthemums.
	fenoxycarb -- IGR 0.6 A (PT 2100 Preclude)	NA -- Aerosol	NA -- Aerosol	See product label	Apply aerosol using a rate of 1 lb per 1,500 to 3,000 ft^2. Treat as late in the day as possible. Building should be vented before reentry.
	fenpropathrin -- PY 30 EC (Tame)	2/3 tsp	10 2/3 fl oz	See product label	Can be used with 1/3 lb per 100 gallons acephate (Orthene Turf, Tree and Ornamental 75 SP).
	fluvalinate -- PY 22.3 F (Mavrik)	1/8 to 5/8 tsp	2 to 10 fl oz	See product label	Also labeled as a plant dip.
	horticultural oil -- Other 98.8 EC (Ultra Fine, Sun Spray)	2 1/2 to 5 Tbsp	1 to 2 gallons	4 hours	Oil is a contact killer of pests; thorough coverage is required. May cause foliar injury if sprayed during extremely humid conditions.
	imidacloprid -- Other 1 G (Marathon)	NA -- Granular	NA -- Granular	12 hours	Can be incorporated into substrate at 3 lb per yd^3; broadcast over flats at 9 to 15 oz per 1,000 ft^2; or surface applied to pots (rate varies with pot size -- use 1/3 tsp (1.3 grams) per 6" pot).
	kinoprene -- IGR 65.1 EC (Enstar II)	1/2 to 2/3 tsp	5 to 10 fl oz	4 hours	
	lambda-cyhalothrin -- PY 10 WSP (Topcide)	—	1 to 4 packets	24 hours	
	lindane -- CH 25 WP (Lindane)	1 Tbsp	1 lb	See product label	
	malathion -- OP 25 WP (Malathion) 57 EC (Malathion)	2 tsp 2 tsp	40 oz 2 pt	See product label See product label	Malathion may injure begonia, crassula, ferns, petunia, orchids, pansy, African violet, gloxinia, some red carnations, and some rose varieties.
	methiocarb -- CA 75 WP (Mesurol)	2 to 4 tsp	1 to 2 lb	24 hours	Up to 4 applications per season. Not for use in the landscape.
	naled -- OP 58 EC (Dibrom)	NA --Vaporized	NA -- Vaporized	24 hours	Apply on steam pipes at rate of 1 fl oz per 10,000 ft^3 of greenhouse volume. Then heat pipes to 160 °F. Naled will corrode pipes with continued use. Naled vapor treatment may injure 'White Butterfly' and 'Golden Rapture' rose, 'Pink Champagne' chrysanthemums, wandering jew, poinsettias, and Dutchman's pipe.

Table 14-3 (Continued)

Insect or Related Pest	Pesticide, Classification, and Formulation†	Amount of Formulation per Gallon of Spray	Amount per 100 Gallons of Water	Minimum Interval Between Application	Precautions and Remarks
Aphids, continued	nicotine sulfate -- Other 40 EC (Nicotine Sulfate)	1 1/2 tsp	1 1/2 pt	See product label	Nicotine sulfate will damage young chrysanthemums and lilies.
	oxamyl -- CA 24 L (Vydate L) 10 G (Oxamyl)	2 to 4 tsp NA -- Granules	2 to 4 pt NA -- Granules	See product label See product label	Ornamental uses of Vydate L are being phased out. Apply 22 to 30 oz of Oxamyl granules per 1,000 ft². Equates into 1/8 tsp per 6" to 10" azalea-depth pot.
	permethrin -- PY 36.8 EC (Astro)	1/4 to 1/2 tsp	4 to 8 fl oz	24 hours	
	pyrethrins -- BO 6 EC (Pyrenone) 0.5 A (PT 1100 Pyrethrum)	1/8 to 1 1/2 tsp NA -- Aerosol	2 to 12 fl oz NA -- Aerosol	See product label See product label	Pyrethrins are not recommended for use directly on open blooms or on bracts showing color. Apply PT 1100 using a rate of 1 lb per 3,000 ft². Treat as late in the day as possible. Ventilate prior to reentry.
	resmethrin -- PY 24.3 EC (Resmethrin) 1 A (PT 1200 TR Resmethrin & PT 1200 DS Resmethrin)	1 tsp NA -- Aerosols	1 pt NA -- Aerosols	See product label See product label See product label	Apply PT 1200 TR using a rate of 1 lb per 1,500 to 3,000 ft². Treat as late in the day as possible. Ventilate prior to reentry. Apply PT 1200 DS using a release rate of 5 to 10 seconds per 100 ft².
	soap -- Other 25 EC (Olympic Insecticidal) 49 EC (M-Pede) 49.5 EC (Olympic Insecticidal)	2 1/2 to 5 fl oz 1 1/4 to 2 1/2 fl oz 2 1/2 fl oz	— 1 to 2 gal 2 gal	See product label 12 hours 12 hours	Do not apply more than two times consecutively; too frequent of applications can cause foliar discoloration.
	sulfotepp -- OP 15 A (Dithio, Plantfume 103)	NA -- Smoke Generator	NA -- Smoke Generator	See product label	Consult product label for recommended rates.
Armyworms	acephate -- OP 75 SP (Orthene Turf, Tree and Ornamental) 3 A (PT 300 TR Orthene & PT 1300 DS Orthene)	1 to 2 tsp NA -- Aerosols	1/3 to 2/3 lb NA -- Aerosols	See product label See product label	See aphids section for comments.
	azadirachtin -- BO 0.3 F (Margosan-O) 3 EC (Azatin EC) 4.5 EC (Neemazad)	2 1/2 to 5 tsp 1/2 tsp 1/8 to 1/4 tsp	40 to 80 fl oz 8 fl oz 2 1/4 to 4 1/2 fl oz	4 hours 4 hours 4 hours	
	Bacillus thuringiensis -- MI 3.2 AS (Dipel) 15 AS (Victory)	1 to 2 tsp 2 1/2 to 5 tsp	1 to 2 pt 40 to 80 fl oz	4 hours 4 hours	
	bifenthrin -- PY 10 WP (Talstar) 7.9 F (Talstar) 0.5 A (PT 1800 Attain)	1 to 5 tsp 1/2 to 2 1/2 tsp NA -- Aerosol	6.4 to 32 oz 2 to 10 fl oz NA -- Aerosol	See product label See product label See product label	See aphids section for comments.
	cyfluthrin -- PY 20 WP (Decathlon)	1/4 tsp	1.9 oz	See product label	
	diflubenzuron -- IGR 25 WSP (Adept)	—	4 to 8 packets	12 hours	Apply at a volume of 1 gallon of final solution per 200 ft².
	fluvalinate -- PY 22.3 F (Mavrik)	1/8 to 5/8 tsp	2 to 10 fl oz	See product label	See aphids section for comments.
	lambda-cyhalothrin -- PY 10 WSP (Topcide)	—	1 to 4 packets	24 hours	

Table 14-3 (Continued)

Insect or Related Pest	Pesticide, Classification, and Formulation†	Amount of Formulation per Gallon of Spray	Amount per 100 Gallons of Water	Minimum Interval Between Application	Precautions and Remarks
Armyworms continued	pyrethrins -- BO 6 EC (Pyrenone) 0.5 A (PT 1100 Pyrethrum)	1/8 to 1 1/2 tsp NA -- Aerosol	2 to 12 fl oz NA -- Aerosol	See product label See product label	See aphids section for comments.
	resmethrin -- PY 24.3 EC (Resmethrin) 1 A (PT 1200 TR Resmethrin & PT 1200 DS Resmethrin)	1 tsp NA -- Aerosols	1 pt NA -- Aerosols	See product label See product label See product label	See aphids section for comments.
Beet Armyworm	chlorpyrifos -- OP 20 ME (PT 1325 DuraGuard)	1/2 to 1 Tbsp	1 1/2 to 3 pt	12 hours	
	fenpropathrin -- PY 30 EC (Tame)	2/3 tsp	10 2/3 fl oz	See product label	See aphids section for comments.
	permethrin -- PY 36.8 EC (Astro)	1/4 to 1/2 tsp	4 to 8 fl oz	24 hours	
Broad Mite	bifenthrin -- PY 10 WP (Talstar) 7.9 F (Talstar) 0.5 A (PT 1800 Attain)	1 to 5 tsp 1/2 to 2 1/2 tsp NA -- Aerosol	6.4 to 32 oz 2 to 10 fl oz NA -- Aerosol	See product label See product label See product label	See aphids section for comments.
	dienochlor -- CH 50 WP (Pentac) 38 F (Pentac)	1 tsp 1/2 tsp	8 oz 8 fl oz	See product label See product label	Repeat pentac applications in 5 to 14 days for effective control.
	lambda-cyhalothrin -- PY 10 WSP (Topcide)	—	2 to 4 packets	24 hours	
	pyridaben -- DI 75 WP (Sanmite)	—	2 to 4 oz	12 hours	Do not exceed 10.5 oz of product per acre per application.
Brown Soft Scale	bendiocarb -- CA 76 WP (Dycarb, Ficam, Turcam)	3/4 Tbsp	5 1/2 oz	See product label	Avoid excessive runoff.
	bifenthrin -- PY 10 WP (Talstar) 7.9 F (Talstar) 0.5 A (PT 1800 Attain)	1 to 5 tsp 1/2 to 2 1/2 tsp NA -- Aerosol	6.4 to 32 oz 2 to 10 fl oz NA -- Aerosol	See product label See product label See product label	See aphids section for comments.
	lambda-cyhalothrin -- PY 10 WSP (Topcide)	—	2 to 4 packets	24 hours	
	sulfotepp -- OP 15 A (Dithio, Plantfume 103)	NA -- Smoke Generator	NA -- Smoke Generator	See product label	See aphids section for comments.
Cabbage Looper	acephate -- OP 75 SP (Orthene Turf, Tree and Ornamental) 3 A (PT 300 TR Orthene & PT 1300 DS Orthene)	1 to 2 tsp NA -- Aerosols	1/3 to 2/3 lb NA -- Aerosols	See product label See product label	See aphids section for comments.
	azadirachtin -- BO 0.3 F (Margosan-O) 3 EC (Azatin EC) 4.5 EC (Neemazad)	2 1/2 to 5 tsp 1/2 tsp 1/8 to 1/4 tsp	40 to 80 fl oz 8 fl oz 2 1/4 to 4 1/2 fl oz	4 hours 4 hours 4 hours	
	Bacillus thuringiensis -- MI 3.2 AS (Dipel) 15 AS (Victory)	1 to 2 tsp 2 1/2 to 5 tsp	1 to 2 pt 40 to 80 fl oz	4 hours 4 hours	
	chlorpyrifos -- OP 20 ME (PT 1325 DuraGuard)	1/2 to 1 Tbsp	1 1/2 to 3 pt	12 hours	
	cyfluthrin -- PY 20 WP (Decathlon)	1/4 tsp	1.9 oz	See product label	

Insect or Related Pest	Pesticide, Classification, and Formulation†	Amount of Formulation per Gallon of Spray	Amount per 100 Gallons of Water	Minimum Interval Between Application	Precautions and Remarks
Cabbage Looper, continued	diazinon -- OP 23 EC (Knox-Out)	1 to 2 Tbsp	3 to 6 pt	See product label	See aphids section for comments.
	dichlorvos -- OP 5 A (Vapona)	NA -- Smoke Generator	NA -- Smoke Generator	See product label	See aphids section for comments.
	fluvalinate -- PY 22.3 F (Mavrik)	1/8 to 5/8 tsp	2 to 10 fl oz	See product label	See aphids section for comments.
	lambda-cyhalothrin -- PY 10 WSP (Topcide)	—	1 to 4 packets	24 hours	
	naled -- OP 58 EC (Dibrom)	NA --Vaporized	NA -- Vaporized	24 hours	See aphids section for comments.
	permethrin -- PY 36.8 EC (Astro)	1/4 to 1/2 tsp	4 to 8 fl oz	24 hours	
	pyrethrins -- BO 6 EC (Pyrenone) 0.5 A (PT 1100 Pyrethrum)	1/8 to 1 1/2 tsp NA -- Aerosol	2 to 12 fl oz NA -- Aerosol	See product label See product label	See aphids section for comments.
	resmethrin -- PY 24.3 EC (Resmethrin) 1 A (PT 1200 TR Resmethrin & PT 1200 DS Resmethrin)	1 tsp NA -- Aerosols	1 pt NA -- Aerosols	See product label See product label See product label	See aphids section for comments.
Caterpillars	*Beauveria bassiana* -- MI 7.16 EC (Naturalis-O)	2 tsp to 2 Tbsp	30 to 100 fl oz	4 hours	See aphids section for comments.
	azadirachtin -- BO 4.5 EC (Neemazad)	1/8 to 1/4 tsp	2 1/4 to 4 1/2 fl oz	4 hours	
	soap -- Other 25 EC (Olympic Insecticidal)	2 1/2 to 5 fl oz	—	See product label	See aphids section for comments.
Chrysanthemum Gall Midge	lindane -- CH 25 WP (Lindane)	1 Tbsp	1 lb	See product label	For use on chrysanthemums only. Weekly applications may be needed.
Crickets	chlorpyrifos -- OP 20 ME (PT 1325 DuraGuard)	1/2 to 1 Tbsp	1 1/2 to 3 pt	12 hours	
	cyfluthrin -- PY 20 WP (Decathlon)	1/4 tsp	1.9 oz	See product label	
	lambda-cyhalothrin -- PY 10 WSP (Topcide)	—	1 to 4 packets	24 hours	
	pyrethrins -- BO 6 EC (Pyrenone) 0.5 A (PT 1100 Pyrethrum)	1/8 to 1 1/2 tsp NA -- Aerosol	2 to 12 fl oz NA -- Aerosol	See product label See product label	See aphids section for comments.
	soap -- Other 25 EC (Olympic Insecticidal) 49 EC (M-Pede) 49.5 EC (Olympic Insecticidal)	2 1/2 to 5 fl oz 1 1/4 to 2 1/2 fl oz 2 1/2 fl oz	— 1 to 2 gal 2 gal	See product label 12 hours 12 hours	See aphids section for comments.
Cucumber Beetle	*Beauveria bassiana* -- MI 7.16 EC (Naturalis-O)	2 tsp to 2 Tbsp	30 to 100 fl oz	4 hours	See aphids section for comments.
	bifenthrin -- PY 10 WP (Talstar) 7.9 F (Talstar) 0.5 A (PT 1800 Attain)	1 to 5 tsp 1/2 to 2 1/2 tsp NA -- Aerosol	6.4 to 32 oz 2 to 10 fl oz NA -- Aerosol	See product label See product label See product label	See aphids section for comments.

Insect or Related Pest	Pesticide, Classification, and Formulation†	Amount of Formulation per Gallon of Spray	Amount per 100 Gallons of Water	Minimum Interval Between Application	Precautions and Remarks
Cucumber Beetle, continued	chlorpyrifos -- OP 20 ME (PT 1325 DuraGuard)	1/2 to 1 Tbsp	1 1/2 to 3 pt	12 hours	
	fluvalinate -- PY 22.3 F (Mavrik)	1/8 to 5/8 tsp	2 to 10 fl oz	See product label	See aphids section for comments.
	pyrethrins -- BO 6 EC (Pyrenone) 0.5 A (PT 1100 Pyrethrum)	1/8 to 1 1/2 tsp NA -- Aerosol	2 to 12 fl oz NA -- Aerosol	See product label See product label	See aphids section for comments.
	resmethrin -- PY 24.3 EC (Resmethrin) 1 A (PT 1200 TR Resmethrin & PT 1200 DS Resmethrin)	1 tsp NA -- Aerosols	1 pt NA -- Aerosols	See product label See product label See product label	See aphids section for comments.
	soap -- Other 25 EC (Olympic Insecticidal)	2 1/2 to 5 fl oz	—	See product label	See aphids section for comments.
Cutworms	acephate -- OP 75 SP (Orthene Turf, Tree and Ornamental) 3 A (PT 300 TR Orthene & PT 1300 DS Orthene)	1 to 2 tsp NA -- Aerosols	1/3 to 2/3 lb NA -- Aerosols	See product label See product label	See aphids section for comments.
	azadirachtin -- BO 0.3 F (Margosan-O) 3 EC (Azatin EC)	2 1/2 to 5 tsp 1/2 tsp	40 to 80 fl oz 8 fl oz	4 hours 4 hours	
	Bacillus thuringiensis -- MI 3.2 AS (Dipel) 15 AS (Victory)	1 to 2 tsp 2 1/2 to 5 tsp	1 to 2 pt 40 to 80 fl oz	4 hours 4 hours	
	bifenthrin -- PY 10 WP (Talstar) 7.9 F (Talstar) 0.5 A (PT 1800 Attain)	1 to 5 tsp 1/2 to 2 1/2 tsp NA -- Aerosol	6.4 to 32 oz 2 to 10 fl oz NA -- Aerosol	See product label See product label See product label	See aphids section for comments.
	cyfluthrin -- PY 20 WP (Decathlon)	1/4 tsp	1.9 oz	See product label	
	fluvalinate -- PY 22.3 F (Mavrik)	1/8 to 5/8 tsp	2 to 10 fl oz	See product label	See aphids section for comments.
	lambda-cyhalothrin -- PY 10 WSP (Topcide)	—	1 to 4 packets	24 hours	
	pyrethrins -- BO 6 EC (Pyrenone) 0.5 A (PT 1100 Pyrethrum)	1/8 to 1 1/2 tsp NA -- Aerosol	2 to 12 fl oz NA -- Aerosol	See product label See product label	See aphids section for comments.
Cyclamen Mite	endosulfan -- CH 24.2 EC (Thiodan) 50 WP (Thiodan) 5 A (Thiodan)	2 tsp 1 Tbsp NA -- Smoke Generator	2 pt 1 lb NA -- Smoke Generator	24 hours 24 hours See product label	See aphids section for comments.
Earwigs	soap -- Other 49 EC (M-Pede) 49.5 EC (Olympic Insecticidal)	1 1/4 to 2 1/2 fl oz 2 1/2 fl oz	1 to 2 gal 2 gal	12 hours 12 hours	See aphids section for comments.
Fungus Gnats	azadirachtin -- BO 0.3 F (Margosan-O) 3 EC (Azatin EC)	2 1/2 to 5 tsp 1/2 tsp	40 to 80 fl oz 8 fl oz	4 hours 4 hours	
	Bacillus thuringiensis -- MI 3.2 AS (Dipel) 15 AS (Victory)	1 to 2 tsp 2 1/2 to 5 tsp	1 to 2 pt 40 to 80 fl oz	4 hours 4 hours	

Insect or Related Pest	Pesticide, Classification, and Formulation†	Amount of Formulation per Gallon of Spray	Amount per 100 Gallons of Water	Minimum Interval Between Application	Precautions and Remarks
Fungus Gnats, continued	bifenthrin -- PY 10 WP (Talstar) 7.9 F (Talstar) 0.5 A (PT 1800 Attain)	 1 to 5 tsp 1/2 to 2 1/2 tsp NA -- Aerosol	 6.4 to 32 oz 2 to 10 fl oz NA -- Aerosol	 See product label See product label See product label	See aphids section for comments.
	chlorpyrifos -- OP 20 ME (PT 1325 DuraGuard)	1/2 to 1 Tbsp	1 1/2 to 3 pt	12 hours	
	cyfluthrin -- PY 20 WP (Decathlon)	1/4 tsp	1.9 oz	See product label	
	diazinon -- OP 23 EC (Knox-Out)	1 to 2 Tbsp	3 to 6 pt	See product label	See aphids section for comments.
	diflubenzuron -- IGR 25 WSP (Adept)	—	2 packets	12 hours	Apply at a volume of 1 to 3 gallons of final solution per 100 ft^2.
	fenoxycarb -- IGR 25 WP (Precision)	1/2 tsp	4 oz	See product label	
	kinoprene -- IGR 65.1 EC (Enstar II)	1/2 to 2/3 tsp	5 to 10 fl oz	4 hours	Enstar II is for larvae control only.
	nematodes (*steinernema carpocapsae*) -- Other 17 AS (Exhibit)	—	—	See product label	Apply 1 gallon per 10,000 ft^2 for larvae control. Make three applications, each 7 days apart for best results.
	oxamyl -- CA 24 L (Vydate L) 10 G (Oxamyl)	 2 to 4 tsp NA -- Granules	 2 to 4 pt NA -- Granules	 See product label See product label	See aphids section for comments.
	permethrin -- PY 36.8 EC (Astro)	1/4 to 1/2 tsp	4 to 8 fl oz	24 hours	See beet armyworm for comments.
	pyrethrins -- BO 6 EC (Pyrenone) 0.5 A (PT 1100 Pyrethrum)	 1/8 to 1 1/2 tsp NA -- Aerosol	 2 to 12 fl oz NA -- Aerosol	 See product label See product label	See aphids section for comments.
	resmethrin -- PY 24.3 EC (Resmethrin) 1 A (PT 1200 TR Resmethrin & PT 1200 DS Resmethrin)	 1 tsp NA -- Aerosols	 1 pt NA -- Aerosols	 See product label See product label See product label	See aphids section for comments.
	soap -- Other 25 EC (Olympic Insecticidal)	2 1/2 to 5 fl oz	—	See product label	See aphids section for comments.
Hemispherical Scale	bendiocarb -- CA 76 WP (Dycarb, Ficam, Turcam)	3/4 Tbsp	5 1/2 oz	See product label	Avoid excessive runoff.
Leafhoppers	*Beauveria bassiana* -- MI 7.16 EC (Naturalis-O)	2 tsp to 2 Tbsp	30 to 100 fl oz	4 hours	See aphids section for comments.
	bifenthrin -- PY 10 WP (Talstar) 7.9 F (Talstar) 0.5 A (PT 1800 Attain)	 1 to 5 tsp 1/2 to 2 1/2 tsp NA -- Aerosol	 6.4 to 32 oz 2 to 10 fl oz NA -- Aerosol	 See product label See product label See product label	See aphids section for comments.
	chlorpyrifos -- OP 20 ME (PT 1325 DuraGuard)	1/2 to 1 Tbsp	1 1/2 to 3 pt	12 hours	
	cyfluthrin -- PY 20 WP (Decathlon)	1/4 tsp	1.9 oz	See product label	
	fenpropathrin -- PY 30 EC (Tame)	1/3 to 2/3 tsp	5 1/3 to 10 2/3 fl oz	See product label	See aphids section for comments.

Insect or Related Pest	Pesticide, Classification, and Formulation†	Amount of Formulation per Gallon of Spray	Amount per 100 Gallons of Water	Minimum Interval Between Application	Precautions and Remarks
Leafhoppers, continued	lambda-cyhalothrin -- PY 10 WSP (Topcide)	—	1 to 4 packets	24 hours	
	malathion -- OP 25 WP (Malathion) 57 EC (Malathion)	 2 tsp 2 tsp	 40 oz 2 pt	See product label See product label	See aphids section for comments.
	permethrin -- PY 36.8 EC (Astro)	1/4 to 1/2 tsp	4 to 8 fl oz	24 hours	See beet armyworm for comments.
	pyrethrins -- BO 6 EC (Pyrenone) 0.5 A (PT 1100 Pyrethrum)	 1/8 to 1 1/2 tsp NA -- Aerosol	 2 to 12 fl oz NA -- Aerosol	See product label See product label	See aphids section for comments.
	resmethrin -- PY 24.3 EC (Resmethrin) 1 A (PT 1200 TR Resmethrin & PT 1200 DS Resmethrin)	1 tsp NA -- Aerosols	1 pt NA -- Aerosols	See product label See product label See product label	See aphids section for comments.
	soap -- Other 25 EC (Olympic Insecticidal) 49.5 EC (Olympic Insecticidal)	2 1/2 to 5 fl oz 2 1/2 fl oz	— 2 gal	See product label 12 hours	See aphids section for comments.
Leafminers	abamectin B1 -- Other 2 EC (Avid)	1/2 tsp	8 fl oz	See product label	Do not use on ferns or conifers.
	acephate -- OP 75 SP (Orthene Turf, Tree and Ornamental) 3 A (PT 300 TR Orthene & PT 1300 DS Orthene)	1 to 2 tsp NA -- Aerosols	1/3 to 2/3 lb NA -- Aerosols	See product label See product label	See aphids section for comments.
	azadirachtin -- BO 0.3 F (Margosan-O) 3 EC (Azatin EC) 4.5 EC (Neemazad)	 2 1/2 to 5 tsp 1/2 tsp 1/8 to 1/4 tsp	 40 to 80 fl oz 8 fl oz 2 1/4 to 4 1/2 fl oz	 4 hours 4 hours 4 hours	
	bendiocarb -- CA 76 WP (Dycarb, Ficam, Turcam)	3/4 Tbsp	5 1/2 oz	See product label	Avoid excessive runoff.
	bifenthrin -- PY 10 WP (Talstar) 7.9 F (Talstar) 0.5 A (PT 1800 Attain)	 1 to 5 tsp 1/2 to 2 1/2 tsp NA -- Aerosol	 6.4 to 32 oz 2 to 10 fl oz NA -- Aerosol	See product label See product label See product label	See aphids section for comments.
	chlorpyrifos -- OP 20 ME (PT 1325 DuraGuard)	1/2 to 1 Tbsp	1 1/2 to 3 pt	12 hours	
	cyromazine -- IGR 75 WP (Citation)	—	2 2/3 oz	See product label	Must wear protective clothing when harvesting. Labeled for chrysanthemum only.
	diazinon -- OP 23 EC (Knox-Out)	1 to 2 Tbsp	3 to 6 pt	See product label	See aphids section for comments.
	dichlorvos -- OP 5 A (Vapona)	NA -- Smoke Generator	NA -- Smoke Generator	See product label	See aphids section for comments.
	malathion -- OP 25 WP (Malathion) 57 EC (Malathion)	 2 tsp 2 tsp	 40 oz 2 pt	See product label See product label	See aphids section for comments.
	oxamyl -- CA 24 L (Vydate L) 10 G (Oxamyl)	 2 to 4 tsp NA -- Granules	 2 to 4 pt NA -- Granules	See product label See product label	See aphids section for comments.

Table 14-3 *(Continued)*

Insect or Related Pest	Pesticide, Classification, and Formulation†	Amount of Formulation per Gallon of Spray	Amount per 100 Gallons of Water	Minimum Interval Between Application	Precautions and Remarks
Leafminers, continued	permethrin -- PY 36.8 EC (Astro)	1/4 to 1/2 tsp	4 to 8 fl oz	24 hours	See beet armyworm for comments.
Leaf Rollers	acephate -- OP 75 SP (Orthene Turf, Tree and Ornamental) 3 A (PT 300 TR Orthene & PT 1300 DS Orthene)	1 to 2 tsp NA -- Aerosols	1/3 to 2/3 lb NA -- Aerosols	See product label See product label	See aphids section for comments.
	azadirachtin -- BO 0.3 F (Margosan-O) 3 EC (Azatin EC)	2 1/2 to 5 tsp 1/2 tsp	40 to 80 fl oz 8 fl oz	4 hours 4 hours	
	bifenthrin -- PY 10 WP (Talstar) 7.9 F (Talstar) 0.5 A (PT 1800 Attain)	1 to 5 tsp 1/2 to 2 1/2 tsp NA -- Aerosol	6.4 to 32 oz 2 to 10 fl oz NA -- Aerosol	See product label See product label See product label	See aphids section for comments.
	cyfluthrin -- PY 20 WP (Decathlon)	1/4 tsp	1.9 oz	See product label	
	diazinon -- OP 23 EC (Knox-Out)	1 to 2 Tbsp	3 to 6 pt	See product label	See aphids section for comments.
	lambda-cyhalothrin -- PY 10 WSP (Topcide)	—	1 to 4 packets	24 hours	
	naled -- OP 58 EC (Dibrom)	NA --Vaporized	NA -- Vaporized	24 hours	See aphids section for comments.
	pyrethrins -- BO 6 EC (Pyrenone)	1/8 to 1 1/2 tsp	2 to 12 fl oz	See product label	See aphids section for comments.
	resmethrin -- PY 24.3 EC (Resmethrin) 1 A (PT 1200 TR Resmethrin & PT 1200 DS Resmethrin)	1 tsp NA -- Aerosols	1 pt NA -- Aerosols	See product label See product label See product label	See aphids section for comments.
Mealybugs	acephate -- OP 75 SP (Orthene Turf, Tree and Ornamental) 3 A (PT 300 TR Orthene & PT 1300 DS Orthene)	1 to 2 tsp NA -- Aerosols	1/3 to 2/3 lb NA -- Aerosols	See product label See product label	See aphids section for comments.
	azadirachtin -- BO 0.3 F (Margosan-O) 3 EC (Azatin EC) 4.5 EC (Neemazad)	2 1/2 to 5 tsp 1/2 tsp 1/8 to 1/4 tsp	40 to 80 fl oz 8 fl oz 2 1/4 to 4 1/2 fl oz	4 hours 4 hours 4 hours	
	bendiocarb -- CA 76 WP (Dycarb, Ficam, Turcam)	3/4 Tbsp	5 1/2 oz	See product label	Avoid excessive runoff.
	bifenthrin -- PY 10 WP (Talstar) 7.9 F (Talstar) 0.5 A (PT 1800 Attain)	1 to 5 tsp 1/2 to 2 1/2 tsp NA -- Aerosol	6.4 to 32 oz 2 to 10 fl oz NA -- Aerosol	See product label See product label See product label	See aphids section for comments.
	chlorpyrifos -- OP 20 ME (PT 1325 DuraGuard)	1/2 to 1 Tbsp	1 1/2 to 3 pt	12 hours	
	cyfluthrin -- PY 20 WP (Decathlon)	1/4 tsp	1.9 oz	See product label	
	diazinon -- OP 23 EC (Knox-Out)	1 to 2 Tbsp	3 to 6 pt	See product label	See aphids section for comments.

Insect or Related Pest	Pesticide, Classification, and Formulation†	Amount of Formulation per Gallon of Spray	Amount per 100 Gallons of Water	Minimum Interval Between Application	Precautions and Remarks
Mealybugs, continued	dichlorvos -- OP 81 EC (Vapona) 5 A (Vapona)	 — NA -- Smoke Generator	 — NA -- Smoke Generator	 See product label See product label	The 81 EC product is for use on African violets and begonias only. Vaporize 1 fl oz per 10,000 ft^3. Ventilate thoroughly after two hours post application. See aphids section for comments on the 5 A product.
	fenpropathrin -- PY 30 EC (Tame)	2/3 tsp	10 2/3 fl oz	See product label	See aphids section for comments.
	fluvalinate -- PY 22.3 F (Mavrik)	1/8 to 5/8 tsp	2 to 10 fl oz	See product label	See aphids section for comments.
	horticultural oil -- Other 98.8 EC (Ultra Fine, Sun Spray)	2 1/2 to 5 Tbsp	1 to 2 gallons	4 hours	See aphids section for comments.
	imidacloprid -- Other 1 G (Marathon)	NA -- Granular	NA -- Granular	12 hours	See aphids section for comments.
	kinoprene -- IGR 65.1 EC (Enstar II)	1/2 to 2/3 tsp	5 to 10 fl oz	4 hours	
	lambda-cyhalothrin -- PY 10 WSP (Topcide)	—	2 to 4 packets	24 hours	
	malathion -- OP 25 WP (Malathion) 57 EC (Malathion)	 2 tsp 2 tsp	 40 oz 2 pt	 See product label See product label	See aphids section for comments.
	naled -- OP 58 EC (Dibrom)	NA --Vaporized	NA -- Vaporized	24 hours	See aphids section for comments.
	oxamyl -- CA 24 L (Vydate L) 10 G (Oxamyl)	 2 to 4 tsp NA -- Granules	 2 to 4 pt NA -- Granules	 See product label See product label	See aphids section for comments.
	permethrin -- PY 36.8 EC (Astro)	1/4 to 1/2 tsp	4 to 8 fl oz	24 hours	See beet armyworm for comments.
	pyrethrins -- BO 6 EC (Pyrenone) 0.5 A (PT 1100 Pyrethrum)	 1/8 to 1 1/2 tsp NA -- Aerosol	 2 to 12 fl oz NA -- Aerosol	 See product label See product label	See aphids section for comments.
	resmethrin -- PY 24.3 EC (Resmethrin) 1 A (PT 1200 TR Resmethrin & PT 1200 DS Resmethrin)	 1 tsp NA -- Aerosols	 1 pt NA -- Aerosols	 See product label See product label See product label	See aphids section for comments.
	soap -- Other 25 EC (Olympic Insecticidal) 49 EC (M-Pede) 49.5 EC (Olympic Insecticidal)	 2 1/2 to 5 fl oz 1 1/4 to 2 1/2 fl oz 2 1/2 fl oz	 — 1 to 2 gal 2 gal	 See product label 12 hours 12 hours	See aphids section for comments.
	sulfotepp -- OP 15 A (Dithio, Plantfume 103)	NA -- Smoke Generator	NA -- Smoke Generator	See product label	See aphids section for comments.
Millipedes	malathion -- OP 25 WP (Malathion) 57 EC (Malathion)	 2 tsp 2 tsp	 40 oz 2 pt	 See product label See product label	See aphids section for comments.
Mites	*Beauveria bassiana* -- MI 7.16 EC (Naturalis-O)	2 tsp to 2 Tbsp	30 to 100 fl oz	4 hours	See aphids section for comments.
	diazinon -- OP 23 EC (Knox-Out)	1 to 2 Tbsp	3 to 6 pt	See product label	See aphids section for comments.

Table 14-3 (Continued)

Insect or Related Pest	Pesticide, Classification, and Formulation†	Amount of Formulation per Gallon of Spray	Amount per 100 Gallons of Water	Minimum Interval Between Application	Precautions and Remarks
Mites, continued	dicofol -- CH 35 WP (Kelthane 35) 50 WP (Kelthane T/O)	 1 1/2 Tbsp 1 Tbsp	 1 to 1 1/3 lb 1/2 to 1 lb	 12 hours 12 hours	
	fenpropathrin -- PY 30 EC (Tame)	1/2 to 1 tsp	10 2/3 fl oz	See product label	See aphids section for comments.
	fluvalinate -- PY 22.3 F (Mavrik)	1/8 to 5/8 tsp	2 to 10 fl oz	See product label	See aphids section for comments.
	methiocarb -- CA 75 WP (Mesurol)	2 to 4 tsp	1 to 2 lb	24 hours	See aphids section for comments.
	soap -- Other 25 EC (Olympic Insecticidal) 49.5 EC (Olympic Insecticidal)	 2 1/2 to 5 fl oz 2 1/2 fl oz	 — 2 gal	 See product label 12 hours	See aphids section for comments.
Moths	chlorpyrifos -- OP 22.5 EC (Dursban Pro)	1/2 to 1 Tbsp	1 1/2 to 3 pt	12 hours	
Plant Bugs	bifenthrin -- PY 10 WP (Talstar) 7.9 F (Talstar) 0.5 A (PT 1800 Attain)	 1 to 5 tsp 1/2 to 2 1/2 tsp NA -- Aerosol	 6.4 to 32 oz 2 to 10 fl oz NA -- Aerosol	 See product label See product label See product label	See aphids section for comments.
	cyfluthrin -- PY 20 WP (Decathlon)	1/4 tsp	1.9 oz	See product label	
	fluvalinate -- PY 22.3 F (Mavrik)	1/8 to 5/8 tsp	2 to 10 fl oz	See product label	See aphids section for comments.
	lambda-cyhalothrin -- PY 10 WSP (Topcide)	—	1 to 4 packets	24 hours	
	resmethrin -- PY 24.3 EC (Resmethrin) 1 A (PT 1200 TR Resmethrin & PT 1200 DS Resmethrin)	 1 tsp NA -- Aerosols	 1 pt NA -- Aerosols	 See product label See product label See product label	See aphids section for comments.
	soap -- Other 25 EC (Olympic Insecticidal) 49.5 EC (Olympic Insecticidal)	 2 1/2 to 5 fl oz 2 1/2 fl oz	 — 2 gal	 See product label 12 hours	See aphids section for comments.
Scale Insects	acephate -- OP 75 SP (Orthene Turf, Tree and Ornamental) 3 A (PT 300 TR Orthene & PT 1300 DS Orthene)	 1 to 2 tsp NA -- Aerosols	 1/3 to 2/3 lb NA -- Aerosols	 See product label See product label	See aphids section for comments.
	azadirachtin -- BO 3 EC (Azatin EC)	1/2 tsp	8 fl oz	4 hours	
	bifenthrin -- PY 0.5 A (PT 1800 Attain)	NA -- Aerosol	NA -- Aerosol	See product label	See aphids section for comments.
	chlorpyrifos -- OP 20 ME (PT 1325 DuraGuard)	1/2 to 1 Tbsp	1 1/2 to 3 pt	12 hours	
	cyfluthrin -- PY 20 WP (Decathlon)	1/4 tsp	1.9 oz	See product label	
	diazinon -- OP 23 EC (Knox-Out)	1 to 2 Tbsp	3 to 6 pt	See product label	Diazinon should not be applied directly to open blooms; it is not labeled for use on poinsettias.
	fenoxycarb -- IGR 25 WP (Precision) 0.6 A (PT 2100 Preclude)	 1/2 tsp NA -- Aerosol	 4 oz NA -- Aerosol	 See product label See product label	See aphids section for comments.

Insect or Related Pest	Pesticide, Classification, and Formulation†	Amount of Formulation per Gallon of Spray	Amount per 100 Gallons of Water	Minimum Interval Between Application	Precautions and Remarks
Scale Insects, continued	horticultural oil -- Other 98.8 EC (Ultra Fine, Sun Spray)	2 1/2 to 5 Tbsp	1 to 2 gallons	4 hours	See aphids section for comments.
	kinoprene -- IGR 65.1 EC (Enstar II)	1/2 to 2/3 tsp	5 to 10 fl oz	4 hours	
	lambda-cyhalothrin -- PY 10 WSP (Topcide)	—	1 to 4 packets	24 hours	
	oxamyl -- CA 24 L (Vydate L) 10 G (Oxamyl)	2 to 4 tsp NA -- Granules	2 to 4 pt NA -- Granules	See product label See product label	See aphids section for comments.
	pyrethrins -- BO 6 EC (Pyrenone) 0.5 A (PT 1100 Pyrethrum)	1/8 to 1 1/2 tsp NA -- Aerosol	2 to 12 fl oz NA -- Aerosol	See product label See product label	See aphids section for comments.
	soap -- Other 25 EC (Olympic Insecticidal) 49 EC (M-Pede) 49.5 EC (Olympic Insecticidal)	2 1/2 to 5 fl oz 1 1/4 to 2 1/2 fl oz 2 1/2 fl oz	— 1 to 2 gal 2 gal	See product label 12 hours 12 hours	See aphids section for comments.
Shore Flies	diflubenzuron -- IGR 25 WSP (Adept)	—	2 packets	12 hours	Apply at a volume of 1 to 3 gallons of final solution per 100 ft^2.
	fenoxycarb -- IGR 25 WP (Precision)	1/2 tsp	4 oz	See product label	
Slugs and Snails	metaldehyde -- Other 4 P (Deadline Bullets and Granules) 5 P (Snarol)	NA -- Pelleted bait NA -- Pelleted bait	NA -- Pelleted bait NA -- Pelleted bait	See product label See product label	More than one application usually necessary. Follow label directions for application rates.
	methiocarb -- CA 75 WP (Mesurol)	8 tsp	4 lb	24 hours	See aphids section for comments.
Sowbugs	cyfluthrin -- PY 20 WP (Decathlon)	1/4 tsp	1.9 oz	See product label	
	lindane -- CH 25 WP (Lindane)	1 Tbsp	1 lb	See product label	Apply 1 gallon of spray per 150 ft^2. Repeat applications every 7 to 10 days may be required for effective control.
	malathion -- OP 25 WP (Malathion) 57 EC (Malathion)	2 tsp 2 tsp	40 oz 2 pt	See product label See product label	See aphids section for comments.
	resmethrin -- PY 1 A (PT 1200 TR Resmethrin & PT 1200 DS Resmethrin)	NA -- Aerosols	NA -- Aerosols	See product label See product label	See aphids section for comments.
Spider Mites	abamectin B_1 -- Other 2 EC (Avid)	1/4 tsp	4 fl oz	See product label	Do not use on ferns or conifers.
	bifenthrin -- PY 10 WP (Talstar) 7.9 F (Talstar) 0.5 A (PT 1800 Attain)	1 to 5 tsp 1/2 to 2 1/2 tsp NA -- Aerosol	6.4 to 32 oz 2 to 10 fl oz NA -- Aerosol	See product label See product label See product label	See aphids section for comments.
	chlorpyrifos -- OP 20 ME (PT 1325 DuraGuard)	1/2 to 1 Tbsp	1 1/2 to 3 pt	12 hours	
	dichlorvos -- OP 5 A (Vapona)	NA -- Smoke Generator	NA -- Smoke Generator	See product label	See aphids section for comments.

Table 14-3 *(Continued)*

Insect or Related Pest	Pesticide, Classification, and Formulation†	Amount of Formulation per Gallon of Spray	Amount per 100 Gallons of Water	Minimum Interval Between Application	Precautions and Remarks
Spider Mites, continued	dicofol -- CH 35 WP (Kelthane 35) 50 WP (Kelthane T/O)	 1 1/2 Tbsp 1 Tbsp	 1 to 1 1/3 lb 1/2 to 1 lb	 See product label See product label	
	dienochlor -- CH 50 WP (Pentac) 38 F (Pentac)	 1 tsp 1/2 tsp	 8 oz 8 fl oz	 See product label See product label	Repeat pentac applications in 5 to 14 days for effective control.
	fluvalinate -- PY 22.3 F (Mavrik)	1/8 to 5/8 tsp	2 to 10 fl oz	See product label	See aphids section for comments.
	horticultural oil -- Other 98.8 EC (Ultra Fine, Sun Spray)	2 1/2 to 5 Tbsp	1 to 2 gallons	4 hours	See aphids section for comments.
	lambda-cyhalothrin -- PY 10 WSP (Topcide)	—	2 to 4 packets	24 hours	
	naled -- OP 58 EC (Dibrom)	NA --Vaporized	NA -- Vaporized	24 hours	See aphids section for comments.
	oxamyl -- CA 24 L (Vydate L) 10 G (Oxamyl)	 2 to 4 tsp NA -- Granules	 2 to 4 pt NA -- Granules	 See product label See product label	See aphids section for comments.
	soap -- Other 25 EC (Olympic Insecticidal) 49 EC (M-Pede) 49.5 EC (Olympic Insecticidal)	 2 1/2 to 5 fl oz 1 1/4 to 2 1/2 fl oz 2 1/2 fl oz	 — 1 to 2 gal 2 gal	 See product label 12 hours 12 hours	See aphids section for comments.
Spittlebugs	cyfluthrin -- PY 20 WP (Decathlon)	1/4 tsp	1.9 oz	See product label	
Spittlebugs, continued	resmethrin -- PY 24.3 EC (Resmethrin)	1 tsp	1 pt	See product label	See aphids section for comments.
	soap -- Other 49 EC (Insecticidal, M-Pede)	2 1/2 fl oz	7 1/2 qt	12 hours	See aphids section for comments.
Silverleaf (Sweetpotato) Whitefly	azadirachtin -- BO 0.3 F (Margosan-O) 3 EC (Azatin EC)	 5 tsp 2/3 to 1 tsp	 80 fl oz 10 to 16 fl oz	 4 hours 4 hours	
	fenpropathrin -- PY 30 EC (Tame)	1/2 to 1 tsp	10 2/3 fl oz	See product label	See aphids section for comments.
Thrips	acephate -- OP 75 SP (Orthene Turf, Tree and Ornamental) 3 A (PT 300 TR Orthene & PT 1300 DS Orthene)	 1 to 2 tsp NA -- Aerosols	 1/3 to 2/3 lb NA -- Aerosols	 See product label See product label	See aphids section for comments.
	azadirachtin -- BO 0.3 F (Margosan-O 4.5 EC (Neemazad)	 2 1/2 to 5 tsp 1/8 to 1/4 tsp	 40 to 80 fl oz 2 1/4 to 4 1/2 fl oz	 4 hours 4 hours	
	Beauveria bassiana -- MI 7.16 EC (Naturalis-O)	2 tsp to 2 Tbsp	30 to 100 fl oz	4 hours	See aphids section for comments.
	bendiocarb -- CA 76 WP (Dycarb, Ficam, Turcam)	1 1/2 tsp	12 to 20 oz	See product label	Avoid excessive runoff.
	bifenthrin -- PY 10 WP (Talstar) 7.9 F (Talstar) 0.5 A (PT 1800 Attain)	 1 to 5 tsp 1/2 to 2 1/2 tsp NA -- Aerosol	 6.4 to 32 oz 2 to 10 fl oz NA -- Aerosol	 See product label See product label See product label	Apply aerosol using a rate of 1 lb per 1,500 to 3,000 ft^2. Treat as late in the day as possible. Building should be vented before reentry.
	chlorpyrifos -- OP 20 ME (PT 1325 DuraGuard)	1/2 to 1 Tbsp	1 1/2 to 3 pt	12 hours	

Table 14-3 (*Continued*)

Insect or Related Pest	Pesticide, Classification, and Formulation†	Amount of Formulation per Gallon of Spray	Amount per 100 Gallons of Water	Minimum Interval Between Application	Precautions and Remarks
Thrips, continued	cyfluthrin -- PY 20 WP (Decathlon)	1/4 tsp	1.9 oz	See product label	
	diazinon -- OP 23 EC (Knox-Out)	1 to 2 Tbsp	3 to 6 pt	See product label	Diazinon should not be applied directly to open blooms; it is not labeled for use on poinsettias.
	fluvalinate -- PY 22.3 F (Mavrik)	1/8 to 5/8 tsp	2 to 10 fl oz	See product label	See aphids section for comments.
	horticultural oil -- Other 98.8 EC (Ultra Fine, Sun Spray)	2 1/2 to 5 Tbsp	1 to 2 gallons	4 hours	See aphids section for comments.
	lambda-cyhalothrin -- PY 10 WSP (Topcide)	—	1 to 4 packets	24 hours	
	lindane -- CH 25 WP (Lindane)	1 Tbsp	1 lb	See product label	Apply 1 gallon of spray per 150 ft^2. Repeat applications every 7 to 10 days may be required for effective control.
	malathion -- OP 25 WP (Malathion) 57 EC (Malathion)	 2 tsp 2 tsp	 40 oz 2 pt	 See product label See product label	See aphids section for comments.
	oxamyl -- CA 24 L (Vydate L) 10 G (Oxamyl)	 2 to 4 tsp NA -- Granules	 2 to 4 pt NA -- Granules	 See product label See product label	See aphids section for comments.
	pyrethrins -- BO 6 EC (Pyrenone)	1/8 to 1 1/2 tsp	2 to 12 fl oz	See product label	See aphids section for comments.
	resmethrin -- PY 24.3 EC (Resmethrin) 1 A (PT 1200 TR Resmethrin & PT 1200 DS Resmethrin)	 1 tsp NA -- Aerosols	 1 pt NA -- Aerosols	 See product label See product label	See aphids section for comments.
	sulfotepp -- OP 15 A (Dithio, Plantfume 103)	NA -- Smoke Generator	NA -- Smoke Generator	See product label	See aphids section for comments.
Twospotted Spider Mite	fenpropathrin -- PY 30 EC (Tame)	1/2 to 1 tsp	10 2/3 fl oz	See product label	See aphids section for comments.
	fluvalinate -- PY 22.3 F (Mavrik)	1/8 to 5/8 tsp	2 to 10 fl oz	See product label	See aphids section for comments.
	pyridaben -- DI 75 WP (Sanmite)	—	2 to 4 oz	12 hours	See broad mite section for comments.
Western Flower Thrips	acephate -- OP 75 SP (Orthene Turf, Tree and Ornamental)	1 to 2 tsp	1/3 to 2/3 lb	See product label	See aphids section for comments.
	azadirachtin -- BO 3 EC (Azatin EC)	3/4 to 1 tsp	12 to 16 fl oz	4 hours	
	fomentate hydrochloride -- CA 92 SP (Carzol)	1 tsp	8 oz	24 hours	Local use registration in North Carolina.
Whiteflies	acephate -- OP 75 SP (Orthene Turf, Tree and Ornamental) 3 A (PT 300 TR Orthene & PT 1300 DS Orthene)	 1 to 2 tsp NA -- Aerosols	 1/3 to 2/3 lb NA -- Aerosols	 See product label See product label	See aphids section for comments.

Table 14-3 (Continued)

Insect or Related Pest	Pesticide, Classification, and Formulation†	Amount of Formulation per Gallon of Spray	Amount per 100 Gallons of Water	Minimum Interval Between Application	Precautions and Remarks
Whiteflies, continued	azadirachtin -- BO 0.3 F (Margosan-O) 3 EC (Azatin EC) 4.5 EC (Neemazad)	 2 1/2 to 5 tsp 1/2 tsp 1/8 to 1/4 tsp	 40 to 80 fl oz 8 fl oz 2 1/4 to 4 1/2 fl oz	 4 hours 4 hours 4 hours	
	Beauveria bassiana -- MI 7.16 EC (Naturalis-O)	2 tsp to 2 Tbsp	30 to 100 fl oz	4 hours	See aphids section for comments.
	bendiocarb -- CA 76 WP (Dycarb, Ficam, Turcam)	3/4 tsp	5 1/2 oz	See product label	See aphids section for comments.
	bifenthrin -- PY 10 WP (Talstar) 7.9 F (Talstar) 0.5 A (PT 1800 Attain)	 1 to 5 tsp 1/2 to 2 1/2 tsp NA -- Aerosol	 6.4 to 32 oz 2 to 10 fl oz NA -- Aerosol	 See product label See product label See product label	See aphids section for comments.
	chlorpyrifos -- OP 20 ME (PT 1325 DuraGuard)	1/2 to 1 Tbsp	1 1/2 to 3 pt	12 hours	
	cyfluthrin -- PY 20 WP (Decathlon)	1/4 tsp	1.9 oz	See product label	
	diazinon -- OP 23 EC (Knox-Out)	1 to 2 Tbsp	3 to 6 pt	See product label	See aphids section for comments.
	dichlorvos -- OP 5 A (Vapona)	NA -- Smoke Generator	NA -- Smoke Generator	See product label See product label	See aphids section for comments.
	diflubenzuron -- IGR 25 WSP (Adept)	—	2 packets	12 hours	Apply at a volume of 1 gallon of final solution per 200 ft^2.
	endosulfan -- CH 24.2 EC (Thiodan) 50 WP (Thiodan) 5 A (Thiodan)	 2 tsp 1 Tbsp NA -- Smoke Generator	 2 pt 1 lb NA -- Smoke Generator	 24 hours 24 hours See product label	See aphids section for comments.
	fenoxycarb -- IGR 0.6 A (PT 2100 Preclude)	NA -- Aerosol	NA -- Aerosol	See product label	See aphids section for comments.
	fenpropathrin -- PY 30 EC (Tame)	2/3 tsp	10 2/3 fl oz	See product label	See aphids section for comments.
	fluvalinate -- PY 22.3 F (Mavrik)	1/8 to 5/8 tsp	2 to 10 fl oz	See product label	See aphids section for comments.
	horticultural oil -- Other 98.8 EC (Ultra Fine, Sun Spray)	2 1/2 to 5 Tbsp	1 to 2 gallons	4 hours	See aphids section for comments.
	imidacloprid -- Other 1 G (Marathon)	NA -- Granular	NA -- Granular	12 hours	See aphids section for comments.
	kinoprene -- IGR 65.1 EC (Enstar II)	1/2 to 2/3 tsp	5 to 10 fl oz	4 hours	
	lambda-cyhalothrin -- PY 10 WSP (Topcide)	—	2 to 4 packets	24 hours	
	naled -- OP 58 EC (Dibrom)	NA --Vaporized	NA -- Vaporized	24 hours	See aphids section for comments.
	oxamyl -- CA 24 L (Vydate L) 10 G (Oxamyl)	 2 to 4 tsp NA -- Granules	 2 to 4 pt NA -- Granules	 See product label See product label	See aphids section for comments.
	permethrin -- PY 36.8 EC (Astro)	1/4 to 1/2 tsp	4 to 8 fl oz	24 hours	

Table 14-3 *(continued)*

Insect or Related Pest	Pesticide, Classification, and Formulation†	Amount of Formulation per Gallon of Spray	Amount per 100 Gallons of Water	Minimum Interval Between Application	Precautions and Remarks
Whiteflies, continued	pyrethrins -- BO				
	6 EC (Pyrenone)	1/8 to 1 1/2 tsp	2 to 12 fl oz	See product label	See aphids section for comments.
	0.5 A (PT 1100 Pyrethrum)	NA -- Aerosol	NA -- Aerosol	See product label	
	pyridaben -- DI				
	75 WP (Sanmite)	—	4 to 6 oz	12 hours	See broad mite section for comments.
	resmethrin -- PY				
	24.3 EC (Resmethrin)	1 tsp	1 pt	See product label	See aphids section for comments.
	1 A (PT 1200 TR Resmethrin & PT 1200 DS Resmethrin)	NA -- Aerosols	NA -- Aerosols	See product label	
				See product label	
	soap -- Other				
	25 EC (Olympic Insecticidal)	2 1/2 to 5 fl oz	—	See product label	See aphids section for comments.
	49 EC (M-Pede)	1 1/4 to 2 1/2 fl oz	1 to 2 gal	12 hours	
	49.5 EC (Olympic Insecticidal)	2 1/2 fl oz	2 gal	12 hours	
	sulfotepp -- OP				
	15 A (Dithio, Plantfume 103)	NA -- Smoke Generator	NA -- Smoke Generator	See product label	See aphids section for comments.

When the possibility of resistance is a problem, one pesticide should be selected for the program against an insect. This pesticide should be used for at least one generation. Since it is difficult to determine when a generation ends, it is safest to use the pesticide for the length of time of two or three generations. For western flower thrips during high temperatures, this time frame would be two to three weeks. Then, a pesticide from a different chemical class should be selected for the next time period (see Table 14-3 for chemical classes). In this way, insects that developed resistance to the first pesticide will likely be eradicated by the second pesticide. The third pesticide should be from yet another chemical class. By this point, it should be possible to safely return to the first pesticide. Care should be taken not to mix or alternate pesticides since simultaneous resistance to a number of pesticides can occur, thus reducing the grower's ability to control the pest.

One further point that helps to avoid resistance is to use pesticides with nonspecific modes of action. Many pesticides adversely affect a single process in the insect. It is possible for a mutant to occur with an alternative to that process such that it is not affected by that pesticide. Insecticidal soap is thought to kill in part by suffocation. Many processes in the insect would have to be altered to impart

Table 14-4 Insect Control for Greenhouse Vegetables.[1]

Commodity	Insect	Insecticide and Formulation	Amount of Formulation	Minimum Interval (Days) Between Last Application and Harvest	Precautions and Remarks
CUCUMBER	Aphid	dichlorvos (DDVP) 10 A	Follow label directions	1 day	
		malathion (various) 10 A 57 EC 25 WP	 1 lb/50,000 cu ft 1 qt/100 gal water 4 lb/100 gal water	 1 day 1 day 1 day	Apply as needed in the closed greenhouse in air above the plants. Spray when the temperature is 70° to 85°F. Keep ventilator closed for 2 hr or overnight. Ventilate before reentry. Hazardous to honey bees.
		nicotine sulfate (various)	4.5 oz/50,000 cu ft	1 day	Foliage sprays may be used.
		insecticidal soap (M-Pede) 49 EC	2 tbsp/gal water	0	
	Cabbage looper	*Bacillus thuringiensis* (various)	0.5 to 1 lb or 3 pt/ 100 gal water	—	
	Cucumber beetles	methoxychlor (Marlate) 10 A 50 WP 25 EC	 1 lb/50,000 cu ft 2 tbsp/gal water 2 tsp/gal water	 7 days 7 days 7 days	
	Spider mite	tetradifon (Tedion 400)	0.5 pt/100 gal water	—	Not more than 3 applications during the fruiting season. Effective against mite eggs.
		soap (insecticidal) 49 EC	2 tbsp/gal wat		
	Whitefly and leafminer	malathion (various) 10 A 50 WP 25 WP	 1 lb/50,000 cu ft 1 qt/100 gal water 4 lb/100 gal water	 1 day 1 day 1 day	For details see Cucumber — Aphid
		insecticidal soap (M-Pede) 49 EC Pyrellin EC *Beauveria bassiana* (Mycotrol WP)	 2 tbsp/gal water 1 to 2 pt 1/4 lb / 20 gal water	 0 0 0	May be used alone or in combination. Acts as an exciter. Apply when whiteflies observed. Repeat in 4- to 5-day intervals.
LETTUCE	Aphid, leafminer, and whitefly	Pyrethrins and PBO (Pyrenone)	12 oz/20 gal water	0	May be used alone or tank mixed with a companion insecticide (see label for details).
		malathion (various) 10 A 57 EC 25 WP	 1 lb/50,000 cu ft 1 qt/100 gal water 4 lb/100 gal water	 10 days 14 days 14 days	Do not use endosulfan or naled on lettuce.
		dichlorvos (Vapona, DDVP) 10 A	1 oz/3,000 cu ft	1 day	
		insecticidal soap (M-Pede) 49 EC Pyrellin EC *Beauveria bassiana* (Mycotrol WP)	 2 tbsp/gal water 1 to 2 pt 1/4 lb/20 gal water	 0 0 0	May be used alone or in combination. Acts as an exciter. Apply when whiteflies observed. Repeat in 4- to 5-day intervals.
	Cabbage looper	malathion (various) 10 A 57 EC 25 WP	 1 lb/50,000 cu ft 1 qt/100 gal water 4 lb/100 gal water	 10 days 14 days 14 days	
		Bacillus thuringiensis Javelin WG	0.5 to 1.25/100 gal water	0	
	Spider mite	dichlorvos (Vapona, DDVP) 10 A	1 oz/3,000 cu ft	1 day	
		insecticidal soap (M-Pede) 49 EC	2 tbsp/gal water	0	
TOMATO	Aphid	dichlorvos (DDVP) 10A	Follow label directions	1 day	
		endosulfan (Thiodan, Phaser) 10 A 50 WP 3 EC	 1 lb/50,000 cu ft 1 lb/100 gal water 1 qt/100 gal water	 15 hr 2 days 2 days	Make sure the greenhouse is tightly closed, then apply in the air above the plants. The optimum temperature for application is 70° to 80°F. Keep greenhouse closed for at least 2 hr. Highly toxic—use with caution. Ventilate before reentry. Do not exceed 5 applications of endosulfan per year.
		malathion (various) 10 A 57 EC 25 WP	 1 lb/50,000 cu ft 1 qt/100 gal water 4 lb/100 gal water	 15 hr 1 day 1 day	Apply as needed in the closed greenhouse in the air above the plants. Spray when the temperature is 70° to 85°F. Keep ventilator closed for 2 hr. Hazardous to honey bees.
		nicotine sulfate (various)	4.5 oz/50,000 cu ft	1 day	Foliage sprays may be used.

Commodity	Insect	Insecticide and Formulation	Amount of Formulation	Minimum Interval (Days) Between Last Application and Harvest	Precautions and Remarks
		insecticidal soap (M-Pede) 49 EC	2 tbsp/gal water	0	
		Pyrellin EC	1 to 2 pt	0	May be used alone or in combination. Acts as an exciter.
		Beauveria bassiana (Mycotrol WP)	1/4 lb / 20 gal water	0	Apply when whiteflies observed. Repeat in 4- to 5-day intervals.
	Armyworm	endosulfan (Thiodan, Phaser)			See remarks under Aphids (above).
		10 A	1 lb/50,000 cu ft	15 hr	
		50 WP	1 lb/50,000 cu ft	2 days	Do not exceed 5 applications of endosulfan per year.
		3 EC	1 qt/100 gal water	2 days	
		malathion (various)			See instructions for Aphids (above).
		10 A	1 lb/50,000 cu ft	15 hr	Hazardous to honey bees.
		57 EC	1 qt/100 gal water	1 day	
		25 WP	4 qt/100 gal water	1 day	
		Bacillus thuringiensis			
		Javelin WG	0.5 lb to 1.25 lb/ 100 gal water	0	
		Agree WP	1 to 2 lb	0	
	Cabbage looper	*Bacillus thuringiensis* (various)	0.5 to 1 lb or 2 to 3 pt/ 100 gal water	0	
		Javelin WG	0.5 to 1.25 lb/ 100 gal of water	0	
		Agree WP	1 to 2 lb	0	
	Climbing cutworm	See Armyworm			
	Leafminer	malathion (various) 10 A	1 lb/50,000 cu ft	15 hr	See remarks—TOMATO-Aphid.
		diazinon (Diazinon, Spectracide) AG 500, 50 WP	4 to 8 oz/100 gal water	3 days	Keep ventilators closed for 2 hr or overnight. Plant injury may result if labeling directions are not followed. For use by members of N.C. Greenhouse Vegetable Growers Association, Inc. only.
	Millipedes and crickets	malathion (various) 5 D	Follow label directions		Apply to soil at base of plants. Do not contaminate fruit.
	Slug	metaldehyde (Metason) bait	Follow label directions		Apply to soil surface around plants. Do not contaminate edible parts.
	Spider mite	naled (Dibrom)			See remarks—TOMATO-Aphid.
		60 EC	5 fl oz/50,000 cu ft	1 day	
		60 EC	1 pt/100 gal water	1 day	
		insecticidal soap (M-Pede) 49 EC	2 tbsp/gal water	0	
	Thrips	methoxychlor (Marlate) 10 A	0.5 lb/50,000 cu ft	7 days	
		Beauveria bassiana (Mycotrol WP)	1/4 lb / 20 gal water	0	Apply when whiteflies observed. Repeat in 4- to 5-day intervals.
	Tomato fruitworm	See Armyworm			
	Whitefly	dicholorovs (DDVP, Vapona)	Follow label directions	24 hr	Sanitation, use of yellow sticky traps, and *Encarsia* parasites are encouraged.
		malathion (various)			Apply as needed in the closed greenhouse in the air above the plants. Spray when the temperature is 70° to 80°F. Sprays every second day may be needed. Keep ventilator closed for 2 hr. Hazardous to honey bees.
		10 A	1 lb/50,000 cu ft	15 hr	
		57 EC	1 qt/100 gal water	1 day	
		25 WP	4 lb/100 gal water	1 day	
		insecticidal soap (M-Pede) 49 EC	2 tbsp/gal water	0	
		pyrellin (pyrethrins, rotenone and cube resin)	2 tsp/gal water	0	Pyrellins may be used on many vegetables and herbs for a wide range of insect pests. Read and follow label directions.
		Pyrethrins and PBO (Pyrenone)	12 oz/ 20 gal water	0	May be used alone or tank mixed with a companion insecticide. (See label for details).
		Beauveria bassiana (Mycotrol WP)	1/4 lb / 20 gal water	0	Apply when whiteflies observed. Repeat in 4- to 5-day intervals.

[1]From NCSU College of Agriculture and Life Sciences (1997)

resistance to this mode of action. Thus, it is not likely that resistance would soon appear for this pesticide.

Further Pesticide Considerations

Timing

The length of time between pesticide applications is very important. It depends upon (1) the residual life of the pesticide, (2) the life cycle of the pest, and (3) the method of kill of the pesticide.

When killing insects with a seven-day life cycle by aerosol or smoke insecticides that kill adults but not eggs, it is necessary to treat at six-day intervals until the insect is eliminated. Usually, three treatments are sufficient. Figure 14-17 shows this sequence of action. The first application (day 1) kills most of the insects but not the eggs. The eggs begin hatching immediately after treatment because no insecticidal residue is left by aerosols or smokes. Six days later, before these insects have matured to the stage when they can lay eggs, a second application of insecticide kills them. Insects that escaped the first application are also killed, but they have had a chance to lay a few eggs. These eggs hatch after day 6, and on day 12, before the new insects have laid eggs, they are killed by the third insecticide application, thus ending the population of insects.

If a spraying interval longer than the life cycle of the insect is used, eggs will always be present and the insect population will not be eliminated. Generally, life cycles increase in length with decreasing temperatures. An insecticide that leaves a residue capable of killing may be applied less frequently because it continues to kill for part of the interval between applications.

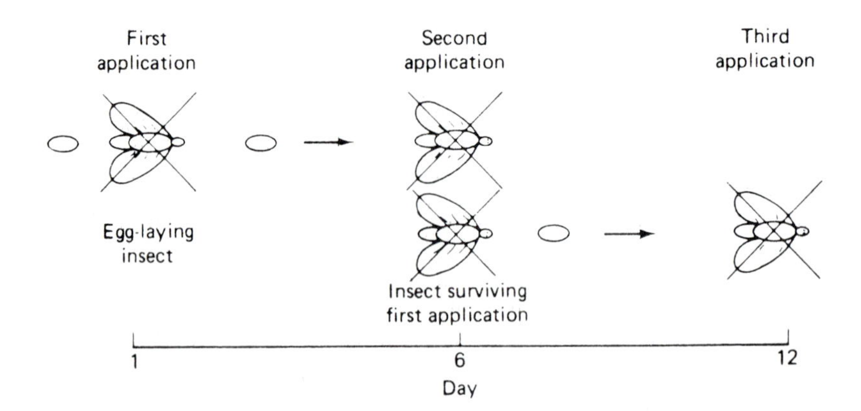

Figure 14-17 **An illustration of the destruction of a population of insects having a seven-day life cycle by means of aerosol or smoke with no residual effect.**

Pesticide application frequencies of five to seven days are common for the eradication of many types of insects and mites. Even a frequency as short as three days may be necessary for eradicating spider mites and whiteflies during hot weather. Systemic insecticides are to be considered separately since their long residual effectiveness eliminates the need for periodic application.

Pesticide Compatibility

Do not mix pesticides until you are certain that they are safe together. Sometimes, two pesticides that are safe to use on a crop individually become toxic to the crop when mixed. Other times, mixing two pesticides reduces their effectiveness because of precipitation or clumping. Further mixing precautions are to be found on the actual pesticide labels.

As a general rule, avoid mixing different kinds of formulations, such as wettable powders and emulsifiable concentrates. Avoid mixing most pesticides in alkaline solutions (pH above 7.0), such as the alkaline fertilizers. Never mix herbicides (weed killers) with any other pesticides. Use separate equipment for herbicides. An herbicide residue in any insecticide spray could cause extensive damage to a crop.

Plant Toxicity

When a pesticide is injurious to a plant species as a whole, recommendations will not be found on the label for that crop. There are, however, pesticides labeled for use on crops where there are a few cultivars that are injured by the pesticide. Lists of cultivars injured by specific pesticides can be found in some pesticide recommendations and also in the "Precautions and Remarks" column of Table 14-3. Whenever you are not certain about the plant toxicity of a pesticide, you should apply it to a small number of plants and wait one week to determine whether or not it is safe.

Shelf Life

Some pesticides are dated as to their effective life; many are not. Generally, after one year they should be inspected for symptoms of deterioration. The effectiveness of some pesticides is reduced or lost, while others may become toxic to plants. Emulsifying agents in EC formulations often become toxic. To help guard against this problem, pesticides should be dated when purchased and properly inventoried to ensure use of the older materials first. The visual symptoms listed in Table 14-5 indicate the need to dispose of a material.

Pesticide Safety

Most greenhouse firms find it necessary to apply pesticides at some point in time even when they are using a biological control program. A number of rules must

Table 14-5 **Visual Symptoms of Various Pesticide Formulations That Indicate It Is Time to Stop Using and Dispose of Pesticide**

Formulation	*Visual Evidence of Deterioration*
Emulsifiable concentrates	When milky coloration does not occur with the addition of water and when sludge is present or any separation of components is evident in the container.
Oil sprays	When milky coloration does not occur with the addition of water.
Wettable powders	When lumping occurs and the powder will not suspend in water.
Dusts	Excessive lumping.
Granulars	Excessive lumping.
Aerosols	Generally effective until the opening of the aerosol dispenser becomes obstructed.

be followed for protection of personnel, the crop, and the environment. Growers should contact their state or governmental agency of pesticide enforcement for a copy of the regulations.

Worker Protection Standard (WPS)

Individual countries have their own sets of regulations governing the use of pesticides. In the United States, the EPA established the Worker Protection Standard (WPS) in 1992 to protect agricultural workers and pesticide handlers. A *worker* is anyone who (1) is employed (including self-employed) for any type of compensation, and (2) is doing tasks, such as harvesting, weeding, or watering, relating to the production of agricultural plants. A *pesticide handler* is anyone who (1) is employed (including self-employed) for any type of compensation by an agricultural establishment or a commercial pesticide-handling establishment that uses pesticides in the production of agricultural plants on a farm, forest, nursery, or greenhouse, and (2) is doing any of the following tasks:

a. Mixing, loading, transferring, or applying pesticides.

b. Handling opened containers of pesticides.

c. Acting as a flagger.

d. Cleaning, handling, adjusting, or repairing the parts of mixing, loading, or application equipment that may contain pesticide residues.

e. Assisting with the application of pesticides, including incorporating the pesticide into the soil after application has occurred.

f. Entering a greenhouse or other enclosed area, after application and before the inhalation exposure level listed on the product labeling has been reached or one of the WPS ventilation criteria has been met, to (1) operate ventilation equipment, (2) adjust or remove coverings, such as tarps, used in fumigation, or (3) check air concentration levels.

g. Entering a treated area outdoors after application of any soil fumigant to adjust or remove soil coverings, such as tarps.

h. Performing tasks as a crop advisor (1) during any pesticide application, (2) before any inhalation exposure level or ventilation criteria listed in the labeling have been reached or one of the WPS ventilation criteria has been met, or (3) during any restricted-entry interval.

i. Disposing of pesticides or pesticide containers.

The standard was amended in April 1995 and again in June 1996. While each state must abide by the WPS rules, they may also apply more stringent rules. Major responsibilities of agricultural employers under the WPS as it now stands are as follows. This list is not complete; thus, one should refer to the whole document before engaging in pesticide use.

Posting of Information Each employer must post and update as necessary the following information at a central location. In addition, employers must tell their employees where this central location is and notify them if there is any change in the posted emergency facility. The information includes the following:

1. An EPA-WPS safety poster.
2. The name, address, and telephone number of the nearest emergency medical facility.
3. The following facts about each pesticide application (from before each application begins until 30 days after the restricted-entry interval [REI]):

   a. product name, EPA registration number, and active ingredient(s)

   b. location and description of the treated area

   c. time and date of application and the REI

The *restricted-entry interval* (REI) is the period of time indicated on the pesticide label during which workers are not allowed to enter the treated area.

Pesticide Safety Training The employer holds responsibility to have each worker and handler trained by a certified applicator or a WPS designated trainer. The certified applicator should use written and/or audiovisual EPA-WPS materials for training workers and handlers. Each worker and handler must be trained within the first five days of employment and at least once every five years thereafter.

Decontamination Sites Employers must establish decontamination sites within 1/4 mile of each worker and handler that contain enough water for routine and emergency whole-body washing and for eye flushing (at least 16 fl oz [0.5 L] eye-flush water per worker or handler), plenty of soap and single-use towels, and a clean coverall. Provide these same supplies where personal protective equipment is removed at the end of a task and at each mixing and loading site.

Emergency Assistance When any handler or worker may have been poisoned or injured by pesticides, the employer must:

1. Promptly make transportation available to an appropriate medical facility.
2. Promptly provide to the victim and to medical personnel:
   a. the product name, EPA registration number, and active ingredient(s)
   b. all first aid and medical information from the label
   c. a description of how the pesticide was used
   d. information about the victim's exposure

Restrictions during Applications The following rules must be followed during pesticide application.

1. Allow only appropriately trained and equipped handlers to be in greenhouses during a pesticide application and until the air concentration level listed on the label is met or, if no such level is listed, until after two hours of ventilation with fans.
2. Do not allow handlers to apply a pesticide so that it contacts, directly or through drift, anyone other than trained and equipped handlers.
3. Make voice or visual contact at least every two hours with anyone handling pesticides labeled with a skull and crossbones.
4. Make sure that a trained handler equipped with labeling-specified personal protective equipment maintains constant voice or visual contact with any handler in a greenhouse who is doing fumigant-related tasks, such as application or air-level monitoring.

Restricted-Entry Intervals (REI) Pesticide labels can carry REI periods from 0 to 72 hours. For this period after pesticide application, only suitably trained and equipped handlers are allowed in the treated area. There is an exception to this rule that allows workers to enter the treated area during the REI period to perform emergency tasks and to irrigate plants. Such early entry is not allowed until four hours after pesticide application and after any inhalation exposure level listed on the product labeling has been reached or any WPS ventilation criteria have been met. Each worker is restricted to eight hours of early-entry exposure in each 24-hour period. Such workers must receive worker training prior to entering the treated area. If they are going to come into contact with the pesticide they must be provided with personal protective equipment as specified on the label for early entry.

Notice about Applications Some pesticide labels require double notification. In this case, workers must be given orally a description of the area to be treated with pesticide, the REI, and a warning not to enter during the REI. In addition, signs must be posted in two languages at the entrances to the greenhouse. If dual notification is not required, signs must be posted at all entrances to the greenhouse. Signs must be posted before pesticide application and removed within three days after the end of the REI, but before workers reenter.

Equipment Safety Pesticide handling equipment must be inspected before each use, and repaired or replaced as needed. Only appropriately trained and equipped handlers are allowed to repair, clean, or adjust pesticide equipment that contains pesticides or residues. A number of accidents have occurred as a result of improperly maintained equipment. Old spray hoses in particular should be replaced before they burst and douse the operator.

Personal Protective Equipment (PPE) *Personal protective equipment* (PPE) is apparel and devices worn to protect the body from contact with pesticides or pesticide residues, including coveralls, chemical-resistant suits, gloves, footwear, aprons, headgear, protective eyewear, and respirators. Employers have the following PPE responsibilities:

1. Provide handlers with the appropriate PPE in clean and operating condition.
2. Make sure that handlers wear the PPE correctly and use it according to the manufacturer's instructions. If a handler wears a respirator, make sure that it fits the wearer correctly.
3. Inspect all PPE before each day of use for leaks, holes, tears, or worn places, and repair or discard any damaged equipment.
4. Provide handlers with clean places away from pesticide-storage and pesticide-use areas to store personal clothing not in use, put on PPE at the start of any exposure period, and take off PPE at the end of any exposure period.
5. Take any necessary steps to prevent heat illness while PPE is being worn.
6. Do not allow any handler to wear home or take home PPE contaminated with pesticides.

Storage Area

In addition to the preceding WPS regulations governing personnel safety, there are other regulations and commonsense rules that will further guarantee safe handling of pesticides. Store pesticides in a well-ventilated, locked closet or room to which children and other unauthorized people have no access. Be sure the temperature remains at 40 to 90°F (4 to 32°C). Place a warning sign on the door indicating that the room contains pesticides. Ideally, the sign should state: "Pesticide Storage, Authorized Personnel Only, In Case of Emergency Call. . . ." A list of the chemical contents of the room should be available. A floor plan of the building showing the location of the pesticide room and its contents should be provided for the local fire department. This will give them a better chance to protect themselves in the event of a fire. You must have a Material Safety Data Sheet (MSDS), which is provided by the manufacturer or supplier, for all hazardous chemicals. Some states have adopted more stringent storage regulations.

Labels

Be sure that labels remain on all containers. Tape or glue them on if they become loose. Read the label before using any pesticide; it contains essential information. To use a pesticide for any other purpose is illegal. The toxicity rating, symptoms of poisoning, and antidote also are generally included. Rates, crops, PPE required, and the REI period are listed as well. Failure to comply with this information could lead to ineffectiveness, injury to the crop, or fines for improper use.

Pesticide Containers

Keep pesticides in their original containers. Never transfer them to another container. If the label is lost from a container and there is any uncertainty about the contents, dispose of the material. Do not guess! Do not reuse pesticide containers. The slightest residue is potentially dangerous. Several cases have occurred in which pesticides were transferred to household containers such as soda bottles, only to be found and drunk by a child.

LD_{50} Values

Categories of toxicity of pesticides have been set forth in the Federal Insecticide, Fungicide, and Rodenticide Act (FIFRA). The toxicity of pesticides is quantitatively expressed as an LD_{50} value either for oral or dermal entry into the body. The former refers to ingestion through the mouth; the latter, to absorption through the skin. In either event, LD_{50} refers to the least dose of pesticide, expressed in milligrams per kilogram (mg/kg) of body weight, that will kill 50 percent of a group of animals tested within 14 days of poisoning. These figures are fairly well applicable to humans. The lower the LD_{50} value, the more poisonous the pesticide, because less of it is required to kill. Pesticides in Category I have acute oral LD_{50} values (to rats) of 0 to 50 ("Highly Toxic") and must bear the words *Danger* and *Poison* as well as a picture of a skull and crossbones on the package. Any time this designation is seen, exercise the ultimate in precaution.

Category II pesticides have acute oral LD_{50} values (to rats) of 50 to 500, are designated "Moderately Toxic," and have the word *Warning* on their label. Category III pesticides have acute oral LD_{50} values (to rats) of 500 and above, are designated "Toxic," and bear the word *Caution* on their label. Note the LD_{50} values of the pesticides you use and be sure that your pesticide handlers are aware of them.

Eating and Smoking

Never smoke, eat, or drink while preparing or applying pesticides. These activities provide a very likely avenue of entry for the pesticide into your body. Keep all foods and beverages out of and away from the pesticide storage room. Do not eat in a greenhouse where plants are routinely treated with pesticides. Set up a lunch and coffee-break area in a safe part of the work building where there is no chance of pesticide contamination.

Disposal of Pesticide Containers

Empty pesticide containers are potentially dangerous because of the difficulty of removing all residue from them. Each state has a set of rules for disposing of them. Contact your cooperative extension service to learn what these rules are.

Summary

1. Insects can constitute a major problem for greenhouse crops. An integrated pest management (IPM) approach should be taken to increase pest control effectiveness and minimize reliance on pesticides. Steps involve eradication of weeds in and around the greenhouse because these can harbor insects and diseases. The greenhouse should be cleaned and root substrate should be pasteurized prior to crop establishment. New plants introduced into the greenhouse should be checked for pests and treated if any are found. A constant surveillance, including the use of sticky cards, should be kept for the appearance of pests in the greenhouse and careful notes should be recorded. The greenhouse environment may need to be altered to tip the advantage away from the pest to the crop. Finally, pest eradication measures may need to be taken. These can involve biological controls or pesticide application.
2. Biological control involves the release of predator insects, parasite insects and related organisms, or pathogenic microorganisms into the crop that seek out and destroy the pests. Several beneficial organisms are commercially available for this use today.
3. The major insects and related pests in greenhouses include aphids, fungus gnats, leaf miners, mealybugs, mites, scale insects, slugs, snails, thrips, whiteflies, and worms.

4. There are eight methods of pesticide application in the greenhouse. High-volume aqueous spray application of pesticides has been the most popular method. Low-volume spray application is similar and involves liquid preparations 10 to 20 times more concentrated than sprays but applied in considerably less volume. Application of dust formulations of pesticides is practiced to a limited degree. High- and low-volume spray and dust applications leave a residue on the plant surface that has a residual effectiveness. Other methods of application, including aerosol, fog, smoke, and volatization, disperse the pesticides throughout the greenhouse atmosphere, killing insects in hard-to-reach crevices and under benches, but have little or no residual activity. Systemic insecticides can be applied to the root substrate or the plant. They are taken up into the plant and result in the death of insects and related pests feeding on the plant. These pesticides have long residual activities and can eliminate the need for other methods of application.
5. Pesticides must be selected from government-approved charts to ensure compliance with safety rules, pest control effectiveness, and avoidance of plant injury.
6. Proper selection of pesticides and timing of applications is important to maximize effectiveness and prevent development of resistance in insects.
7. Human safety should be the major concern on each employer's mind. Safety precautions should include the following:
    a. Compliance with governmental regulations for safe pesticide application. In the United States this is the Worker Protection Standard (WPS).
    b. A labeled, well-ventilated, locked storage area for pesticides, set up and maintained in compliance with governmental regulations.
    c. Labels on all pesticide containers, MSDS sheets for each, and the use of appropriate containers.
    d. Familiarity of each pesticide handler with the toxicity of each pesticide they apply.
    e. No eating or smoking in areas where pesticides are being or have been applied.

References

1. Baker, J. R., ed. 1994. *Insects and related pests of flowers and foliage plants.* North Carolina Coop. Ext. Ser. Bul. AG-136. Raleigh, NC.
2. Baker, J. R., and D. A. Bailey. 1996. Pesticides labeled for greenhouse ornamental insect and related pest control. *North Carolina Commercial Flower Growers Bul.* 41(5).
3. Baker, J. R., and E. A. Shearin. 1995. *Insect screening.* Orn. and Turf IPM Info Note 104. North Carolina Coop. Ext. Ser., Raleigh, NC.
4. Baker, J. R., and E. A. Shearin. 1996. Screen test. *Greenhouse Manager* 13(1):42, 45.
5. Baker, J. R., Bethke, J. A., and E. A. Shearin. 1995. Insect screening. In: Banner, W., and M. Klopmeyer, eds. *New Guinea Impatiens: A Ball Guide,* pp. 155–170. Batavia, IL: Ball Publishing.

6. Bell, M. L., and J. R. Baker. 1997. Choose a greenhouse screen based on its pest exclusion efficiency. *N. C. Flower Growers' Bull.* 42(2):7–13.
7. Cornell Cooperative Extension. 1994. *1995 recommendations for the integrated management of greenhouse florist crops: management of pests and crop growth.* Cornell Coop. Ext., Ithaca, NY.
8. Coyier, D. L., and J. J. Gallian. 1982. Control of powdery mildew on greenhouse-grown roses by volatilization of fungicides. *Plant Disease* 66:842–844.
9. Garber, M. P., W. G. Hudson, K. Bondari, A. R. Chase, R. K. Jones, R. F. Mizell, J. G. Norcini, and R. D. Oetting. 1996. Pest management in the United States greenhouse and nursery industry. *HortTechnology* 6(3):194–221.
10. Hussey, N. W., and N. Scopes. 1985. *Biological pest control: the glasshouse experience.* Ithaca, NY: Cornell Univ. Press.
11. NCSU College of Agriculture and Life Sciences. 1997. *1997 North Carolina agricultural chemicals manual.* College of Agr. and Life Sci., North Carolina State Univ., Raleigh, NC.
12. U.S. Environmental Protection Agency. 1993. *Worker protection standard.* 40CFR Parts 156 and 170. (Available from the Superintendent of Documents, U.S. Government Printing Office, Washington, DC 20402.)
13. Weekman, G. T., ed. 1983. Applying pesticides correctly. USDA and Environmental Protection Agency. North Carolina Agr. Ext. Ser., Raleigh, NC.

 CHAPTER 15

Disease Control

Infectious diseases of greenhouse crops are the downfall of the careless grower. Many diseases cannot be eradicated; at best, they can only be contained. Others cannot be contained, which means that the infected portion of the crop must be quickly identified and removed. As a whole, fungicides and bactericides are not nearly as effective as are insecticides and miticides.

Prevention plays a very important role in disease control. But even with the best of preventive programs, disease organisms will get a foothold in the greenhouse. Some disease organisms are transmitted in soil or groundwater and then are carried in on the bottom of a pot or the sole of a shoe. Some fungi such as the rusts and *Botrytis* produce windblown spores. If host plants are growing near the greenhouse, these spores can be blown into it. Your disease prevention program depends upon that of other greenhouse ranges if you are purchasing seedlings or cuttings or, in some cases, even seed. Numerous pathogens are readily transported on these plant materials. Knowing the inevitability of disease, the manager must be careful to check daily for its presence in the same manner that he or she watches for insects and checks the need for water. The value of IPM (covered in Chapter 14) pertains equally well to disease control.

Much of what has been discussed for pest control is apropos of this discussion on control of disease organisms. The same application equipment is used, including sprayers, dusters, foggers, and smokes. Surfactants are important in WP formulations. LD_{50} values are assigned to fungicides, bactericides, and nematicides as well. In general, all of the safety rules governing insecticides and miticides pertain equally well to fungicides, bactericides, and nematicides.

Diseases of Greenhouse Crops

There are numerous pathogenic diseases of greenhouse crops. They come under four general categories: viruses, bacteria, fungi, and nematodes (Figure 15-1). It is important to know the characteristics of each, as well as the life cycles of specific pathogens, in order to determine how to control them. Some require free water on the plant foliage in order to develop. Others require a very wet root substrate or appear exclusively on purchased cuttings. Each of these requirements suggests a very effective way of controlling the pathogen.

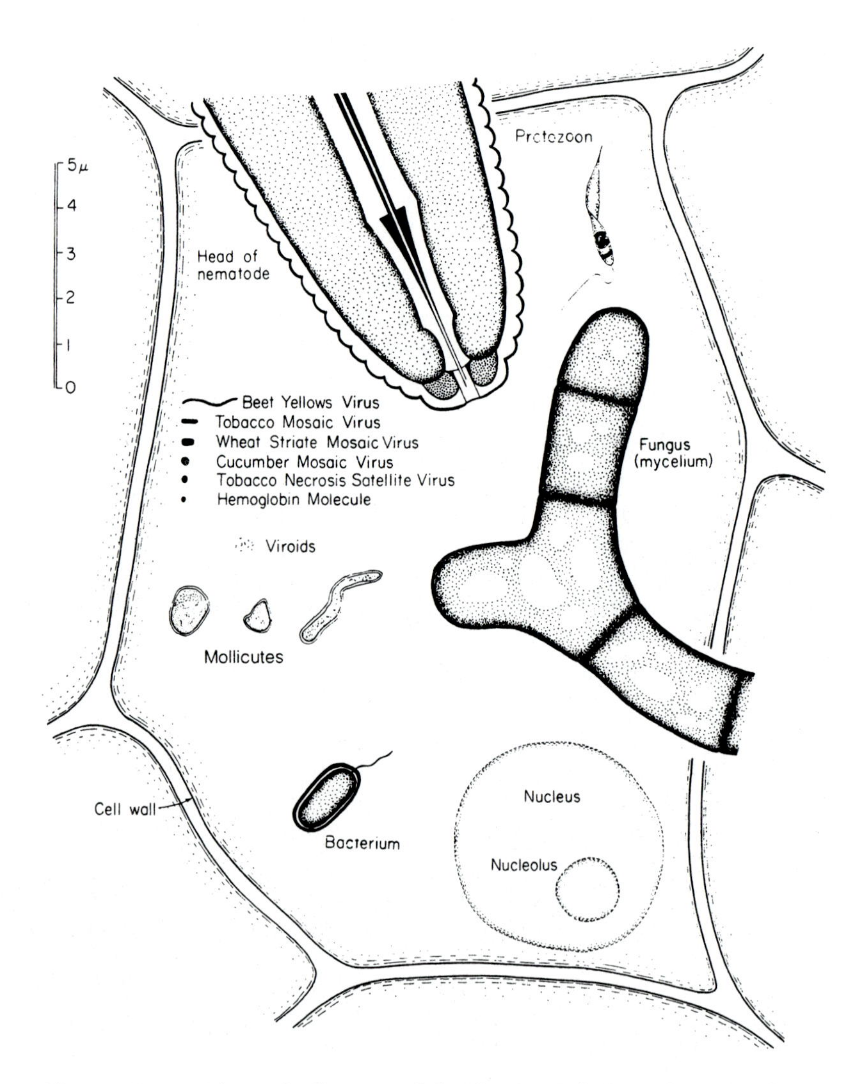

Figure 15-1 **Schematic diagram of the shapes and sizes of certain plant pathogens in relation to a plant cell. (*From G. N. Agrios,* Plant Pathology, 4th ed. *Academic Press, New York, 1997.*)**

Aside from the introductory paragraphs, the following accounts of viruses, fungi, and nematodes are taken verbatim from the writing of Daughtrey and Horst of the Department of Plant Pathology at Cornell University, as presented in the 1995 recommendations for integrated management of greenhouse florist crops (Cornell Cooperative Extension, 1994).

Viruses. Viruses are submicroscopic infectious agents that generally consist of particles composed of protein surrounding genetic material (RNA or DNA). Because plants do not produce antibodies, they neither recover from a virus infection nor become immune. Once a plant is infected, it may remain infected for life, even though the symptoms of disease become masked or the plants grow out of it under certain conditions. Thus, perennial plants and vegetatively propagated greenhouse plants carry the virus from one crop to the next with continuing loss to the disease.

Symptoms. The most common symptom of virus infection is stunting or dwarfing. Leaves may also show distinctive signs, most commonly color changes. Leaves may show spots, streaks, blotches, and rings of light green, yellow, white, brown, or black, or they may develop uniform yellow or orange coloration. Leaves also may change in size or shape, either puckering or developing rolled margins. Flowers may be dwarfed, deformed, streaked, faded, colored green instead of their usual color, or even changed into leafy structures. These are only a few of the more obvious symptoms caused by virus or viruslike causal agents. [See Figure 15-2.]

Spread. Generally, viruses are not transmitted through seed although some, such as tomato ring spot and tobacco ring spot, which affect geraniums, are seed transmitted. A bedding plant crop grown from seed can suffer serious loss if a virus disease

Figure 15-2 Mosaic virus of rose with its characteristic symptoms of leaf puckering and yellow discoloration. (*From R. K. Jones, Department of Plant Pathology, North Carolina State University, Raleigh, NC 27695-7616.*)

(such as tomato spotted wilt virus) is introduced by an insect vector and has an efficient means of spread. The next year, however, the crop will again start clean.

Although viruses can spread unaided from cell to cell in one plant, they require active assistance to spread from one growing plant to another as well as a wound through which to enter the plant. Most frequently, viruses are spread by insects feeding on a healthy plant after feeding on an infected one, by grafting with a scion from an infected plant, or by using infected stock plants as a source of cuttings.

Insect control can be critical to virus control in greenhouses. The current difficulty in controlling western flower thrips, for example, creates a potential danger of widespread impatiens necrotic spot virus infections.

Indexing Program. The use of pathogen-free propagating material is extremely important in any disease control program. Virus indexing is used to eliminate viral pathogens from propagative material of chrysanthemums, carnations, geraniums, orchids, lilies, hydrangeas, and foliage plants. Culture- and virus-indexed plant material is initially free of the internal pathogens for which the index is set up to check; however, such plant material is not disease resistant and requires a growing medium free of the designated pathogens and good cultural practices if the full potential of healthy plants is to be realized.

Virus indexing makes use of indicator plants or serological assays for each specific virus since many cultivars can act as "sleepers," that is, virus carriers that show no external symptoms. Sleeper varieties or cultivars are a tremendous threat since they can be responsible for a large amount of virus spread before the grower realizes there is a serious problem. When the plants are about to flower, the seriousness of the problem is realized in uneven plant growth and flowering time along with a reduction in flower quality.

Diseases of a few crops for which indexing programs have been developed are listed below.

Chrysanthemum

- Viruses
  - Chrysanthemum stunt
  - Chrysanthemum mosaics
  - Chrysanthemum aspermy
  - Chrysanthemum chlorotic mottle
  - Tomato spotted wilt
- Vascular wilts
  - Verticillium wilt
  - Bacterial blight

Carnation

- Viruses
  - Carnation mottle
  - Carnation ring spot
  - Carnation mosaic
  - Carnation streak
  - Carnation etch-ring
  - Necrotic fleck
- Vascular wilts
  - Fusarium wilt
  - Phialophora wilt
  - Bacterial wilt
  - Slow wilt

Geranium

- Viruses
  - Tomato ring spot
  - Tobacco ring spot
  - Pelargomium flower break
- Vascular wilts
  - Bacterial blight
  - Verticillium wilt

Impatiens Necrotic Spot Virus (INSV). Currently INSV, previously called tomato spotted wilt virus (TSWV), is the most common and most damaging virus in the greenhouse industry. [See Figure 15-3.] The virus has an extremely broad host range, and its insect vector, the western flower thrips, is widespread and hard to control. TSWV is vectored by this thrips also. [For control suggestions, see Chapter 14.] INSV is also a problem on foliage plants as well as most crops other than roses and poinsettia. Geraniums have only rarely shown symptoms of the disease. Tuberous dahlias and chrysanthemums are sometimes affected by the closely related virus TSWV.

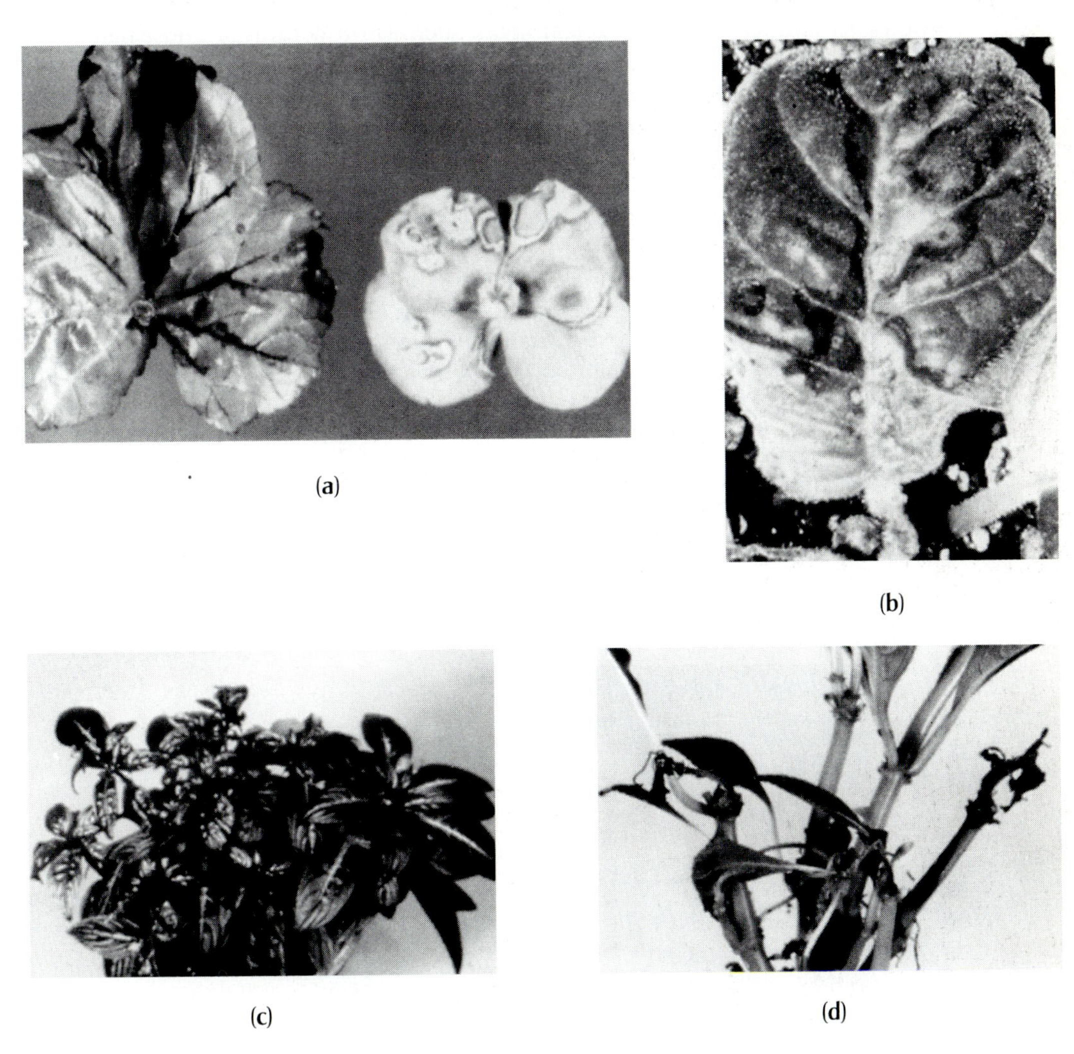

(a) (b) (c) (d)

Figure 15-3 Some of the more common symptoms produced by impatiens necrotic spot virus. (a) Elatior begonia with concentric rings in petals and veinal necrosis in leaves. (b) Gloxinia leaf displaying chlorosis and necrosis. (c) New Guinea impatiens plant showing a common pattern of infection with only one sector of the plant affected. Foliage on the left side is stunted and distorted. (d) Exacum plant with lesions on stems and young leaves. (*Photos courtesy of R. K. Jones and J. W. Moyer, Department of Plant Pathology, North Carolina State University, Raleigh, NC 27695-7616.*)

Bacteria

Bacteria are single-celled microorganisms. Bacterial diseases are difficult to control. A few bactericides exist. Control is primarily through prevention and elimination of infected plants. There are not as many bacterial diseases (Figure 15-4) as fungal diseases. Some of the more common bacterial diseases are bacterial wilt of carnation (*Pseudomonas caryophylli*); bacterial blight (stem rot and leaf spot) of geranium (*Xanthomonas pelargoni*); soft rot of cuttings, corms, and bulbs (*Erwinia chrysanthemi*); bacterial leaf spots such as on geranium and English ivy (*Xanthomonas hederae*); fasciation on crops including carnation, chrysanthemum, geranium, and petunia (*Corynebacterium fascians*); and crown gall on rose, chrysanthemum, and geranium (*Agrobacterium tumefasciens*). The following description is from Daughtrey and Horst.

> **Bacterial Blight of Geranium**. Bacterial blight is caused by *Xanthomonas pelargonii,* which can cause leaf spots as well as systemic infections in geraniums. Leaf symptoms are either an overall tiny spotting or a wedge-shaped yellow area often followed by leaf wilting. The disease can cause black dieback of growing points and stem cankers at the base of the petioles. In hot humid weather, the bacteria spread from infected leaves into the stem, becoming systemic and killing the plant.
>
> Zonal and ivy geraniums are most likely to develop symptoms of this disease; a few cases of leaf spots on Regal geraniums have been observed. Hardy geranium species may be a source of bacteria that can cause disease on greenhouse crops of Pelargonium species. Hybrid "seed" geraniums can become badly diseased if they are grown with an infested cutting crop. Plants in other families than the *Geraniaceae* are not susceptible.

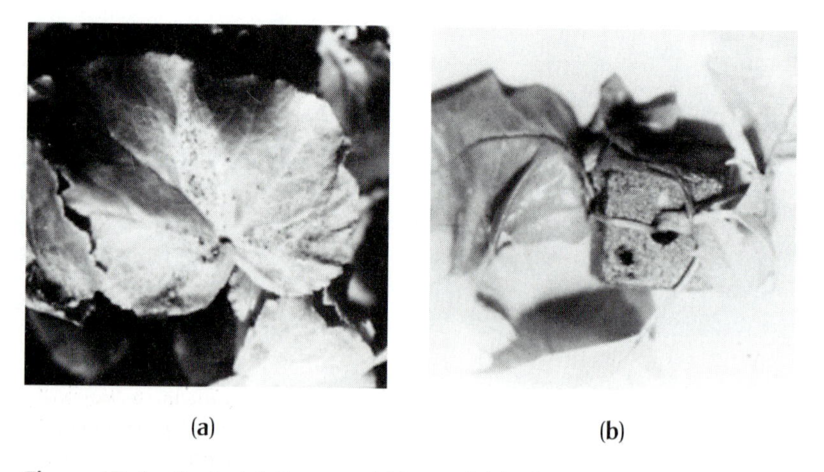

(a) (b)

Figure 15-4 Bacterial diseases: (a) bacterial leaf spot on Rieger begonia and (b) bacterial soft rot of the stem of a poinsettia cutting. (*From D. L. Strider, Department of Plant Pathology, North Carolina State University, Raleigh, NC 27695-7616.*)

Xanthomonas-free material for cuttings is assured through careful culture indexing. Culture indexing is performed by specialists and involves removing thin slices obtained aseptically from the base of a cutting and placing the slices in a nutrient medium. Cultures of nutrient media showing any fungus or bacterial growth are discarded along with the cuttings from which the slices were removed.

Cultural Control. Grow culture-indexed cuttings only, and grow stock plants using individual tube watering systems. The organism is easily spread by splashing water. Subirrigation may spread the disease from root system to root system. Keep stock from different suppliers separate, and grow seedling geraniums separate from cutting crops. Do not hang ivy geraniums over a bench or floor crop of geraniums. Do not grow hardy (perennial) geranium species near greenhouse crops of pelargoniums. Rogue out symptomatic plants immediately.

Fungi

The fungal diseases are the most numerous and lend themselves best to control measures (Figure 15-5). Fungal organisms are much more complex than bacteria. They are multicellular organisms, often comprising several tissues. Some of the more important categories as described by Daughtrey and Horst follow.

Powdery Mildew. Powdery mildew, one of the most easily recognized of all plant diseases, is characterized by the presence of a whitish, powdery mildew growth on the surfaces of leaves, stems, and sometimes petals. The fungal threads and the spores that develop on short, erect branches are visible with a strong hand lens. Under some conditions, however, the threads are so sparse that the mildew can be detected only by examination under strong light with a good lens or dissecting microscope. In some cases, the mildew development is limited to small areas in which the leaf cells are killed and turn black.

The mildew spores are easily detached from the sporophores and carried by air currents to surrounding plants where they initiate new infections. On some plants, such as grape ivy, rose, and delphinium, the young foliage and stems often become severely distorted in addition to being covered by the whitish mildew growth. Seriously affected plants may be of little value as cut flowers or potted plants.

Poinsettia crops were affected by a new powdery mildew disease in 1992. The disease developed rapidly during the fall when bracts were already showing color. While scouting for whiteflies on poinsettias, also watch for powdery mildew colonies on the upper or lower surface of older leaves. At times a yellow spot on the upper leaf surface may indicate a mildew colony growing on the undersurface. Pick all affected leaves and initiate fungicide treatment immediately.

Bioenvironmental Control. Unlike the spores of nearly all other fungi, powdery mildew spores can germinate and initiate infections at humidity levels far below those commonly encountered in the greenhouse. Development of mildew following infection, however, may be more rapid and luxurious at higher humidities. As a deterrent to mildew in greenhouses, ventilation and heating should be adjusted to avoid high-humidity conditions. Irrigate plants early in the day. Heat at least one hour before sunset, and provide adequate ventilation. Horizontal air flow systems assist in powdery mildew management.

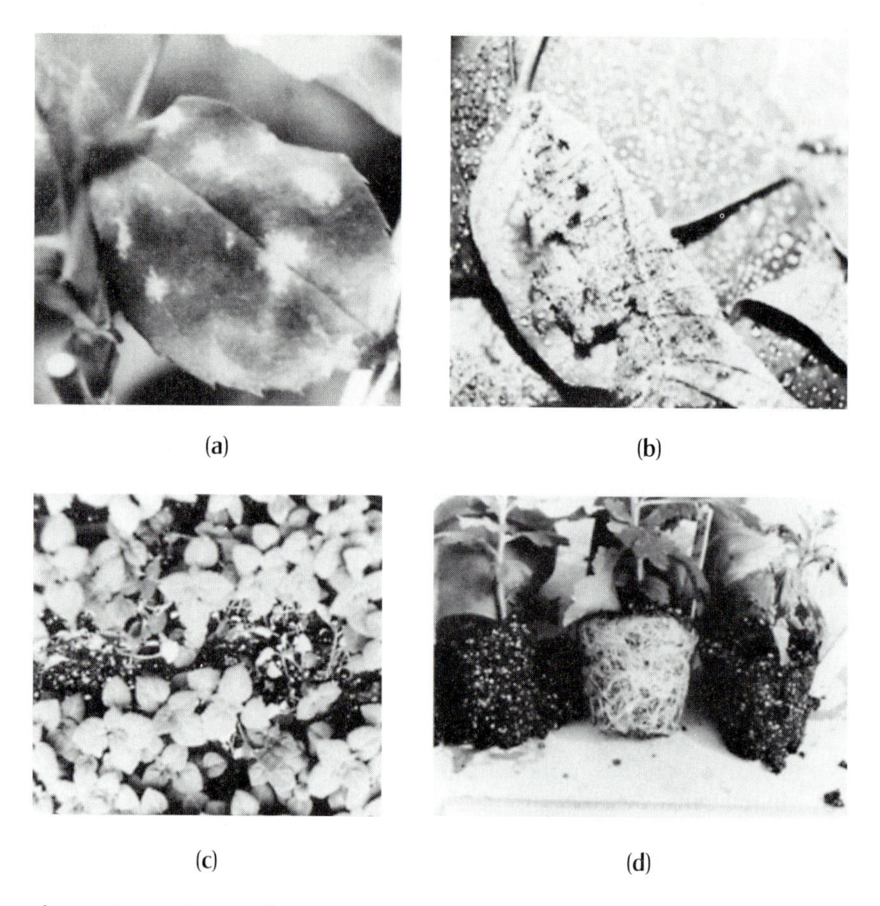

Figure 15-5 **Fungal diseases: (a) powdery mildew on rose; (b) *Botrytis* blight in the spore-forming stage on a poinsettia leaf under mist propagation; (c) damping-off of salvia seedlings; and (d) *Pythium* root rot of chrysanthemum (*left and right*) and normal plant (*center*).** (*From R. K. Jones and D. L. Strider, Department of Plant Pathology, North Carolina State University, Raleigh, NC 27695-7616.*)

Chemical Control. Under some conditions, fungicides for mildew control are essential. Systemic and nonsystemic protectant materials are available for spray application [see Tables 15-2 and 15-3]. In addition, sulfur can be vaporized by painting a slurry of sulfur in water on steam heating pipes or by heating pure sulfur in vaporizers. Regular use of sulfur as a preventive will usually keep powdery mildew from becoming a serious problem.

Botrytis Blight. The common gray mold fungus, *Botrytis cinerea,* attacks a wide variety of ornamental plants, probably causing more losses than any other single pathogen. The fungus causes a brown rotting and blighting of affected tissues. It commonly attacks the stems of geranium stock plants and wounds on cuttings. As a result of *Botrytis* infection, very small seedlings can be rotted, snapdragon, zinnia, exacum, or lisianthus stems can be girdled, and petal tissues of many plants, including carnations, chrysanthemums, roses, azaleas, and geraniums, can be spotted and ruined. The fungus is usually identified by the development of fuzzy grayish spore masses over the surface of the rotted tissues, although such sporulation will not develop under dry conditions.

Spores of *Botrytis* are produced on distinctive dark-colored, hairlike sporophores and are readily dislodged and carried by air currents to new plant surfaces. The spores will not germinate and produce new infections, however, except when in contact with water, whether from splashing, condensation, or exudation. Only tender tissues (seedlings, petals), weakened tissues (stubs left in taking cuttings, tissues infected by powdery mildew), injured tissues (bases of cuttings), or old and dead tissues are attacked on most crops. Active, healthy tissues, other than petals, seldom are invaded. Petals shed from crops in hanging baskets may encourage *Botrytis* leaf infections on the crops below.

Bioenvironmental Control. Because high humidity is required for spore production and actual condensation is necessary for spore germination and infection, *Botrytis* can usually be controlled under glass by avoiding splashing and by heating and ventilating to prevent any condensation on the plant surfaces. Because the fungus readily attacks old or dead tissues and produces tremendous quantities of airborne spores, the importance of strict sanitation cannot be overemphasized. All old blossoms and dead leaves should be removed, and all fallen leaves and plant debris on or under the benches should be gathered and burned.

Chemical Control. Fungicides may be required under some greenhouse conditions, especially with highly susceptible crops such as exacum, geranium, poinsettia, and fuchsia [see Tables 15-2 and 15-3].

Root Rot Diseases. *Rhizoctonia, Phytophthora,* and *Pythium* not only cause damping-off of seedlings but together with *Thielaviopsis* are very important in causing root and basal stem rots of older plants. These fungi are common inhabitants of soil and attack a wide range of plants. They are spread by the mechanical transfer of mycelia, sclerotia, or resting spores in infested soil particles (on flats, tools, pots, baskets, or in the end of the watering hose) or infected plant tissue.

Whereas sanitation measures are effective against all the root rot fungi, fungicides are more specific in their control benefits. The most important control measures are (1) the use of a light, well-drained soil mix; (2) thorough pasteurization of the mix as well as disinfestation of the containers, tools, and benches that come in contact with the plants [see Table 15-1]; (3) the use of clean plants; (4) the enforcement of a sound sanitation program; and (5) the use of supplementary soil treatments with chemicals to minimize recontamination [see Tables 15-2 and 15-3].

Pythium Root Rot. *Pythium* causes a dark brown to black wet rot that makes roots soften and disintegrate. It typically attacks below the soil surface and may extend up into the base of the stem.

Bioenvironmental Control. Pythium is favored by cool, wet, poorly drained soils. Using a well-drained mix with sufficient air pore space and avoiding excessive levels of ammonium or soluble salts will minimize *Pythium* losses.

Rhizoctonia Root Rot. *Rhizoctonia* causes a drier root or stem rot. Affected tissues are brown or tan. It is favored by an intermediate range of moisture, neither too wet nor too dry. Cankers formed by *Rhizoctonia* usually appear at the soil line; roots are also sometimes affected in peat-lite mixes.

Bioenvironmental Control. *Rhizoctonia* disease is often favored by warm temperatures, so losses will typically occur during spring bedding plant production and summer pot plant propagation.

Thielaviopsis Root Rot. *Thielaviopsis* causes a drier stem lesion than *Rhizoctonia,* one that soon turns black because a large number of black spores of the fungus are produced in the lesion. It may also cause a very black root rot, but this is likely to occur only in mixes containing soil. In recent years, however, losses have been seen in pansy and vinca bedding plant crops started as plugs. High pH and poor drainage in the plug tray would encourage *Thielaviopsis* root rot. Losses have also occurred in hanging baskets of fuchsia grown at high pH (6.5 to 7.0), and in poinsettia crops.

Bioenvironmental Control. The disease is not a problem in soil adjusted to pH 4.5 to 5.0.

Damping-Off Disease. Damping-off of seedlings, which is caused mostly by fungi, can be a complex of several diseases occurring separately or simultaneously. Most commonly, either *Rhizoctonia* or *Pythium* is involved. *Botrytis, Sclerotinia,* and *Alternaria* are also occasionally responsible for damping-off.

Preemergence Infection. Seed decay before germination or rot of seedlings before emergence is commonly caused by a water mold, usually *Pythium* or sometimes *Phytophthora.*

Postemergence Infection. Rot developing at the soil line after emergence, which causes the seedling to topple, is most commonly caused by *Rhizoctonia.* This is the conspicuous type of damping-off most frequently reported by growers. Older seedlings may be infected at the soil surface and yet remain upright. Transplanted seedlings remain hard and stunted and eventually die. In some cases, water molds (such as *Pythium*) invade the rootlets at the tips and progress upward to the stem, whereupon the plant dies.

Cultural Control. For all practical purposes, *Rhizoctonia* and *Pythium* do not have an airborne stage. Therefore, spread of both fungi depends upon the mechanical transfer of mycelia, sclerotia, or resting spores in infested soil particles (on flats, tools, baskets, or in the end of the watering hose) or infected plant tissue. Thus, if soil or another medium is steamed or chemically treated and care is taken to prevent recontamination, damping-off should be of little significance. Sowing seed in a layer of screened sphagnum, vermiculite, perlite, peat-lite mix, or other sterilized material also helps. However, some peat moss used in peat-lite mixes may carry these pathogens, and the seed itself may occassionally carry damping-off pathogens.

Chemical Control. Fungicide-treated seed is available for some crops. To avoid plant injury, it is best to rely on careful sanitation practices rather than on fungicide drenches to protect crops until after seedling emergence. Preplant mix incorporation of granular fungicide formulations may lead to phytotoxicity problems with some seedling species, particularly if ingredients are not well distributed through the soil mix.

If experience has shown that particular plant species are plagued by damping-off, make spot applications of appropriate fungicides to just those species [see Tables 15-2 and 15-3].

If preventative fungicide treatments are made to all crops, pay careful attention to the appropriate dosage delivery for the size of the container and recognize that fungicides labeled for ornamentals are often not registered for use on vegetable seedlings.

Verticillium Disease. *Verticillium* is a fungus capable of infecting a wide variety of ornamental plants; some of the more important are chrysanthemums, China asters, snapdragons, roses, geraniums, and begonias. Symptoms vary with the host.

Snapdragons can appear completely healthy until blossoms develop; then the foliage can suddenly wilt completely. The conductive tissues of some varieties can turn brown or purple, particularly the woody stem tissues.

With chrysanthemums, there is usually a marginal wilting of the leaves, followed by chlorosis and eventually death and browning of the leaves, which remain attached and hang down against the stem. These symptoms commonly develop at first on only one side of the plant and only after blossom buds have formed. Young, vigorous plants usually remain symptomless.

The buds on one or two branches of red-flowered varieties of greenhouse roses turn blue and fail to open; the leaves and the green stem tissues may become mottled, and when the stem is shaken, the leaves fall from the plant, and the stem dies. Additional shoots can develop from basal buds and go through the same sequence, though eventually a shoot may remain healthy. Usually there is no vascular discoloration.

With semituberous-rooted begonias, some yellowing of leaf margins can occur, but the most distinctive symptom is the development of an extremely shiny lower leaf surface.

The symptoms thus are quite variable, but the most characteristic ones are one-sided development, wilting and yellowing of leaf margins progressing upward from the lowest leaves, lack of leaf and stem lesions, and normal-appearing roots.

The fungus causing the disease invades the soil and may persist there for many years. Initial infection usually occurs through normal roots, and the fungus grows upward through the water-conducting (xylem) tissues. Infected plants of some types (for example, chrysanthemums) are usually not killed by the fungus and, during periods of rapid vegetative growth, can appear symptomless.

Cultural Control. Cuttings taken from symptomless diseased plants can carry the fungus internally and introduce the disease to new areas. Obtain planting stock only from a reliable dealer, and purchase chrysanthemums and geraniums from propagators who culture index all nucleus stock (see Bacterial Blight of Geranium for a description of culture indexing).

Chemical Control. Plant only in soilless mixes or in soils that have been steamed or treated with chloropicrin to eliminate *Verticillium.*

Nematode Diseases

Nematodes are very small, round worms sometimes called eelworms. For the most part, they cannot be seen by the unaided eye. Nematodes are present in essentially all soils, but many of the numerous types are not harmful. Nematode assays of soil samples will quite often reveal the presence of harmful types. It is not until a large population of these types builds up that crop injury occurs (Figure 15-6). Often, this buildup does not occur because of natural predators in the soil. Nematodes are rarely a problem in soilless root substrates. They should not be a problem in soil-based root substrate if it is pasteurized. The following description of nematodes is by Daughtrey and Horst.

> **Root-Knot Nematode.** Root-knot nematodes may cause plants to appear stunted and unthrifty and to wilt on warm days. When the root system is examined, galls are generally conspicuous and easily recognized. On some crops, root-knot nematodes may cause crop loss even when only a few galls are evident. The presence of root-knot nematodes may also increase the amount of plant injury from bacterial and fungal diseases, or it may break the resistance of plants to these diseases.
>
> Galled plants will not perform as well as healthy ones, but adequate moisture and fertility may mask the difference in vigor between nematode-infested and healthy plants.
>
> Six kinds of root-knot nematodes are recognized in the United States today. Only the northern root-knot nematode, *Meloidogyne hapla,* survives outdoors in northern states. Thus, the other five kinds are usually shipped in on plant material. The host ranges and host-parasite relationships may vary, but all have essentially the same life history.

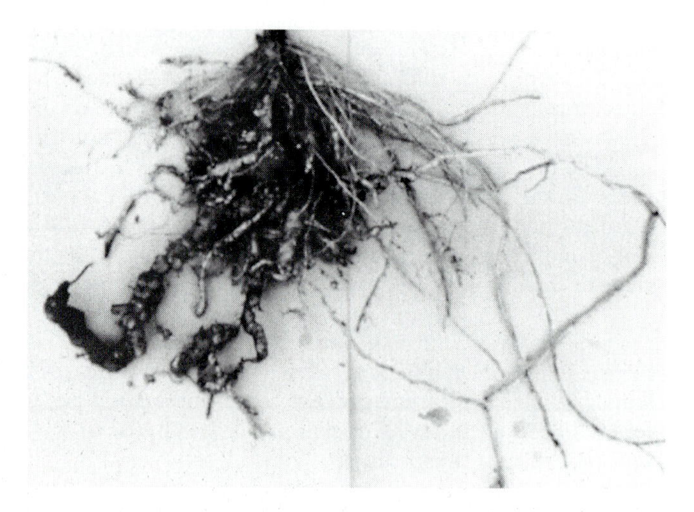

Figure 15-6 Symptoms of root-knot nematode on tomato roots. (*From R. K. Jones, Department of Plant Pathology, North Carolina State University, Raleigh, NC 27695-7616.*)

Eggs of *Meloidogyne* are about twice as long as they are wide. They are usually found in a gelatinous mass about the posterior end of the female. Eggs hatch into small, slender worms (larvae) about 1/50 inch (0.5 mm) long. The larvae migrate through the soil seeking new roots, which they enter near the tip. Once inside the root, with its head located in what will become the vascular cylinder, the nematode does not change position. Stimulated by the nematode's saliva, nearby root cells develop into giant cells, which provide nourishment. Other cells adjacent to the nematode enlarge and increase in number, forming the familiar gall or knot. After the giant cells are functioning, the nematode goes through three molts (shedding of cuticle), becomes an adult female, and starts the cycle over. A female can lay as many as 2,000 eggs during her life, but the average is probably 200 to 500.

The temperature of the soil is critical in the development of the nematode. It takes about 17 days at 29°C (85°F) for females to develop from infective larvae to egg-laying adults, 21 to 30 days at 24°C (76°F), and 57 days at 16°C (60°F). Females fail to reach maturity at temperatures above 33°C (92°F) or below 15°C (59°F).

Spread within a greenhouse occurs through movement of infested soil or plant debris by workers, water, and possibly wind. Migration of larvae through the soil is limited to perhaps a few feet (1 m) per year.

There is no known cure for root-knot nematodes. With continued care, infected bed or bench plants can produce a good crop. Discard infected potted plants carefully to prevent spreading the nematode. Preplanting treatments of steam or fumigants effectively eliminate nematodes from soil, but be sure that infested crop residues are thoroughly decomposed.

Other Nematodes Affecting Roots. Other root-attacking nematodes can cause chlorosis and stunted and unthrifty growth of aboveground parts of the plant. Affected roots may be shortened, thickened, excessively branched to the point of becoming matted, and occasionally killed.

Foliar Nematodes. Leaf (foliar) nematodes cause deformity of young growth, leaf spots, and defoliation. The spots are first discernible on the lower leaf surface as yellowish or brownish areas, which eventually turn almost black. Although the lesions are small at first, with favorable temperature and moisture they may spread until much of the leaf is destroyed. Unlike other nematodes, foliar nematodes do not persist in the soil in the absence of living host-crop tissues.

On chrysanthemum plants, the leaf veins retard the spread of the nematodes through the leaf, causing the lesions to be V-shaped or angular. Infection begins on the lower leaves and progresses upwards.

On peperomia, gloxinia, African violet, and Elatior begonias, the lesions are less definite in outline and infection may occur on any leaf.

Disease Prevention

The value of root substrate pasteurization, discussed in Chapter 7, can readily be seen. It also should be apparent that no resistance to pathogen reinfection is imparted to the root substrate by pasteurization. A total program of sanitation is needed. Learn to integrate disease control into all the cultural operations of your business.

Sterilize Pots and Other Containers

All materials that have contacted another crop must be sterilized before coming into contact with a newly pasteurized root substrate. In this list of materials are tools, flats, pots, wire or plastic supports for plants, and watering systems. Presented in Table 15-1 are some useful disinfectants and items that can be sterilized with each. Since bleaches will discolor clothing, an apron should be worn during their use.

Sterilize Potting Benches

Too often, because potting benches are used for work and storage, tools, motors, and supplies accumulate there. Dirt begins to settle between and around items, and soon it is impossible to sterilize the benches. Keep potting benches clear. Sweep them off after each use with a broom maintained just for that purpose. Do not use a broom that is also used on the floor. At the end of each week, it is a good idea to swab the benches off with a disinfectant or bleach.

Table 15-1 **Common Disinfectants Used in the Greenhouse for Sterilizing Tools, Containers, and Other Materials That Come into Contact with Pasteurized Root Substrate**

Disinfectant	*Rate*	*Application*
Sodium hypochlorite (household bleach)	1 part bleach (5.25% sodium hypochlorite) to 9 parts water	Tools—dip to wet all parts.
		Water system pipes—thoroughly wet with a sponge or rag.
		Flexible water tubes—soak until all interior surfaces are wet for one minute and rinse with water.
		Plastic or wire plant supports—syringe (a proportioner may be used).
Disinfectants (such as Green-Shield, Physan 20, Triathlon)	see label rates	Thoroughly wet hard, nonporous surfaces such as floors, walls, benches, tools, pots, and flats and wait for the prescribed time.
Copper naphthenate	2% solution in Stoddard solvent (VarSol)	Wooden benches and flats—paint or dip and wait until dry to use.

Isolate Root Substrate Storage

Bins for holding root substrate components should be high enough so that groundwater does not flow into them. If they are in a location where dust from crop areas can blow into them, they should be covered.

Avoid Foreign Soil

People will sometimes naturally put one foot on the side wall of a bench when they are talking or surveying a greenhouse. This should not be done because soil on the bottom of their shoes will be scraped off into the bench. Such soil may have originated in another greenhouse range or in someone's garden. The chance of a disease organism being in it is quite high.

Foreign soil will undoubtedly be deposited on the greenhouse floor. One easy way in which it is spread to the bench is by the end of the watering hose. It is a natural impulse to drop the hose on the ground when one is done with it. A hose on the ground picks up soil, which is later flushed off all along the bench during watering. If the floor is contaminated, the whole crop may quickly become infected. A simple, inexpensive broom-handle clip can be nailed to the side of the bench near the water faucet, or a hook can be fashioned out of heavy wire. When the water is turned off, the hose end is hung up.

Clean Up Debris

Sometimes, a crop may become its own worst enemy. In a weakened state, it is more susceptible to a number of pathogens. Plant parts should never be cast where a disease organism developing in them can produce inoculum for the crop. Poorly rooted cuttings should never be thrown under the bench. When pinching the tops from plants or removing the lateral flower buds (disbudding), never throw these plant parts on the floor. Strap a cloth pouch around your waist for holding plant tissue. Disbud pot mums over a container into which the buds can fall (Figure 15-7). Do not expect employees to always go the extra yard for disease prevention. Devise systems whereby they can operate in accordance with sound principles of sanitation with little or no extra effort.

At the end of each crop, there is usually a percentage of plants or flowers left that do not meet the market standards. If another crop is not coming in directly behind the first one, it is tempting to leave rejected plants in the bench until all hope of sale is gone. Clean these plants out immediately! They are weak and will become even weaker as they are neglected. They constitute likely tissue to be infected and, once infected, are an excellent source of infection for other crops. During spring and fall, when evening condensation is common on plants, old flowers invariably become infected with *Botrytis.* It is nearly impossible to eradicate in old flowers.

Plant debris should be placed in a compost pile or dump far enough away

Figure 15-7 **A disbudding stand holds the pot mum at working height. As disbuds are removed and dropped, a funnel directs them into a basket. Plant tissue discarded on the floor provides a good host for disease organisms.**

from the greenhouse so that soilborne microorganisms, windblown spores, and insects contained in it cannot make their way back to the greenhouse.

Clean Stock

When you purchase seedlings or cuttings, you are relying on the sanitation program of another business. Select your plant sources by their reputation. An occasional pathogen or pest problem can come along with plants from the best of propagators. Inspect each lot of plants. If disease is present, isolate these plants and discard or treat them immediately.

If you propagate your own cuttings, maintain a careful disease prevention program on the stock plants. Carefully inspect stock plants to avoid taking cuttings from any infected plants. If a knife is used for removing cuttings, periodically dip it in rubbing alcohol to sterilize it. Transport the cuttings in clean containers and work on a sterilized surface. If there is any doubt, keep the cuttings on clean newspaper.

Environmental Control

The life cycles of pathogens as discussed earlier suggest cultural procedures for reducing their incidence. Diseases such as *Botrytis* blight depend upon free water on the plant surface for spores to germinate. Free water often occurs as condensation. Warm air holds more water than cold air. During warm fall and spring days, the air picks up moisture. Evenings at these times of year are generally cold. As the air cools, its moisture-holding capacity drops until the dew point is reached and water begins to condense on any solid surface.

Two steps can be taken to combat condensation on foliage. During periods of high relative humidity in the greenhouse, the ventilators or exhaust fans can be used to exchange the warm moist air in the greenhouse with cooler air from outside. The moisture-holding capacity of air increases with increasing temperature. Thus, the relative humidity of the cold entering air will fall as it is heated, resulting in dry air within the greenhouse. The heat loss is small since the mass of the exhausted air is small relative to the combined mass of the greenhouse structure, plants, root substrate, floor, and so on, which all hold heat inside the greenhouse.

A time clock is the least expensive system. It can be used to exchange a single volume of air in the late afternoon when the lower night temperature begins and possibly one or more times during the night. It is particularly important to exchange air at the point when the warm day temperature is switched to the cooler night temperature. If not, the relative humidity of the warm moist air from the day will increase as it cools toward the lower night temperature setting. In so doing, it might reach the dew point where condensation settles out on cool surfaces such as leaves.

However, time clocks do not provide for careful control of humidity, nor are they always energy efficient. A computer will provide the best control. Responding to a signal from a humidity sensor, the computer can activate venting equipment at any time that air humidity needs to be lowered and for only the period of time required to lower the relative humidity to a set point. In this way, air is not exhausted during nights when the relative humidity remains satisfactorily low. Also, on humid nights when more ventilation is required than would have been anticipated for a time clock–controlled system, it will be applied.

Moving air in a closed greenhouse is the second step that can be taken to eliminate water on plant surfaces. The horizontal airflow system or the overhead polyethylene convection-tube system will minimize temperature differentials and cold spots where condensation is likely to occur.

Mildew is encouraged by high humidity. Ventilating during the early stages of heating and maintaining good air circulation will help control it. Humidity can further be reduced by watering early in the day when the warm air can absorb moisture from wet surfaces.

Spread of root-rot and damping-off pathogens depends on mechanical transfer of the root substrate in which they reside. Automatic watering helps because it minimizes the splashing and lateral transfer of substrate that are associated with hand watering. It also eliminates the use of a nozzle, which may periodically touch the root substrate along the bench.

A number of root-rot and damping-off pathogens are enhanced by high root substrate moisture levels. A well-drained root substrate should always be used. Water should be applied only as needed. Watering too frequently encourages development of these diseases.

Weed Control

Weeds harbor insects and disease. The importance of eliminating weeds in and around the greenhouse was already stressed in Chapter 14. Various weeds can serve as a host for several pathogens that infect greenhouse crops.

Fungicide and Bactericide Recommendations

The primary approach to disease control should be an IPM program. When preventive measures fail to hold the line and pesticides are required, surveillance will indicate the specific areas to be treated and the number of applications required. Diseases attacking specific greenhouse floral crops and their control recommendations are listed by crop in Table 15-2 (developed by R. K. Jones). Similar recommendations for greenhouse vegetable crops are found in Table 15-3 (developed by F. J. Louws and P. B. Shoemaker). These tables have been reproduced from the 1997 *North Carolina Agricultural Chemicals Manual.*

Specific crops that are particularly susceptible to soil-borne root-rot and damping-off diseases and that are grown in an area where such diseases have been a problem can be protected by the use of fungicidal drenches. Truban is effective against *Pythium* and *Phytophthora,* Aliette against *Phytophthora,* Subdue against *Pythium* and *Phytophthora,* Thiophanate-methyl (Cleary's 3336) against *Rhizoctonia* and *Thielaviopsis,* and Terraclor against *Rhizoctonia.* These fungicides are mixed in water and are applied to the root substrate in the fashion of a normal watering. Combinations are often used to give a broader spectrum of protection. Cleary's 3336 plus Truban or Cleary's 3336 plus Subdue are popular combinations. Banrot is a commercially available combination similar to Cleary's 3336 plus Truban. Upon planting, a new crop is drenched with one of these. Bedding plants, Easter lily, and poinsettia, are some crops for which this practice is popular.

Table 15-2 Commercial Floral Crop Disease Control[1]

Crop	Disease	Pesticide and Formulation	Rate of Formulation	Schedule and Remarks
ANY CROP	Powdery mildew	fenarimol (Rubigan AS)	4 to 12 oz per 100 gal	Spray every 7 to 14 days.
AFRICAN VIOLET	Phytophthora rot and Phythium root rot	etridiazole (Truban 30 WP)	3 to 10 oz/100 gal/400 sq ft or 0.5 pt/6-in. pot	Apply additional water immediately after application. Repeat at 4-to 8-week intervals if needed.
		metalaxyl (Subdue 2E)	0.5 to 2 fl oz/100 gal/ 400 to 800 sq ft	Repeat every 1 to 2 months. Do not apply rates of 1.6 to 2 fl oz/100 gal more often than once every 6 weeks.
	Botrytis blight	thiophanate methyl (Cleary 3336 or Domain)	See label.	Spray every 7 to 14 days.
		mancozeb 80% W	1.5 lb/100 gal	Spray every 7 to 10 days.
		iprodione (Chipco 26019)	1 to 2 lb/100 gal	Spray every 7 to 14 days.
		vinclozalin (Ormalin)	0.5 to 1 lb/100 gal	Spray every 7 to 14 days.
		chlorothalonil (Daconil, Termil)	1.5 lb/100 gal 3.5 oz/10,000 cu ft	Treat every 7 to 14 days.
	Powdery mildew	Systhane	4 oz/100 gal	Spray every 10 to 14 days.
		triadimefon (Strike)	2 to 4 oz/100 gal	Spray every 30 days.
ASTER	Rust	Systhane	4 oz/100 gal	Spray every 10 to 14 days.
AZALEA	Ovulinia petal blight	Systhane	4 oz/100 gal	Spray when flower starts to show color.
		triadimefon (Strike)	8 to 16 oz/100 gal	Make one application as first flower buds show color. Spray later varieties as they show color.
	Leaf gall	ferbam 76 W	2 tsp/gal	Spray just before leaves unroll in spring and 10 days later.
	Phytophthora root rot	etridiazole (Truban 30 WP)	10 oz/100 gal/400 sq ft or 0.5 pt/6-inch pot	Water-in immediately after application. Repeat at 4- to 12-week intervals. Effective for disease prevention.
		metalaxyl (Subdue 2 E)	1 to 2.4 oz/100 gal for 400 to 800 sq ft	Repeat applications every 2 to 4 months.
		fosetyl-Al (Aliette 80 WDG)	1 to 2 lb/1,000 sq ft or 0.5 to 1.5 pt/sq ft	Drench monthly.
	Phomopsis die-back	thiophanate methyl (Cleary's 3336 or Domain)	See label.	Spray older plants immediately after any summer pruning. Prune out any dead branches.
	Phytophthora root rot and nematodes	methyl bromide	2 lb/100 sq ft	Apply under tarp. Wait 1 to 3 weeks to plant.
	Powdery mildew	Systhane	4 oz/100 gal	Every 10 to 14 days.
		triadimefon (Strike)	2 to 4 oz/100 gal	Spray as needed.
	Rhizoctonia web blight	thiophanate methyl (Cleary's 3336 or Domain)	See label.	Spray weekly in rooting bed or every 7 to 14 days during July and August outside.
		iprodione (Chipco 26019 50 WP)	1 to 2 lb/100 gal	Apply as a foliar spray during July and August every 7 to 14 days.
		chlorothalonil (Daconil 2787 W85)	1.5 lb/100 gal 1 tbsp/gal	Apply at first sign of disease, or in mid-July, and repeat at 7- to 14-day intervals until early September.
BEGONIA	Botrytis blight	See African Violet.		
	Powdery mildew	Systhane	4 oz/100 gal	Spray every 10 to 14 days.
		triadimefon (Strike)	2 to 4 oz/100 gal	Spray every 7 to 14 days as needed. Outdoors only.
		dinocap (Karathane 19.5 WP)	4 oz/100 gal	Spray every 7 to 14 days as needed. Apply with caution when temperature is above 85°F.
		sulfur (dust)	Follow label instructions.	Do not apply sulfur when temperature is above 90°F.
	Rhizoctonia stem rot	thiophanate methyl (Cleary's 3336 or Domain)	See label.	Drench at first sign of disease. Repeat at 10- to 14-day intervals as needed.
		iprodione (Chipco 26019 50 WP)	1 to 2 lb/100 gal	Apply as a foliar spray during July and August every 7 to 14 days.
	Pythium root rot	metalaxyl (Subdue 2 E)	0.25 to 2 fl oz/100 gal	Apply at 1- to 2-month intervals.
CALENDULA	Powdery mildew	propiconazole (Banner 14%)	5 fl oz/100 gal	Apply every 30 days.

[1]From NCSU College of Agriculture and Life Sciences (1997).

Table 15-2 (continued)

Crop	Disease	Pesticide and Formulation	Rate of Formulation	Schedule and Remarks
CARNATION	Alternaria blight	anilazine (Dyrene 50 WP)	1 lb/100 gal	Spray every 7 to 14 days after disease begins to appear.
	Gray mold (Botrytis)	chlorothalonil (Daconil 2787 75 WP)	1 lb/100 gal 2 tsp/gal	Spray every 7 to 14 days.
		thiophanate methyl (Cleary's 3336 or Domain)	See label	
		chlorothalonil (Termil)	3.5 oz/10,000 cu ft	Fumigate every 7 to 14 days.
		iprodione (Chipco 26019 50 WP)	1 to 2 lb/100 gal	Spray every 7 to 14 days.
	Pythium and Phytophthora root rot	etridazole (Truban 30 WP)	4 to 6 oz/100 gal to 800 sq ft 1 tsp/4 gal to 30 sq ft	Apply at planting. Apply additional water immediately after application. Repeat at 4- to 8-week intervals if needed.
		metalaxyl (Subdue 2 E)	0.5 to 2 oz/100 gal to 800/sq ft	Start at transplanting and repeat at 1- to 2-month intervals if needed.
	Fusarium wilt	steam	180° F for 30 min under tarp	Use disease-free plants.
CHRYSANTHEMUM	Stemphylium ray speck	chlorothalonil (Daconil 2787 75 WP)	1 lb/100 gal 2 tsp/gal	Spray every 7 to 14 days.
		chlorothalonil (Termil)	3.5 oz/10,000 cu ft	Fumigate every 7 to 14 days.
	Gray mold	chlorothalonil (Daconil 2787 75 WP)	1 lb/100 gal 2 tsp/gal	Spray every 7 to 14 days after disease first appears.
		mancozeb 80% W	1.5 lb/100 gal	Spray every 7 to 10 days.
		thiophanate methyl (Cleary's 3336 or Domain)	See label	Spray every 7 to 14 days after disease first appears.
		chlorothalonil (Termil)	3.5 oz/10,000 cu ft	Fumigate every 7 to 14 days after disease first appears.
		iprodione (Chipco 26019)	1 to 2 lb/100 gal	Apply every 7 to 14 days.
	Pythium root rot	etridiazole (Truban 30 WP)	4 to 6 oz/100 gal to 800 sq ft 1 tsp/4 gal to 30 sq ft	Apply at planting. Apply additional water immediately after application. Repeat at 4- to 8-week intervals if needed.
		metalaxyl (Subdue 2 E)	0.5 to 2 fl oz 100/gal 400 to 800 sq ft	Repeat at 1- to 2-month intervals. Do not use 1.6 to 2 fl oz rates more often than every 2 weeks.
	Pythium, Fusarium, and root parasitic nematodes	methyl bromide	1 to 2 lb/100 sq ft	Apply under cover 10 to 14 days before planting.
		steam	180° F for 30 min under tarp	
	Rhizoctonia stem rot	thiophanate methyl (Cleary's 3336 or Domain)	See label	Drench immediately after planting. Use 0.5 pt per 6-in. pot.
		iprodione (Chipco 26019)	1 to 2 lb/100 gal	Spray every 7 to 14 days.
	Phythium and Rhizoctonia root rot	etridiazole + thiophanate methyl (Banrot 40 WP)	6 to 12 oz/100 gal/400 sq ft; 6 to 12 oz/100 gal (0.5 pt/6-in. pot)	Repeat at 4- to 12-week intervals.
	Fusarium wilt	thiophanate methyl (Cleary's 3336 50 WP)	1 lb/100 gal, use 1 to 2 pt/sq ft	Apply as drench immediately after transplanting.
CINERARIA	Powdery mildew	triadimefon (Strike)	1 to 2 oz/50 gal	Spray every 30 days.
COLUMBINE	Botrytis blight Rhizoctonia stem rot	iprodione (Chipco 26019)	1 to 2 lb/100 gal	Spray every 7 to 14 days.
CYCLAMEN	Botrytis blight	iprodione (Chipco 26019)	1 to 2 lb/100 gal	Spray every 7 to 14 days.
		mancozeb 80% W	1.5 lb/100 gal	Spray every 7 to 10 days.
		thiophanate methyl (Cleary's 3336 or Domain)	See label	Spray every 7 to 10 days.
		thiophanate methyl (Cleary 3336 WP)	1.5 lb/100 gal	Spray every 7 to 10 days.
		vinclozalin (Ornalin)	0.5 to 1 lb/100 gal	Spray every 7 to 14 days.
DAHLIA	Powdery mildew	Systhane	4 oz/100 gal	Spray every 10 to 14 days.
DELPHINIUM	Powdery mildew	propiconazole (Banner)	5 fl oz/100 gal	Apply every 10 to 14 days.
		Systhane	4 oz/100 gal	Spray every 10 to 14 days.
EXACUM	Botrytis stem rot	thiophanate methyl (Cleary's 3336 or Domain)	See label	Drench plants just after potting and repeat every 7 to 10 days until plants begin to bloom.
		chlorothalonil (Daconil 2787 75 WP)	1.5 lb/100 gal 1 tbsp/gal	

Crop	Disease	Pesticide and Formulation	Rate of Formulation	Schedule and Remarks
GERANIUM	Botrytis blight	See Chrysanthemum		
	Pythium blackleg	etridiazole (Truban 30 WP)	4 to 6 oz/100 gal to 800 sq ft 1 tsp/4 gal to 30 sq ft	Apply at planting. Apply additional water immediately after application. Repeat at 4- to 8-week intervals if needed.
		metalaxyl (Subdue 2 E)	0.5 to 2 fl oz/100 gal to 1 to 2 pt/sq ft	Repeat every 1 to 2 months. Do not use 1.6 to 2 fl oz rates more than every 6 weeks.
	Rust	Systhane	4 oz/100 gal	Spray every 10 to 14 days.
GERBERA	Powdery mildew	triadimefon (Strike)	1 to 2 oz/50 gal	Spray every 30 days.
GLADIOLUS	Curvularia leaf spot	mancozeb 80 WP	1.5 to 2 lb/100 gal 2.5 to 3 tsp/gal	Same as for Botrytis.
	Fusarium corm rot	Busan 75 EC	1 pt/100 gal	Soak corms 15 min. Prestorage plus preplant.
		thiophanate methyl (Cleary's 3336 or Domain)	12 to 16 oz/100 gal soak for 30 min	
GLOXINA	See African violet			
GRAPE IVY	Powdery mildew	triadimefon (Strike)	1 to 2 oz/50 gal	Spray every 30 days.
IMPATIENS	Rhizoctonia stem rot	thiophanate methyl (Cleary's 3336 or Domain) chlorothalonil (Daconil 2787)	See label	Drench at first sign of disease and repeat at 14-day intervals as needed.
	Alternaria leaf spot	mancozeb	1.5 to 2 lb/100 gal	Spray as needed
INTERIORSCAPE PLANTS	Pythium and Phytophthora root rot	metalaxyl (Subdue 2 E)	0.5 to 1.25 oz/100 gal	Drench every 2 to 3 months. Check label for crops.
IRIS	leaf spot	chlorothalonil (Daconil 2787 75 WP) thiophanate methyl (Cleary's 3336 or Domain)	1.5 lb/100 gal 1 tbsp/gal 16 oz/100 gal	Spray every 10 to 14 days in spring. Spray at first sign of disease. Repeat at 14 day intervals.
	Pythium root rot	metalaxyl (Subdue 2 E)	0.25 to 2 fl oz/100 gal	Drench every 1 to 2 months.
KALANCHOE	Powdery mildew	See African violet		
LILY (EASTER)	Gray mold Botrytis blight	thiophanate methyl (Cleary's 3336 or Domain)	See label	Spray every 7 to 14 days.
		chlorothalonil (Termil)	3.5 oz/10,000 cu ft	Fumigate every 7 to 14 days. Do not apply when foliage is wet or temperature is above 75° F.
	bulb rot (Fusarium)	thiophanate methyl (Cleary's 3336 WP)	12 to 16 oz/100 gal	Soak bulbs 30 min before planting.
	Rhizoctonia root rot	See Chrysanthemum		
	Pythium root rot	etridiazole (Truban 30 WP)	3 to 10 oz/100 gal	Drench immediately after planting. Use 0.5 pt per 6-in. pot.
		metalaxyl (Subdue 2 E)	0.5 to 2 fl oz/100 gal/400 to 800 sq ft	Repeat every 1 to 2 months. Do not use 1.6 to 2 fl oz rates more often than every 6 weeks.
	Pythium andRhizoctonia root rot	See Chrysanthemum		
NARCISSUS	Basal rot (Prestorage and/or preplanting)	thiabendazole (Mertect 140-F)	30 fl oz/100 gal	Soak 15 to 30 min 24 to 48 hr after digging.
PANSY	Black root rot *Thielaviopsis sp.*	Terraguard 50 W	2 to 4 oz/100 gal	Drench every 3 to 4 weeks.
		thiophanate methyl	See label	Drench.
	Pythium root rot	metalaxyl (Subdue 2 E)	0.25 to 2 fl oz/100 gal	Drench every 1 to 2 months.
	Anthracnose leaf spot	mancozeb 80 WP	1.5 lb/100 gal	Spray as needed
PHLOX	Powdery mildew	triadimefon (Strike)	2 to 4 oz/100 gal	Outdoor use only.

Crop	Disease	Pesticide and Formulation	Rate of Formulation	Schedule and Remarks
POINSETTIA	Botrytis blight	thiophanate methyl (Cleary's 3336 or Domain)	See label	Spray every 7 to 14 days.
		chlorothalonil (Termil)	3.5 oz/10,000 cu ft	Fumigate every 7 to 14 days. Do not apply when foliage is wet or temperature is above 75° F. For foliage only.
		chlorothalonil (Daconil 2787 F)	2 pt/100 gal	Spray every 7 to 14 days as needed. Discontinue applications prior to bract formation.
		iprodione (Chipco 26019 50 WP)	1 to 2 lb/100 gal	Spray every 7 to 14 days as needed.
	Rhizoctonia root rot and Thielaviopsis root rot	thiophanate methyl (Cleary's 3336 or Domain)	See label.	Drench immediately after planting. Use 0.5 pt per 6-in. pot.
	Pythium root rot	etridiazole (Truban 30 WP)	3 to 10 oz/100 gal	Drench immediately after planting. Use 0.5 pt per 6-in. pot.
		metalaxyl (Subdue 2 E)	0.5 to 2 fl oz/100 gal/400 to 800 sq ft	
	Pythium and Rhizoctonia root rot	See Chrysanthemum		
	Alternaria leaf blight	chlorothalonil (Daconil 2787F)	2 pt/100 gal	Spray as a preventative.
		mancozeb 80 WP	1.5 to 2 lb/100gal	
ROSES (Greenhouse)	Downy mildew	mancozeb 80 W	1.5 lb/100 gal	Spray every 7 days as needed.
	Powdery mildew	Terraguard 50 W	4 to 16 oz/100 gal	Every 7 to 14 days.
		triforine (Funginex 18.2% EC)	12 to 18 fl oz/100 gal	Apply every 7 to 10 days as needed.
		dodemorph (Milban 39% EC)	32 oz/100 gal	Spray every 10 to 14 days. For commercial use only. Some varieties may be sensitive.
		triadimefon (Strike)	1 to 2 oz/50 gal	Apply to runoff every 30 days.
		fenarimol (Rubigan AS)	4 to 12 oz/100 gal	Spray every 7 to 14 days.
		piperalin (Pipron LC)	0.25 pt/100 gal	Spray at first sign of disease. Make one application as eradicant.
SNAPDRAGON	Pythium Phytophthora	metalaxyl (Subdue 2 E)	0.25 to 2 fl oz/100 gal	Apply every 1 to 2 months.
	Rust or Cercospora leafspot	mancozeb 80 WP	1.5 lb/100 gal 1 tbsp/gal	Spray every 10 to 14 days beginning at first appearance of disease. If severe disease develops spray every 7 days.
	Rust	propiconazole (Banner 14.3%)	10 fl oz/100 gal	Apply every 7 to 10 days.
	Botrytis blight	**See African Violet.**		
	Damping-off Preplant	steam	160° F for 30 minutes	Before planting under tarp.
		methyl bromide	2 lb/100 sq ft	Before planting, wait 7 days after removing tarp.
		etridiazole + thiophanate methyl (Banrot 8 G)	Mix 8 oz/cu yd of soil mixture	
	Damping-off Postplant	etridiazole + thiophanate methyl (Banrot 40 WP)	4 to 8 oz/100 gal/ 800 sq ft	Re-treat at 4- to 8-week intervals.
	Powdery mildew, rust	Systhane	4 oz/100 gal	Spray every 10 to 14 days.
		triadimefon (Strike)	2 to 4 oz/100 gal	

Crop	Disease	Pesticide and Formulation	Rate of Formulation	Schedule and Remarks
SPATHPHYLLUM	Cylindrocladium root rot	Terraguard 50 W	4 to 8 oz/100 gal	
	Myrothecium petiole rot	Terraguard 50 W	4 to 8 oz/100 gal	Foliar spray
SWEET WILLIAM	Leaf spot	propiconazole	6 fl oz/100 gal	Spray at first sign of disease and repeat at 14-day intervals.
SYNGONIUM	Myrothecium petiole rot	Terraguard 50 W	4 to 8 oz/100 gal	Foliar spray
TULIP BULBS	Blue mold (Penicillium)	thiophanate methyl (Cleary's 3336)	12 to 16 oz/100 gal	Dip for 30 min. Water temperature 26 to 28° C.
		propiconazole (Banner 14.3%)	24 to 48 oz/50 gal	Dip for 15 min. Water temperature should be 26 to 28° C.
VERBENA	Pythium root rot	metalaxyl (Subdue 2 E)	0.25 to 2 fl oz/100 gal	Apply every 1 to 2 months.
VINCA	Pythium Phytophthora	metalaxyl (Subdue 2 E) (Subdue 2 G)	0.25 to 2 fl oz/100 gal 15 to 30 oz/1,000 sq ft	Apply every 1 to 2 months.
ZINNIA	Cercospora leaf spot	Systhane	4 oz/100 gal	Spray every 10 to 14 days.
	Powdery mildew	Systhane	4 oz/100 gal	Spray every 10 to 14 days.
		triadimefon (Strike)	2 to 4 oz/100 gal 1 tsp/gal	Spray as needed.
Greenhouse flowering crops such as CHRYSANTHEMUM GERANIUM POINSETTIA SNAPDRAGON and ANNUALS	Pythium root rot	etridiazole (Truban 30 WP)	0.5 lb/100 gal	Drench as disease appears. Use disease-free seeds. Sani- tation and prevention are most important.
		metalaxyl (Subdue 2 E)	0.5 to 2 fl oz/100 gal 1 to 2 pt/ sq ft	Repeat every 1 to 2 months. Do not use 1.6 to 2 fl oz rates more frequently than every 6 weeks.
	Rhizoctonia root rot	thiophanate methyl (Cleary's 3336 or Domain)	See label	
		iprodione (Chipco 26019 50 WP)	0.4 lb/100 gal	Apply 1 to 2 pt/sq ft
	Pythium and Rhizoctonia	Banrot	Follow label instructions	
Flowering Annuals except SALVIA and CARNATION	Damping-off	methyl bromide	2 lb/100 sq ft under tarp	10 to 14 days before planting. Follow cautions on label.
Flowering Annuals	Damping-off	steam	160° F for 30 min under tarp	Plant immediately after soil cools.
Flowering Ornamentals	Botrytis	thiophanate methyl (Cleary's 3336 WP)	0.75 lb/100 gal	Spray as needed.

Table 15-3 Disease Control Schedule for Greenhouse Vegetable Crops[1]

Commodity	Disease	Material[3]	Formulation Rate/100 gal	Minimum Days		Schedule and Remarks
				Harv.	Reentry	
GREENHOUSE	Sanitation	Solarization	140°F, 4 to 8 hr for 7 days	—	—	Close up greenhouse during hottest and sunniest part of summer for at least 1 week. Greenhouse must reach at least 140°F each day. Remove debris and heat sensitive materials and keep greenhouse and contents moist; will not control pests 0.5 in.or deeper in soil; not effective against TMV.
		Added heat	180°F for 30 min	—	—	Remove all debris and heat-sensitive materials. Keep house and contents warm.
		methyl bromide 98%	3 lb/1,000 cu ft	—	—	Clean out greenhouse, moisten interior, close tightly, treat for 24 hr at 65°F or higher, and ventilate.
SOIL	Fusarium crown and root rot[2] Wilts: Bacterial[1] Fusarium[1] Verticillium[1] Nematodes[1]	Steam, Vapam, or chloropicrin	—	—	7 to 21	Preplant soil treatment. See table on sanitizing greenhouses and plant beds.
BEANS (dry)	Leaf spots	mancozeb 80 W	1.5 to 2 lb/acre	30	1	Spray first appearance of leaf spot or downy mildew. Not effective against powdery mildew. Approximate equivalencies: (1.5 lb/acre = 6.8 grams/gal) (1.5 lb/acre = 2.5 tsp/gal)
BROCCOLI BRUSSELS SPROUTS CAULIFLOWER				7	1	
CABBAGE KOHLRABI				7	1	
CUCURBITS				5	1	
EGGPLANT				5	1	
KALE				10	1	
LETTUCE ENDIVE				10	1	
MELONS				5	1	
ONIONS			2 to 3 lb/acre	7	1	
CUCUMBERS	Sclerotinia, white mold	Botran 75W	1.3 lb/100 gal	14	1	Apply to diseased areas.
LETTUCE (leaf)	Botrytis	Botran 75W	2.6 lb/100 gal	14	1	Spray 7 days after transplanting and when half mature.
RHUBARB	Botrytis	Botran 75W	1.3 lb/100 gal	3	1	Start weekly sprays at bud emergence.
TOMATO	Botrytis Early blight Gray leafspot Late blight Leaf mold[1]	chlorothalonil — 20% smoke generator (Exotherm termil)	3.5-oz can per 1,000 sq ft	0	1	Smoke generator. Start program before disease appearance. Repeat treatments at weekly intervals.
	Botrytis	Botran 75W	1 lb	10	1	Spray stems up to 24 in. on 7-day schedule.
	Bacterial hollow stem	Clorox	1 pt/10 pt water	0	—	Leave 1-in. stumps when pruning. Dip knife in Clorox solution. Don't handle wet plants. Copper fungicide sprays may help.
		copper fungicides	See label	0	1	

[1]From NCSU College of Agriculture and Life Sciences (1997).

Table 15-3 (continued)

Commodity	Disease	Material[3]	Formulation Rate/100 gal	Minimum Days		Schedule and Remarks
				Harv.	Reentry	
TOMATO (continued)	Anthracnose Early and late blights Gray leaf spot Septoria leaf spot	mancozeb 4 F	1.2 to 2.4 qt/ 100 gal water	5	1	Spray when disease first appears. Repeat weekly. Do not use mancozeb on tender, young plants.
		mancozeb 80 W	1.5 to 3 lb/100 gal water	5	1	
		maneb 75 W	1.5 to 2 lb/100 gal	5	1	
	Sclerotinia	—	—	—	—	Benlate as used for control of Botrytis should give some control. Soil treatment effective.
	Southern blight *(Sclerotium rolfsii)*	Terraclor L	0.2% solution, 0.5 pt per plant	—	—	Cover soil at base of plant at transplanting.
	TMV[1,2] (on seed)	sodium hypo—chlorite 5.25% solution (Clorox)	2 pt in 8 pt of water			Use 1 gal of mix per pound of seed. Wash seed in it for 40 min, provide continuous agitation; air dry promptly.
	Viruses	Milk (skim)	—	—	0	Dip hands before handling plants. (See Plant Pathology Information Note 186).
	Fusarium crown and root rot[2] Wilts: Bacterial[1] Fusarium[1] Verticillium[1] Nematodes[1]	Steam, Vapam, or chloropicrin	—	—	7 to 21	Preplant soil treatment. See table on sanitizing greenhouses and plant beds.
	Southern blight	Terraclor 75 W	2.2 lb	—	—	Use 0.5 pt/plant. transplant use only.

[1] Resistance available.

[2] Use sanitation, seed treatment.

[3] Other formulations may be available.

Summary

1. Categories of pathogens causing plant diseases are viruses, bacteria, fungi, and nematodes.
2. Viruses are the smallest of the pathogenic organisms infecting plants and are similar in size and chemistry to the genetic material (DNA) contained in the nuclei of plant and animal cells. There are no pesticides for controlling viruses, and plants neither recover nor become immune to them. Control depends upon procurement of virus-free plants, insect control, and disposal of infected plants. The more common symptoms include stunting; distortion of leaves or flowers; and spots, streaks, blotches, or rings on leaves with yellow, white, brown, black, or orange discolorations. Viruses are most commonly spread by specific insects or through vegetative propagation by such means as grafting and cuttings.

3. Bacteria are single-celled microorganisms. Bacterial diseases are difficult to control since only a few bactericides exist. Again, control is mainly through elimination of infected plants as well as pasteurization of root substrate and sterilization of containers and tools. The number of bacterial diseases on greenhouse crops is small in relation to the number of fungal diseases. The more common symptoms of various bacterial diseases are wilting; stem rot; leaf spot; soft rot of cuttings, corms, or bulbs; fasciation; and crown gall.
4. Fungal pathogens constitute a large group of multicellular organisms. They are more successfully controlled than other categories of pathogens because there are a larger number of effective chemicals (fungicides). Pasteurization of root substrate and sterilization of containers and tools also play an important part in control. Some of the most common fungal diseases of greenhouse crops are powdery mildew, *Botrytis* blight, *Verticillium* wilt, and root rots including *Pythium, Rhizoctonia,* and *Thielaviopsis.* Control measures range from the reduction of high humidity around plants and free moisture on plants to the application of fungicides (both topical and systemic).
5. Nematode diseases are caused by small, round worms that are usually not visible to the eye. Nematodes abound in all soils; most are harmless. These pests penetrate plant roots, causing lack of vigor and stunting of the plant, shortened and thickened roots, and chlorosis of the foliage. Root-knot nematodes stimulate the development of giant root cells, which develop into galls or knots. Foliar nematodes infect leaves, causing yellowish or brownish spots and areas that enlarge and turn darker. Leaf death and sometimes abscission follows. Root-knot nematodes can be controlled only by discarding infected plants and pasteurizing the root substrate. Other root-attacking nematodes can be controlled by postplanting root substrate applications of chemicals.
6. The first, and very often the only, line of defense against diseases is prevention. For this reason, an IPM (integrated pest management) program is the most sensible approach. Purchase disease-free plants by dealing with reputable propagators. Periodically pasteurize all root substrate and sterilize growing containers and tools. Prevent weed establishment in and around the outside of greenhouses. Clean up plant debris such as pinched-off plant tops and disbuds. Maintain proper air-circulating equipment, heating and ventilating practices, and watering practices to minimize the occurrence of free water on plants. Above all, keep a constant watch for initial disease development and take appropriate action when it occurs.
7. When disease does get a foothold, follow proper label recommendations for the use of an appropriate bactericide, fungicide, or nematicide. These pesticides fall under the same laws of usage as insecticides, and one must adhere strictly to instructions on the label. The same rules of safety apply, and similar methods of application are used. Insecticides, miticides, and disease-control chemicals are often applied together.

References

In addition to the references listed at the end of Chapter 14, the following are suggested.

1. Baker, K. F., ed. 1957. *The U.C. system for producing healthy container-grown plants.* Univ. of California Agr. Exp. Sta. and Ext. Ser. Manual 23. Univ. of California, Berkeley, CA.
2. Chase, A. R. 1987. *Ornamental foliage plant diseases.* St. Paul, MN: APS Press.
3. Chase, A. R., and T. K. Broschat. 1991. *Diseases and disorders of ornamental palms.* St. Paul, MN: APS Press.
4. Chase, A. R., M. Daughtrey, and G. W. Simone. 1995. *Diseases of annuals and perennials: a ball guide.* Geneva, IL: Ball Publishing.
5. Cornell Cooperative Extension. 1994. *1995 recommendations for the integrated management of greenhouse florist crops: management of pests and crop growth.* Cornell Coop. Ext., Ithaca, NY.
6. Coyier, D. L., and M. K. Roane. 1986. *Rhododendron and azalea diseases.* St. Paul, MN: APS Press.
7. Daughtrey, M., and A. R. Chase. 1992. *Ball field guide to diseases of ornamentals.* Geneva, IL: Ball Publishing.
8. Daughtrey, M. L., R. L. Wick, and J. L. Peterson. 1995. *Compendium of flowering potted plant diseases.* St. Paul, MN: APS Press.
9. Gould, C. J., and R. S. Byther. 1979. *Diseases of* Narcissus. Washington State Univ. Coop. Ext. Bul. 709. Washington State Univ., Pullman, WA.
10. Gould, C. J., and R. S. Byther. 1979. *Diseases of tulips.* Washington State Univ. Coop. Ext. Bul. 711. Washington State Univ., Pullman, WA.
11. Horst, R. K. 1983. *Compendium of rose diseases.* St. Paul, MN: APS Press.
12. Jarvis, W. R. 1992. *Managing diseases in greenhouse crops.* St. Paul, MN: APS Press.
13. Jones, J. B., J. P. Jones, R. E. Stall, and T. A. Zitter. 1991. *Tomato diseases.* St. Paul, MN: APS Press.
14. NCSU College of Agriculture and Life Sciences. 1997. *1997 North Carolina Agricultural Chemicals Manual.* Raleigh, NC, North Carolina State Univ. College of Agriculture and Life Sciences.
15. Powell, C. C., and R. K. Lindquist. 1992. *Ball pest and disease manual.* Geneva, IL: Ball Publishing.
16. Strider, D. L., ed. 1984. *Diseases of floral crops.* New York: Praeger Scientific.

 CHAPTER 16

Postproduction Quality

Fresh flowers and ornamental plants differ from edible agricultural products because the latter are consumed or processed within days of harvest, while the former are expected to maintain an aesthetically pleasing appearance as long as possible. The postproduction period for fresh flowers is referred to as *vase life;* for containerized plants, it is called *shelf life.* Postproduction longevity is determined by conditions during a chain of periods beginning with early plant production and extending through late production, holding, shipping, and retailing. Thus, the decision-making process leading to maximum postproduction quality begins with the crop production planning step prior to planting.

Fresh Flowers

Unlike containerized plants, fresh flowers present a special problem. A fresh flower is still a living specimen even though it has been cut from the plant. Its maximum potential vase life, although acceptable in the marketplace, is short. There are many impinging forces that can interact to reduce fresh-flower vase life—that is, the period of time during which fresh flowers possess aesthetic value. As an industry, we have not been highly successful in preserving the potential life of fresh flowers. As mentioned earlier, some 20 percent of harvested fresh flowers become unmarketable as they move through the market channel (harvesting, packaging, transporting, and selling). A very significant proportion of the remaining flowers are sold in a weakened condition, which leads to consumer dissatisfaction. Fortunately, there are well-known solutions for the bulk of this problem. First, we need to take a look at why there is such a decline in the vase life of fresh flowers.

Vase Life

Cultural Influences Basically, those forces that improve crop quality before and after harvest usually improve vase life. Light intensity is very important. A crop grown under dirty glass or during a period of inclement winter weather, such that light is a limiting factor for photosynthesis, will be low in carbohydrate content. Respiration continues after the flowers are harvested, but little photosynthesis occurs because light is limited in the packing house, the florist shop, and the consumer's home. When carbohydrates are low, respiration is very low and flower senescence (deterioration) occurs. Optimum light intensity during growth of the crop is very important to vase life.

The time of the day when flowers are harvested can be very important for some crops—for instance, rose. Carbohydrates build up during the day through photosynthesis and reach a peak in late afternoon. During the night, carbohydrates are utilized during respiration. Roses cut at 4:30 P.M. were found by Howland (1945) to last longer than those cut at 8:00 A.M.

Temperature also enters into the picture because it influences photosynthesis and respiration, which in turn influence carbohydrate accumulation. During hot periods of the year, crops sensitive to high temperatures, such as carnation and rose, have shorter vase lives because flowers contain low carbohydrate levels. When the temperature is raised to an adversely high level to force earlier flowering, the same problem occurs.

Nutrition of the crop likewise has an effect on flower longevity. Shortages or toxicities of nutrients that retard photosynthesis will reduce vase life. Deficiencies in a number of nutrients, including nitrogen, calcium, magnesium, iron, and manganese, result in a reduction in the chlorophyll content, which in turn reduces photosynthesis. The net result is a low carbohydrate supply for each flower. On the other hand, high levels of nitrogen at flowering time can have an adverse effect on keeping quality.

Diseases and insects reduce the vigor of plants and directly reduce vase life. Diseases also reduce vase life indirectly; injured tissue releases large quantities of ethylene gas, which hastens senescence of fresh flowers.

Cause of Vase-Life Decline Fresh flowers deteriorate for one or more reasons. Five of the most common reasons for early senescence are the following:

1. Inability of stems to absorb water because of blockage.
2. Excessive water loss from the cut flower.
3. A short supply of carbohydrates to support respiration.
4. Presence of diseases.
5. Negative effect of ethylene gas.

Inability to absorb water is a very common reason for premature wilting. The water-conducting tubes in the stem (xylem) become plugged. Bacteria, yeast, and/or fungi living in the water or on the flower foliage proliferate in the containers holding the flowers. These microorganisms and their chemical products plug the stem ends, restricting water absorption. They continue to multiply inside and eventually block the xylem tubes (Figure 16-1). Chemical blockage also can occur. Chemicals that are present in some stems, upon cutting, change into a gumlike material that blocks the end of the stem. This material is suspected to be composed of oxidized tannins in some plants, and in others it is unidentified. A third source of blockage can be air trapped in xylem tubes. When flowers are received at wholesale or retail establishments it is advisable to cut the ends of the stems off under water. Special knives are available for mounting under water in tubs for this purpose. When stems are cut in this fashion, water rather than air enters the end of the stem. After cutting, the stems can be removed immediately from the water. This procedure is particularly effective for rose.

Excessive water loss from flowers can lead to wilting and reduction in quality and vase life. After harvest, flowers should be removed from the field or greenhouse and refrigerated as soon as possible. Leaving the flowers out of water and in warm air or in warm drafts such as from a heater causes considerable damage. Flowers should be in water and under cool temperatures as much as possible from the time they are cut until they reach the final customer.

Low carbohydrates are another reason for flower deterioration. A low carbohydrate supply can occur as a result of improper storage temperature and han-

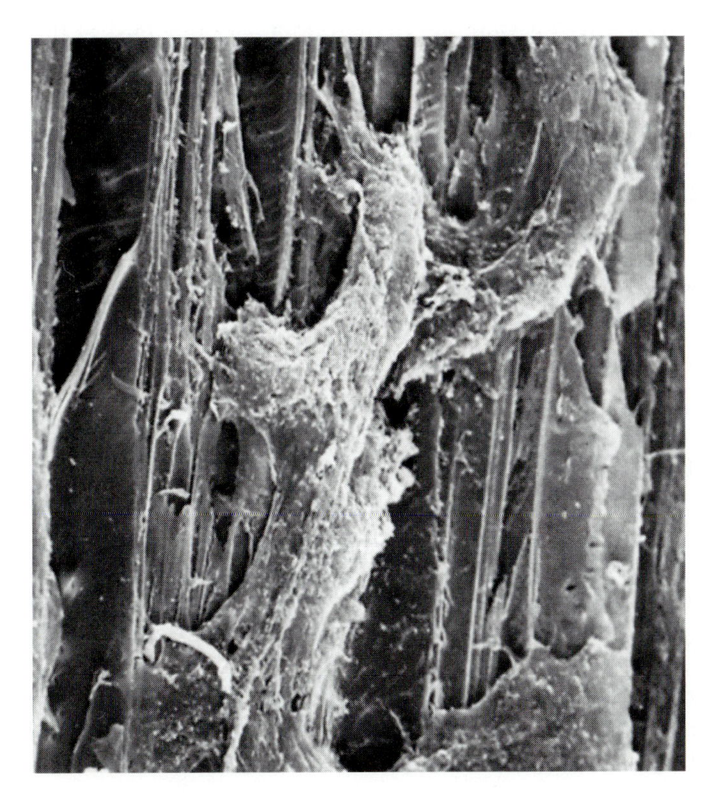

Figure 16-1 A longitudinal section (1,500X magnification) of a rose stem, showing the interior of water-conducting cells and a slime plug blocking some of the cells. Such slime plugs can be composed of microorganisms, particularly bacteria, and solidified compounds from the flower itself. (*Photo courtesy of H. P. Rasmussen, Department of Horticulture, Michigan State University, E. Lansing, MI 48824.*)

dling. Respiration continues to be governed by temperature after harvest. Low temperatures reduce respiration and conserve carbohydrates, thereby prolonging quality and vase life. Each of the many stages in the marketing channel must be watched. Flowers should be placed in cold storage as soon after harvesting as possible. They should be refrigerated during surface transport and during holding periods by the wholesaler and retailer. Serious damage occurs when flowers are left on a heated loading dock at the motor- or airfreight terminal or when they are left sitting in a hot warehouse for a day or so.

The harmful effects of disease and pests, as well as the effects of ethylene, have already been pointed out. Fruits, especially apples, give off large quantities of ethylene gas, making it inadvisable to store lunches containing fruits in coolers. It has already been mentioned that ethylene is evolved from plant tissue, particularly injured and old plant tissue. Coolers should be kept clean of plant debris such as cut stems and leaves that might accumulate on the floor. Old unsalable flowers should be discarded.

Ethylene gas has many deleterious effects. Generally, it causes premature deterioration of flowers. It also causes sleepiness (the upward cupping of petals) of carnation flowers, which gives the flowers an appearance of wilting. This phenomenon is not reversible.

Preservatives for Extending Vase Life Considerable research has been conducted over the past 30 years to find a preservative solution that will combat some of the causes of flower deterioration and reduction of vase life. One of the earliest home remedies called for table sugar (sucrose) plus aspirin, and sometimes a penny was added to the vase to provide copper as a bactericide. Another remedy used carbonated lemon soft drinks containing sugar. There is some value in these remedies, but aspirin is not readily soluble and the penny is essentially insoluble. Hence, the remedies supply sugar but do not control microbial growth.

Floral preservatives perform three functions:

1. They provide sugar (carbohydrate).
2. They supply a bactericide to prevent microbial growth that causes blockage of the water-conductive cells in the stem.
3. They acidify the solution. This function suppresses bacterial development and, through some unknown process, prevents wilting of flowers. It is suspected that the acidity helps prevent chemical blockage.

Various universities and the USDA have developed successful preservatives (Figure 16-2). The most popular preservatives today contain 8-hydroxyquinoline citrate (8-HQC) and sucrose (common table sugar). Listed in Table 16-1 are preservative formulas for five fresh flowers. The 8-HQC is a bactericide and an acidifying agent. Besides suppressing bacterial development and lowering the pH, 8-HQC also prevents chemical blockage, thus aiding in the absorption of water. Sucrose taken up by the stem maintains quality and turgidity and extends vase life by supplementing the carbohydrate supply.

Figure 16-2 **Floral preservative trials on gladiolus flowers. The flower on the left is in water, and the one on the right is in a preservative containing 600 ppm 8-hydroxyquinoline citrate plus 4% sucrose. (*Photo courtesy of F. J. Marousky, USDA, Agricultural Research Service, Bradenton, FL 33505.*)**

There are a number of commercial preservatives on the market, including products such as Floralife, Petalife, Oasis, Rogard, and Everbloom. These work well. One can also purchase 8-HQC under the name *oxine citrate* from florist-supply companies and add sucrose to make the preservatives listed in Table 16-1.

The bactericide 8-HQC is not totally effective in preventing the buildup of bacteria in floral solutions. Chlorine is a very effective bactericide but dissipates

Table 16-1 **Floral Preservative Formulas for Five Fresh Flowers**

	8-HQC			*Sucrose*		
Flower	*oz/10 gal*	*g/L*	*ppm*	*oz/10 gal*	*g/L*	*%*
Gladiolus	0.80	0.6	600	54	40	4
Carnation	0.27	0.2	200	27	20	2
Chrysanthemum[1]	0.27	0.2	200	27	20	2
Rose	0.27	0.2	200	27–42	20–30	2–3
Snapdragon	0.41	0.3	300	20	15	1.5

[1]Use this formula for other flowers in general.

quickly from solution unless provided in a slow-release form. Two slow-release forms sold extensively in products including bleaches, deodorizers, detergents, dishwashing compounds, and swimming-pool additives are DICA (sodium diclorisocyanurate) and DDMH (1,3-dichloro-5,5-dimethylhydantoin). Both are highly effective bactericides for floral preservation of aster, carnation, gladiolus, gypsophila, and rose. Each is used at a concentration of 300 ppm (0.41 oz/10 gal, 0.3 g/L) in the place of 8-HQC (Marousky 1976). DICA or DDMH is used with sucrose at a concentration of 2 percent (27 oz/10 gal, 20 g/L). These chlorine compounds will bleach stems and leaves immersed in the preservative solution. They may also injure outer petals of roses, but this occurs after normal senescence of the flower begins. These disadvantages are outweighed by the superior bactericidal effects of these materials. They are particularly useful for gypsophila, which is exceptionally prone to bacterial buildup.

Floral preservatives are very effective in maintaining quality and extending longevity. On the average, they can double the vase life of cut flowers when compared to water. Snapdragons with a life expectancy of 5 to 6 days last up to 12 days in preservative. The life expectancy of roses can be extended from 3 to 5 days to 7 to 10 days. Carnations with a vase-life expectancy of 5 days, after extensive shipping, have been shown to last 12 days in preservatives.

Refrigerated Storage

The most common system for handling harvested flowers is refrigerated storage, which involves the following sequential steps.

1. Flower stems should be cut with a sharp knife or shears to prevent crushing of stem and water-conduction cells.
2. The cut flowers should be placed in a preservative solution as soon as possible to prevent wilting. The flowers should not be allowed to be out of water while they are waiting to be transferred to the storage or grading rooms. If flowers are cut in the field, buckets containing solution can be brought out

on trailers to hold the harvested flowers. Flowers cut in the greenhouse should not be left in the sun or out of water for more than a few minutes. One person should be assigned to carry these flowers to the grading room or storage cooler immediately. A conveyor system can be used to carry cut flowers to the grading room.

3. As soon as flowers arrive at the storage room, they should be placed in preservative solution inside the refrigerated storage room. If wilted, they should be placed in a warm preservative solution at room temperature until turgid. They should then be placed in the cooler.
4. The temperature of the refrigerated room should be 33 to 40°F (0.5 to 4°C). The lower the temperature the better because the respiration rate falls off with diminishing temperature. Low respiration rates have an effect similar to that resulting from adding sucrose to the preservative solution in that they conserve carbohydrates within the flower. A temperature range of 35 to 40°F (2 to 4°C) is usually encountered in flower coolers. Special attention should be paid to some flowers, such as orchids and gardenias, which cannot withstand low temperatures. If cattleya orchids are stored below 50°F (10°C), they will show signs similar to frost injury (petal browning).
5. Air should be gently circulated inside the cooler only to the extent necessary to ensure uniform temperatures in all areas. Unprotected flowers placed in a direct airstream will be desiccated. Flowers immediately adjacent to a cooling coil may freeze even though the air temperature is above freezing. Since the coil itself is below the freezing point, radiant heat is lost from the flower to the coil, and the flower can be colder than the surrounding air.
6. Potential sources of ethylene gas should be avoided by keeping fruit and vegetables out of the cooler. Discard old flowers. Wash the inside of the cooler periodically.
7. Replace the preservative solution at two- to seven-day intervals. The preservative should be checked periodically for bacterial growth, which is apparent when the solution becomes cloudy. In spite of the bactericides in preservatives, microorganisms will develop and need to be eliminated periodically. To accomplish this, wash the buckets with a disinfectant such as bleach.

Refrigerated storage goes beyond this point, but from here on it becomes difficult to ensure that it is carried out properly. The flowers are sold to a wholesaler who in turn sells them to a retail shop. These people should continue to preserve the quality you have worked hard to maintain. The wholesaler and retailer should hold the flowers under refrigeration as you have. Whenever possible, flowers should be transported under refrigeration. Needless delays at shipping terminals should be avoided. Instruct the wholesaler and retailer to cut ½ inch (1.3 cm) from the base of the stems whenever it has been necessary to leave the flowers out of water for a period of time, and then to place them in warm water at a cool air temperature to prevent the ends of the stems from drying out and restricting water movement.

Many growers, wholesalers, and retailers are of the opinion that these procedures, particularly the use of floral preservatives, are not necessary. Undoubtedly, they have partial evidence to support their view. However, if they looked at the whole market channel, they would realize they are wrong. It is too late for the retailer to get maximum effectiveness from a preservative if the grower or the wholesaler has failed to use one. Flowers left to wilt in the greenhouse while others are cut have already lost a significant portion of their quality and longevity. Precautions taken after this time will have diminished effects and at times may appear to be without effect. Very often, abusive handling is the main culprit in flower deterioration.

Dry Storage

Flowers can be held in refrigerated storage for one to three weeks, depending on the species. Refrigerated storage is more generally used as an aid for maintaining quality as flowers pass through the market channel. Dry storage is used when flowers must be held for periods longer than one to five days. Roses may be held in dry storage up to 18 days, chrysanthemums and carnations up to three weeks, and rooted cuttings of chrysanthemums and carnations for as long as six weeks. Gladioli do not store well.

Flower prices depend to a great degree upon market demand. Prices are high at holidays, but flowers cannot always be scheduled to bloom at each holiday. Dry storage offers a means of holding flowers without deterioration for a high-priced holiday market.

Only the best-quality flowers should be dry stored. Those of poor quality will have a short vase life, if any at all, when they are removed from storage. Flowers should be cut and packaged for storage immediately without being placed in water. Standard cardboard flower boxes are suitable, but a lining of polyethylene film should be placed in them to cover the flowers and seal in moisture (Figure 16-3). Desiccation can be a problem in long-term storage, especially when an absorbent container such as cardboard is used.

A common problem of dry storage is the presence of free water on the flowers, which encourages the development of disease. While flowers freeze only at temperatures below 29°F (–2°C), the free water will freeze at 32°F (0°C). Resulting ice crystals on the petals can be injurious. Boxes and flowers packed at warm temperatures develop condensation (free water) as the plants and air inside are cooled. Because of the polyethylene barrier, the water cannot escape. Disease, enhanced by this moisture, is a common cause of failure in dry storage. Boxes of flowers should be cooled open in a 38-to-40°F (3-to-4°C) cooler and then sealed and placed in a 31°F (–1°C) cooler.

Since most flowers freeze at 27 to 29°F (–3 to –2°C), it is essential that the temperature stay above this point. Flower life expectancy is lessened at 33°F (0.5°C) and drops rapidly at temperatures above this point. Many of the failures of this system have been due to high temperatures or fluctuating temperatures. Since the dry-storage cooler should not be open too often, another cooler is

Figure 16-3 Bunches of pompon chrysanthemums being packed in a polyethylene-lined cardboard carton. The polyethylene will be placed over the flowers, the lid will be placed on the box, and then it will be stored at 31°F (–1°C) for a period of up to three weeks. (*Photo courtesy of F. J. Marousky, USDA, Agricultural Research Service, Bradenton, FL 33505.*)

needed for regular refrigerated storage. The 31°F (–1°C) cooler is often built inside the 35-to-40°F (2-to-4°C) cooler to provide a more uniform temperature. Space should be left between boxes of flowers when they are placed in storage initially. Respiration is occurring, and this produces heat. A large stack of boxes can generate enough heat and provide sufficient insulation to prevent thorough cooling of the inner flowers. Leave space between each stack of boxes and between every other box in a stack to permit the absorption of heat by circulating cool air.

Flowers removed from dry storage need to be hardened. Cut ½ inch (1.3 cm) from the bottom of each stem. Place the flowers in a preservative solution inside a 38-to-40°F (3-to-4°C) cooler. Allow the flowers to become fully turgid before marketing them; this will take 12 to 24 hours. With proper handling, dry-stored flowers should have reasonable quality and the same longevity as fresh flowers. Poor temperature control or disease will decrease quality and longevity.

Dry storage is used only to a limited degree by the industry and works best with chrysanthemum. Chrysanthemum, carnation, and rose are the crops to which it is primarily applied. Much more potential exists here than is being realized. The main reason for its low level of acceptance has probably been failures due to inept handling of the system.

Bud Harvesting

Bud harvesting is a procedure that is used infrequently but is fairly well proven and has a tremendous potential. Carnation and chrysanthemum can be harvested and shipped in the bud stage, which cuts down greatly on their volume and hence lowers the cost of shipping. The wholesaler may then store the buds or open them immediately for resale. Once open, the flower has at least the same vase-life potential as a mature harvested flower.

Bud harvesting enables a grower to produce more crops per year in the same greenhouse space. The grower must, of course, either provide space for opening these buds or pass along part of his or her production savings to the wholesaler or retailer who then must provide facilities and time for opening the buds. In any event, there is a significant increase in net return to the grower.

There are other advantages to this system. Buds are more immune to handling injuries and ethylene toxicity, making a higher-quality final product possible. As in the case of mature harvested flowers, buds will dry-store very well, enabling the grower to build up inventory for higher-priced market dates. Bud harvesting is not a new concept for all crops; gladioli, irises, peonies, roses, and tulips, have always been cut in the bud stage.

Carnations are cut when from ½ to 1 inch (13 to 25 mm) of petal color is showing (Figure 16-4). Standard chrysanthemums are cut when the buds are 2 in (51 mm) in diameter. Buds at this stage can be placed directly into dry storage, or they can first be shipped under ice or refrigeration in a box and then be put into dry storage. When needed, buds are removed from the storage box, ½ inch (13 mm) of stem is cut off, and they are placed in a floral preservative solution. The buckets of buds are held in an opening room at 70 to 75°F (21 to 24°C) until the buds are fully open. A low light intensity is provided in the opening room. Carnation buds open in two to three days and chrysanthemum buds in seven to nine days. The open flowers may be held under refrigeration in the preservative solution, or they may be sold directly. The quality and longevity of these flowers have been reported to be superior to that of flowers harvested at maturity.

Bud harvesting is becoming important. Growers who ship flowers great distances, across North America or from other countries, recognize its value and find it necessary to use this system. Greater cooperation among growers, wholesalers, and retailers will foster it even more.

Silver Thiosulfate (STS)

Silver has a bactericidal activity. On the strength of this point, silver has been tested as a component of floral preservatives. Vase life of carnations has been increased with silver (Kofranek and Paul 1972; Halevy and Kofranek 1977). However, when silver is applied in the form of a soluble salt (most commonly silver nitrate), the silver is only slightly absorbed by cut flowers. Most is wasted, and results are limited. It was found in 1978 that silver in the complex form of silver thiosulfate (STS) is readily taken up and translocated in plants and cut flowers (Veen and van de Geijn 1978). In this form, it is very active in preventing post-production problems. Silver acts as an antagonist of ethylene (Beyer 1976), thereby blocking the aging effects of ethylene. Some problems that are reduced

(a)

(b)

Figure 16-4 (a) Carnation buds cut at three stages of maturity: (left) petals just showing, (center) 1/4 inch (6 mm) of petals showing, and (right) 3/4 inch (19 mm) of petals showing. (b) The same buds after three days in an opening solution. The youngest buds will require seven to eight days of total opening time, the intermediate buds four to five days, and the oldest buds three days. The oldest buds are ready for retail use after three days in the opening solution. For greatest efficiency of growing and postproduction handling time, the buds should be harvested when 3/4 inch (19 mm) of petals is showing. (*Photos courtesy of F. J. Marousky, USDA, Agricultural Research Service, Bradenton, FL 33505.*

are curling of carnation petals (sleepiness), abscission of petals and florets of numerous plants, drooping and epinasty (twisting) of poinsettia leaves, and a general rapid decline in cut-flower condition.

STS must be formulated by the grower as follows (Cameron, Reid, and Hickman 1981):

1. Dissolve either 4.25 oz of prismatic sodium thiosulfate pentahydrate ($Na_2S_2O_3 \cdot 5H_2O$) or 2.8 oz of anhydrous sodium thiosulfate ($Na_2S_2O_3$) in 1 pint of water.
2. Dissolve 0.7 oz of silver nitrate ($AgNO_3$) in 1 separate pint of water.
3. Slowly pour the silver nitrate solution into the sodium thiosulfate solution while rapidly stirring. Some acceptable browning of this concentrated STS solution may occur.
4. The concentrated STS solution is diluted before use according to the requirements of the plant being treated. When 1 fl oz of concentrated STS is diluted in 1 gal water, the concentration becomes 108 ppm silver, which is equivalent to 332 ppm (1 mmol) STS.

The metric equivalent to this preparation is as follows:

1. Dissolve 100 g of prismatic sodium thiosulfate pentahydrate or 63 g of anhydrous sodium thiosulfate in 500 mL of water.
2. Dissolve 17 g of silver nitrate in another 500 mL of water.
3. Mix the silver nitrate solution into the sodium thiosulfate solution.
4. Use 10 mL of the concentrated STS solution in each liter of water to achieve a concentration of 108 ppm silver, which is equivalent to 332 ppm (1 mmol) STS.

An exciting use for STS has been for extending the vase life of carnations (Reid et al. 1980). Studies have shown that vase life can be doubled. The concentration of STS to use depends on the length of time stems are left in it. Reid and Staby (1981) developed a graph relating the required concentration for any given desired exposure time. Three effective alternatives from their work include a 10-minute pulse with 4 mmol STS at room temperature, a one-hour pulse with 2 mmol STS at room temperature, or an overnight (20-hour) treatment with 1 mmol STS in a cool room at 32 to 35°F (0 to 2°C).

STS treatment shows promise of increasing vase life of a number of other cut flowers, including delphinium, dendrobium orchid, enchantment lily, gerbera, *Matthiola* (stock), and snapdragon. The vase lives of gladiolus and rose are not improved by STS; however, the quality of gladiolus flowers is improved. STS used in combination with floral preservatives further increases vase life beyond that of the floral preservative or the STS alone.

The Future

We can look for some very fascinating developments in the future with regard to postproduction handling. Two systems in particular show promise. For some time now, fruit has been stored in "controlled atmosphere" (CA) systems. Crisp, fresh apples stored from September to June are the products of this system. The merits of the system in relation to flower storage are being tested. In such a system, flowers would be stored at low temperatures in an atmosphere very low in oxygen (1 to 3 percent) and high in CO_2 (2 to 5 percent). The low oxygen and high CO_2 levels further reduce the rate of respiration.

Hypobaric (low-pressure) storage is a much newer idea. Fruits, vegetables, or flowers are placed in a sealed chamber, and a vacuum is established in the chamber down to a pressure of about 50 mm of mercury (6.7 kPa). The chamber is constructed to maintain a vacuum, yet allow fresh air to be swept through it. The chamber is also cooled to a low temperature. As the chamber is evacuated, the oxygen content is reduced to a level that sharply reduces respiration. Ethylene gas evolved by the plant tissue is quickly removed by the flowing air.

Containerized Plants

Production Phase

Overall Production There is a general rule of thumb that applies to the three categories of containerized plants, which include flowering pot plants, bedding plants, and green plants. The rule states that those environmental and cultural conditions that ensure best production quality also ensure the best postproduction keeping quality. Conditions include temperature, light, nutrition, water, and CO_2.

When a sub-optimum level of light is supplied during production, plants accumulate a lower level of carbohydrates. It is important that there be an extra supply of carbohydrates in the plant to provide energy during dark or dimly lighted times. Such times occur during shipment to market, when the plant is held in the retail establishment, and possibly if it is placed in a poorly lighted consumer environment. Low-carbohydrate plants in these situations may abort flower buds, develop smaller and paler flowers, or develop chlorotic leaves that later drop off.

Care should be taken during production to meet the degree of shoot compactness specified for each crop grown. This may be accomplished through chemical growth regulators or cultural procedures. Compact plants are less susceptible to mechanical damage during handling in the market channel. Also, compact plants tend to transpire less water. This reduces the likelihood of injurious levels of drying during the market stage.

Plug trays for growing plug seedlings and flats for bedding plants are available in various depths, all with the same surface area. Deeper cells in these containers favor maintenance of postproduction quality. The deeper cells hold more root substrate; thus, they hold more water. There is less chance of drying while in the market channel.

Toning during Late Production All categories of containerized plants are less subject to quality deterioration in the market and consumer environments if they have been toned at the end of the production period. *Toning* is a process in which a reversible stress is applied to a plant for the purpose of increasing its tolerance of mechanical and environmental adversities. Less-than-adequate levels of fertilizer, water, heat, or light can be used to tone a plant. The stress is applied at the end of the production period, usually during the last two weeks. A longer period of toning is necessary for green plants that have long production times. Toned plants withstand mechanical stress, drying, extreme cold, and heat better than untoned plants. They also last longer in the market and consumer environments.

The most effective toning stress is a shortage of nutrition. Nitrogen appears to have the predominant effect among all of the nutrients. However, the stress is generally applied by cutting back application of all fertilizer. If the nutritional stress is too great, postproduction quality will be impaired. Foliage may be chlorotic, flowers may be smaller, and shelf life may be reduced. In order to apply the correct level of stress, it is important to know the nutrient status of the substrate two weeks before market date so that an appropriate shift in fertilization can be made. Fertigation usually results in much higher nutrient levels in the substrate at the end of the crop than weekly fertilizer application does. When substrate analysis indicates that the available levels of nutrients, particularly nitrogen, are at or above the high end of the optimum range, fertilization can be discontinued for the last two weeks of production. If substrate nutrient levels are in the middle of the optimum range, fertilization can be cut in half for the last two weeks. This may be done by reducing the fertilizer concentration to one-half or by applying the full-strength fertilizer at half the frequency. If nutrient levels in the substrate are at the low end of optimum, it may be sufficient to stop fertilization for the last week only. In the unlikely event that the crop is receiving insufficient nutrition when the last two weeks start, it would be best to continue fertilization up to market date.

Water stress is perhaps the second most popular method for toning containerized plants. As the frequency of water application to a crop increases, the percentage of water in the plant increases, often without any increase in dry-matter content. Plants in this state are said to be *soft*. Soft plants wilt quickly in periods of full light, in drying wind, or in adverse settings such as a sidewalk or a parking lot at a garden center. Toned plants stand up to these hostile situations much better. As in the case of fertilizer toning, the degree of water stress is important. Plants should be allowed to wilt moderately before each watering. They should not be allowed to wilt to the point where root or leaf tips burn. This extra level of water stress is applied to bedding plants after the visible-bud stage.

Low-temperature toning is used for flowering pot plants and bedding plants but generally not for green plants. Green plants are, for the most part, tropical or subtropical plants that perform best at warm night temperatures of 60 to 70°F

(16 to 21°C). For most species, it is best not to lower the temperature below this range. When the temperature is lowered for the last two to three weeks of production, flowering pot plants develop deeper-colored flowers and sometimes larger flowers. These plants typically last longer in the consumer environment. Depending on the plant species, the temperature can be lowered by 5 to 10°F (3 to 6°C). Care must be taken not to lower the temperature too much with white-flowered plants, because some can develop a pink tint at lower temperatures. Bedding plants grown at lower temperatures after the visible-bud stage are more resistant to mishandling, high transpiration situations, and light frosts. According to Armitage (1993), bedding plants fall into two categories in terms of tolerance to low temperatures, as seen in Table 16-2. Group A plants are toned at 50 to 55°F (10 to 13°C); group B plants are toned at 58 to 62°F (14 to 17°C).

Low-light toning is not practiced for bedding plants. Greenhouse production of bedding plants encompasses the early portion of the life cycle of these plants when high photosynthetic rates are needed to drive growth. After the market stage, these plants will go to gardens where light conditions will be even higher than in the greenhouse. A period of restricted light at the end of the production period could lead to undesirably tall and weak plants, chlorosis of leaves, and poor adaptation to the outside environment upon transplanting. Flowering pot plants are likewise not light toned. If done, this would lead to low carbohydrate levels that would limit shelf life.

Low-light toning is restricted mainly to green plants. Many green plants originate in brighter environments than those into which they will be placed in consumer buildings. They will grow best at these higher light intensities in the production greenhouses or shade houses. However, if they are shifted directly from high to low light situations, leaves will become chlorotic and possibly will abscise. It is necessary to lower the light intensity near the end of production.

Table 16-2 **Temperatures Suitable for Toning Bedding Plants**[1]

Group A 50 to 55°F (10 to 13°C)	*Group B 58 to 62°F (14 to 17°C)*
ageratum	begonia
alyssum	celosia
calendula	coleus
dianthus	impatiens
marigold	pepper
pansy	tomato
petunia	vinca (catharanthus)
perennials (all)	zinnia
phlox	
salvia	
snapdragon	
torenia	

[1]From Armitage (1993).

The length of this period depends on the species and size of plant and can extend from six weeks to six months or longer. Water and fertilizer toning are not substitutes for light toning in green plants. All three toning systems are customarily used for these plants.

Silver Thiosulfate (STS) Foliar sprays of STS in combination with 0.1 percent of the spreader-sticker Tween 20 have been shown to prevent abscission of flowers or flower parts from several pot-plant crops. A common problem of seedling geraniums has been "shatter," the premature dropping of petals. Shatter is particularly troublesome during shipping and marketing. It has been shown that sprays of 0.5 mmol STS to entire geranium plants or 2.0 mmol STS to only the buds of other plants result in complete prevention of petal drop three weeks later, when plants are placed in a market environment for six days, while plants sprayed with water drop florets continuously during the six days (Cameron and Reid 1983). Sprays should be timed to coincide with the appearance of color in the first florets (Miranda and Carlson 1981). There is an indication that geranium plants sprayed with STS are more susceptible to *Pythium* root rot. Calceolaria plants that were sprayed to runoff with 0.5 mmol STS and, one week later, moved to a laboratory where they were exposed to 1 ppm ethylene for two days or to four days of simulated transport at 77°F (25°C) in the dark show dramatic improvement over plants sprayed with water. Flower drop in plants treated with STS can be reduced from 91 percent to 36 percent in the ethylene test and from 83 percent to 22 percent in the transport test (Cameron and Reid 1983). Bracteole drop from bougainvillea plants brought on by water stress can be greatly minimized by sprays of 0.5 mmol STS (Cameron and Reid 1983). Zygocactus are also very prone to flower drop during transportation and marketing. Sprays of 2.0 mmol STS to plants in the tight-bud stage almost completely prevent flower drop during exposure 0.5 ppm ethylene for two days or to four days in the dark at 80°F (27°C) (Cameron and Reid 1981; Cameron, Reid, and Hickman 1981).

Poinsettia plants placed in sleeves for shipping purposes can develop droopy bracts and epinastic (twisted) leaves. This apparently results from a buildup in the sleeve of ethylene released from the mechanically stressed plant tissues (Sacalis 1977). These problems have been significantly reduced by spraying plants with a 3 mmol solution of silver ions 24 hours before placing them in the sleeves (Saltveit and Larson 1981). These problems, however, can be minimized by using more porous paper or fiber rather than solid plastic sleeves and by using the sleeves for the least time possible (Staby, Thompson, and Kofranek 1979). Complete correction can usually be obtained by allowing plants to stand for a period of time after sleeve removal and before selling.

Shipping

A level of 0.1 ppm ethylene is injurious to plants. A level of 10 ppm can cause death in a matter of hours. Ethylene symptoms include leaf chlorosis; abscission of leaves, buds, and flowers; flower abortion; wilting; and epinasty. Sources of ethylene during packing and shipping can include exhaust from fossil-fuel engines, plant debris, and the plants themselves, particularly if they are damaged. Gasoline-engine exhaust can contain from approximately 100 to over 200 ppm ethylene, depending on the condition of the engine, while diesel engines produce about 60 ppm.

A number of steps can be taken to lower ethylene levels. Old and damaged foliage and flowers should be removed from plants before packing them for shipment. This will eliminate a major source of ethylene and will make the plant more marketable. Old plants and leaves should be removed from packing and holding areas and shipping trucks. Again, these are a large source of ethylene. Ethylene in engine exhaust can be minimized by using carts with electric motors rather than fossil-fuel motors for moving plants, turning off truck engines whenever possible, and providing good ventilation in the packing and holding areas. Many plants are shipped in cartons. Less ethylene will build up around plants if ventilated cartons are used. Whenever possible, the truck compartment should be ventilated.

In addition to removal of ethylene itself and ethylene-generating tissue, steps can be taken to reduce the deleterious effect of the remaining ethylene. Both ethylene production and the sensitivity of plants to ethylene diminish as temperature decreases. At 40°F (4°C), many flowering pot plants do not show symptoms of ethylene toxicity. Holding and shipping spaces should be set at as low a temperature as is tolerated by the plant types being handled.

Low temperature also plays a valuable role by reducing the respiration rate of plants. When plants are in the dark or at a low light intensity, net photosynthesis does not occur. Under these conditions, a high rate of respiration would result in rapid depletion of the carbohydrate stores in the plant. The consequence would be a short shelf life. Temperatures during shipping should be as low as possible without causing injury. The temperatures listed in Table 16-2 for toning bedding plants are the same low temperatures that can be used for shipping. Flowering pot plants have minimum shipping temperatures in the range of 40 to 60°F (4 to 16°C), depending on species. Cold-sensitive plants such as poinsettia and hibiscus have a minimum temperature of 50°F (10°C).

Ideally, plants should be lighted during shipping. This is generally not possible inside trucks, because bedding plants are carried on closely spaced shelves and flowering pot plants are often boxed. The adverse effects of low light are chlorosis, leaf drop, pale flower color, floral abortion, and tall thin plants. These problems diminish at lower temperatures. When a truck without cooling is used for shipping, it is advisable to insulate the truck, avoid parking it in the sun for prolonged periods, and travel in the early morning or late afternoon as much as possible. Bedding plants can be shipped successfully in darkness for a period of two days, and green plants for seven days, when precautions are taken to lower temperature.

Retail Environment

Environmental conditions in the retail area should allow for a low level of net photosynthesis. Under these conditions, plants are able to maintain carbohydrate levels and grow slowly. If the light level is too low, respiration will outpace photosynthesis, resulting in a depletion of carbohydrates. At the very least, consumer longevity will be shortened. With time, these plants can develop chlorotic leaves, exhibit paler flower colors, and drop leaves and buds. On the other hand, if conditions are conducive to rapid growth, bedding plants will outgrow their space in flats and will stretch. Flowering pot plants will mature through much of their postproduction stage that was intended to be enjoyed by the consumer. Green plants will lose the level of acclimatization achieved at the end of the production phase and fail to adjust to the consumer environment. A heavily shaded greenhouse constitutes a very desirable retail environment. A 75 percent shade level may be necessary to reduce light to the required level to prevent rapid growth during brighter months. Less shade can be used in the winter. A light level of 500 to 1,000 fc (5.4 to 10.8 klux) is desirable for green plants (Blessington and Collins 1993) and a level of 250 to 700 fc (2.7 to 7.5 klux) serves well for bedding plants (Nelson and Carlson 1987).

Temperature is an important factor in the retail area. If it can be held low, growth will be contained and high light levels will not be a detriment. Consumer comfort must be balanced with plant needs. When plants are displayed in a greenhouse it should be possible to maintain a low temperature. Hard goods, information, and the cashier can be housed in a retail building at a comfortable temperature. When plants are in the retail building, the lowest acceptable temperature for consumers should be maintained. Green plants differ from bedding and flowering pot plants in respect to retail conditions. They perform well in a retail area at temperature, light, and relative-humidity conditions that will exist in the subsequent consumer environment.

Toxic levels of ethylene must be avoided in the retail area. This is a particular problem for bedding plants, which are often displayed outdoors. Since parking lots are usually an integral part of retail settings, auto exhaust is a problem. A few steps can be taken to reduce ethylene exposure. Keep plants as far removed from vehicular traffic paths as sound product promotion practice will allow. Display the more tolerant species closest to these traffic paths. See Table 16-3 for a list of bedding plants ranked according to their sensitivity to ethylene. Ensure that there is good natural air movement in the plant area.

Table 16-3 Sensitivity of Bedding Plants to Ethylene Based on Exposure of 1 ppm for 3, 6, 12, 24, or 48 Hours[1]

Insensitive	*Moderate*	*High*
alyssum	begonia	geranium (petal shatter)
calendula	coleus	impatiens
marigold	petunia	salvia
ornamental pepper		snapdragon
zinnia		

[1]From Armitage (1993).

Fertilizer is usually not applied during the retail phase. It could stimulate growth that is not desired. Care is taken to restrict water so that soft leaf expansion is avoided. Water is applied at the onset of stress.

Summary

1. The five common factors that reduce vase life of harvested fresh flowers are the inability of stems to absorb water due to blockage by microbial organisms or solidified chemicals; excessive water loss from the cut flower; a low supply of carbohydrates to support respiration; the presence of pathogenic diseases; and the buildup of ethylene gas derived particularly from fruit or from injured and deteriorating plant material.
2. Maximum vase-life potential of fresh flowers is achieved by producing high-quality flowers rich in carbohydrates; preventing the occurrence of diseases before and after harvest; harvesting at the proper stage; placing flowers in a floral preservative immediately upon cutting and maintaining them in such a preservative throughout the marketing channel; keeping cut flowers at a low temperature of 33 to 40°F (1 to 4°C) whenever possible during handling and marketing; and preventing the buildup of ethylene gas around stored flowers by removing injured and old plant material and keeping fruit out of the storage cooler.
3. Floral preservatives can double the vase life of fresh flowers when compared to water. Generally, they provide sugar to supplement the carbohydrate supply in the flower, a bactericide, and an acidifying agent that suppresses bacterial development in the storage water. Sucrose (table sugar) is a common source of sugar, and 8-hydroxyquinoline citrate is often used as the bactericide and acidifying agent.
4. Refrigerated storage of fresh flowers in preservative solution works well for periods of a few days, as is often necessary during the market period. Dry storage of flowers can be used when they must be held for a longer period—up to 18 days for roses and three weeks for chrysanthemums and carnations. Freshly cut, turgid flowers are placed in polyethylene-lined cartons without water, and the cartons are held in a 31°F (–1°C) cooler. Flowers cut in this manner have essentially the same longevity after removal as flowers handled in the more conventional manner.
5. Some flowers, such as roses, irises, and tulips, have traditionally been cut in the bud stage, while others, such as carnations and chrysanthemums, are customarily harvested in an open stage. These latter crops can also be harvested in the bud stage with a considerable reduction in the culture time required in the greenhouse. Harvested buds may be shipped and/or stored at this stage. Ultimately, they are placed in an opening solution, similar to a floral preservative, at room temperature. Carnation buds open in two to three days, and chrysanthemum buds open in seven to nine days. A savings in pro-

duction expense and shipping cost can be realized, and less flower injury is sustained during shipping.

6. Silver thiosulfate (STS) has found application as a fresh flower preservative for carnation, delphinium, dendrobium orchid, enchantment lily, gerbera, *Matthiola*, and snapdragon. The vase life of carnations can be doubled. In this form, silver effectively moves into the flower and blocks the effects of ethylene. Research shows that abscission of flowers or floral parts can be delayed by sprays of STS for a number of crops including bougainvillea, calceolaria, geranium, and zygocactus. There is an indication, however, that STS sprays can render geranium plants susceptible to *Pythium* root rot.
7. Postproduction longevity and quality of containerized plants is increased during production by those factors that enhance production quality.
8. Toning (hardening) plants during the later phase of production increases postproduction quality. Reductions in fertilization, water frequency, and temperature during the last two weeks benefit bedding plants and flowering pot plants. Lower fertilization, light intensity, and watering frequency during the last six weeks to six months acclimatizes green plants so that they adjust better to the consumer environment.
9. It is important that containerized plants be shipped at a cool temperature to prevent weak etiolated growth and reduce sensitivity to ethylene. It is advisable to provide ventilation in the handling area, the truck, and packing cartons to minimize ethylene concentrations.
10. Light and temperature in the retail area for bedding plants and flowering pot plants should be high enough for a low net photosynthetic rate to prevent depletion of carbohydrate reserves in these plants. These factors should not be higher because rapid growth would cause stretching of bedding plants and loss of shelf life in flowering pot plants. Green plants need to be held at the same light level that was used at the end of production for acclimatization. Fertilizer is generally not applied to any plants during the retail phase and water is applied only as stress appears.

References

References with an asterisk before them lend themselves well to a general overview of the subject.

*1. Armitage, A. M. 1993. *Bedding plants: prolonging shelf performance*. Batavia, IL: Ball Publishing.
2. Ball, V., ed. 1985. *The Ball Red Book*, 14th ed. Reston, VA: Reston Publishing.
3. Beyer, E., Jr. 1976. A potent inhibitor of ethylene action in plants. *Plant Physiol.* 58:268–271.
*4. Blessington, T. M., and P. C. Collins. 1993. *Foliage plants: prolonging quality*. Batavia, IL: Ball Publishing.

5. Boodley, J. W., and J. W. White. 1969. Post-harvest life. In Mastalerz, J. W., and R. W. Langhans, eds. *Roses: a manual on the culture, management, diseases, insects, economics and breeding of greenhouse roses*, pp. 78–92. Pennsylvania Flower Growers' Association, New York State Flower Growers' Association, Inc., and Roses, Inc.
6. Cameron, A. C., and M. S. Reid. 1981. The use of silver thiosulfate anionic complex as a foliar spray to prevent flower abscission of zygocactus. *HortScience* 16:761–762.
7. Cameron, A. C., and M. S. Reid. 1983. Use of silver thiosulfate to prevent flower abscission from potted plants. *Scientia Hort.* 19:373–378.
8. Cameron, A. C., M. S. Reid, and G. W. Hickman. 1981. Using STS to prevent flower shattering in potted flower plants—a progress report. *Flower and Nursery Report for Commercial Growers* (Fall 1981). Univ. of California Coop. Ext. Ser.
9. Carpenter, W. J., and D. R. Dilley. 1975. *Investigations to extend cut flower longevity.* Michigan Agr. Exp. Sta. Res. Rep. 263.
10. Fjeld, T., and E. Stromme, eds. Sixth international symposium on postharvest physiology of ornamental plants. *Acta Hort.* No. 405.
11. Halevy, A. H., and A. M. Kofranek. 1977. Silver treatment of carnation flowers for reducing ethylene damage and extending longevity. *J. Amer. Soc. Hort. Sci.* 102:76–77.
12. Howland, J. E. 1945. A study of the keeping quality of cut roses. *Amer. Rose Annual* 30:51–66.
13. Kofranek, A. M., and A. H. Halevy. 1972. Conditions for opening cut chrysanthemum flower buds. *J. Amer. Soc. Hort. Sci.* 97:578–584.
14. Kofranek, A. M., and J. L. Paul. 1972. Silver impregnated stems aid carnation flower longevity. *Florists' Rev.* 151(3913):24–25.
15. Marousky, F. J. 1970. New methods for improving keeping quality for gladiolus, roses and chrysanthemums. *Florists' Rev.* 145 (3770):67, 116–119.
16. Marousky, F. J. 1976. *Control of bacteria in vase water and quality of cut flowers as influenced by sodium dichloroisocyanurate, 1,3-dichloro-5,5-dimethylhydantoin, and sucrose.* ARS-S-115. USDA, Washington, DC.
17. Mastalerz, J. W. 1969. Low temperature dry storage. In Mastalerz, J. W., and R. W. Langhans, eds. *Roses: a manual on the culture, management, diseases, insects, economics and breeding of greenhouse roses*, pp. 150–156. Pennsylvania Flower Growers' Association, New York State Flower Growers' Association, Inc., and Roses, Inc.
18. Miranda, R., and W. H. Carlson. 1981. How to stop petal shattering in hybrid seed geranium. *Grower Talks* 45(7):18–22.
*19. Nell, T. A. 1993. *Flowering potted plants: prolonging shelf performance.* Batavia, IL: Ball Publishing.
20. Nelson, L. J., and W. Carlson. 1987. Improve the marketability of bedding plants. *Greenhouse Grower* 5(3):84–85.
21. Reid, M. S., J. L. Paul, M. B. Farhoomand, A. M. Kofranek, and G. L. Staby. 1980. Pulse treatments with silver thiosulfate complex extend the vase life of cut carnations. *J. Amer. Soc. Hort. Sci.* 105:25–27.
22. Reid, M. S., and G. L. Staby. 1981. "Super" carnations—a concept. *Canadian Florist, Greenhouse and Nursery* 76(1):40, 42, 44, 46, 48.
23. Robertson, J. L., and G. L. Staby. 1976. Economic feasibility of once-over bud harvest of standard chrysanthemums. *HortScience* 11:159–160.
24. Sacalis, J. N. 1977. Epinasty and ethylene evolution in petioles of sleeved poinsettia plants. *HortScience* 12:388.
*25. Sacalis, J. N. 1993. *Cut flowers: prolonging freshness.* Batavia, IL: Ball Publishing.
26. Saltveit, M. E., ed. International symposium on the physiological basis of post-harvest technologies. *Acta Hort.* No. 343.

27. Saltveit, M. E., and R. A. Larson. 1981. Reducing leaf epinasty in mechanically stressed poinsettia plants. *J. Amer. Soc. Hort. Sci.* 106:156–159.
28. Sass, P., ed. 1993. International symposium on postharvest treatment of horticultural crops. *Acta Hort.* No. 368.
*29. Staby, G. L., J. L. Robertson, D. C. Kiplinger, and C. A. Conover. 1976. *Proc. National Floricultural Conference on Commodity Handling.* Ohio Florists' Association, 2001 Fyffe Ct., Columbus, OH 43210.
30. Staby, G. L., J. F. Thompson, and A. M. Kofranek. 1979. Post-harvest characteristics of poinsettias as influenced by handling and storage procedures. *Florists' Rev.* 165:86–87, 136–139.
31. Veen, H. 1983. Silver thiosulfate: an experimental tool in plant science. *Scientia Hort.* 20:211–224.
32. Veen, H., and S. C. van de Geijn. 1978. Mobility and ionic form of silver as related to longevity of cut carnations. *Planta* 140:93–96.

 CHAPTER 17

Marketing

There is as much science, technology, and art applied to floral crops after harvest as before. Storage, packaging, transportation, design, advertising, marketing, and servicing can all be involved in flower or plant handling after harvest. The input can be sufficiently great to justifiably raise the final retail price to several times the level of the wholesale price. Failure to properly market a crop can negate the efforts that have gone into producing a quality crop.

Marketing actually begins with the planning of the crop. It entails a market-demand evaluation to ensure that the correct crops, sizes, colors, and so forth are grown to meet market needs. Cultural schedules are developed to finish the crop at a potentially profitable time. All too often, growers become concerned entirely with maximizing the use of bench space and lose sight of the market demand and selling price of the crop. Once a crop is properly planned, the more obvious steps of marketing begin at harvest time.

Packaging

It requires skill to harvest fresh flowers at the proper stage of maturity. This topic is covered in detail in books on flower crop production and will not be discussed here. It is sufficient to note that, with the exception of roses, there is a degree of latitude in the stage of development at which the flower must be cut. The exact stage depends on the length of time and type of handling in the market channel. European and Colombian carnations are often harvested in a tight stage (with guard petals upright) to facilitate shipping and lengthen vase life. Carnations grown for local consumption are generally harvested open (with guard petals horizontal or lower) to minimize handling time. Standard chrysanthemums can be similarly harvested in the bud stage, as described in Chapter 16.

Floral products are packaged in conventional unit sizes. Roses and carnations are packaged in bunches of 25, while the number is 10 for standard chrysanthemums, daffodils, gladioli, irises, snapdragons, tulips, and most other fresh flowers. Pompon chrysanthemums are bunched according to weight, with 9 oz (255 g) being common. Generally, the stems are 30 in (76 cm) long and no fewer than five stems are included in a bunch. The weight of the bunch varies with different growers. Bunches of fresh flowers are often placed in a plastic sleeve to protect the blooms, and the stem ends are bound with a rubber band or string. Bunches so wrapped are placed in cardboard containers for refrigerated storage or shipping. Colors are not mixed in individual bunches, nor are the types of flowers mixed within a carton. Different-colored bunches are conventionally mixed within the carton in whatever proportion they are produced. The wholesaler and retailer are expected to take them in this ratio to guarantee a market for all. Communication is required between grower and wholesaler for the system to work.

Pot plants are usually sold individually to full-service florists. Some of the larger suppliers of mass markets package pot plants in cardboard cartons in varietal proportion so that they can be more easily stacked in trucks, handled, and inventoried, especially if they are distributed from the central warehouse of a chain store.

Pot plants are often placed in a plastic or paper sleeve just prior to shipping. The sleeve compacts the foliage, reducing the amount of valuable shipping space required by each plant. It also protects the plant from damage during handling. Some growers take advantage of the sleeve as a strategic place to advertise their company and to offer cultural suggestions.

Packaging will play a very important role in the future, particularly for those growers servicing the mass market. Plants are already being marketed in complete enclosures that nearly eliminate evaporation and the need to water them during the period of marketing. The enclosure consists of clear plastic for viewing the plant and for transmission of light for photosynthesis. Some use cardboard for the frame, a handle, and a place to advertise. Fresh flowers also may be packaged to prolong shelf life. Such packaging enables control of the atmosphere within and extends life. The location of the packaging step (grower, wholesaler, or retailer) will depend on the comparative costs and returns of the alternatives. Packaging is a fertile area that the grower should consider.

Grades and Standards

There has been considerable controversy over grading. Opponents cite hidden factors such as the increased cost of handling. Proponents see grading as a means of discouraging poor quality in the marketplace and achieving financial renumeration for quality. It could also go a long way toward nurturing consumer satisfaction.

Most fresh-flower producers use a grading system. One problem, however, is the diversification of grading systems among growers and even the shifting of standards by an individual grower as average flower sizes change throughout the year. If grades could be standardized for all growers, it would be a great benefit for wholesalers and retailers. Ultimately, what benefits the market system and the consumer usually brings benefits to the grower.

Standardized grading could give both the marketer and the consumer a means for judging and demanding the quality they are willing to pay for. It would give the grower a tangible objective and measuring stick for achieving a better product. Greater consumer satisfaction should lead to increased product demand. Higher-quality production and handling would help reduce flower loss in the market channel, which could be helpful in reducing the final selling price of flowers. Obviously, marketers must get involved in this aspect also.

Traditionally, rose growers have graded their product. Nearly all growers have graded by stem length. Increments of 2, 3, or 4 in (51, 76, or 100 mm) have been used to separate grades, with the most common being 3 in (76 mm) and beginning at a minimum length of 9 in (23 cm). Flowers with weak stems, blemished foliage, off-color blooms, or bullhead blooms are sold as a utility grade. Recently the Floral Marketing Association (FMA) in cooperation with the Society of American Florists (SAF) has developed grades and standards for 21 types of fresh flowers (FMA/SAF 1996). An example set of standards developed for carnations by the SAF is presented in Table 17-1.

Green plants in particular should be graded to protect the consumer as well as the grower of quality plants. Green plants are grown in a favorable environment relatively rich in nutrients and sunlight. They are then utilized, hopefully for many years, in a rather marginal indoor environment. A period of acclimati-

Table 17-1 Society of American Florists (SAF) Standards for Carnation Grades[1]

	Blue Grade (Fancy) (1)	*Red Grade (Standard) (2)*	*Green Grade (Short) (3)*
Minimum length[2]	22 in (56 cm)	17 in (43 cm)	12 in (30 cm)
Minimum flower diameter[3]	Tight[4]—2 in (51 mm)	1¾ in (44 mm)	No requirement
	Fairly tight—2 in (64 mm)	2¼ in (57 mm)	
	Open—3 in (76 mm)	2¾ in (70 mm)	

[1]Flowers in the blue, red, and green grades should be full, symmetrical, and free of insects, disease, and mechanical injury; they should also be free of bloom defects such as slab side, bullhead, blow heads, singles, sleepy appearance, splits, and discoloration. The stems should be of sufficient strength so that they do not deviate more than 30° from the horizontal plane when held 1 inch (2.5 cm) above the minimum length of the grade with the natural curvature down. Any flowers with these defects are either sold at a lower price or discarded.

[2]Length is measured from the top of the bloom to the cut end of the stem.

[3]The flower diameter is the greatest dimension of the petals measured through the center of the bloom. The guard petals of open blooms are held horizontally when size is determined.

[4]*Tight:* guard petals are up, center petals are up but fluffed. *Fairly tight:* guard petals are horizontal, center petals are up and fluffed. *Open:* guard petals are horizontal or lower, center petals are out or down.

zation must be provided by the grower for these plants to make the transition successfully. Acclimatization can be costly since it entails a period of slow growth when nutrients and light are reduced. Growers who do not acclimatize their crops may realize a profit in the short run, but in the long run the industry is hurt by consumer dissatisfaction. Standards for green plants should take into account such handling so that it is encouraged and rewarded. As early as 1977, a set of standards existed for green plants (Gaines 1977). Grades and standards were recently completed for about 20 green-plant crops by the FMA in association with the Florida Nursery Growers Association (FNGA) (FMA/FNGA 1995).

Standards did not exist for flowering pot plants until recently. In 1987, the FMA and the SAF made the decision to promote the development of grades and standards. The first set of guidelines was released in 1989 and covered azalea, chrysanthemum, lily, and poinsettia. Today, there are grades and standards for 20 pot crops (FMA/SAF 1995).

The standards are voluntary, and it is hoped that they will never be absolutely binding. Differences in taste exist across regions. Desired pot-plant height is taller in the western regions of the United States than in the eastern regions. The consumer preference is for an 18-to-20-inch (46-to-51-cm) pot mum measured from the base of the pot to the top of the plant in California; on the East Coast, a 14-to-16-inch (36-to-41-cm) plant is preferred. Included in the standards are height, minimum width of the top of the plant, number and developmental stage of buds, condition of foliage and roots, strength of stems, and the presence of a care tag. Separate standards are developed within each species of plant for various grades in accordance with pot size. For example, poinsettia grades include "small" for 4-to-4.5-inch (10-to-11-cm) pots, "medium" for 5-to-5.5-inch (13-to-14-cm) pots, "large" for 6-to-6.5-inch (15-to-17-cm) pots, "extra large" for 7-to-8.5-inch (18-to-22-cm) pots, and "jumbo" for 10-inch (25-cm) pots.

When the grower has finished grading and packing fresh flowers or pot plants, it is generally his or her task to ship them to the wholesaler or retailer. Green plants are an exception, as they are often shipped by the grower at the retailer's expense or are picked up directly by the wholesaler or retailer.

The Market System

Consumers exist wherever people live—in cities, towns, and villages scattered throughout the states and provinces. Floral production, however, is more centralized. This is particularly true of the fresh-flower industry. The heaviest concentration of gladioli comes from Florida; carnations, from Colorado and California; and spray-type chrysanthemums, from Florida and California. The trend is not as well established for flowering plants, but the vast majority of green plants come from Florida, California, and Texas. Under such circumstances, a complex marketing system is necessary (Figure 17-1).

The marketing system serves the functions of gathering together the various floral products of many diverse growers, bringing these within reach of con-

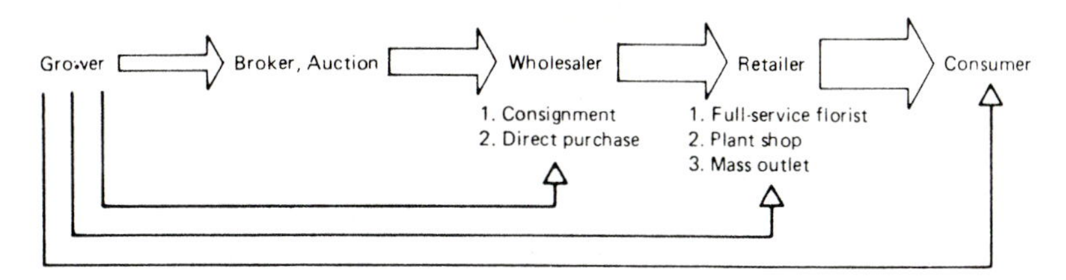

Figure 17-1 **Channels through which fresh flowers flow from the grower to the final consumer.**

sumers both close to and distant from the producers, and developing a consumer awareness and desire to purchase the floral products.

Fresh Flowers

The floral marketing system has several components. There are a number of possible channels within the system. Fresh flowers pass through the most extensive channel, thus providing us with a good overview of the whole system. In The Netherlands, fresh flowers and flowering plants are customarily brought to an auction, where they are purchased by wholesalers. Such auctions have appeared in the United States on a limited scale in the past 20 years. The broker quite often serves as an alternative to the auction in the United States. In this situation, a brokerage firm purchases flowers or plants from growers to fill orders it has received from wholesalers or large retail florists. Most often, the broker is located in the region of concentrated production, so that crops can be examined to match them to the various grade and quality levels demanded by clients. The wholesalers served are scattered at considerable distances from the production area. It is this gap between grower and wholesaler that creates the need for auctions and brokers. The broker system is popular in the fresh-flower production areas of California as well as in the green plant production areas of Florida and California.

Traditionally, fresh flowers in the United States have passed directly from the grower to the wholesaler. Often, the wholesaler is a commission wholesaler, one who takes flowers on consignment. This means that the grower is paid for flowers that the wholesaler sells but not for those that he or she fails to market. The commission wholesaler sells the flowers at a wholesale price and then takes a commission of about 25 percent from this price, returning the remainder to the grower.

Recently, a trend has been developing for wholesalers to buy flowers outright from the grower or broker. This system is more expedient where flowers are mass produced in one area and wholesaled a great distance away. Although the wholesaler appears to assume all the risk, this is not the case. Flower losses can be reflected in lower subsequent returns to the grower or higher prices for the consumer. The latter affects consumer demand, which hurts the grower.

Wholesalers sell flowers to retailers. Some retailers travel to the wholesale house to make their purchases; others are serviced by trucks operated by the wholesaler. Wholesale florists quite often stock supplies needed by retail florists, such as ribbon, net, vases, and wreaths, which are used in the daily operation of a full-service retail flower shop. The inventory may be larger, including plastic flower arrangements and giftware to be sold directly by the retail florist.

Most sales are made over the phone by salespeople employed by the wholesale florist. When the orders are filled, the remaining space on the truck is filled with flowers and merchandise that will probably be sold along the route (Figure 17-2). Each florist is generally serviced twice a week, often by more than one wholesaler. These truck routes serve a very valuable role for retail florists who

Figure 17-2 **The interior of a wholesaler's truck used for delivering fresh flowers and pot plants to full-service retail florists. Although it is not common, a small refrigerated room for fresh flowers is located at the forward part of this truck.**

are located in remote areas. The wholesale florist likewise plays a valuable role for the retail florist near transportation facilities. The wholesaler brings together hundreds of items from numerous sources for the retailer's use. This saves the retailer considerable time and expense, as well as the problem of overstocking on items that must be purchased in case lots and soon become outdated. The wholesaler makes it his or her business to keep abreast of changing tastes in supplies, which further benefits the retailer.

There are various types of retailers, as anyone who purchases flowers or plants is well aware. Traditionally, we think of the full-service florist shop, where there is a designer to arrange flowers and delivery service is available. Plant shops and flower boutiques are becoming more numerous. These are cash-and-carry outlets where plants and flowers may be purchased. Sometimes, simple arrangements of flowers are offered, mainly for home decoration, but delivery service is not provided. A number of full-service florists operate such shops as well. Mass-market retail outlets have become a very large business. These are the cash-and-carry stands, wagons, and minishops that are located in high-traffic areas within supermarkets, department stores, discount stores, shopping malls, and airports, as well as on street corners. The markup is generally 30 to 40 percent. Therefore, when a plant sells for \$1.00, the mass-market outlet retains \$0.30 to \$0.40, while the grower receives \$0.60 to \$0.70.

A subtle difference (often misunderstood) exists between two systems for setting the retail price. *Markup* refers to a percentage of the retail price, while *mark-on* relates to a percentage of the wholesale price. To illustrate the difference, assume that you as grower receive \$1.00 each for your pot plants. One retail outlet using a 33 percent markup charges its customers \$1.50 per plant: \$1.00 (wholesale price) ÷ (1 – 0.33) = \$1.49. A second outlet using a 33 percent mark-on charges \$1.33 per plant: 1.33 × \$1.00 (wholesale price) = \$1.33.

Flowering Pot Plants

Flowering pot plants are sold through auctions in The Netherlands. As a rule, in the United States they are sold directly to retail outlets by growers. Flowering pot-plant growers are generally situated near population centers. Long-distance transportation does not enter into the picture as extensively as in the case of fresh-flowers. The typical flowering pot-plant grower would operate one or more trucks for delivery purposes and have his or her own sales department. Plants are generally delivered within a radius of one working day's travel. There has been a feeling in the past that pot plants are too heavy to transport the distances that fresh flowers are shipped. To a degree this is true, since pot plants cannot generally be shipped by air, as many fresh flowers are. However, in recent years, some very large pot-plant ranges have developed to supply the mass market. Such ranges successfully deliver plants 500 or more miles (800 km) by truck. Success is dependent upon having fewer delivery points per truck, larger orders per stop, and stores that can accommodate large trucks. Insulated trucks are used that are heated in winter and cooled in summer.

Green Plants

Green plants are produced most extensively in Florida, California, and Texas, which dictates the need for a marketing system as described for fresh flowers. Many of these plants pass through brokers to wholesalers. A significant group of green-plant wholesalers turns out to be growers operating in close proximity to the retail market. While the basic line of green plants is the tropicals, which in spite of transportation are most economically produced in subtropical regions, there are some green plants, such as ferns, that can be produced economically in close proximity to the market. The local grower-wholesaler business combination allows the flexibility needed to develop this potential.

Wholesalers truck their plants from Florida or California to their greenhouse range, where they are held until they can be marketed to retailers along their truck routes. The greenhouse is necessary for holding these plants, since a considerable length of time may pass before some are sold. The greenhouse also affords an opportunity to supplement the line of plants with some that can be produced more profitably than they can be bought. There are yet other plants that are purchased in early stages of growth and are finished locally. The grower-wholesaler business combination is working out well for the green-plant industry.

Direct Sales

Locally produced fresh flowers are in high demand when they are produced continually throughout the year and at a high level of quality. There is some effort on the part of retailers to trade directly with such sources. With a modest effort, a grower can establish a market without passing through the wholesaler. This system is used particularly by smaller growers. New growers of fresh flowers, and more often of pot plants, sometimes sell directly to the final consumer. This allows them to enter into two businesses for little more overhead than that of the growing operation. Funds are generated more quickly this way. This system works well when the extra labor can be provided by the owner, assuming that there is not a more profitable use for his or her labor at the time. This is at best a temporary system, and soon a decision must be made as to whether to operate one or both businesses.

A number of large full-service retail outlets and some mass-market retailers operate their own production ranges. Care must be taken to keep separate records on each business, lest one should exist at the expense of the other. Quite often, the retail outlet is the more successful of the two, in which case it might be better to purchase flowers and plants from another source.

Flower Auctions

Flower auctions exist where there is concentrated production some distance away from the retail market. The Dutch flower auctions are the most famous (Figure 17-3). Flower production concentrated in a few regions of a small nation,

(a)

(b)

Figure 17-3 Exterior and interior views of the United Flower Auctions in Aalsmeer, The Netherlands. This is the largest flower auction in the world, with approximately 4,000 members. Such auctions serve as a distribution channel between growers and wholesalers. (*Photos courtesy of United Flower Auctions Aalsmeer, Aalsmeer, The Netherlands.*)

The Netherlands, supplies retail markets throughout the world. To make the distribution system efficient, growers send fresh flowers and plants to an auction where they are purchased by the wholesalers, who in turn distribute them to retailers throughout the world. This adds an extra link between the grower and the wholesaler in the distribution chain illustrated in Figure 17-1, but in so doing it brings wholesalers into contact with hundreds of growers who would otherwise be unreachable.

Flower auctions in The Netherlands and Canada also exist because of the practices of some wholesalers who play growers off against one another. Growers by definition are in a weak market position because of the perishability of their product. By uniting in a producers' cooperative, individual growers strengthen their market position.

Dutch auctions charge the grower/members a commission of about 5 percent. The buyer pays about 0.3 percent service costs. This does not necessarily increase the retail price over that in a system without a flower auction, since the job of the wholesaler is made more efficient by the auction.

A typical auction functions as follows. Flowers or plants are delivered by the grower to the auction, where they are set out on display. Early in the day, wholesalers peruse each lot to assess quality and condition. In so doing, they decide which lots they wish to purchase and how much they are willing to pay. Later in the morning, wholesalers take their assigned seats in the auction room, as pictured in Figure 17-4. Each lot of plants is brought one at a time before the wholesalers. The clock at the front of the room indicates the identification number of the grower of the plants and the lot number of the plants. An auction employee holds the plants for all to see, and the auctioneer gives a brief assessment of the plants. The sale begins when the clock pointer, set on 100, is released and begins its descent in price. The value of each unit on the clock's scale of 0 to 100 is denoted on the clock—whether it be 1, 5, 10, 25, or 100 Dutch cents. When the pointer comes down to the price a wholesaler is intending to pay, he or she presses a desk button that stops the clock and electrically records his or her identification number and the price on the clock at that point. The wholesaler finalizes the sale by indicating the quantity of the lot he or she wishes to purchase. In a matter of a few hours, all of the day's sales are made.

This system works rapidly. As many as 700 transactions can be made per clock per hour. The interests of both seller and buyer are served. If the buyer waits for an exceptionally low price, he or she may lose the chance to purchase the plants desired. If he or she bids too soon, a needlessly high price is paid. Each lot of plants is judged independently, and its price is established accordingly. The principle of supply and demand expresses itself in this system.

When sales are finished, flowers or plants are moved from their display area to loading docks where the trucks of the various wholesalers are waiting. Even this process is often mechanized. Carts are loaded according to purchaser and are moved automatically along tracks to the loading area. When the day ends, the auction house is ready to repeat its cycle.

Three auctions have opened in Canada in Montreal, Toronto, and Vancouver, and two have opened in the United States on Long Island and in San Diego.

Figure 17-4 **One of many auction rooms at the United Flower Auctions in Aalsmeer, The Netherlands, where over two billion flowers and plants are sold each year. (*Photo courtesy of United Flower Auctions Aalsmeer, Aalsmeer, The Netherlands.*)**

Concentrated production at a distance from scattered markets played a role in the establishment of these auctions, particularly the Toronto and San Diego auctions. Wholesalers and retailers alike make purchases in these auctions.

Advertising

The need for advertising varies. A grower who sells to one wholesaler or to a few wholesalers will generally have little motivation to advertise. The grower of a centralized crop such as green plants in Florida or fresh flowers in California will probably be interested in new wholesale outlets. This grower often advertises in the various florist trade papers.

The retailer has the greatest need for advertising. Unfortunately, cost may be a deterrent. Those who do advertise generally find it profitable. Newspaper ads are most commonly used. Radio spots are also valuable, particularly toward the weekend and in connection with a gardening program. Television has been used by some and can have a far-reaching effect when done properly. Mailing lists have provided a very successful avenue of communication with the consuming public for many retailers.

It is not the intent of this book to take more than a cursory look at retail marketing. While the major burden of advertising rests on the retailer, the grower is not without obligation. The allied supply industry—growers, wholesalers, and retailers—are all parts of one system that culminates in the sale of floral products to the consumer. It has been demonstrated in the floral industry that advertising effectively increases the demand for these products. This ultimately benefits all segments of the industry; thus, all should share in the advertising program. Shared advertising is often practiced in other businesses. The Coca-Cola sign, so often used to display the name of a restaurant, is paid for in part by The Coca-Cola Company. Advertisements for a given product, regardless of the retail outlet, will carry the same logo (sketch, picture, and so on). The logo is developed and provided at the expense of the producer. The advertising cost for many items presented by the local supermarket in its newspaper ads is borne by the producer of the products.

There are national and international advertising programs in the floral industry. Floral wire services collect a percentage of the gross wire sales of their member retail florists and use these funds for wide-range advertisement. Individual retail florists expend additional funds for local advertising. Through the centralized program of the wire houses, expensive but highly effective advertising media can be used. National television and major magazine ads are procured. Billboard space is contracted. Consumer information literature is underwritten, such as the booklet *Professional Guide to Green Plants*, sponsored by the Florists' Transworld Delivery Association.

The closest the floral industry comes to a properly shared advertising program is seen in PromoFlor and is the efforts of the American Florists' Marketing Council (AFMC) of the Society of American Florists (SAF). This latter organization carries on a national advertising program with funds derived from all segments of the floral industry on a voluntary basis. The AFMC runs advertisements in national magazines and newspapers such as *U.S. News & World Report, Redbook, Sports Illustrated*, and *The Wall Street Journal*, as well as radio spot ads on the ABC, CBS, and NBC television networks. In addition, they prepare and offer at cost in-store display banners, newspaper advertisement mats ready to submit to the newspaper once the retailer's name and address are inserted, radio spot scripts, and truck and billboard signs. The overall program is having a positive effect on increasing the floral market but needs to be much larger, in light of the potential market.

Floral growers have an obligation to share the overall marketing responsibility of the industry. There are several things they can do:

1. Financially support cooperative advertising programs such as the AFMC.
2. Explore the possibility of and, when warranted, work with wholesalers and retailers in local promotional programs.
3. Establish communications with the wholesale and retail segments of the industry through membership in their organizations, attendance at their conventions, and reading of their literature.

Much of the potential of the floral industry is dependent upon a greater degree of cooperation among the diverse businesses making it up. A major problem in the industry today is lack of unity, as seen in separate grower, wholesaler, and retailer organizations, meetings, literature, and attitudes. Such disunity can hurt even at the individual grower's level. Wire services periodically feature specific fresh flowers and plants in their promotional programs. Grower alerts are issued long in advance of the promotion date, but many growers are not tuned in. This has a negative effect in the marketplace, since the promotional item falls into short supply and prices rise adversely. It can be disadvantageous to the growers, who find themselves heavy on nonpromotional items and short on those in demand. Through interindustry communication, it should be possible to use promotional programs as a means for coping with inadvertent overproduction and periods of low market demand and for establishing consumer demand for products and product forms, thus rendering larger profit to the grower, greater ease of handling in the market channel, and increased consumer satisfaction.

There are other ways in which growers can play a role in the overall promotional or advertising program. They must concern themselves with consumer satisfaction. This can be done by selecting plant varieties that stand up best in the region in which they are marketed. Fuchsias are beautiful almost anywhere in the spring, but are a disappointment to consumers in hot climates when the heat of summer arrives. Such sales should be discouraged, and in their place crops adapted to the situation should be promoted. The grower has a responsibility to make such decisions and to educate the retailer. It is the further responsibility of the grower to produce plants of high quality, free of insects and disease. Whether consumers relate plant failure to the grower or to themselves, the main effect is the erosion of the desire to make a subsequent purchase.

The consuming public has an underlying desire for information. This is often as important as the product itself. The grower should provide identification and cultural information with each unit sold. Plastic stakes are available with such information for many types of pot plants. If they are not available, one could have such stakes made, attach an information sheet to the plant, or have the information printed on the plastic sleeve if one is used.

The grower's responsibility to educate does not stop here. He or she must pass information along to the retailer as to how the product is to be handled during marketing. The grower should also supply information that the retailer can pass along to the consumer. This responsibility is particularly important in the mass-market channels, where some merchandisers have little experience in handling plants. Some larger growers supplying mass markets have found it advantageous to work with management in chain stores in training their produce managers to properly handle floral products.

There are no binding laws forcing growers to participate in advertising or promotional programs. Advertising as discussed thus far falls into two categories: (1) brand-name advertising, in which the advertising firm is directly promoting its own products; and (2) generic advertising.

In brand-name advertising, for example, Nelson the Florist advertises poinsettias for Christmas so that the townspeople will buy from him rather than from the supermarket. The effects of such advertising are relatively easy to evaluate. The fact that most retailers and wholesalers engage in it is testimony to its success.

Generic advertising promotes flowers and plants in general without reference to any brand names. Its purpose is to expand the total market. The PromoFlor and AFMC programs are examples of this type of advertising or promotion. It is difficult to evaluate the usefulness of such advertising, since the effects are indirect. A large producer servicing the mass market over an expansive region will probably sense an effect and feel that the expenditure returns a profit. A smaller grower, particularly one selling to full-service retail florists, may not feel that it is profitable. This is a business decision that must be made by each firm; however, too few businesses have realistically considered generic advertising. It would be better for the floriculture industry as a whole if more businesses were involved.

Summary

1. Packaging of fresh flowers has been standardized by convention. A set number or weight of flowers constitutes a bunch. Bunches are shipped in cardboard cartons, the number contained within depending upon the grade. Pot plants were once customarily sold individually and often in plastic sleeves; however, it is common today for pot plants to be shipped in cardboard cartons and to be sold in the multiple contained within a carton. The number in a carton depends on the pot size, the type of plant, and the grower. Pot-plant packaging is new and is not standardized.
2. Grades and standards exist for some fresh flowers and are being established for pot plants. Grading is a voluntary program that is practiced by many growers. It would be advantageous to the floral industry and the consumer if all growers adhered to a single grading system. Today, many systems are in use.
3. Fresh flowers are usually purchased from growers by wholesalers, who in turn sell them to retailers. Brokers and auctions are playing an increasing role in moving fresh flowers from growers to wholesalers.
4. Flowering pot plants are generally sold directly by growers to retailers. Green pot plants often pass through a broker and a wholesaler en route to the retailer.
5. Flower auctions are popular in Europe and exist to a limited degree in Canada and the United States. They serve well as a channel between growers and wholesalers when production is concentrated and located at a considerable distance from the retail market.
6. Advertising, as in any other business, is critical to the floral industry. Any increase in consumer demand has the potential to benefit all segments of this industry. While the heaviest investment in advertising is made by retailers, the burden is shared by wholesalers, growers, and allied trades as well. PromoFlor as well as The American Florists' Marketing Council (AFMC) of the Society of American Florists (SAF) carries out a promotional program enhanced by voluntary contributions from all of the industry. Considerably

more promotional effort must be made by the floriculture industry before it comes up to the standards of most other industries. Growers can do their part by supporting existing national promotional programs, joining in local promotional programs with wholesalers and retailers, communicating more extensively with wholesaling and retailing groups, supplying technical information to retailers to aid them in handling floral products and better advising the consumer, and producing high-quality plants well acclimated to the consumer's environment so that satisfaction is guaranteed.

References

Numerous popular and academic books exist on marketing. As a student, one should consider a course in marketing essential.

1. Berninger, L. M. 1982. *Profitable garden center management*, 2nd ed. Reston, VA: Reston Publishing.
2. FMA/FNGA. 1995. *Recommended grades and standards for foliage plants.* Floral Marketing Association and Florida Foliage Growers Association. (Available from FMA, 1500 Casho Mill Rd., P.O. Box 6036, Newark, DE 19714-6036.)
3. FMA/SAF. 1995. *Recommended grades and standards for potted plants.* Floral Marketing Association and Soc. of Amer. Florists. (Available from FMA, 1500 Casho Mill Rd., P.O. Box 6036, Newark, DE 19714-6036.)
4. FMA/SAF. 1996. *Recommended grades and standards for fresh cut flowers.* Floral Marketing Association and Soc. of Amer. Florists. (Available from FMA, 1500 Casho Mill Rd., P.O. Box 6036, Newark, DE 19714-6036.)
5. Gaines, R. L. 1977. *Guidelines to foliage plant specifications for interior use.* Apopka, FL: Florida Foliage Association.
6. Laurie, A., D. C. Kiplinger, and K. S. Nelson. 1979. *Commercial flower-forcing*, 8th ed., chap. 14. New York: McGraw-Hill.
7. Nichols, R., and G. Sheard, eds. 1975. Post harvest physiology of cut flowers. *Acta Hort.* No. 41.
8. Pfahl, P. B. 1973. *The retail florist business*, 2nd ed. Danville, IL: Interstate Printers and Publishers.
9. SAF Grades and Standards Committee. *Standard grades for carnations.* Soc. of Amer. Florists, 901 N. Washington St., Alexandria, VA 22314.
10. Staby, G. L., J. L. Robertson, D. C. Kiplinger, and C. A. Conover. 1976. *Proc. National Floricultural Conference on Commodity Handling.* Ohio Florists' Association, 2001 Fyffe Ct., Columbus, OH 43210.

 CHAPTER 18

Business Management

Management and labor are distinctly different activities. *Management* is the directing of labor, time, and materials. *Labor* is the execution of plans that have been developed. The owner of small greenhouses often finds it necessary to be a laborer as well as a manager. This is all right as long as he or she never loses sight of the need to manage. Without proper management, an owner-manager expends a great deal of effort with little return, the attitude of the labor force deteriorates, and the business fails to meet its goals. This situation is unstable and ultimately leads to failure.

Management efforts must be applied to planning the expenditure of labor, time, and materials, and must allocate these expenditures properly to crop production and marketing. As the greenhouse range grows in size, the integration of production and the subsequent marketing of this production become complex; responsibilities such as purchasing materials, handling billing and payments, bookkeeping, and even correspondence in general become great enough to distract the manager from production and marketing operations. At this point, a business affairs office with its own staff is warranted, as well as additional managerial personnel.

Business Structure

Managers themselves must be properly organized and managed in order for a business to succeed. The general manager is at the top of the managerial ladder. He or she is responsible for all departments. The manager of each department answers directly to the general manager. It is important that each department manager be held responsible for his or her assignment and the work of employees below him or her. At the same time, each employee should answer to one person only.

The labor of a greenhouse production business falls into four general categories (Figure 18-1).

1. The production department oversees the efforts directly involved in producing crops.
2. The marketing department solicits orders, packages the crop, and delivers and performs whatever other services might be required at the point of sale.
3. The engineering department maintains the physical plant and equipment. It has the further task of custom-building facilities and equipment such as benches in the greenhouse, racks for the trucks, and so on.
4. The business affairs department handles matters such as recordkeeping for tax and cost-accounting purposes, billing, purchasing, and payroll administration.

Full responsibility for the business is held by the owner. The owner of a small business often fills the various management roles as well. In this case, he or she serves two roles and is entitled to the manager's salary and whatever profits are left over. The small business has few employees, and direct communication can be maintained between each employee and the owner. Although the functions of the four departments exist, separate managers are not required. The owner is the general manager and also the manager of each department.

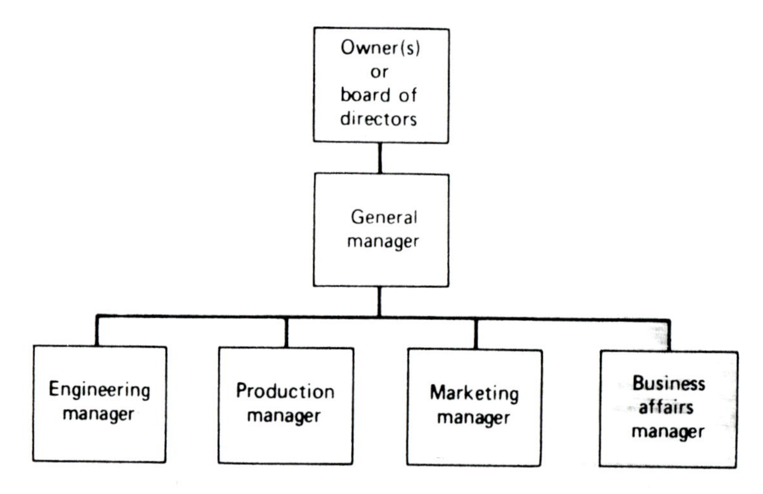

Figure 18-1 A typical managerial structure of a greenhouse business. The owner(s), in the case of a single proprietorship or partnership, or the board of directors of a corporation, carry the full responsibility for the business. The managers of each of four departments take orders directly from a general manager, who takes orders from the owner. All labor within a department answers directly to the management of that department. The four common departments are engineering (repairs and construction), production, marketing, and business affairs.

When the business grows in size, it becomes impractical for the owner to manage all functions. The owner assesses his or her talents and interests and continues to manage one or more departments. Personnel are employed to manage the other departments. For example, the general manager of the company might continue to serve as manager of the engineering and marketing departments. The general manager hires a business affairs manager, who at first manages and carries out the tasks of this department alone and later supervises a number of subsequently hired clerks. The general manager also hires a production manager to manage the labor force involved in growing the crops. The two new managers take their orders directly from the general manager, who is the owner in this case. Each employee receives his or her orders from his or her department manager and not from the general manager.

Further expansions may increase the workload of the general manager to the point where the owner serves only in that capacity. Engineering and marketing managers may then be hired. Perhaps at this time, the size of the production department has become too great for a single manager to effectively handle. Submanagers (often called growers) might be hired to manage production sections. The submanagers take orders from the production manager, and, in turn, the employees under them take orders only from them. Production sections are defined by logic. If the greenhouse business is situated in two locations, each might constitute a section. If pot plants and fresh flowers are grown, it is logical to place each in its own section since the physical facilities (bench type and arrangement, planting area, and so on) differ.

The owner may enter into other business ventures, such as wholesale marketing, retailing, or even an unrelated business. The establishment of another business may possibly require all of his or her attention, in which case the owner should relinquish the role of general manager of the greenhouse business. At this point, he or she no longer draws a managerial salary but does receive the profits.

Two or more owners in partnership must organize in such a manner that the business is commanded with a single voice. Duality of command is constantly open to discrepancy and undermines productivity. One partner might serve as general manager, while the other serves below him or her as manager of one or more departments. In another arrangement, one owner may be a silent partner; the other, a general manager. Both partners share in the profits according to their initial agreement, but the managing partner generally draws in addition a salary in accordance with his or her managerial input. There are many ways to organize a partnership, but the important point is that a unified line of command is established and that the agreement between partners is in writing.

The owners of a corporation are the stockholders. A number of greenhouse businesses are corporations. Obviously, each owner cannot be allowed to give orders. Stockholders' meetings are periodically convened to decide the objectives and methods of operation of the corporation. A board of directors is set up to represent all the stockholders. The board of directors communicates in a single voice with the general manager of the company.

A proper balance of management and labor must be achieved. Management is costly, as students of floriculture well know, since it is these positions they ultimately seek. But indiscriminate reduction of the management force can be even more costly because it leads to a breakdown in communication and inefficiency of resource utilization. The owner must watch the profit-and-loss statement and weigh it against the operational efficiency of his or her business to determine the proper time to adjust the management force. Presented in Table 18-1 are the average number of people in various management and labor positions, other than production labor, for six categories of flowering pot-plant firms (Brumfield et al. 1981). These firms encompass three sizes (20,000, 100,000, and 400,000 ft^2; 1,860, 9,290, and 37,160 m^2) and two market channels (mass-market and full-service florists).

Table 18-1 **Typical Number of People in Various Management Positions in Each of Three Greenhouse Firm Sizes Producing Pot Plants for Either the Mass Market or Full-Service Flower Shops[1]**

Firm Size/ Management Position	*Mass Market*	*Flower Shop*
20,000-ft^2 firm size:		
General manager	1	1
Salespeople	0.80	1.33
Secretary-bookkeeper	0.40	0.50
100,000-ft^2 firm size:		
General manager	1	1
Production-maintenance-labor manager	1	1
Growers	1	2
Salespeople	1.33	2
Secretary-bookkeeper	0.75	1
400,000-ft^2 firm size:		
General manager	1	1
Assistant general manager		1
Production manager	1	1
Growers	5	6
Labor manager	1	
Maintenance manager		1
Maintenance people	2	3
Sales manager	1	1
Salespeople	6	9
Secretary-bookkeeper	1.5	3.5

[1]From Brumfield et al. (1981).

Labor Management

Labor management begins with personal management. To manage others, one must manage one's own life in such a way as to develop traits of leadership. Leadership, coupled with financial and other personal inducements, should provide the motivation needed for workers to carry out their job assignments.

Leadership

A leader is one who through practice comfortably exhibits traits of self-motivation and perseverance that will provide the impetus to stimulate activity when there is a resistance to move ahead. He or she must provide the motivation when the task becomes wearisome and success seems out of reach. A leader must naturally maintain traits of integrity and justice. These qualities gain respect, without which leadership cannot exist. A just system gives the worker a sense of security and an inducement to render an honest day's service for an honest day's wage. Justice calls for setting aside personal prejudices and relationships to see that each individual is judged on his or her own merits. It calls for a just reward where it is earned, and constructive criticism or assistance where it is needed. A leader must have empathy, for it is only through compassion and understanding that the myriad of gaps in personality and position of the workers can be bridged. Unless some common ground can be found, communication is not possible.

Elements of Success

The manager who has the qualities of leadership must organize his or her efforts in such a way as to achieve success. Success depends upon a goal, a plan, faith, and perseverance.

A Goal The manager must establish the success of the business as his or her primary goal. This can be spelled out in many ways. It may be a monetary figure that the business should achieve by some point in time, a volume of production for the existing greenhouse area, a prespecified lower level of crop loss, a projected level of quality, an expansion in business size, or the adoption of new crops into the production scheme. If the manager is not the owner, it is important that his or her goals coincide with those of the owner. A business in the free-enterprise system can afford to reimburse employees financially only in proportion to their contributions toward the financial goal of the business.

When the manager sets a goal, it must be explicit—as in a dollar value of income, a certain number of pots, or a definite percentage of flowers in the premium grade. When a specific quantity is set as the goal, the manager is able to determine where he or she is in relation to the goal. With this knowledge, all efforts can be apportioned to reach the goal. The goal is very important. Without it, a course of action cannot be plotted. Some years ago, contestants of quiz shows were interviewed 10 years after they had won large sums of money. They were asked to show their positive achievements as a result of the unexpected sum of money that had entered their lives. More than 9 out of 10 failed to show any lasting value derived from the money. On the contrary, many found themselves in a poorer lot of life as a result. Loss of initiative, complacency, inability to adapt back to a lower material level of life when the funds were depleted, and other such problems took their toll. Divorce was a common result. The problem stemmed from acquisition of money without a goal to direct its proper use. It is unusual for a person to be able to properly manage that which he or she is incapable of earning. If the contestants had had a realistic goal and a plan for using the acquired money, it is very likely that it would have improved their lot in life.

A Plan A goal alone is not enough to bring success. Just as one does not drive to an unfamiliar destination without a road map, a goal is not attained without a plan. Plans are sometimes demanded of us, as in the case of arranging a loan. Lending agencies demand a *pro forma* (balance sheet) as part of a loan application to see how the money is to be used and what the chances of success and repayment are. This same planning should enter into all operations of the greenhouse range.

Before a plan is drafted, the objective should be researched. The manager should obtain literature in order to establish a background. He or she should communicate with other firms that have successfully undertaken this objective. Suppliers of materials required for the objective are another good source of information. Finally, the manager should integrate this knowledge into his or her own experience and logic and sit down to formulate a plan.

The plan must include a timetable. A reasonable timetable is a weapon against procrastination. The plan should earmark primary objectives within the goal for periodic review and secondary objectives for setting up daily work plans. Each objective must have a stated date of accomplishment.

Crop production depends upon a number of straightforward operations performed on a precise schedule. A good program has the misleading appearance of monotony. A chrysanthemum range, for instance, has its recurring three-month cycle of planting, pinching, lighting, shading, disbudding, watering, fertilizing, spraying, and harvesting. A new manager can quickly settle into a state of boredom and, in the absence of a plan, can begin to perform operations late or miss them altogether. Monotony is quickly replaced with the almost impossible task of saving the crop. The astute manager seeks to establish a simple culture plan that meets all the needs of the plant on schedule. He or she then satisfies the need for adventure and creative outlet through an appreciation for the plant and an attention to detail.

Even the best plans do not hold up forever. Most change begins subtly. The keen manager develops a sense for where insects or disease might first appear and the ability to recognize the infinitesimal changes in plant appearance signaling encroaching disorder. In short, he or she heads off conflict before the need arises to fight a battle. This ever-changing challenge can be met only when the gross physical requirements of the crop are guaranteed in a plan.

Faith Goals and plans require a degree of faith. If the manager harbors doubt in his or her ability to accomplish the plan, it tends to feed on itself and grow in his or her mind. The feeling is inadvertently transmitted to the workers, who will magnify it and reflect it back. Doubt is self-destructive and can be countered only by faith.

Everyone has doubts at one time or another. Doubt can be minimized by practicing an attitude of positive thinking. Doubts that still exist can be disposed of through a process of *auto-suggestion.* We form impressions from everything we do, see, hear, or feel. These impressions may be negative or positive. There is no middle ground. Information received by our conscious faculties feeds into our subconscious. Our minds are at work day and night gathering evidence to support conclusions we have drawn.

When we entertain the idea of failure, we begin to see evidence around us that would suggest failure. When we anticipate success, our minds tend to blank out evidence that would suggest failure and recognize evidence supporting impending success. Our faith in success is thereby strengthened. This in turn causes us to gravitate toward an environment in which answers exist for the needs of our goal; thus, it nurtures our ability to draft a plan.

Our attitudes are readily communicated even without speech. An attitude of direction and self-confidence attracts other positive-thinking people. This is important because one rarely solves all of his or her problems independently. Each member of a group contributes a different perspective and additional information. The greenhouse manager should never "go it alone." He or she should seek a relationship with positive-thinking people at the state university and among the management of other greenhouses, as well as with allied tradespeople, community business people, and civic groups. All conceivable types of information and perspectives eventually come to bear on greenhouse management.

Perseverance Besides a goal, a plan, and faith, success requires perseverance. Very often, a person's first plan will fail. One who gives up at this point will not be a successful manager. Each apparent failure has a lesson contained in it that points the way to an improved plan. People who press on after repeated failures invariably succeed.

The effective manager comes to learn that success is a journey and not a destination. While a plan for the culture of a crop must be simple and precise, it must also be continually altered to accommodate changing cultivars, climate, market dates, automation, and so forth. Maintenance of faith requires constant practice of positive thinking. Above all, to realize viable goals, a manager must continuously develop a perspective on the business firm he or she serves, the floriculture industry, the labor force, and the needs of society as a whole.

The Manager-Employee Relationship

Assuming that the labor force has an adequate level of skill and motivational potential, its accomplishments will depend upon the manager. The labor force must be aware of the managerial structure, know the goals toward which they work, be delegated sufficient authority to accomplish their jobs, understand the system by which they are to be evaluated, and be assured of recognition for their efforts.

Management Structure As previously discussed, each employee must answer to only one superior. Such a system gives continuity to the chain of command so that the firm's goals are not altered or diluted. It also minimizes confusion in the minds of employees, so that each can more fully apply himself or herself to the task. To maintain such a system, each employee should be made aware of the overall structure of management. Although the structure demands that each employee take orders from his or her immediate superior, there should also be a system for higher appeal in the event that an employee feels unfairly treated by a superior. Such a system guards the employee against unfair treatment and at the same time allows the firm to identify improper management.

Goals People seek to improve their self-esteem. Some attain self-esteem by serving in a managerial role; others, by implementing plans of the firm that they deem to be of value. In either event, the employee seeks to have a part in a worthwhile goal. No matter how mundane one's role may be, each employee can relate to the overall mission if he or she is properly motivated.

Therefore, each individual should be made aware of the goals in which he or she participates, their values to the firm or to society, and the importance of his or her part in the plan. This is an easy task for the floriculture manager because the product is one that brings pleasure to people. It enriches their lives, improves the human environment, helps to heal wounds, adds to the pitch of emotional experiences, and expresses sentiments more aptly than words. Flowers, like their artistic and recreational counterparts, bring added meaning to life beyond that of survival.

The manager should point out the goals of the firm. A properly motivated employee takes pride in the growth of the firm for which he or she works. Greenhouse owners have found to their surprise a spirit of exhilaration among employees during periods of greenhouse expansion. One would think that in such a period of added stress, just the opposite mood would take over; however, a feeling of accomplishment prevails because workers are participating in goal setting. A firm that is living off its depreciation (running into the ground) is one in which management of the labor force is difficult, if not impossible.

Finally, the manager must clearly inform each employee of the task he or she is to achieve as part of the overall goal. The task should be spelled out in detail, and a deadline should be given. It is important that the manager have the employee repeat the work assignment to eliminate any misconceptions. No doubt should exist in the employee's mind as to what is expected. This puts the employee in a position to apply all of his or her resources directly to accomplishing the task. Any doubt in the objective will dilute an employee's efforts.

Where the chain of command is long, involving perhaps the owner, the general manager, the production manager, and several subproduction managers, it is wise to post a long-range set of production plans for periodic reference by all concerned. Some growers have devised graphical ways of doing this, as seen in Figure 18-2. Such a visible plan aids managers in maintaining their responsibilities and in briefing their employees.

Delegation of Authority Without some level of authority, an employee cannot organize his or her own activity. The manager must always assume full responsibility for tasks performed by those under his or her command, but must also delegate authority to subordinates. Such authority may cover decisions of priority, purchases, and labor assistance. This authority permits employees to organize their efforts for greater efficiency and to proceed without minute-to-minute supervision of the manager.

Delegation of authority becomes more important as the number of employees answering to a manager increases. It often turns out to be one of the most difficult roles for the manager to perform. It can be difficult to relinquish authority over applications of growth regulators or pesticides when the stakes are so high and the responsibility for avoiding error still rests on the manager. If such authority is not relinquished, however, the manager may not be able to attend to responsibilities of even higher priority.

Evaluation Employees should know at the outset how performance will be evaluated. Such a system gives employees a chance to gauge their performance. They have a chance to improve performance before being reprimanded and to pace themselves. The evaluation system is equally valuable for the manager because it provides a means for guiding the professional development of the employee and thereby further assuring that his or her responsibilities as a manager are met.

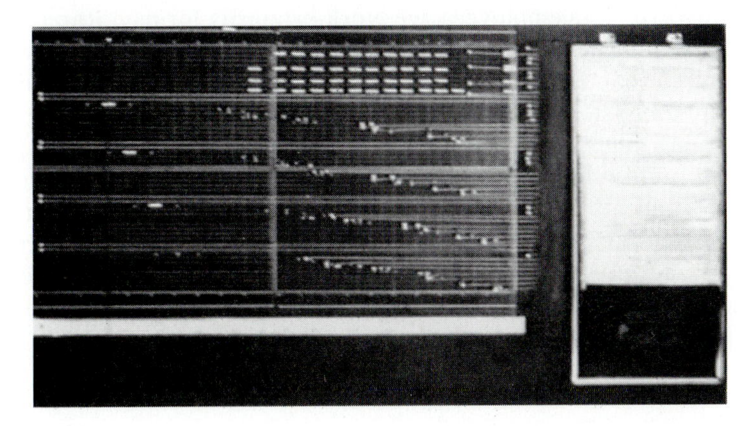

Figure 18-2 A production schedule chart in the management office of the Royal Carnations Co. in Bogotá, Colombia. Each row represents a production area and has holes for each week of the year. Numbered pegs are inserted in holes for the weeks when an operation is to be executed for that production area. The number refers to the operation, and a detailed set of instructions for the operation can be found in the pages at the left side of the board.

A set of standards for the employee's work should be established. In the case of disbudding pot mums, for example, this might include the loss of not more than one terminal bud for every three pots, a time allocation of three minutes per pot, no unnecessary breakage of foliage, placement of all disbuds in a receptacle, and orderly replacement of the pot and watering system after disbudding.

The method of evaluation should be set. A person planting cuttings might be given a set of labels bearing his or her name so that one can be placed in each section in which this person plants. Periodically, the manager checks the planting operation, complimenting those who have performed well and correcting those who have not.

It is advisable to set up a periodic meeting with each employee to review performance. This serves as a reminder to the employee to give some attention to his or her efforts, and it gives the manager an easy means to communicate a judgment without unduly alarming the employee.

Some managers reserve comments for times when job performance is poor. Some employees may understand that no comment is a vote of confidence, but most are unable to respond to this system. A periodic evaluation system circumvents the problem of lack of recognition.

Sometimes, it helps to provide an overall record of accomplishment. For example, a grower might post a chart in the center of the flower production area. Data could be added weekly to graphs showing the total production for the week, the proportion of the total represented in each flower grade, and the quantity of loss due to neglect on the part of labor. This system would provide all employees with a continuous evaluation of the range as a whole and would tend to encourage higher performance.

Reward As employees, we work primarily for pay. Rarely do we feel that we have enough. Increased pay is a stimulus for improved performance, but increases must be handled equitably. If poor evaluation of performance or favoritism enters into the system of pay raises, it has a negative effect on performance.

At times, financial recognition of superior service is not possible. This does not nullify the need for an evaluation-and-reward system. People are social beings and as such are very concerned about recognition. The manager must make it his or her business to notice good performance and to express appropriate appreciation. At the same time, the manager must be certain to notice and help correct poor performance. Both are an integral part of the system for encouraging performance.

Working Conditions

Working conditions are as important as the manager-employee relationship for encouraging good performance. Consider your own feelings when walking along a street in a town with no trees or plantings and with noisy traffic passing a few feet away versus walking along a pedestrian mall landscaped with lawns and planters and overhung by trees. Without realizing it, many greenhouse ranges develop into a harsh, repelling environment that brings about negative feelings in the employees. How much stimulation is there to plant seedlings neatly and at the precise depth when all around are weeds, trash, and unrepaired greenhouses?

Facilities The greenhouses, headhouse, restrooms, and surroundings should be orderly and clean. It was pointed out earlier that this is an important part of insect and disease control. It is also important to proper management. A harmonious environment suggests a state of finesse, which can be achieved with a little encouragement by the manager.

A job is not finished until it is cleaned up. Tools, empty cartons, and so forth should always be in their proper places. Greenhouse aisles, the headhouse, and areas around the greenhouse should be clear. Aside from the negative messages that such messes impart, they also present hazards and a physical barrier to efficient operation.

There should be a program of preventive maintenance for all equipment to ensure that jobs will always be done on schedule. A little paint on a tank before it rusts, grease on a bearing before it freezes, or a tune-up on a rototiller before it stops will prevent breakdowns that could snowball into a stoppage of many other operations.

Each human has an internal rhythm. When the pace of his or her work is geared to this rhythm, efforts are minimized and productivity is maximized. Disruptions in the form of ambiguous orders, undue changes in orders, and equipment breakdown break the work momentum. It is fatiguing and depressing to the employee.

Work facilities should be respectable. Human dignity dictates that bathroom facilities be provided. If the very being of an individual does not command respect, why should his or her productivity be any different? A pleasant area for eating and taking breaks also should be provided. A brief repose at midmorning, noon, and midafternoon benefits the firm as well as the employee. A tired employee is not productive.

There are many other aspects of the physical facilities that warrant attention if the manager simply puts himself or herself in the position of the employee. Worthwhile improvements are those that prevent needless fatigue and facilitate work efficiency. Rubber mats on the floor and, under some circumstances, chairs are an asset to progress. Convenient centralization of tools and supplies also increases efficiency. The range layout as a whole should be formulated with efficiency in mind. A flat site and ridge-and-furrow greenhouses rather than separate structures permit automation and minimize walking effort. Service buildings on the north side, midway along the greenhouses, minimize travel distance. Permanently plumbed pesticide lines, local steam outlets for pasteurization, and central fertilizer proportioning equipment improve efficiency. Conveyor systems for moving flowers, pot plants, and supplies to or from production areas should be considered.

Product Quality The demand for low-quality products is small. The profitability of such production is low at best. Some years ago, it was stated in the

Florist and Nursery Exchange that Larry Taylor of Denver Wholesale Florist found on a year-round average that it took 1.8 standard-grade (second) carnations, 3.5 short-grade blooms, 7.5 design-grade blooms, or 11 split-calyx blooms to equal the profit of one fancy-grade (first) carnation.

Aside from market price, product quality is important to personnel management. It affects the same principle of the employee relating to the firm. When one knows that he or she is part of a quality production scheme, the incentive exists to try to meet these standards in his or her own work.

Education Most people take pleasure in learning. It is flattering to an employee when the firm thinks enough of him or her to provide an education along with the job. Actually, there is a mutual advantage, since employees who understand the *why* and *what* of their tasks have the potential to be better workers. They are in a position to reason out better ways of doing the job and how to solve a problem when the job is not proceeding smoothly.

Education in a small firm need not consist of anything more than the manager talking with employees as they work. They should be given an appreciation of the various cultural procedures involved in a crop and how they interrelate. Employees should be aware of the quality standards required by the market. They should know the problems that can arise from mistakes such as insect or disease establishment, excessively high or low temperatures, improper photoperiod control, incorrect planting depth, nutritional disorders, and overwatering. Worthwhile employees welcome such knowledge and use it to better themselves within the firm and to assist the manager in meeting his or her responsibilities.

Larger firms, in addition to the procedure just discussed, sometimes make use of training sessions for their employees. These may be held on the premises of the firm and be conducted by management within the firm or by instructors hired from outside. Outside services are available for topics such as management and marketing. Visits from university personnel, allied trade representatives, or competitive greenhouse operators can be a valuable source of information. If possible, the general manager should arrange an opportunity for key personnel to meet with such individuals. One cannot help but be impressed by the professional manner in which Colombian growers receive such visitors. Preparation is apparent in the complete involvement of the management staff and the organized quest for information by each staff member.

Numerous meetings are sponsored each year by industry organizations, state universities, and other state and federal agencies. These meetings are an excellent educational opportunity for the owner and his or her key employees. Many state universities with horticultural or plant science departments conduct annual or semiannual floricultural short courses of one to three days' duration. Commodity groups such as Roses, Inc. and the Professional Plant Growers' Association sponsor meetings. The Society of American Florists, the Produce Marketing Association, and various wire services hold meetings with topics ranging in scope from the grower to the wholesaler to the retailer. Many other wholesaler and retailer associations also sponsor meetings.

Most of the organizations mentioned publish newsletters containing current floral news items as well as articles on technical subjects. Growers should definitely get on the mailing list of the horticulture department at their local state university. They should join their local flower growers' association as well as a national growers' association. Many growers join associations in other states as a means of expanding their sources of information and ideas. Information derived from these organizations should be passed down through the firm by one or another of the methods discussed earlier.

Production Management

Record Keeping

A grower who does not keep records is committed to repeating the same errors over and over. Every business must keep records for income tax purposes. With a little more thought and effort, a set of records can be developed for cost accounting purposes. *Cost accounting* is a system for assessing the costs of conducting a business. The costs of each input—labor, utilities, and materials—are determined and compared to a reasonable proposal of costs. The overall profitability of the business is then determined.

To know at the end of a year that a business made a profit is not enough. Some crops may have been profitable, while others were marketed at a loss. Certain grades of fresh flowers or sizes of pot plants may have contributed little or nothing to the profit. One market channel may have been more profitable than another. These differences must be known, or else the poorer alternatives may be allowed to increase out of proportion to the better alternatives. Cost accounting provides a tool for comparing the cost of producing units. The units may be different crops, different sizes of a given crop (bedding plants in $1^{3}/_{4}$-inch versus 3-inch pots), different market dates for a given crop, or different methods for producing and marketing a crop.

Cultural Records Before growing a crop, one should decide which records to keep. One set will be financial, including costs of items such as plants, containers, root substrate, labor, utilities, and so forth. The other set of records is cultural in nature. Cultural records are maintained for the purposes of (1) providing a plan for duplicating successful crops, and (2) giving an accounting from which the cause of errors in the culture of the crop can be determined and then corrected in the next crop.

Long before a crop is planted, a cultural schedule should be written, listing dates and labor budgets for operations such as root substrate preparation, planting, syringing, fertilization, pesticide application, pinching, pruning, chemical growth regulation, disbudding, harvesting, and cleaning up. This cultural schedule should be maintained in the general manager's office. The information should be duplicated on a cultural schedule record sheet to be hung in the greenhouse at the location of the crop being grown (Figure 18-3). The cultural schedule record sheet serves as a daily reminder to the production manager as to the various operations that must be performed.

When each operation is performed, the date is entered on the cultural schedule record sheet in the greenhouse and the name of the performing employee is entered. Should an unscheduled operation or an alteration in a scheduled operation be necessary, a description of the operation is entered in the record. At the end of each day, the entries are verified and initialed by the manager overseeing the operations.

Plant Environment Records A second set of culture-related records contains the plant environment records, including temperatures inside and outside

Greenhouse section		Benches	Crop	Cultivar	
IV		9-15	Cut Mums	Nob Hill	
Date scheduled	Date accomplished	Operation		Employee	Mgr. initials
3-7		Plant 7'' x 8''			
3-7		Fertilize, half strength			
3-7		Start lighting at night			
3-14		Fertilize and spray			
3-21		Fertilize and spray			
3-28		Pinch			
3-28		Fertilize and spray			
4-4		Fertilize and spray			
4-11		Fertilize and spray			
4-18		Fertilize and spray			
4-18		Start shading			
4-19		Prune plants back to 2 or 3 shoots			
4-25		Fertilize and spray			
5-2		Fertilize and spray			
5-9		Fertilize and spray			
5-16		Fertilize and spray			
5-23		Fertilize and spray			
5-30		Fertilize and spray			
5-30		Disbud			
6-6		Fertilize and spray			
6-13		Spray			
6-20		Spray			
6-24		Harvest			

Figure 18-3 A typical cultural schedule record placed at the end of a bench in the greenhouse. All planned cultural operations are entered on the record prior to planting the crop. As the operations are performed, the actual dates, any changes in descriptions, and the names of employees performing the operations are entered. The manager verifies the record with his or her initials.

the greenhouse, solar radiation, root substrate nutrient analyses, foliar analyses, insect and disease occurrence, and visual observations. Temperature should be recorded in the greenhouse to determine whether the desired temperatures have been maintained. Deviations from both low-temperature and high-temperature phases have an adverse effect on plant growth and prevent efficient use of energy. Such records give an assessment of the quality of the heating and cooling equipment and indicate breakdowns. They can prevent the erroneous conclusion, when the crop matures at the wrong time, that the schedule is incorrect. To change the cultural schedule in this case would only lead to a second mistimed crop. Recording thermometers with a seven-day record are available for this purpose. This is one more task at which a computerized greenhouse-control system excels. The computer can log temperature, relative humidity, light intensity, and any other parameter received from a sensor. The computer goes a step further by being able to tabulate or graph such data.

Inside temperatures give an indication of the condition of the temperature-control equipment, while outside temperatures are a reflection of fuel and electrical consumption. A record of outside temperatures can be obtained from your local branch of the National Weather Service. Data are gathered at several points in each state and are published. Greenhouse climate-control computer systems are available with outside climate-monitoring stations. Climate factors monitored include temperature, light intensity, rain, wind speed, and wind direction. For more sophisticated cost-analysis programs, the grower may want a record of daily minimum and maximum temperatures, heating degree days, and solar radiation. The winter-temperature and heating-degree-days data are useful in determining whether the fuel bill for one winter will be representative of successive years. These data also can be used to determine what proportion of the total fuel bill to allot to each crop grown during the heating season. This is important information for cost accounting. Summer-temperature data can be used in a like manner. Extremes in summer temperature result in crop delay and poor quality. Such records permit proper assessment of the blame.

Solar-radiation values indicate the amount of light reaching the earth's surface and thereby indicate when light is a limiting factor. When growth is limited by insufficient light, increases in temperature, fertilization, or CO_2 level are ineffective and a waste of money. Records such as those for solar radiation give an indication of which factor is limiting growth, thereby enabling the grower to decide whether alteration of environmental factors will be profitable.

Periodic root substrate tests and foliar analysis reports should be saved chronologically by crop. These are also valuable in determining factors that limit growth and thereby explaining exceptionally good or poor growth. As mentioned in Chapter 9, these records are also used for establishing the fertilization program itself.

All states have agricultural extension agents who can identify insect and disease problems. Some states have insect and disease clinics where samples can be sent for identification. Whenever an insect or disease problem is identified, it should be reported in the plant environment records. Again, these factors explain poor quality and yield.

Production Records The third set of culture-related records needed by the general manager are the production records. These records are gathered throughout the growth period of the crop. The production manager should assess the condition of each crop weekly and enter this assessment into the production record. For a crop such as gloxinia, the width of a dozen typical plants might be measured and the average value entered into the record. The average height of a chrysanthemum crop could be measured and recorded. Visual observations also should be recorded, considering factors such as form, leaf color, leaf size, stem thickness, and appearance of chlorosis or necrosis.

These types of information allow for the comparison of the present crop with previous crops. A problem such as phosphorus deficiency, which is not apparent to the eye in early stages, can be identified through smaller-than-normal growth measurements. By looking at a poor crop in retrospect, the grower can identify the stage of growth when trouble first occurred. The cultural records and plant environment records can then be checked to find the cause of the problem. In most cases, when the cause can be found, it can be corrected in subsequent crops.

The production record should also include the number of blooms or pots harvested, the date, and the grade or quality. These records are needed for cost accounting and are used in the same manner as the growth measurements described earlier.

Financial Records Just as cultural records must be gathered so that cultural mistakes can be identified, assessed, and corrected, financial records must be collected for improving the procedures for conducting business.

Income. Income should be recorded by crop. It is important to further subdivide income by date of sale, market outlet, and grade of product. Such a breakdown allows for comparison of relative profitability of season, market outlets, and grades.

Expenses. All inputs into the production and marketing of each crop must be identified. Each input is then quantified in a monetary value and entered as an expense. Some expenses are easily identified with a given crop, such as cuttings, pots, planting labor, disbudding labor, and trucking to market. These are known as *variable costs* because their magnitude varies with the size of the crop and from one crop to another. Other costs are known as *fixed costs* because they will continue even when production stops. Examples are interest on the loan for buildings and equipment, taxes, insurance, and management salaries. There are yet other costs that appear to fall between fixed and variable. These are *semifixed costs,* which increase as production increases but are not directly related to the number of units produced. Fuel, electricity, and lower-level management costs are examples of semifixed costs. They increase with increasing production but are not directly related to a pot of mums or a bunch of roses.

Variable costs permit the most sensitive cost analysis. The method of record keeping sometimes determines whether an expense can be treated as variable, fixed, or semifixed. Labor will be at best a semifixed cost if only the total number of hours worked per week is recorded. When the number of hours expended

Table 18-2 Codes and Descriptions of Production Labor

Code	Labor Task	Code	Labor Task
1	Sterilization of benches	10	Shifting plants
2	Set up watering system	11	Black cloth shading
3	Set up lights/shade	12	Pesticide application
4	Planting	13	Pruning
5	Moving plants to benches	14	Weed control
6	Watering	15	Disbudding
7	Fertilization	16	Staking and tying
8	Pinching	17	Cutting flowers
9	Growth regulator application	18	Cleaning up after crop

on each crop is recorded, it can be treated as a variable cost since the labor per unit of production can be determined. Labor in the former case would be divided by the total area in production, regardless of whether the crop was pot mums or poinsettias. Since each square foot of production area carries the same expense, a comparison of the cost of production or the profitability of pot mums and poinsettias would not take into account labor, one of the largest expenses. A comparison in the latter case, where labor was accurately related to each crop, would permit an accurate comparison of the crops.

Labor could be even further identified by operation. An example of various crop production labor tasks is seen in Table 18-2. Each production manager or grower is required to fill out a time sheet at the end of each day, identifying the quantity of time they spent on the labor assigned to them, the crop, and the labor operation category (Figure 18-4). From these sheets, the total labor input for each operation for each crop can be calculated.

This type of variable expense record permits comparisons within a crop. Alternative methods may be studied, such as manual pinching of azalea versus chemical pinching. Such a record also indicates where the greatest expenses lie, so that the possibilities for their reduction can be studied.

Fixed Expenses. Detailed expense records by crop are very important but are not always possible. The first letters of each of the five common fixed expenses (depreciation, interest, repairs, taxes, and insurance) spell *DIRTI;* these expenses are known as the "DIRTI Five."

Depreciation is a means of allocating the costs of fixed assets such as buildings, vehicles, machinery, and so on. For purposes of cost analysis within the firm, the cost of an asset is depreciated over the useful life expectancy of the asset. If a glass greenhouse is anticipated to serve a useful function for 20 years, then for straight-line depreciation its purchase price is divided by 20 to determine the annual depreciation cost. This allows one to allocate the proper cost of greenhouses to a given crop. The loan for the greenhouse may be for only 10 years. In this case, the amount of annual repayment of the principal on the loan is twice the amount of the depreciation. Depreciation is generally calculated over a

Task Code	Employee	Crop	Hours

Figure 18-4 **A daily time sheet to be filled out by the production manager or grower, indicating the quantity of labor expended on each crop by labor employees under the manager's command.**

shorter period of time than its useful life for income tax purposes. This permits a greater tax break in early years when the dollar is worth more.

Depreciation is not a savings fund to be used for replacing equipment or buildings. It is money already spent for the original equipment and buildings. As such, it is a business expense that can be deducted from sales revenue for determining taxable income. A separate fund should be set aside for asset replacement, if desired.

Interest is the cost to the business for using money to set up and run the business. The interest cost exists whether the money was borrowed from a commercial lending institution or provided by the owner. If the owner puts up the money to establish the business, he or she must expect to receive interest for this money from the business. Otherwise, the interest that could have been made by investing the money elsewhere is lost. This is known as an *opportunity cost.* To get an accurate picture of profitability, interest on all borrowed money must be entered as an expense.

Repairs to facilities and equipment will be required periodically. Maintenance also falls into this category since it is a logical expense for keeping repairs realistic.

Property taxes are a fixed expense. They include taxes paid to municipal, county, and state governments.

Insurance on facilities and equipment is a fixed expense. Labor-related insurance is not included here but comes under the expense category of labor.

Other fixed expenses include management salaries, the services of accountants and attorneys, travel to technical meetings or business appointments relating to the business in general, organizational dues, contributions, entertainment, and office expenses. Of these, management salaries bear close scrutiny.

When the owner is the manager, there is a temptation to consider management salary as profit. This should not be done, because it interferes with cost accounting. An unrealistically low cost of production emerges and cannot be maintained as the range expands beyond the size that the owner can manage alone. The manager should consider what a hired manager would be paid to do the job and enter it as an expense. The owner should realize that an employee could be hired to replace him or her, thereby freeing the owner to derive income from another endeavor. This type of thinking might set the stage for a more profitable use of time.

The lower levels of management come close to a variable expense. Growers are often assigned to specific jobs. As greenhouses are added, more growers are hired. If a direct relationship can be established between crops and grower input, then their salaries can be considered a variable expense.

Variable Expenses. These expenses are the easiest to identify. Each increases directly as the number of units produced increases. In this category are labor, plants, seeds, and growing supplies such as pots, root substrate, labels, pesticides, growth regulators, and fertilizers.

Fuel can be a variable expense if some measure of consumption is kept relative to crops. The weekly number of heating degree days could be used to determine the fuel consumption attributed to each week of the heating season. Within each week, the fuel cost could be proportioned to each crop according to area occupied and inside temperature maintained. This is too large a bill to be treated as a fixed expense.

Electricity could be treated in the same manner. The greater part of total consumption may be accounted for during the summer by cooling and during the winter by photoperiodic lighting.

Sales expenses should take into account packaging materials (sleeves, labels, and cartons), packaging labor, delivery labor, and sales labor. In a large operation where separate office personnel and telephones are used for sales, these expenses can also be listed under sales and can possibly be relegated to crops.

Labor has already been discussed. As indicated, labor of production and marketing should be treated as a variable expense as much as possible.

Expense Comparison. It is difficult to state a general cost of production or marketing. There is a different set of figures for each crop. There are also great differences among categories of growers. A small business may have a relatively high labor bill but low equipment expenses compared to a large automated business. The relative relationship of each expense to the total can be seen in the circle graph in Figure 18-5. This is a composite of a study of several categories of fresh-flower and pot-plant growers in Wisconsin and Michigan. The study indicated that the sales expense for the growers surveyed is low and would probably run at least 25 percent for the industry as a whole.

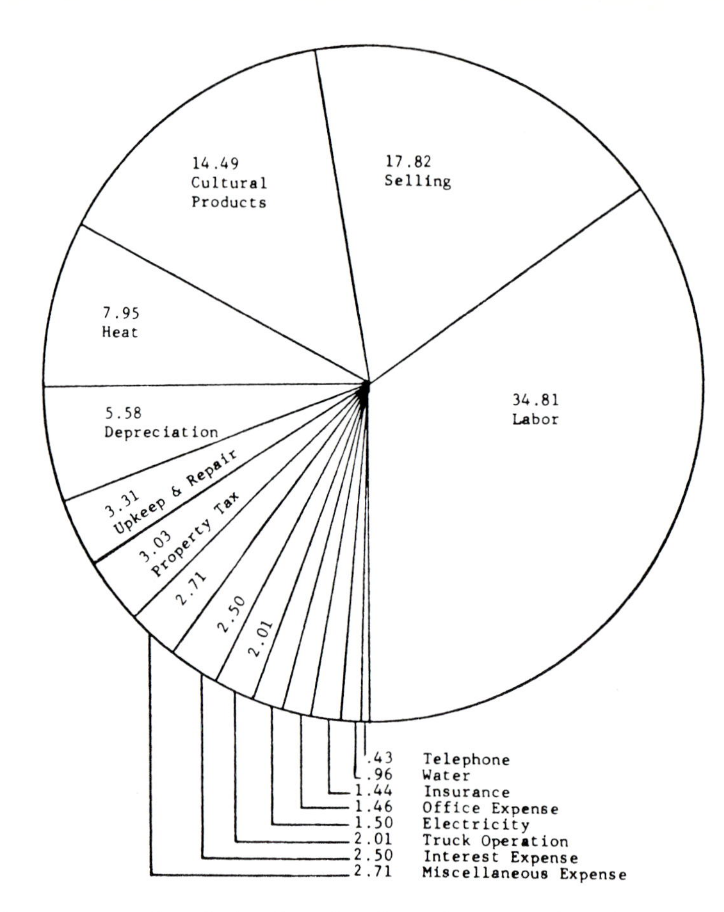

Figure 18-5 The relationship of each production and marketing expense to the total expense. (From Grimmer [1975].)

Cost Analysis

The records gathered thus far are in two general categories—cultural and financial. The cultural records can set a consideration of the financial records into perspective. Before the accounting office settles down to a cost analysis of a crop that has shown no profit, the cultural records should be studied. Failure may be due to one of several diverse factors, such as managerial error in executing the cultural schedule, uncontrolled disease, or boiler failure. In any case, assessment and correction of the problem are cultural considerations. An analysis of expenses and revenue will do little to shed light on the problem.

Profit or Loss Assessment After the cultural records have been studied, and barring any unduly large cultural problem, cost analysis becomes an important operation. A cost-analysis statement (Table 18-3) is developed for each crop to

Table 18-3 **Cost Analysis Sheet for an Individual Crop Listing Expenses, Revenues, and Profit or Loss**

EXPENSES

Fixed and Semifixed

Depreciation (facilities and equipment)	________
Interest	________
Repairs and maintenance (facilities and equipment)	________
Taxes	________
Insurance	________
Office expenses	________
Telephone	________
Accounting and legal fees	________
Travel and dues	________
Management salaries	________
Automotive	________
Miscellaneous	________

Variable

Seed	________
Cultural supplies	________
Labor	________
Fuel	________
Electricity	________
Sales costs	________

Total Expenses ________

REVENUES

Grade	Units	Unit Price	Total

Total Revenues ________

PROFIT OR LOSS ________

determine the profitability of each. The statement further serves to pinpoint the causes of expense and sources of revenue.

Fixed expenses are commonly determined on a per-square-foot-of-bench basis, and the crop is assessed according to the area it occupies. Variable expenses are assigned directly to the crop. Revenue may be identified by flower grade for fresh-flower crops. If more than one market channel is used, and pricing differs, revenue is entered according to market channel. Cost analysis by crop permits the grower to determine which crops are most profitable at various times of the year. It gives him or her an opportunity to determine the most profitable market dates in spite of seasonal changes in expenses such as fuel. It can be a difficult task to determine workable combinations and rotations of crops for greenhouse culture. Without this type of cost analysis, serious error may be made.

In determining the profitability of a crop, one must add in the fixed costs associated with vacant bench space that is associated with the crop. This may occur during the early culture of a crop such as bedding plants, when plants are in the seedling or cutting stages. Vacant space is held for spacing of plants as they develop. The rotation in which bedding plants and poinsettias are grown makes use of greenhouse space in the spring and fall but leaves considerable space open in the summer. Often, this space cannot be fully utilized because the market demand for floral products is low at this time and the length of time between the bedding-plant and poinsettia crops is too short to grow many crops. Fixed expenses continue and must be met for the empty space during this period.

Planning for Profit Increase Profit depends on the management of expenses and the production of revenue. Many things can be done to increase revenue. Many methods have already been introduced in reference to various greenhouse operations.

The market price of fresh flowers varies with supply and demand; however, there are well-established annual patterns. A price curve can be developed from market reports given in weekly trade papers. Quite often, land-grant university libraries maintain some of these. The firm should subscribe to such literature and develop a record of its own. Flowering pot-plant prices are fairly stable, but demand varies considerably. There are times when part of the crop cannot be sold. Even if this were not the case, the percentage of pots not sold due to form or condition would tend to decrease during a period of peak demand. Whether the increased demand affects the price or the percentage of the crop sold, the result is an increase in revenue.

Crop rotations should be designed to maximize bench utilization, as long as they do not result in crops of low profitability. The grower with a bedding plant–poinsettia rotation might look into propagating poinsettia cuttings during the summer. If this is not profitable for the grower's own use alone, the selling of cuttings might be considered. Even though this is the same plant that he or she grows in the fall, the grower must consider it as a separate crop for cost-analysis purposes. The sales price of the cuttings used within the business is the price the business would normally pay for these. Profits for the propagation business are then calculated from this point.

Greenhouse productivity also can be maximized by the physical arrangement of the growing space. Benches running the length of the greenhouse will occupy about 67 percent of the floor space, while peninsular benches may cover 75 percent of available space. The former arrangement is often more practical for fresh flowers, but a pot-plant grower will benefit more from the latter arrangement. Movable benches that make use of some of the aisles also should be considered. The system of growing pot plants on a paved floor offers even a greater advantage for increasing revenue. Hanging-basket plants grown over aisles permit nearly 100 percent use of greenhouse space. Increased use of greenhouse space is important because it reduces the fixed costs per unit of growing space. This concept should be pursued as long as it is not offset by another input such as labor. Conceivably, the aisles could become too narrow or the blocks of plants too wide to allow for efficient use of labor. Such relationships can be established through cost analysis of trial crops.

Revenue can be increased by producing higher-quality products if one is not currently meeting the market standards. There is always a demand for quality. Low-quality produce is generally dumped first. Even for flowering pot plants, there is a range of prices. A grower of high-quality plants can usually command a higher price for his or her products. This is most pronounced for fresh flowers, which are priced according to grade. The cultural records should suggest factors in the greenhouse environment that are limiting growth, as well as errors in the cultural schedule. A fresh coat of paint, reflective material on the north wall, clean glass, and proper plant spacing can improve light intensity, which is often limiting in the winter. Repair or replacement of heating and cooling equipment, injection of CO_2 into the greenhouse atmosphere, better sanitation, improved drainage of the root substrate, and many other factors discussed in this book should be considered for increasing quality. There is little profit in the floriculture industry for low-quality producers.

Cutting fresh flowers in the bud stage shortens the length of time the crop is in the bench, thus permitting a higher volume of production. Greenhouse space is replaced by opening room space in the grower's service building or, even more preferably, the marketplace. The substitution of space and gain in time have been shown to be profitable.

Revenue is tied into the choice of a market channel. Prices can vary among the different outlets. The speed of payment is quite variable. Most large mass-marketing chains have a reputation for prompt payment; uncollected debts are not common. A rapid cash flow permits growers to minimize the amount of money to be borrowed and thereby minimizes interest expense. Slow receipt of payment and bad debts are a real problem for some growers.

Raising prices to increase revenue usually does not work for the flower grower. A modest increase might be supported by exceptional quality or superior service. Any extensive increase opens the door for competitive growers. To a degree, flower growers are price takers rather than price setters.

Another step in improving profits is that of managing (reducing) expenses, which can be a very elusive challenge. Growers tend to react to sudden changes and overlook other more important, subtle changes. A good system of cost analysis can prevent this problem.

Professor A. O. Voigt (1976) of Pennsylvania State University presented an interesting analysis of a grower's situation. Even though the monetary figures are out of date, the principle is of value. The fuel cost for producing a 6-inch (15-cm) poinsettia rose $0.065 from a cost of $0.06 in 1972 to a cost of $0.125 in 1974. The industry as a whole was deeply alarmed at the doubling in fuel costs, and many felt that such prices might eventually put them out of business. Strong support developed for research into methods of energy conservation and alternative energy sources—and justifiably so. But what about the other expenses? While the cost of fuel doubled by rising $0.065, the grower's total production and marketing costs rose $0.54 in the same period. The cost of cuttings alone increased by a larger amount than did the fuel, and the total labor bill increased by $0.07 per pot.

Fuel prices are a real challenge for the grower, but they are not the only problem. Considerable attention should be given to labor, marketing expenses, and cultural supplies, which are larger expenses that are also rising rapidly (see Figure 18-5). The cost of supplies can be significantly reduced by ordering in large quantities to take advantage of volume discounts. A common complaint of the companies supplying greenhouse materials (seed, pesticides, fertilizers, pots, and so on) is that growers tend to buy as needed. Discounts of 5 to 15 percent are passed up by failing to consolidate orders to cover three-to-four-month operating periods. This is the length of time of a typical crop. The discounts are possible in part because supply companies may have the order shipped direct from the manufacturer rather than from their own warehouse. Extra handling and warehousing costs are avoided.

Worse than the loss of discounts are the additional shipping charges. Small orders may be assessed at minimum freight rates. A large order could be shipped for the same freight cost. Whole truckloads or carloads have a lower per-unit freight rate than partial loads. This may be seen in a higher rate or in a single fixed charge for a partial load. For heavy materials, the freight bill can equal the cost of the materials. Freight savings are worthwhile but often overlooked.

Cash discounts are often overlooked because of their small apparent size. A 2 percent discount for payment within 20 days equates to better than 36 percent annual interest. There are slightly more than 18 20-day periods in a year, and within each period 2 percent is collected. Even though the interest is collected only once on a given purchase, a greenhouse firm makes purchases throughout the year. If the firm's cash flow is low, such that the cash discount is rarely received, then the firm may be paying 36 percent more for materials than is necessary.

Each firm needs to calculate the impact on profit of a 1 percent shift in each expense factor entering into the production of a crop, as presented in Table 18-4. Armed with such a change-sensitivity chart, the progressive grower is in a position to pinpoint the production operation most critical to profits and then to do something about improving the condition of this operation.

Table 18-4 **Sensitivity of Changes in Economic Profit to 1 Percent Increases in Some Factors of Production of Chrysanthemums in 4-Inch (10-cm) Pots in a Plastic Greenhouse Operation Model[1]**

Factor (Increased by 1 Percent)	*Economic Profit (Change in Percent)*
Price received per plant ($)	+26.90
Yield (% of marketable crop)	+24.63
Production materials expense ($)	–7.17
Production labor (person-hours)	–5.25
Labor wage rate ($)	–5.25
Equipment investment ($)	–4.18
Management costs ($)	–3.93
Cuttings ($)	–3.81
Packaging costs ($)	–2.28
Fuel ($)	–2.18
Interest rate on the money (%)	–0.97

[1]From Kirschling and Jensen (1974).

The point cannot be emphasized too strongly that successful ownership or management of a greenhouse business for someone else depends as much upon a knowledge of business principles as it does upon a technical knowledge of crop production. A person entering or already in the greenhouse business should consider supplementing his or her knowledge with courses in accounting, personnel management, business law, and marketing.

Computer Assistance

Most people agree with the principles of record keeping and cost analysis set forth in this chapter; however, few greenhouse firms have an adequate system. Greenhouse managers feel that it is difficult to devise and establish an adequate record-keeping system and that it is time-consuming to maintain one. This has, in the past, been partly true.

Today, however, the computer greatly lessens the burden. The computer is actually a file and a calculator. The cultural, plant-environment, production, and financial records described earlier in this chapter can be entered into a computer at any frequency, rather than onto papers in a file drawer. The computer has the capability of recalling any specific records with minimal operator input. One might wish to contrast the variable costs of producing pot mums versus gloxinias.

Records for the years in which they have been collected can be immediately retrieved. Singling out an individual year, averaging several years, or expressing the variable costs on a per-pot or per-square-foot-of-bench-area basis are among numerous options that can be obtained in moments with simple commands. Data may be tabulated or presented in graphic form. Analyses of the retrieved data can be easily performed to assist in management decisions. All of these operations could be performed without a computer, but the time required would generally discourage it.

The computer gives even greater assistance in that planned cultural operations and greenhouse space allocations can be entered prior to planting. The computer can then be used to generate work schedules for each day. At the end of a year, unused bench space can be plotted out to study plans for better use of this space.

Word-processing software programs can make the daily work of the office staff more efficient. Form letters need to be typed only once. Correspondence with similar content can have only the differences typed. While many greenhouse firms have computerized their business office affairs, far fewer have computerized monitoring and control of the plant environment. The day is rapidly approaching when it will be necessary for greenhouse firms to have two computer systems—one for the business office and a second for greenhouse environmental control.

Summary

1. There are four general categories of operations within the greenhouse business: crop production, marketing, engineering, and business affairs. Each department has a manager, and over all four is a single general manager. The general manager answers to the owners. In a large business, there may be submanagers in each department. In a small business, the owner may serve as general manager as well as manager of each department.
2. A manager's ability to govern the affairs of others depends upon an ability to manage himself or herself. Managers must conduct themselves in a manner that fosters leadership. This includes self-motivation, perseverance, integrity, and a sense of justice. Further, they must operate according to the rules of success. These entail establishment of worthwhile goals, development of precise plans, unfaltering faith in their ability to achieve the goals, and relentless perseverance.
3. An effective manager-employee relationship calls for several situations. Employees must be aware of the management structure so that they know clearly the source of orders and to whom they are to answer. Each employee must know the goals of the firm and the specific portion of each that he or she is to perform. Employees must have sufficient authority to accomplish their assignments. Finally, there must be a system of evaluation by which the em-

ployee's performance is appraised, and there must be a just system of reward or constructive criticism.

4. Working conditions have a bearing on labor management. The physical facilities should be neat to encourage orderly work. Proper bathroom and eating facilities demonstrate respect for the employee and increase the chance that such respect is transferred by the employee to his or her job. Preventive maintenance and timely repair of equipment and facilities reduce discord in the employee's efforts. Efficient arrangement of work areas, supplies, and equipment averts needless expenditures of energy. Production of quality crops evokes a spirit of pride, which the employee transfers to his or her work. Most employees look upon an educational experience provided by the employer as a desirable benefit. Such provision also has great benefit to the firm because it improves the employee's ability to carry out his or her mission.
5. Production management is absolutely dependent upon recordkeeping, for without it the firm is destined to repeat errors. Cultural records serve to identify causes of cultural errors and are used as the framework for improved crop production plans. Financial records, including sources of revenue and expense, provide the tools for analysis of marketing and business affairs in general.
6. The use of records goes well beyond the realm of income tax returns and analysis of past performance. Records are used for profit planning. Profits depend on management of expenses and production of revenue. Records provide the basis for cost analysis of alternative production operations, which in turn has a bearing on expenses. They provide the basis for assessing the effect that increases in selling price have on profits. Most important, such analyses identify the relative effects of various expenses and revenues on profit, thereby providing a system of priorities in the effort to obtain increased profits.

References

1. Bange, G. A., E. E. Bender, and G. A. Stevens. 1972. *Planning and accounting for profit in floriculture.* Univ. of Maryland Agr. Exp. Sta. MP 806.
2. Boyd, R. M., T. D. Phillips, T. M. Blessington, and S. P. Myers. 1982. *Costs of producing selected floricultural crops.* Mississippi Agr. and Forestry Exp. Sta. AEN Res. Rep. 133.
3. Brumfield, R. G., P. V. Nelson, A. J. Coutu, D. H. Willits, and R. S. Sowell. 1981. *Overhead costs of greenhouse firms differentiated by size of firm and market channel.* North Carolina Agr. Res. Ser. Tech. Bul. 269.
4. Covey, S. R. 1989. *The seven habits of highly successful people.* New York: Simon & Schuster.
5. Grimmer, W. W. 1975. *Greenhouse cost accounting.* Gateway Technical Institute, 3520 30th Ave., Kenosha, WI.
6. Hemphill, B. 1997. *Taming the paper tiger*. New York: Random House
7. Hill, N. 1960. *Think and Grow Rich.* New York: Columbine.
8. Kirschling, P. J., and F. E. Jensen. 1974. *Profitability of pot chrysanthemum production under plastic greenhouses.* New Jersey Agr. Exp. Sta. Bul. 835.

9. Montigaud, J. C. 1992. Twelfth international symposium on horticultural economics. *Acta Hort.* No. 340.
10. Nelson, K. S. 1973. *Greenhouse management for flower and plant production.* Danville, IL: Interstate Printers and Publishers.
11. Peale, N. V. 1974. *You can if you think you can.* Greenwich, CT: Fawcett Publications.
12. Perry, D. B., and J. L. Robertson. 1980. *An economic evaluation of energy conservation investments for greenhouses.* Ohio Agr. Res. and Devel. Center, Res. Bul. 1114.
13. Reynolds, R. K., and W. R. Luckham. 1979. *Business management techniques for nurserymen.* Reston, VA: Environmental Design Press.
14. Schwartz, D. J. 1965. *The magic of thinking big.* Englewood Cliffs, NJ: Prentice-Hall.
15. Voigt, A. O. 1976. Are growers confusing energy problems with problems caused by poor marketing? *Florists' Rev.* 158:23, 78–80.

Glossary

Abortion The partial or complete arrest of a developing tissue, as in embryos, buds, and so forth.

Abscission The separation of leaves, flowers, fruits, or other plant parts from the plant, generally following the formation of a separation layer of cells.

Actinomycetes A group of microorganisms apparently intermediate between bacteria and fungi and classified as either.

Aerated steam pasteurization Using a mixture of air and steam that is adjusted to a temperature below that of steam (212°F, 100°C) for pasteurizing root media.

Aeroponics A system for growing plants with their roots suspended in air. Water and nutrients are misted onto the roots.

Apical dominance The suppression of lateral shoot development by the apical bud (shoot tip).

Asset Any item of value or resource. Assets of a business include cash, amounts owed to the business by its customers for goods and services sold to them on credit, merchandise held for sale by the business, supplies, equipment, buildings, and land.

Auxin A group of hormones that induces growth through cell elongation.

Azalea pot A pot with equivalent inside rim diameter of but only three-quarters the depth of a standard pot.

Bactericide An agent or preparation used for killing bacteria.

Bacterium (plural *bacteria*) A unicellular plant that lacks chlorophyll and multiplies by fission.

Bedding plants A wide range of plants that are propagated and cultured through the initial stages of growth by commercial growers and are then sold for use in outdoor flower and vegetable gardens.

Blindness The condition of a plant stem evidenced when the bud stops developing. It is a frequent problem of roses during low-light periods.

Blown head A bloom that is excessively open.

Bluing The objectionable development of a blue pigment in flower petals, usually after harvest.

Boiler horsepower A quantity of heat equal to 33,475 Btu.

Bract A more or less modified leaf subtending a flower or belonging to an inflorescence.

Bracteole A secondary bract, as one upon the pedicel of a flower.

Btu (British thermal unit) The amount of heat required to raise the temperature of 1 pound of water 1°F at or near its point of maximum density.

Bulk density The mass per unit bulk volume. For example, bulk density of a soil-based medium in a dry state might be 70 pounds per cubic foot.

Bullhead A flower whose short petals, particularly at the center, give it a blunt, broad appearance. Also, a flower whose excess number of petals gives it a blunt, broad appearance.

Calyx A term referring to the sepals collectively. It is the first of the series of floral parts and is usually green and leaflike but may be colored like the petals.

Cambium A zone or cylinder of meristematic (dividing) cells located between xylem and phloem tissues in plants. The cambium cells divide to form new xylem and phloem cells.

Cation exchange capacity (CEC) A measure of the ability of an absorbing material such as a root medium to hold exchangeable cations such as various fertilizer nutrients including ammonium nitrogen, potassium, calcium, magnesium, iron, manganese, zinc, and copper. It is generally measured in milliequivalents per 100 cubic centimeters (me/100 cc) of dry absorbing material, and a value of 6–15 me/100 cc is considered ample for greenhouse root media. A root medium with low CEC does not retain nutrients well and consequently must be fertilized often.

CEC *See* Cation exchange capacity.

Central heat system A heating system in which the heat used for heating one or several greenhouses is generated in a single location in one or more boilers.

Chelate A chemical complex that will hold or bind a metal. Metals that are commonly chelated for agricultural use are less subject to tie-up in adverse root-media environments. These metals include iron, manganese, zinc, and copper.

Chloropicrin Tear gas. A chemical used for pasteurizing greenhouse root media. It is not as popular as methyl bromide but can be used in carnation root media.

Chloroplast A specialized body (organelle) in the cytoplasm of some plant cells that contains chlorophyll.

Chlorosis The state in which normally green plant tissue is lighter green and possibly yellow due to the loss of chlorophyll or the failure of chlorophyll to form.

Chord A support member of the greenhouse frame that is under tension.

Clay A mineral component of soils consisting of particles less than 0.002 mm in diameter.

Closed cultural system Any method for growing plants in which the nutrient solution is recirculated. Nutrients are not allowed to leach from the pot or bench to the ground.

CO_2 The chemical formula for the gas carbon dioxide.

Conduction heat loss Heat loss by transmission through a barrier such as the covering of a greenhouse.

Conidiophore A specialized hypha on which one or more conidia are produced.

Conidium (plural *conidia*) An asexual fungus spore formed from the end of a conidiophore.

Container capacity The maximum amount of water a root medium can hold against the force of gravity when this root medium is in a container that has open drainage holes in its base.

Convection heat loss Loss of heat from the greenhouse as it moves in air convection currents to the greenhouse covering, then through the covering by conduction, and finally away from the outside of the covering.

Convection heater A heater that does not contain a heat exchanger. Heat leaves the heater in the smoke. The smoke is carried the length of the greenhouse in a pipe that serves as an exchanger as heat passes through its walls to the greenhouse air.

Corporation A legal entity, separate and distinct from the persons (stockholders or shareholders) who own it. The corporation has all the rights and responsibilities of a person and may buy, own, and sell property; sue and be sued; and enter into contracts with both outsiders and its own shareholders. The most important advantage of the corporate form is its responsibility for its own acts and debts and the freedom of its owners from liability for either.

Cost accounting The use of the cost data of producing a given product for the purpose of assessing and controlling those costs. Since a knowledge of costs and controlling costs is vital to good management, a large greenhouse firm often engages the services of a cost accountant.

Critical night length The length of darkness less than which a short-night plant or more than which a long-night plant will undergo a photoperiodic response. The critical night length varies with plant species and even sometimes with cultivars within a species.

Cross-fluted cellulose pad An evaporative cooling pad composed of laminated sheets of fluted (corrugated) cellulose impregnated with insoluble antirot salts, rigidifying saturants, and wetting agents. Pores are oriented diagonally through the pad in two directions, crossing each other.

Crown bud A flower bud whose development has ceased. It sometimes develops the appearance of a crown. Generally, this cessation of development breaks apical dominance, resulting in the development of side shoots. Crown buds may be caused by excessively low or high temperatures or in long-night plants by a series of short nights while the flower bud is developing.

Cultivar A cultivated variety. A cultivar usually has less variation within it than does a botanical variety.

Curtain wall The nontransparent lower portion of the side walls of a greenhouse.

Cuticle A nonliving waxy layer covering all plant cells that are in contact with air. Although this layer protects plant cells from drying, water and nutrients can slowly penetrate it, as in the case of foliar fertilization.

Cutting The portion of a plant removed for the purpose of asexual propagation. It may be part of a stem, a leaf, or part of a root, depending on the species of plant to be propagated. Commercial cultivars of chrysanthemums, for example, are propagated by removing terminal stem pieces and placing the lower inch of them in a rooting medium in a moist environment to induce new root formation.

Cyclic lighting An alternative method of applying light during the night to achieve the photoperiodic effect of long days. The customary lighting period is divided into a number of subperiods, each comprised of a duration of light followed by darkness. The total duration of light can be reduced by as much as 80 percent. Where three hours of light are customarily applied, six consecutive cycles of 5 minutes of light and 25 minutes of darkness can be substituted, thereby reducing electrical consumption greatly.

Damping-off A disease caused by a number of fungi, mainly *Pythium, Rhizoctonia,* and *Phytophthora*. The symptoms include decay of seeds prior to germination; rot of seedlings before emergence from the root medium; and development of stem rot at the soil line after emergence, causing seedlings to topple.

Day-neutral plant A plant that does not respond to the relative lengths of light and darkness in the daily cycle.

Depreciation Decline in value of an asset due to such factors as wear or obsolescence.

Desiccation The process of drying. Desiccation of plants results from a lack of water. High levels of soluble salts in the root medium cause desiccation of roots by preventing water from entering the roots.

Detergent *See* Surfactant.

DIF The difference between day and night temperature computed by subtracting the night temperature from the day temperature. By controlling DIF, the length of stem internodes (overall plant height) can be controlled. For many crops, the lower the DIF value is, the shorter the plant will be.

Disbudding The process of removing flower buds from a plant stem, generally to improve the size of the remaining bud or buds. In most cases, the terminal flower bud is retained and all of the lateral (side) flower buds are removed.

Disease A plant is said to be diseased when it develops a different appearance or changes physiologically from the normally accepted state. These differences are called symptoms. Disease can be caused by unfavorable environmental conditions such as temperature extremes, insects, or pathogenic organisms such as nematodes, fungi, bacteria, or viruses.

Distribution tube A clear plastic tube with holes along either side that is installed along the length of a greenhouse to provide uniform distribution of air within the greenhouse.

Dry matter That portion of the plant remaining after water has been driven off. For purposes of foliar analysis, leaves are generally dried for one day at a temperature of 158°F (70°C).

Eave A component of the greenhouse frame to which the side wall and roof are connected.

Ebb-and-flow system A cultural system in which containerized plants are grown in a watertight bench top. When watering is required, nutrient solution is pumped into the bench to a depth of 0.5–0.75 inch (13–19 mm). The solution is drawn into the root medium by capillarity. When the pot is thoroughly wet, after about 10–15 minutes, the nutrient solution is drained from the bench to a holding tank where it is stored until needed by the crop again. This is a closed cultural system.

Employee One who works for wages or salary in the service of an employer.

Employer One who employs another individual.

Emulsifiable concentrate (EC) A liquid pesticide preparation in which the pesticide is dissolved in oil and that contains an emulsifying agent to render the oil miscible in water.

Emulsifying agent A chemical that when added to two immiscible liquids renders them miscible.

Epinasty That state in which the more vigorous growth of the upper surface of an organ (as in an unfolding leaf) causes a downward curvature.

Equinox The two times of the year when day and night are of equal length everywhere on the earth. The sun is closest to the equator. The vernal equinox occurs about March 21; the autumnal equinox, about September 23.

Even-span greenhouse A greenhouse both of whose roof slopes are of equal length and angle.

Excelsior pad A pad comprised of curled shreds of wood, generally aspen wood, that is used for evaporative cooling of greenhouses.

Facultative long- and short-night plants Plants that do not require a night length longer or shorter than a given critical length for a response to occur, but that will respond faster if the dark period is longer or shorter respectively than a critical length.

Fan-and-pad cooling A system for cooling greenhouses used during the warm months of the year. Warm air expelled through exhaust fans in one wall is replaced by air entering through wet pads on the opposite wall. The entering air is cooled by the evaporation of water in the pad.

Fan–tube cooling A system for cooling greenhouses used during the cool months of the year. Cold air entering through a louver high in the gable of the greenhouse is directed along the length of the greenhouse through a clear plastic distribution tube. Pairs of holes spaced equidistant along the length of the tube's opposite vertical walls permit uniform air distribution throughout the greenhouse.

Fasciation A malformation in plant stems resulting in an enlarged and flattened stem, as if several stems were fused.

Fertigation The combined application of watering and fertilizer such that fertilizer solution is applied every time the plants require water.

Fertilizer proportioner (Also known as a *fertilizer injector*.) Equipment used to inject concentrated fertilizer solution into a water line to result in a desired dilution prior to plant application.

Fixation The process or processes in a soil by which certain chemical elements essential for plant growth are converted from a soluble or an exchangeable form to a much less soluble or to a nonexchangeable form.

Fixed costs Costs of conducting business that are not directly related to the number or type of items produced. Interest on a greenhouse mortgage, for example, is fixed because it remains unchanged if poinsettias are grown rather than azaleas, or even if no crop is grown.

Flat A container used in greenhouses and nurseries for purposes such as germinating seeds or for holding several small plant containers. Flats are commonly constructed from wood or plastic. They are variable in size but commonly approximate 21 inches (53 cm) long by 11 inches (28 cm) wide by 2.5 inches (6 cm) deep.

Floramull® A white water-absorbing synthetic urea formaldehyde type of resin produced by BASF Corporation. It is an amendment used in root media for its high water-holding capacity; it holds water to the extent of approximately 50 percent of its volume.

Floor heating Application of heat in or near the floor of a greenhouse.

Floriculture The art and science of growing and utilizing those plants valued for their aesthetic characteristics other than woody plants used in outdoor landscape.

Flowering plants Greenhouse crop plants grown in a pot and sold in the flowering state.

Fog cooling A system for cooling greenhouses in which fog is generated inside the greenhouse. As the fog droplets evaporate,. heat is absorbed, thus cooling the air.

Foot-candle (fc) A unit of illumination equal to the direct illumination on a surface everywhere 1 foot from a uniform point source of 1 international candle. It is equivalent to 10.76 lux.

Forced-air heater A heater containing a heat source, a heat exchanger, and a fan for expelling the heated air.

Fresh flowers Flowers marketed subsequent to being cut from commercial crops.

Fritted nutrients Nutrients, usually potassium or micronutrients, contained in a solid, finely ground glass powder. The glass slowly dissolves in the root medium, releasing nutrients over an extended period of time.

FRP (fiberglass-reinforced plastic) A type of panel used as the transparent covering on some greenhouses.

Fungicide An agent or preparation used for killing fungi.

Fungus An undifferentiated plant lacking chlorophyll and conductive tissues.

Gibberellins A category of hormones that stimulate growth through cell division or elongation or both.

Glasshouse A term used more commonly in Europe to designate a structure used for growing plants that has a transparent cover and an artificial heat source. The equivalent American term is *greenhouse*.

Gravelculture A system for growing plants in a root substrate consisting exclusively of gravel.

Gravitational force In reference to soil water it is the force of the earth's gravitational pull on water in the soil that causes water to move downward in the soil profile.

Greenhouse A structure used for growing plants that has a transparent covering and an artificial heat source.

Greenhouse range A term referring collectively to two or more greenhouses at a single location that belong to the same business entity.

Green plants Commercial crop plants grown in a pot and sold primarily for the aesthetic value of their foliage.

Headhouse A work-building in close proximity to or attached to a greenhouse. This facility might be used for purposes such as a workshop, storage area, pesticide room, potting area, or eating area. This building may also be referred to as a *service building.*

Herbicide A chemical used for killing weeds.

Hormone An organic substance produced in one part of the plant and translocated to another part where in small concentrations it regulates growth and development.

Horticulture The art and science of growing fruits, vegetables, flowers, and woody ornamentals as well as spice, medicinal, and beverage plants.

Host plant A plant that is invaded by a parasite and from which the parasite obtains its nutrients.

Humus The relatively stable fraction of the soil organic matter remaining after the major portion of added plant and animal residues have decomposed.

Hydroponics The culture of plants in a root substrate consisting exclusively of water and dissolved nutrients.

Hypha (plural *hyphae*) A single branch of the mycelium that makes up the body of a fungus.

IAA (indole-3-acetic acid) A naturally occurring auxin produced in apical meristems of both roots and shoots.

IBA (indole-3-butyric acid) A synthetically produced auxin.

Infiltration heat loss Loss of heated air from the greenhouse through cracks.

Inoculum The pathogen or its parts that can cause disease; that portion of individual pathogens that are brought into contact with the host.

Insecticide An agent or preparation used for killing insects.

Internode The portion of a plant stem between two nodes. The node is the portion of the stem where one or more leaves are attached.

Interveinal Pertaining to the space between the vascular tissue (veins) on a leaf.

IPM (integrated pest management) A holistic approach for managing pests including insects and related animals, pathogenic diseases, and weeds. Components of the program can contain restriction of pest entry into the cultural area, establishment of environmental conditions unfavorable to the pest at hand, biological control, and, only when necessary, the use of chemical pesticides.

Lap sealant A clear sealing material forced into the space formed between two overlapping panes of glass.

Larva The immature, wingless, and often wormlike form in which some insects hatch from the egg and in which they remain through increase in size and other minor changes until they assume the pupa or chrysalis stage.

Leaching percentage The percentage of water or nutrient solution applied to a pot or bench of root substrate that drains from the bottom of that container.

Lean-to greenhouse A greenhouse built against the side of another structure such that it has only one sloping roof.

Loam A textural class name for soils having reasonably balanced amounts of sand, silt, and clay. Loam soils can contain 7–28 percent clay, 28–50 percent silt, and less than 52 percent sand.

Logo A word, slogan, or sketch used to convey a thought. A logo is often used in advertising programs—for example, *Say It with Flowers* or the cougar on a Mercury automobile advertisement.

Long-night plant A plant that undergoes photoperiodic response, such as flowering, only when the night length is greater than a critical length.

Lumen The unit of light equal to the light emitted in a unit solid angle by a uniform point source of 1 international candle.

Lux The international unit of illumination, being the direct illumination on a surface that is everywhere 1 meter from a uniform point source of 1 international candle. It is equal to 1 lumen per square meter, or 0.0929 foot-candles.

Management The making of decisions that affect the profitability of a business.

Mark-on The percentage of the wholesale price added on to the wholesale price in order to cover overhead and profit and to arrive at a retail price.

Markup The percentage of the retail price added on to the wholesale price to cover overhead and profit. An item purchased for $1 and selling for $2 has a markup of 50 percent and a mark-on of 100 percent.

Mass marketing In the field of floriculture, the sale of floral products through high-traffic outlets such as supermarkets, discount stores, department stores, sidewalk stands, and shops in shopping malls.

Matrix force The attraction of the soil (matrix) for water. This force is responsible for the adsorption and capillarity of water in soil.

Meristem A tissue composed of embryonic, unspecialized cells actively or potentially involved in cell division. An apical meristem is a meristem located at the apex (tip) of a shoot or root.

Methyl bromide A chemical commonly used for pasteurizing greenhouse root media. It should not be used in carnation root media.

Miticide An agent or preparation used for killing mites.

Mycelium The hypha or hyphae that make up the body of a fungus. Mycelium are the microscopic threadlike strands that make up the body of a fungus.

NAA (naphthalene acetic acid) A synthetically produced auxin.

Necrosis The state of being dead and discolored.

Nematicide An agent or preparation used for killing nematodes.

NFT (nutrient film technique) The culture of plants in a system where a thin film (a few millimeters deep) of nutrient solution is circulated through a trough that also contains the plant roots. It is a specialized form of hydroponics.

Open cultural system Any system for growing plants in which nutrient solution is allowed to pass through the root zone and out into the environment.

Opportunity cost The value of other opportunities (alternatives) given up in order to produce or consume any good; that which must be forfeited when alternative *A* is abandoned in order to pursue alternative *B*.

Organelle One of several types of small structures within plant or animal cells that is bounded by a membrane. The chloroplast is one type of organelle in which photosynthesis occurs.

Ovipositor A prominent structure projecting from the posterior end of females of some insects that is used to deposit eggs.

Pasteurization The selective destruction of some, but not all, living microorganisms. Root media are pasteurized to eliminate harmful disease organisms and to retain the beneficial microorganisms.

Pathogen An entity (fungus, bacterium, nematode, virus) that can incite disease.

Peat The organic remains of plants that have accumulated in places where decay has been retarded by excessively wet conditions. There are many types of peat, some desirable and others not, used for greenhouse root media.

Peat humus Peat that is at an advanced stage of decomposition in which the original plant remains are not identifiable. It is not generally a desirable form of peat for greenhouse root media because of its rapid rate of decomposition and its occasionally high rate of ammonium nitrogen release.

Peat moss Peat consisting predominantly of slightly humified (decomposed) *Sphagnum* moss species. Horticultural peat moss contains over 75 percent sphagnum moss.

Pedicel Stem of one flower in a cluster.

Perimeter heating system A row of heating pipe or pipes just inside the perimeter walls of a greenhouse.

Perlite A siliceous volcanic rock that is crushed and heated to 1,800°F to cause it to expand into lightweight (about 6 pounds per cubic foot) particles with closed air-filled cells. Perlite is used as a substitute for sand when a lightweight root medium is desired.

Pesticide An agent or preparation used for killing living organisms that are a nuisance or are harmful to crops.

Petiole The stalk or stemlike portion of a leaf.

Photoperiodism The response of a plant or animal to the relative length of day and night. The response in plants can take on many forms, including flowering, changes in leaf shape or internode length, and bulb or tuber formation.

Photosynthesis The manufacture of carbohydrate from carbon dioxide and water in the presence of chlorophyll, using light energy and releasing oxygen.

Phytotoxic Toxic to plants.

Pinching Removal of the top of a vegetative plant stem in order to cause it to form several branches.

Plug seedlings Seedlings produced and contained in a small cohesive volume of root medium. This unit of root medium is known as a *plug*.

Polyethylene A plastic material used in the greenhouse industry in the form of thin films for covering greenhouses. It is an inexpensive substitute for glass. Generally, two layers are used—an outer layer 6 mils (6 one-thousandths of an inch) thick and an inner layer either 4 (0.10 mm) or 6 (0.15 mm) mils thick.

Pompon chrysanthemums A term used in this book to denote the chrysanthemum cultivars grown with several flowers on each stem. The term *spray chrysanthemum* is more commonly used.

Potable Drinkable.

Precipitation The process whereby a dissolved substance comes out of solution to form a solid. The solid substance is a *precipitate*.

Pressuring fan A fan in the end of the clear plastic greenhouse distribution tube that forces heater air, exterior cold air, or interior warm air through the tube, depending on whether the system is being used for heating, cooling, or air circulation, respectively.

Proprietorship A business owned by a single individual.

Pupa The intermediate, usually quiescent, stage assumed by many insects after the larval stage and maintained until the adult stage.

Purlin A component of the greenhouse frame running the length of the greenhouse just below the roof covering that connects the trusses together.

PVC (polyvinyl chloride) A plastic material available in corrugated sheets. This material was used for covering greenhouses during the 1960s, but, because of its rapid deterioration from ultraviolet light, it has virtually disappeared from the greenhouse industry.

Radiant heat loss The radiation of heat from a warm body, such as plants in a greenhouse, to a cooler body, such as the covering on the greenhouse or the sky and earth outside.

Rafter A frame component spanning the space between the eave and the ridge. Unlike a sash bar, glass is not attached to it.

Reglaze To replace the glass or the glazing compound that seals the glass on a greenhouse.

Respiration Those biochemical processes in the plant or animal that result in the consumption of oxygen and carbohydrate, the evolution of carbon dioxide, and the release of energy. Respiration has the reverse effect of photosynthesis.

Revenue Income; return from investment.

Ridge A component of the greenhouse frame to which the upper portion of the two roof slopes are connected.

Ridge-and-furrow greenhouses Two or more greenhouses connected to each other along their length at the eave. In this case, the eave becomes a gutter, or furrow. The common side wall is eliminated in each greenhouse. Such greenhouses are less expensive to heat and easier to automate than an equivalent area of separate greenhouses.

Rock wool A fibrous material used for thermal and acoustical insulation as well as a root medium for plants. It is made from melted rock that can be basalt or limestone, sometimes in combination with iron slag. The hot liquid is spun and cooled into long fibers. The fibers may be formed into granules for use in potting media, blocks for plant propagation, or large slabs for growing finished crops of vegetables and fresh flowers.

Root medium See *Root substrate.*

Root substrate (also called *root medium)* A suitable substrate in which plant roots can grow. It consists of one or more mineral and/or organic components mixed together. This term is most commonly used in the greenhouse and nursery circles of agriculture.

Sand A soil mineral particle measuring 0.05–2.0 mm in diameter.

Sandculture The culture of plants in a root substrate consisting exclusively of sand.

Sash bar The bar to which glass is attached in a greenhouse.

Senescence The process of growing old; aging.

Sepal One of the components of the calyx.

Shatter When used to describe a floral condition, this term refers to the dropping or abscission of petals.

Short-night plant A plant that undergoes a photoperiodic response, such as flowering, only when the night length is less than a critical length.

Sill The portion of the greenhouse that rests on the curtain wall and to which the side wall sash bars are attached.

Silt A mineral component of soils consisting of particles measuring 0.002–0.05 mm in diameter.

Slab-side (also known as *cling-side*) A flower that has failed to open symmetrically. The petals on part of the circumference are still straight up, while the remaining petals have opened in a normal fashion.

Sleepiness A condition in flowers in which petals curve upward, giving the appearance of a wilted condition. It is commonly caused by ethylene gas after harvest.

Soil The upper, heavily weathered layer of the earth's crust that supports plant life. It is a mixture of mineral and organic materials.

Soil structure The combination of primary soil particles into secondary aggregate particles.

Solenoid valve An electrically activated valve that controls the flow of gases or liquids. Such valves can be activated by a time clock to control the flow of water in automated greenhouse watering systems.

Solstice The two times of the year when the sun is farthest from the equator (closest to the poles). In the Northern Hemisphere, the summer solstice occurs about June 22 and the winter solstice about December 22.

Split A flower having a split calyx, in which the petals protrude from the split. It is a common problem of carnations.

Spore The reproductive unit of fungi consisting of one or more cells; analogous to the seed of green plants.

Sporophore A hypha or fruiting structure bearing spores.

Spreader-sticker *See* Surfactant.

Standard chrysanthemums Cultivars of chrysanthemum customarily grown with one large flower on each stem.

Sterilization The destruction of all living organisms. Greenhouse tools and growing containers are periodically sterilized to eliminate harmful organisms including pathogenic diseases, insects, nematodes, and weeds.

Strap leaves Leaves whose margins are partially or completely missing such that the leaf is narrower than normal, often resembling a strap.

Strut A support member of the greenhouse frame that is under a compression force.

Surfactant A chemical used to alter the surface properties of liquids. Surfactants are added to pesticide sprays to reduce the surface tension over the plant leaf surface. Without a surfactant, complete coverage of the leaf surface often is not achieved. Surfactants used for this purpose include *spreader-stickers, wetting agents*, and *detergents*. Surfactants are also used to enhance the initial wetting of root media containing relatively dry peat moss, which tends to be waxy and water-repellant.

Symphillid A small, translucent to white, many-legged arthropod that ranges up to 1/4 inch (6 mm) in length and that feeds on the roots of plants.

Systemic Spreading internally throughout the plant body. Some pesticides are systemic, as are some pathogens.

Texture The relative proportion of various sizes of mineral particles in a given soil or root medium.

Transpiration The loss of water from plant tissue in the form of vapor.

Tropism A growth response or bending toward or away from a stimulus. Geotropism is in response to gravity; roots grow toward, and shoots away from, the center of the earth's gravity. Phototropism is in response to light; shoots tend to grow toward light.

Trough culture A closed system for growing pot plants in which a single row of pots is placed in a watertight trough arranged on a slight incline. Nutrient solution is pumped to the high end of the trough. It flows by gravity around the bases of the pots to the low end where it is channeled to a holding tank. Solution is drawn into the root medium by capillarity. The solution is reused each time the crop requires watering.

Truss A compound component of the greenhouse frame spanning the width of the greenhouse and consisting of rafters, chords, and struts that are welded or bolted together.

Ultralow volume When used to describe pesticide application to plants, this term refers to the application of a very low volume of liquid pesticide formulation per unit area. When less than 5 gallons is used per acre (47 1/ha), the application is referred to as *low volume*; below 1 gallon per acre (9.5 1/ha), it is referred to as *ultralow volume*.

Uneven-span greenhouse A greenhouse with one roof slope longer than the other, generally for the purpose of adaptation to a hillside.

Unit heater A forced-air heater. Unit heaters are usually mounted overhead in a greenhouse. They may contain a firebox or receive heat in the form of steam or hot water from a boiler elsewhere.

Variable cost A cost that increases proportionately with each additional unit produced and ceases if no units are produced. The cost of pots, root media, and plants are variable costs; the mortgage on the greenhouse range is not a variable cost but rather a fixed cost because it continues even if no plants are produced.

Vascular tissue Tissue in the root, stem, leaf, or flower stem including phloem for conducting organic substances throughout the plant, xylem for conducting water and nutrients primarily from the roots to the shoot, and supporting fiber cells. Vascular tissue in leaves is often called *veins*.

Vase life The length of time that a cut flower retains its aesthetic value after it has been placed on display.

Veinal Pertaining to the vascular tissue (veins) or the tissue immediately above the vascular tissue in a leaf.

Ventilator A glazed panel attached to the greenhouse with hinges that permit opening for ventilation purposes.

Vermiculite A micaceous mineral that exfoliates (expands by separation of the many layers composing it) when heated. It is used in the expanded state as a lightweight component of greenhouse root media. Its desirable properties include a light bulk density of 7–10 pounds per cubic foot, a relatively high cation exchange capacity of 19–23 me/100 g, and a high water-holding capacity.

Vermiculaponics The culture of plants in a root substrate consisting exclusively of vermiculite.

Wettable powder In floriculture, an agricultural chemical formulated generally in talc or dry clay. It is suspended in water by continual mixing and is applied as a spray or root-media drench.

Wetting agent *See* Surfactant.

Whole-firm recirculation A closed circuit system encompassing an entire firm. Effluent from all benches or beds is channeled to a treatment pond. There, it is often treated for pathogens, analyzed, and nutritionally altered prior to being recirculated through the crops.

Witch's broom A symptom of boron deficiency in a plant. A witch's broom consists of a large number of shortened plant stems situated parallel and close to one another to give the appearance of the straws in a broom.

Xylem A tissue in the plant that transports water and nutrients upward from the roots to the foliage. Cells connected from end to end form xylem tubes. Vessels are the predominant xylem cells in flowering plants and have open ends. Tracheids predominate in the conifer (pines, etc.) xylem; rather than having open ends, they have pits along their sides connecting to adjacent tracheid cells. Vessel and tracheid cells are nonliving at the time they carry out the function of water and nutrient transport.

Index

A

B

C

D

E

F

G

H

N

Q

T

U

V

W

Z